Advanced Materials and Chemical Engineering

Lei Guo
https://orcid.org/0000-0001-7849-9583
Tongren University, China

Yan Tan
https://orcid.org/0009-0003-9778-5851
Tongren University, China

Amir Mahmoud Makin Adam
Tongren University, China

IGI Global
Scientific Publishing
Publishing Tomorrow's Research Today

Vice President of Editorial Melissa Wagner
Director of Acquisitions Mikaela Felty
Director of Book Development Jocelynn Hessler
Production Manager Mike Brehm
Cover Design Jose Rosado

Published in the United States of America by
IGI Global Scientific Publishing
701 East Chocolate Avenue
Hershey, PA, 17033, USA
Tel: 717-533-8845 | Fax: 717-533-7115
Website: https://www.igi-global.com E-mail: cust@igi-global.com

Library of Congress Cataloging-in-Publication Data

LCCN: 2026003613 (CIP Data Pending)
ISBN13: 9798337344027
Isbn13Softcover: 9798337344034
EISBN13: 9798337344041
British Cataloguing in Publication Data
A Cataloguing in Publication record for this book is available from the British Library.

Table of Contents

Detailed Table of Contents

Advanced functional ceramic materials represent a critical category within new functional materials and a national strategic emerging industry. These inorganic non-metallic materials are engineered from highly refined and purified powders, unlike traditional ceramics, to exhibit superior properties including high strength, hardness, temperature resistance, and excellent electrical, magnetic, optical, and biological characteristics. Their defining feature is the ability to facilitate mutual conversion between different forms of energy (e.g., mechanical to electrical, as in piezoelectrics), making them indispensable in modern high-tech fields such as microelectronics, aerospace, biomedicine, and communications. The document elaborates on four primary categories of these ceramics. Electronic Ceramics form the most diverse group, encompassing insulating, capacitor, ferroelectric (piezoelectric), semiconductor, and microwave dielectric types.

Zeolites have proven to be highly effective catalysts in various reactions, including acid, base, and redox catalysis. This chapter discusses the use of zeolitic materials as an affordable and efficient approach to producing green hydrogen. It focuses on three main goals: first, providing a general overview of zeolites and their diverse applications across multiple fields; second, examining the role of zeolitic materials as catalysts in hydrogen production processes; and finally, presenting case studies that highlight the mechanisms behind catalytic reactions during hydrogen generation with zeolites.

The scale of global industrial waste generation has reached alarming levels. As a byproduct of human industrial activities, industrial waste has long been a critical challenge in balancing economic growth with environmental sustainability. China's transition from a "waste disposal" mindset to a "resource utilization" paradigm has become not only an environmental imperative but also a strategic economic opportunity. Beyond environmental and economic benefits, this transformation aligns with the global trend in the circular economy. This chapter provides a systematic introduction to the classification and characteristics of industrial wastes, the historical development of resource utilization in China, typical case studies, policy framework in China, challenges and future prospects. The chapter offers detailed descriptions of various processes, operational parameters, practical outcomes and economic benefits. This chapter is intended for researchers, designers and production specialists engaged in environmental protection and resource recycling.

This chapter provides a comprehensive analysis of globally sustainable manganese production and use, addressing key challenges and strategies across the entire value chain. Manganese, critical for steel, batteries, and alloys, faces issues of resource depletion, environmental degradation, and supply chain vulnerabilities due to geographically concentrated reserves. Sustainable solutions emphasize responsible mining (precision extraction, water management, renewable energy, community engagement), low-carbon smelting (hydrogen reduction, plasma technology, biomass reductants), energy-saving and consumption-reduction electrodeposition process of manganese, sustainability of green manganese materials in end-use applications and circular economy practices. Policies and industry collaboration are highlighted as essential drivers for promoting transparency, ethical sourcing, and green innovation. The integration of life-cycle assessment, eco-design, and advanced recycling technologies is urged to achieve a resilient, low-carbon manganese industry supporting global sustainability goals.

Water pollution now involves complex mixtures (heavy metals, POPs, pharmaceuticals, microplastics) that conventional treatment trains often fail to remove when pollutants are dilute, recalcitrant, or transformation-prone. This chapter reviews material-enabled strategies built on (i) semiconductor photocatalysts, connecting band structure, charge dynamics, and ROS generation to performance gains from doping/defects, plasmonic coupling, and Type-II, Z-scheme, and S-scheme heterojunctions; and (ii) MOFs/COFs as modular porous platforms for high-affinity adsorption, ion exchange, and catalysis toward metals and organic micropollutants. It also links synthesis routes to morphology, stability, scalability, and recoverability (e.g., magnetic composites), integrates case studies on dyes/pharmaceuticals, metal removal, and microplastic mitigation, and critiques translation barriers (cost, long-term durability, fouling, ecotoxicity, regeneration). Future directions emphasize responsive materials, AI-assisted discovery, hybrid process integration, and circular-economy resource recovery.

Biomass-derived fuels and chemicals have attracted considerable attention for their benefits, including economical, ecofriendly, nontoxic, and renewable. This chapter will focused on the preparation and characterization of biomass-derived carbon dots, biomass fuels, and biochar. Part I review details the synthesis of carbon dots from green sources such as plants parts and waste. Renewable sources such as biofuels show a potential usage with low carbon dioxide emission, less polluting and can be created from biomass in the forms of organic waste. Part II aims to review the various sources of biofuels and the methods used in the production, and discuss its advantages and sustainability. Biochar, a carbon-rich material produced from the pyrolysis of organic biomass, has gained significant attention as a potential solution for sustainable green remediation practices. Part III aims delving into the diverse methods of biochar production, highlighting their influence on physical and chemical properties.

This chapter outlines the current state and future prospects of the specialty chemicals sector, defined as high-value, performance-driven compounds characterized by molecular specificity and tailored applications. It begins by contrasting these with commodity chemicals, focusing on attributes like molecular precision, application-specific optimization, and regulatory complexity. The review traces the field's evolution from early polymer innovations to modern advancements in green chemistry, AI, and nanotechnology. Interdisciplinary advances in pharmaceuticals, agriculture, electronics, and environmental remediation are explored, highlighting roles for synthetic biology and advanced materials. Key challenges addressed include scalability, regulatory fragmentation, and sustainability trade-offs. Finally, future opportunities driven by quantum computing, bio-inspired design, and circular economy principles are discussed, emphasizing the need for interdisciplinary collaboration and global governance to promote equitable, sustainable growth.

This chapter analyzes pharmaceutical intermediates as the strategic link between basic feedstocks and APIs. It examines the evolution of synthetic methodologies, highlighting green synthesis, continuous flow chemistry, and biocatalysis as key approaches for process intensification. The text details purification strategies, contrasting traditional crystallization with advanced methods such as preparative chromatography and membrane separation. Furthermore, it emphasizes the "Quality by Design" (QbD) paradigm, integrating impurity profiling to ensure regulatory compliance. Exploring artificial intelligence-driven synthetic route planning and

Preface

INTRODUCTION

The field of advanced materials and chemical engineering stands at the forefront of scientific innovation, underpinning technological progress across energy, healthcare, environmental sustainability, and industrial manufacturing. As global challenges intensify—ranging from climate change and resource scarcity to the demand for high-performance materials—there is an urgent need for interdisciplinary approaches that bridge fundamental science with engineering applications. This edited volume, Advanced Materials and Chemical Engineering, seeks to provide a comprehensive and timely contribution to this evolving landscape.

This book brings together leading scholars and researchers from diverse institutions and countries, reflecting the inherently global and collaborative nature of materials science and chemical engineering. The chapters collectively explore both foundational theories and emerging technologies, emphasizing the integration of advanced materials design, process optimization, and sustainability principles. By combining experimental insights with computational approaches, including artificial intelligence and machine learning, the volume highlights how modern research is reshaping traditional paradigms.

As editors, we have carefully curated contributions that not only present state-of-the-art developments but also critically assess current limitations and future directions. The structure of the book is designed to guide readers from core material innovations to applied systems and industrial-scale considerations. It is our hope that this volume will serve as a valuable reference for researchers, practitioners, and graduate students, while also inspiring new lines of inquiry and collaboration.

CHAPTER OVERVIEW

Chapter 1 introduces advanced ceramic materials, emphasizing their functional properties and critical role in modern high-tech applications. It provides a detailed classification of electronic ceramics and explores their multifunctional capabilities in energy conversion and advanced devices.

Chapter 2 examines zeolitic catalysis as a sustainable pathway for hydrogen production. It highlights the structural versatility of zeolites, their catalytic mechanisms, and their application in green energy systems through illustrative case studies.

Chapter 3 focuses on superhard materials, discussing their synthesis, structural properties, and industrial applications. It also explores cutting-edge developments such as machine learning-assisted material design and the integration of theory with practical implementation.

Chapter 4 presents the fundamentals and technological advancements in energy storage materials. It addresses key challenges such as energy density, safety, and cost, while reviewing innovations that support the transition to renewable energy systems.

Chapter 5 explores multiscale modeling and machine learning approaches to understanding degradation mechanisms in lithium-ion batteries. It emphasizes the importance of integrating modeling techniques across different spatial and temporal scales.

Chapter 6 discusses the application of artificial intelligence and machine learning in materials design. It outlines methodologies for prediction, optimization, and microstructural analysis, while addressing current challenges and future opportunities.

Chapter 7 investigates industrial waste resource utilization, particularly within the context of China's transition toward a circular economy. It provides insights into policy frameworks, technological processes, and economic implications.

Chapter 8 examines sustainable manganese production and use, focusing on environmental challenges, low-carbon technologies, and circular economy strategies across the value chain.

Chapter 9 addresses advanced materials for water pollution control, including photocatalysts and porous materials such as MOFs and COFs. It also discusses scalability, environmental impact, and future innovation pathways.

Chapter 10 explores biomass resource utilization for fuels and chemicals. It reviews biomass-derived materials such as carbon dots, biofuels, and biochar, highlighting their role in sustainable energy and environmental remediation.

Chapter 11 provides an overview of specialty chemicals, discussing their evolution, applications, and future prospects in areas such as pharmaceuticals, agriculture, and advanced materials.

Chapter 12 focuses on the synthesis and purification of pharmaceutical intermediates, emphasizing process intensification, quality control, and the integration of modern technologies such as continuous flow chemistry and AI.

Chapter 13 examines recent advances in anticancer drug screening, analyzing traditional and emerging techniques while addressing challenges in clinical translation and drug development efficiency.

Chapter 14 presents a comprehensive study of traditional Chinese medicine, exploring its historical foundations, modern scientific integration, and role in global healthcare systems.

CONCLUSION

This volume reflects the dynamic and rapidly evolving nature of advanced materials and chemical engineering. By bringing together diverse perspectives and research methodologies, it underscores the importance of interdisciplinary collaboration in addressing complex global challenges. The integration of sustainability, digital technologies, and innovative material design emerges as a unifying theme throughout the book.

We believe that the contributions in this collection not only advance current knowledge but also provide a roadmap for future research and industrial application. The emphasis on bridging theory and practice is particularly critical in ensuring that scientific discoveries translate into tangible societal benefits. From energy systems and environmental protection to healthcare and industrial innovation, the impact of these advancements is far-reaching.

As editors, we extend our sincere gratitude to all contributing authors for their dedication and scholarly excellence. We also thank the reviewers and publishing team for their invaluable support. It is our hope that this book will serve as a meaningful resource and inspire continued exploration in the field of advanced materials and chemical engineering.

Chapter 1
Advanced Ceramic Materials With Functional Properties

Runqing Liu
http://orcid.org/0009-0007-0444-7415
Tongren University, China

Xiaoguang Pan
Inner Mongolia University of Science and Technology, China

Jie Song
Fudan University, China

ABSTRACT

Advanced functional ceramic materials represent a critical category within new functional materials and a national strategic emerging industry. These inorganic non-metallic materials are engineered from highly refined and purified powders, unlike traditional ceramics, to exhibit superior properties including high strength, hardness, temperature resistance, and excellent electrical, magnetic, optical, and biological characteristics. Their defining feature is the ability to facilitate mutual conversion between different forms of energy (e.g., mechanical to electrical, as in piezoelectrics), making them indispensable in modern high-tech fields such as micro-electronics, aerospace, biomedicine, and communications. The document elaborates on four primary categories of these ceramics. Electronic Ceramics form the most diverse group, encompassing insulating, capacitor, ferroelectric (piezoelectric), semiconductor, and microwave dielectric types.

DOI: 10.4018/979-8-3373-4402-7.ch001

INTRODUCTION AND CONCEPTUAL FRAMEWORK OF ADVANCED CERAMICS

Advanced functional ceramic materials represent a pivotal category within new functional materials and are recognized as a national strategic emerging industry (Alallaq & Kashan, 2023). These materials are characterized by their unique electrical, magnetic, optical, superconducting, chemical, and biological properties, which often involve mutual conversion capabilities (e.g., piezoelectric effect). They play a vital role in numerous advanced technical fields, including microelectronics, laser technology, optical fiber communications, sensing, superconducting technology, and space exploration (Best et al., 2008). To clarify the terminology used throughout this document and avoid ambiguity, we establish the following conceptual hierarchy:

Functional Ceramics:This is a broad, umbrella term referring to all ceramic materials engineered to possess specific physical or chemical functions beyond their basic structural role. The core of this concept lies in the material's inherent properties. The primary categories discussed in this review—electronic ceramics, superconducting ceramics, bioceramics, and magnetic ceramics—all fall under this classification (Bhadrakumari et al., 2011).

Smart/Intelligent Ceramics:This term describes an advanced subset and an evolving frontier of functional ceramics. Smart ceramics are defined by their ability to sense changes in the external environment (e.g., stress, temperature, electric field, pH) and respond in a controlled, useful manner, potentially exhibiting adaptive, self-diagnostic, or self-healing capabilities. The core differentiator is the sensing-response intelligent behavior. Examples include piezoelectric ceramics used in sensors and actuators.

The following diagram illustrates the hierarchical relationship between these core concepts:

Figure 1. Classification of Advanced Functional Ceramics

Functional ceramics		
	Electronic ceramics	Piezoelectric, photoelectric, pyroelectric, ferroelectric, insulation
	Superconducting ceramics	Superconductivity, low temperature resistance
	Bioceramics	Good biocompatibility with blood and organs
	Magnetic ceramic	High permeability, coercivity, and hardness

In essence, all smart ceramics are functional ceramics, but not all functional ceramics are inherently "smart".Functional ceramics provide the foundation for specific applications, while smart ceramics represent a higher level of sophistication with dynamic environmental interaction.

Distinct from traditional ceramics, advanced ceramics are "produced from highly refined, purified, or chemically synthesized powder raw materials. Through precise chemical composition control, product structure design, and well-regulated manufacturing technologies, they are processed to obtain ceramics with excellent characteristics." These characteristics include high strength, high hardness, exceptional resistance to high temperatures, wear, and corrosion, alongside excellent electrical, optical properties, chemical stability, and biocompatibility (Bochenek et al., 2022).

With the rapid advancement of modern high-tech industries, advanced functional ceramics have become an indispensable part of the new materials sector (Chen et al., 2019). Their applications extend far beyond traditional industrial fields like machinery, chemical engineering, energy, and environmental protection (Cui et al., 2022). They are increasingly critical in high-tech domains such as aerospace, communication electronics, semiconductor microelectronics, biomedicine, national defense, high-speed rail, and new energy vehicles (Dorozhkin, 2010).

Research Status and Progress of Advanced Ceramic Materials

International Context and Milestones

The global development of advanced ceramics has been significantly propelled by strategic initiatives and breakthroughs in several technologically advanced regions. In the United States, defense and aerospace programs, such as the Integrated High Performance Turbine Engine Technology (IHPTET) program, have been instrumental in driving the development and application of high-temperature structural ceramics and ceramic matrix composites (CMCs) (Guillon, 2021). In Japan, national projects like the Fine Ceramics Research and Development Project in the 1980s established its leadership in electronic ceramics, particularly in multilayer ceramic capacitors (MLCCs) and silicon nitride bearings, emphasizing precision manufacturing and commercialization. In Europe, collaborative frameworks such as the European Union's Framework Programmes for Research and Technological Development have fostered cross-border cooperation in fundamental research on functional ceramics, bioceramics, and sustainable material processing technologies. These international efforts have collectively established a robust foundation of knowledge in powder synthesis, advanced sintering techniques, and microstructure engineering (Kim et al., 2014).

China's Research Trajectory and Achievements within the Global Landscape

The development of advanced ceramics in China began in the 1950s. Since the 1970s, numerous domestic universities and research institutes have focused on this field, achieving a series of innovative results. A landmark achievement was the innovative application of fiber-reinforced ceramic matrix composites in protective frames for strategic missiles and satellite antenna windows, demonstrating capabilities parallel to advanced international programs (Li et al., 2023). Furthermore, Chinese research on the phase diagrams of multi-phase nitrogen ceramics gained significant international influence, and the concept of multiphase composite ceramics contributed to developing numerous new materials with excellent comprehensive properties. Since the 1980s, national scientific research projects have tackled key technical challenges, such as utilizing high-temperature ceramics in critical components for high-efficiency engines-a goal also pursued by research institutions in the U.S., Germany, and Japan. This effort spurred R&D in critical technologies like composition design, grain boundary engineering, and various sintering and molding processes (e.g., pneumatic pressing, hot pressing, hot isostatic pressing, net-size molding) (Li et al., 2022), leading to major achievements in the industrialization of high-precision ceramic preparation technology in China.

Current Global Standing and Converging Trends

Entering the 21st century, China's advanced ceramic preparation technology has developed rapidly, breaking foreign monopolies and technological blockades across various sintering technologies, molding processes, powder preparation methods, and precision machining. The performance of functional ceramic materials developed by leading institutions, such as transparent wave-transmitting ceramics, laser ceramics, and ultra-high temperature ceramics, is close to or at the international advanced level, with many technologies successfully industrialized. Today, the global R&D focus is increasingly converging on areas such as multi-functional ceramics, nanotechnology integration, and sustainable manufacturing processes. Supported by strong market demand, the advanced ceramic industry continues to show a positive development trend, finding ever-broadening applications in both traditional industries and cutting-edge technological fields, with China playing an increasingly significant role in the global supply chain and innovation ecosystem (Mu et al., 2010).

Table 1. Overview of Key Properties of Representative Ceramic Materials

Material Category	Representative Material	Key Property 1	Key Property 2	Key Property 3	Primary Application Areas
Insulating Device Ceramics	Alumina (Al_2O_3)	Volume Resistivity (Ω·cm): $>10^{14}$	Dielectric Constant (1 MHz): 9-10	Flexural Strength (MPa): 300-400	IC Substrates, Electronic Device Packaging
Capacitor Ceramics	Barium Titanate ($BaTiO_3$)	Dielectric Constant (ε): 1,000 - 12,000	Curie Temperature (T_c): ~120°C	Dielectric Loss (tanδ): <0.01	Multilayer Ceramic Capacitors (MLCC)
Piezoelectric Ceramics	Lead Zirconate Titanate (PZT)	Piezoelectric Coefficient (d_{33}, pC/N): 300-600	Electromechanical Coupling Coefficient (k): 0.5-0.7	Curie Temperature (T): 180-350°C	Sensors, Transducers, Actuators
Microwave Dielectric Ceramics	($MgTiO_3$-$CaTiO_3$)	Dielectric Constant (ε): 20-25	Quality Factor (Q×f, GHz): 50,000-80,000	Temperature Coefficient of Frequency (τf, ppm/°C): ~0	Microwave Resonators, Filters, 5G/6G Communications
Bio-ceramics (Inert)	Yttria-Stabilized Zirconia (Y-TZP)	Fracture Toughness (MPa·m$^{1/2}$): 5-10	Flexural Strength (MPa): 800-1,200	Vickers Hardness (HV): 1,200-1,400	Dental Implants, Artificial Joint Femoral Heads
Bio-ceramics (Active)	Hydroxyapatite (HAp)	Bone Bonding Strength (MPa): ~10-15	Young's Modulus (GPa): 80-110	Degradation Rate: Controlled/Slow	Bone Defect Fillers, Implant Coatings

continued on following page

Table 1. Continued

Material Category	Representative Material	Key Property 1	Key Property 2	Key Property 3	Primary Application Areas
High-Temperature Superconducting Ceramics	Yttrium Barium Copper Oxide (YBCO)	Superconducting Transition Temperature (T , K): ~92	Critical Current Density (J , A/cm^2): >10^6 (77K)	—	Superconducting Magnets, Power Transmission, Filters
Structural Ceramics	Silicon Nitride (Si_3N_4)	Flexural Strength (MPa): 500-1,000	Fracture Toughness (MPa·m$^{1/2}$): 5-7	Thermal Conductivity (W/m·K): 20-30	Ceramic Bearings, Cutting Tools, Engine Components

Electronic Ceramics: Classification, Properties, and Applications

Electronic ceramics refer to the ceramic materials used in the preparation of various electronic components and devices in the electronics industry, which are inorganic non-metallic materials that use artificially refined inorganic powder as raw materials to achieve specific properties through structural design, accurate stoichiometry, appropriate molding methods and firing systems, and are processed to meet the requirements of dimensional accuracy. It can be widely used in energy, household appliances, automobiles, etc. For instance, the global market for multilayer ceramic capacitors (MLCCs), dominated by Japanese companies like Murata and TDK, relies heavily on the dielectric properties of barium titanate-based ceramics, an area where Chinese manufacturers have made significant strides in recent years (Shekhawat et al., 2021). Table 1 presents an overview of the key properties of major advanced ceramic materials, to facilitate a better understanding of the following content.

Classification of electronic ceramics

Electronic ceramics can be divided into five categories according to their function and use: insulating device ceramics, capacitor ceramics, ferroelectric ceramics, semiconductor ceramics and microwave dielectric ceramics.

Figure 2. (a) Insulating device ceramics; (b) Capacitor ceramics; (c) Semiconductor ceramics; (d) Ferroelectric ceramics; (e) Microwave dielectric ceramics

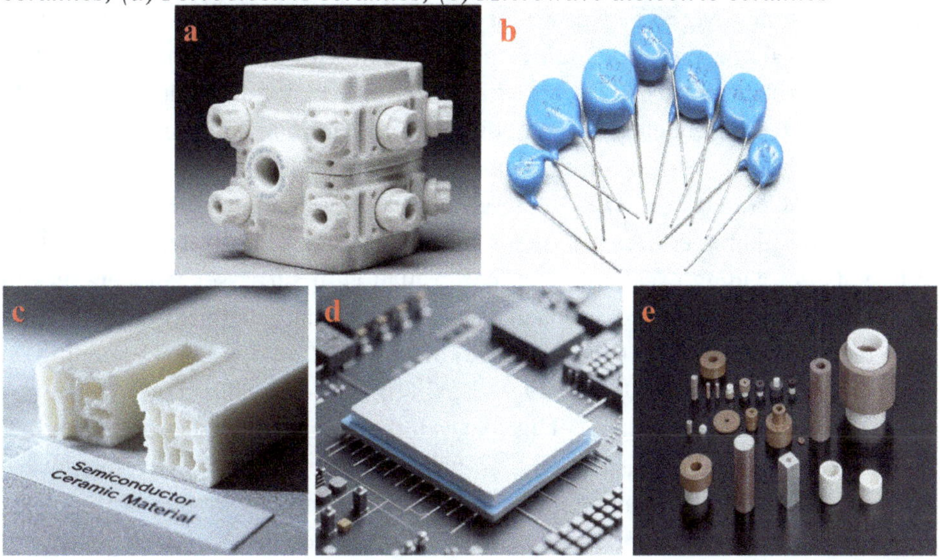

1. Insulating Device Ceramics: The Foundation of Electronic Reliability

Insulating device ceramics serve as the fundamental, non-functional backbone of modern electronics. Their primary role is to provide exceptional electrical isolation, ensuring signal integrity and preventing leakage currents, while simultaneously offering the structural integrity required to support and protect sensitive active components. They are ubiquitously employed as substrates, device packages, seals, and various structural parts in electronic systems, with common applications ranging from simple insulators and coil formers to sophisticated integrated circuit (IC) substrates and high-density interconnect (HDI) boards (Niska et al., 1994).

The performance of these ceramics is dictated by a stringent set of electrical, physical, and chemical property requirements. Electrically, a low dielectric constant (ε) is crucial to minimize parasitic capacitance, which is essential for preserving signal speed and integrity in high-frequency circuits. Similarly, a low dielectric loss tangent (tanδ) is imperative to reduce energy dissipation as heat, thereby improving circuit efficiency, especially in RF applications. High insulation resistivity (ρ) ensures minimal current leakage, maintaining circuit stability, while a high dielectric breakdown strength (E)guarantees reliable operation under high voltage stresses, preventing catastrophic failure. Beyond electrical properties, these materials must exhibit superior mechanical strength, thermal stability, and chemical inertnessto withstand manufacturing processes, thermal cycling, mechanical shock, and harsh operational environments. Widely utilized examples of these critical materials include:

Talc Ceramics ($MgSiO_3$):Known for their cost-effectiveness and excellent high-frequency insulating properties, talc ceramics feature a favorable combination of low dielectric constant and loss. They are a traditional and economical solution, typically employed in radio frequency (RF) circuits, coil skeletons, and insulating substrates where ultra-high thermal or mechanical performance is not the primary driver.

Alumina Ceramics (Al_2O_3): Representing the workhorse of the insulating ceramic family, alumina offers a superior and more balanced portfolio of properties. While also providing excellent electrical insulation, it surpasses talc ceramics in high-temperature resistance, mechanical robustness, and chemical durability. Its high thermal conductivity (for a ceramic) aids in heat dissipation. Depending on the purity (ranging from ~90% to 99.9+% Al_2O_3), its electrical and thermal performance can be precisely tailored. This makes it the dominant choice for demanding applications such as IC packages, high-power device substrates, RF/microwave windows, and structural components in corrosive or high-temperature environments. Its versatility and reliable performance have established it as the industry standard for high-reliability insulating applications.

2. **Capacitor ceramics:** Electronic ceramics used as capacitor dielectrics. This kind of ceramics has the largest amount and the largest variety of specifications. The main ones are high-frequency, low-frequency capacitor ceramic and semiconductor capacitor ceramic (Pattipaka et al., 2024).

High-frequency capacitor ceramics: Based on titanate solid solutions of alkaline earth or rare earth metals, they are used for high-stability and temperature-compensation capacitors in high-frequency circuits due to their stable performance.

Low-Frequency Capacitor Ceramics: Characterized by a very high dielectric constant, they are suitable for bypass, DC blocking, and filtering in low-frequency circuits, though they exhibit higher loss and temperature sensitivity.

Semiconductor capacitor ceramics: A high-performance type utilizing an insulating layer on a semiconducting ceramic grain boundary as the dielectric. This results in an extremely high apparent dielectric constant (up to 10^5), low loss (<1%), and high resistivity, enabling compact, high-reliability capacitors (Randhawa et al., 2024).

3. **Ferroelectric & Piezoelectric Ceramics:** These multifunctional materials are most notably applied for their piezoelectric effect (the conversion between mechanical stress and electrical charge), hence often being called piezoelectric ceramics. Lead zirconate titanate (PZT) is the most common base material. Applications include ultrasonic transducers, resonators, filters, and micro-

actuators. Significant research is focused on developing environmentally friendly lead-free alternatives (Selicani et al., 2025).

4. **Semiconductor Ceramics:** Their electrical conductivity varies significantly in response to external stimuli, making them ideal for various sensors:

Thermistor Ceramics: Include Negative Temperature Coefficient (NTC), Positive Temperature Coefficient (PTC), and Critical Temperature Resistor (CTR) types, used for temperature measurement, control, and compensation.

Photosensitive Ceramics: Exhibit changed conductivity under light illumination, used in optical switches and solar cells.

Gas-Sensitive Ceramics: Their conductivity changes upon exposure to specific gases, enabling applications in leak detection and air quality monitoring.

Humidity-Sensitive Ceramics: Used for humidity measurement and control as their electrical properties are sensitive to water vapor.

5. **Microwave dielectric ceramics:** mainly titanate, niobate, zincate-based ceramics or glass ceramics. It is a kind of high-frequency, low-loss, temperature-stable dielectric material, which has been widely used in microwave resonators, filters, oscillators, phase shifters, microwave capacitors, antennas and microwave substrates, etc., and is the key material for modern microwave communication such as mobile communications, satellite communications, global positioning systems, bluetooth technology and wireless local area networks.

Figure 3. Applications of microwave dielectric ceramics.

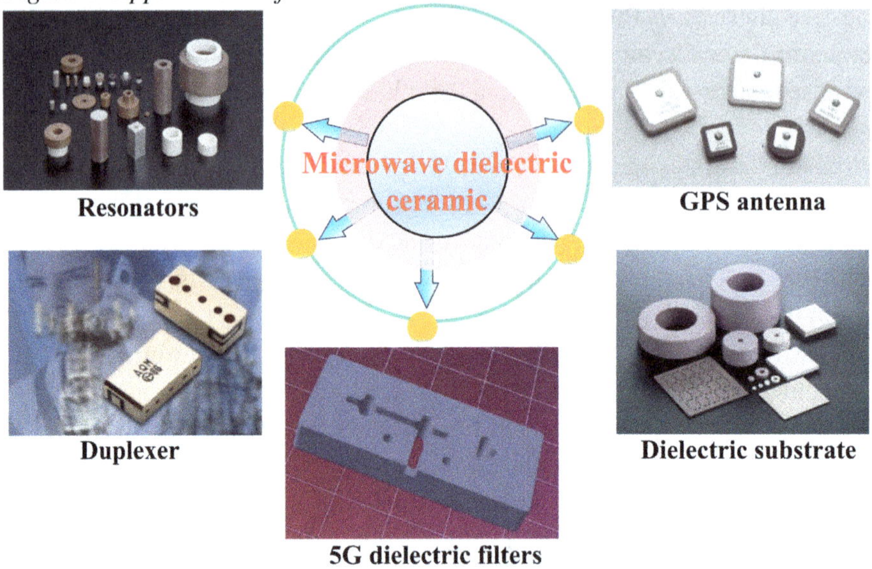

Applications of electronic ceramics materials

The unique properties of the ceramic classes described above directly enable their use in critical high-tech sectors. The following applications demonstrate the translation of material properties into real-world technological solutions.

Optical communication: Components like TOSA (Transmitter Optical Sub-Assembly) and ROSA (Receiver Optical Sub-Assembly), lasers, and optical switches rely on specialized ceramics for their stability and hermeticity. These are essential for backbone networks, data centers, and emerging technologies like LiDAR and laser processing.

Wireless communication systems fundamentally rely on ceramic-based power devices and modules, enabled by the exceptional low-loss properties of microwave dielectric ceramics. In digital mobile communication networks like 5G, ceramic filters and resonators are essential for enabling high-frequency signal transmission with minimal interference and loss, supporting greater bandwidth and network capacity. For critical point-to-point microwave links used in backbone infrastructure, these ceramics ensure stable, long-distance signal integrity. Furthermore, in wireless broadband access networks, ceramic components in radio frequency front-ends and power amplifiers enhance signal efficiency and coverage. Their high quality factor (Q), thermal stability, and tunable permittivity are key to developing compact, high-efficiency, and reliable next-generation communication hardware.

Industrial lasers: used in the packaging of various fiber lasers. Industrial lasers are mainly used in industrial shipbuilding, automobile manufacturing, laser engraving, laser marking, laser cutting, printing roller making, metal and non-metal drilling/cutting/welding, military defense security, medical instruments and equipment, etc.

Consumer electronics:The demand for miniaturization and high performance drives the use of ceramic packages and substrates in smartphones and other portable devices, where they provide excellent heat dissipation, high airtightness, and signal integrity.

Automotive electronics: The harsh under-hood environment requires robust materials. Ceramics are used in sensors (e.g., pressure, level), electronic control units (ECUs), and glow plugs for diesel engines, ensuring reliability and longevity.

Figure 4. Application areas of electronic materials.

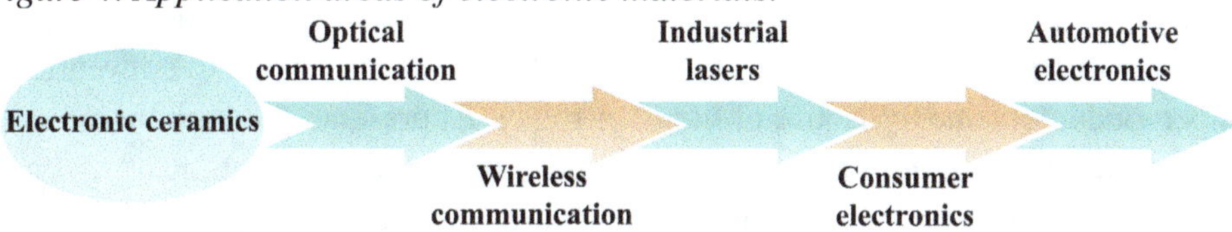

Superconducting Ceramics

Superconducting ceramics refer to ceramic materials that exhibit superconducting properties under specific conditions (low temperatures). It is fully diamagnetic when its resistance drops to zero (Meissner effect). Unlike conventional metal or alloy superconductors (e.g., NbTi, Nb_3Sn), superconducting ceramics typically have higher critical temperatures (Wan et al., 2024). These materials are of interest due to their non-traditional superconducting mechanisms (e.g., high-temperature superconductors) and their potential applications. Since the major breakthrough in the research of superconducting ceramics in the 1980s, the field has attracted significant attention, with China's research in this area reaching the world's advanced level. At present, the application of high-temperature superconducting materials is developing towards high-current applications, electronic applications, diamagnetism and other aspects.

High-temperature superconducting ceramic

High-temperature superconducting ceramic materials are a class of ceramic compounds that exhibit superconducting properties at relatively high temperatures, typically above the boiling point of liquid nitrogen of 77 K (-196 °C). Compared with traditional metal or alloy superconductors (such as Nb_3Sn, critical temperature Tc≈18K), the discovery of high-temperature superconducting ceramic materials breaks through the temperature limit of superconducting applications and has important scientific significance and technical potential.

In 1986, lanthanum-barium-copper-oxygen and yttrium-barium-copper-oxygen high-temperature superconducting materials were successively introduced, which pushed the application of superconductivity technology onto the world stage. Like conventional superconductors, these new materials transmit current without resistance, but the difference between the two is that low-temperature superconducting alloys must be cooled to below -250 C, while high-temperature superconducting ceramics only need to be cooled to -140 °C. Therefore, liquid nitrogen, which is much cheaper than liquid helium and can be supplied in large quantities, can be used to achieve this condition. In many applications, the cost of high-temperature superconducting ceramics can be reduced by 10%~50% compared to conventional superconductors due to the use of liquid nitrogen. At present, although there is still a long way to go before the production and widespread application of low-cost high-temperature superconducting ceramics are realized, and there are still many difficulties in achieving this goal, high-temperature superconducting ceramics have indeed gained real applications from power transmission to microwave technology and radio communication.

Figure 5. High-temperature superconducting ceramic materials

Application prospects of high-temperature superconducting materials

1. Microwave and radio communication filters are a potential application area for high-temperature superconducting ceramic materials. Ybtrium-barium copper oxygen $YBa_2Cu_3O_{7-8}$ (YBCO) filters are used in mobile phone paging stations, and their anti-interference ability is 1000 times higher than that of copper filters, so the combination of YBCO filters and cryogenic amplifiers in paging stations will open up the first substantial market for high-temperature superconducting electronics (Mustafa et al., 2026).

2. The thin film process is the core of microelectronics applications, and the YBCO coated metal strips prepared from it will be possible to be made into superconducting long bands (using flexible metal substrates) for power transmission and enter the field of electrical applications. At the same time, in the preparation of bismuth tie tape, it is rare to find single-strand and multi-strand strips of hundreds to thousands of meters that meet the minimum requirements. Therefore, the high-temperature superconducting materials in large-scale applications in the form of strip (wire) and blocks in underground transmission cables, high-temperature superconducting generators, maglev trains, etc. are also just around the corner.

3. The structural devices used in metrology, sensors, and detectors will also rely on the properties of high-temperature superconducting ceramic materials to grow.

Bioceramics

Bioceramics are a class of ceramic materials used in the fields of medicine and bioengineering, with excellent biocompatibility, chemical stability and mechanical properties (Engstrand, et al.,2015). They are widely used in the fields of human hard tissue repair and replacement, drug delivery, tissue engineering and medical devices. Bioceramics can form a good combination with biological tissues and even promote tissue regeneration, so they have important value in clinical medicine (ValletRegí et al., 2019).

Bioceramics materials as biomedical materials began in the early 18th century. In early 1808, ceramic teeth for dental prostheses were successfully made, and in 1871, hydroxyapatite was synthesized artificially (Han et al., 2020). In 1894, H. Dreeman reported the use of plaster as a bone replacement material. In 1928, Leriche and Policard began researching and applying calcium phosphate as a bone replacement material (Salimi et al., 2020). 1963 was also an important year in the history of bioceramics, when Smith reported the development of a ceramic bone substitute. Due to technical limitations, it was not until 1971 that hydroxyapatite was successfully developed and expanded to clinical applications. In the early 70s of the 20th century, China began to study bioceramics and used them in clinical practice. In 1974, the research on the use of glass-ceramic in artificial joints was carried out; In 1977, alumina ceramics were clinically applied; In 1979, high-purity alumina single crystal was used in clinical practice, and later new bioceramic materials continued to appear and were used in clinical practice. China Shanghai Institute of Ceramics, South China University of Technology, Beijing Institute of Stomatology and other units have conducted in-depth research on bioceramics. The application scope of bioceramics is also gradually expanding, and can now be applied to artificial bone, artificial joints, artificial tooth roots, bone filling materials, bone replacement materials, osseointegration materials, artificial heart valves, artificial tendons, artificial blood vessels, artificial trachea, percutaneous leads can be applied to in vivo medical monitoring, etc (Cavelier et al.,2021). In recent years, with the development of nanotechnology and 3D printing, the application range of bioceramics has been further expanded (Chen et al., 2019).

Classification of bioceramics

1. Traditional bioceramics: Traditional bioceramic materials are mainly divided into the following two categories according to their interaction with biological tissues and their performance characteristics:

Biologically inert ceramic materials: These ceramic materials have little chemical reaction in the body and form a mechanical connection with the surrounding tissue, rather than a chemical bond. They usually form a fibrous tissue wrap around them. In addition, the structure of these materials is relatively stable, the bond force in the molecule is strong, and all have high mechanical strength, wear resistance and chemical stability. It is mainly composed of oxide ceramics, non-oxide ceramics and ceramics. Among them, Al, Mg, Ti and Zr oxides are the most widely used. Typical examples include alumina (Al_2O_3) and zirconia (ZrO_2), which are commonly used in dental restorations and coatings on the surface of artificial joints, and are favored for their high wear resistance and stability (Vaiani et al., 2024).

Bioactive ceramic materials: These ceramic materials can chemically react with surrounding tissues to form chemical bonds, which can be tightly bonded to biological tissues and promote tissue regeneration. Bioactive ceramics include surface bioactive ceramics and bioabsorbent ceramics (biodegradable ceramics). Biosurfactant ceramics usually contain hydroxyl groups and can also be made porous, allowing biological tissues to grow into and bond strongly to their surfaces. The characteristics of bioresorbable ceramics are: they can be partially or fully absorbed, and can induce the growth of new bones in the organism. Bioactive ceramics mainly include: hydroxyapatite and calcium phosphate.

The production process of traditional bioceramics is relatively mature and the cost is low. However, with the development of science and technology, traditional bioceramics are gradually unable to meet the needs of high-end medical applications in terms of performance and biological activity. In order to improve the performance of bioceramics, researchers are exploring new synthesis methods and modification technologies to develop new bioceramic materials with higher bioactivity, stronger mechanical properties and better biocompatibility.

2. Functional bioceramics: Functional bioceramics are a class of bioceramic materials with specific functions, such as antibacterial, anticoagulant, sustained-release drugs, etc (Punj et al., 2021). These materials are designed and prepared through special processes that endow bioceramics with additional properties that make them have a wider range of applications in the medical field. For example, antimicrobial bioceramics can effectively inhibit the growth of bacteria by releasing antibacterial substances such as silver ions, and are widely used in the surface treatment of medical devices. Anticoagulant bioceramics can reduce the formation of thrombosis at the surgical site and improve the success rate of surgery. The R&D and application of

functional bioceramics are of great significance for improving the performance of medical devices and the therapeutic effect of patients. For example, sustained-release drugs bioceramics can encapsulate drugs in materials, and through the interaction between bioceramics and tissues, the slow release of drugs can be realized, thereby improving drug utilization and therapeutic efficacy. In addition, functional bioceramics can also be optimized in terms of mechanical properties, biocompatibility and bioactivity by regulating their microstructure and composition. With the rapid development of nanotechnology, bioengineering and other fields, work remarkable progress has been made in the research and development of bioceramics. New functional bioceramic materials are emerging, such as nanocomposite bioceramics and smart bioceramics, which have great application potential in the biomedical field. Therefore, how to reduce costs and improve biosafety while ensuring performance is the key to the future research and development of functional bioceramics.

Application areas of bioceramic materials

1. Medical devices: Medical devices are one of the main application fields of bioceramics, and bioceramic materials play an important role in medical devices due to their excellent biocompatibility, biodegradability and mechanical properties. In the field of orthopedics, bioceramic materials are widely used in artificial joints, bone plates, bone nails and other implants, which can achieve good fusion with human bones and effectively improve the success rate of surgery. In addition, bioceramics are also widely used in the dental field, such as dental restoration, implants, etc., these products can provide stable support and good aesthetic results (Jayaswal et al., 2010). The application of bioceramics in the field of medical devices is also reflected in its versatility. For example, bioceramics with antimicrobial properties can form a protective film on the surface of the implant, effectively preventing bacterial infection. At the same time, the sustained-release drugs bioceramics can gradually release the drug during the implantation process, reducing the postoperative pain and infection risk of patients. In addition, the research and development of intelligent bioceramic materials has also brought new possibilities in the field of medical devices, such as real-time monitoring of patients physiological conditions through bioceramic implants. With the continuous progress of bioceramic material technology, its application range in the field of medical devices is also expanding (ValletRegí et al., 2011). For example, bioceramic materials can be used in cardiovascular interventional devices such as cardiac stents and endovascular stents to improve the safety and efficacy of surgery. In addition, the application of bioceramics in tumor treatment and drug delivery has also shown great potential. In the future, with the further improvement of the performance of bioceramic materials and the

reduction of costs, their application in the field of medical devices will be more extensive, bringing more benefits to patients.

2. Bone tissue engineering: Bone tissue engineering is one of the important fields of bioceramic application, and bioceramic materials are used as scaffold materials in this field to provide an environment for the growth and differentiation of bone cells. Bioceramic scaffolds have good biocompatibility and biodegradability, and can be gradually replaced by new bone tissue in vivo, thereby promoting bone regeneration. In bone tissue engineering, commonly used bioceramic materials include hydroxyapatite, calcium phosphate, etc., which can mimic the chemical composition of natural bone and provide a suitable attachment and growth environment for bone cells. The application of bioceramics in bone tissue engineering is mainly reflected in the following aspects: firstly, bioceramic scaffolds can be used as temporary scaffolds for bone defect repair to provide support for the growth and vascularization of bone cells; Secondly, bioceramic materials can be combined with bioactive substances such as growth factors and cells to form composite scaffolds to further improve the effect of bone regeneration. Finally, bioceramic scaffolds can also enhance their biological activity and promote rapid healing of bone tissue through surface modification technologies, such as nanocoatings. With the continuous progress of bioceramic material technology, its application in the field of bone tissue engineering is also expanding. For example, new bioceramic materials, such as carbon nanotube composite bioceramics and bioactive glass, provide more options for bone tissue engineering due to their unique properties, such as enhanced mechanical properties and biocompatibility (Engstrand et al., 2015). In addition, the application of bioceramic materials in tissue engineering, drug delivery, and bioprinting has also shown great potential. With the further improvement of the performance of bioceramic materials and the deepening of clinical applications, major breakthroughs are expected to be achieved in the field of bone tissue engineering, bringing new hope for the treatment of patients with bone diseases.

3. Dental applications: Dental applications are an important market for bioceramic materials, and the applications of bioceramics in the dental field mainly include dental restoration, implants, crowns and bridges. Bioceramic materials are widely used in dental implants due to their excellent biocompatibility and mechanical properties. For example, bioceramic implants are able to form a good osseointegration with the alveolar bone to provide stable support for the teeth. In addition, bioceramic materials are also not easy to corrode, not easy to allergies, etc., which makes them have a wide range of application prospects in the field of dentistry. With the continuous development of bioceramic material technology, its application in the field of dentistry is becoming more diverse. New bioceramic materials, such as nano-bioceramics and smart bioceramics, provide more options for dental treatment due to their special properties, such as enhanced mechanical strength, antimicrobial

properties and drug release capabilities. In addition, the application of bioceramic materials in dental cosmetology and dental treatment instruments is also expanding. With the further breakthrough of bioceramic material technology, its application in the field of dentistry will be more extensive, bringing more personalized and efficient oral medical services to patients (Surana et al., 2024).

4. Other applications: Bioceramic materials are also being used in other fields, including burn treatment, skin repair, tissue engineering, etc. In the treatment of burns, bioceramic dressings are able to absorb wound exudate and keep the wound moist, while having antibacterial properties that aid in wound healing. In the field of skin repair, bioceramic materials can be used to make skin substitutes, providing good biocompatibility and mechanical properties to promote the growth of new skin. The application of bioceramics in tissue engineering is mainly reflected in the construction of artificial tissues and organs. By combining bioceramic materials with bioactive substances, cells, etc., biologically active scaffolds can be formed for replacing damaged tissues or organs. For example, in the cardiovascular sector, bioceramics can be used to build artificial blood vessels; In neural tissue engineering, bioceramic scaffolds can promote the growth and regeneration of nerve cells. In addition, bioceramic materials have also shown their unique application value in environmental protection, aerospace, automotive industry and other fields (Hench, 1991). In the field of environmental protection, bioceramics can be used to treat wastewater, waste gas, and soil remediation. In the aerospace and automotive industries, bioceramic materials are used to manufacture high-performance components due to their lightweight, high strength, and high temperature resistance. With the continuous progress of bioceramic material technology, its application potential in more fields is gradually being tapped, providing a new solution for the sustainable development of human society.

Magnetic Ceramic

Magnetic ceramics are a class of inorganic non-metallic functional materials with ferrite as the core, mainly composed of iron oxide compounded with other metals (such as barium, strontium, manganese, zinc, etc.) or rare earth oxides. Its core characteristics include high resistivity, low eddy current loss, excellent chemical stability and high temperature resistance, and are widely used in electronics, machinery, energy, medical and other fields (Sanjabi et al., 2025). With the continuous progress of science and technology, the research and development of magnetic ceramic materials have achieved remarkable results, and their properties and functions are increasingly rich. At present, the magnetic ceramics industry has become one of the important strategic emerging industries in the world.

The development of magnetic ceramics began in the 30s of the 20th century, and its industrial production was driven by radar and communication needs during World War II. In the 50s of the 20th century, the development of garnet-type ferrite (such as $Y_3Fe_5O_{12}$) marked the maturity of the material system. China's magnetic ceramic industry originated in the 50s of the 20th century, with the acceleration of the country's industrialization process, magnetic ceramic materials have gradually been widely used in electronics, automobiles, energy and other fields. From the initial simple magnetic cores and magnetic components to today's precision magnetic ceramic devices, China's magnetic ceramics industry has experienced a process from imitation to innovation. During this period, the strong support of national policies, the continuous breakthrough of scientific research institutions and the continuous improvement of enterprise technology have jointly promoted the rapid development of the industry. In the 80s of the 20th century, China's magnetic ceramics industry began to enter a stage of rapid development, and domestic enterprises have introduced foreign advanced technology to produce internationally competitive magnetic ceramic products. During this period, the application of magnetic ceramics in electronic information technology, automobile industry, energy utilization and other aspects has been significantly improved, and the market demand has been expanding. At the same time, domestic scientific research institutions have increased their investment in the basic theoretical research of magnetic ceramics, which has laid a solid foundation for the technological innovation of the industry. Entering the 21st century, China's magnetic ceramics industry has entered a new stage of development. With the continuous emergence of new materials and new processes, the performance of magnetic ceramics has been greatly improved. In particular, major breakthroughs have been made in the research of high-performance magnetic ceramic materials and nano-magnetic ceramic materials.

Classification and characteristics of magnetic ceramics

According to their performance and application, magnetic ceramic materials can be mainly divided into three categories: soft magnetic ceramics, hard magnetic ceramics and microwave magnetic ceramics.

1. Soft magnetic ceramics have good permeability and low loss characteristics, and are suitable for electronic components such as power supply filters and transformers. The characteristics of soft magnetic ceramics include: high permeability, low loss, good temperature stability and chemical stability. This kind of products are widely used in electronics, automobiles, energy and other fields, especially in

the electronics industry, soft magnetic ceramics are the core materials of key components such as power filters and transformers (Vassilios et al., 2004).

2. Hard magnetic ceramics have high permeability and magnetic saturation strength, and are suitable for magnetic components such as motors and sensors. Hard magnetic ceramic materials are characterized by: high permeability, high magnetic saturation strength, good mechanical strength and corrosion resistance, suitable for manufacturing motors, sensors and other magnetic components (Tavoosi et al., 2023).

3. Microwave magnetic ceramics have excellent microwave performance and are suitable for microwave devices such as radar and satellite communications. The characteristics of microwave magnetic ceramics include: high dielectric constant, low loss, good temperature stability and chemical stability, which are suitable for the manufacture of microwave devices (Kempa et al., 2024).

With the continuous progress of science and technology, the performance and application range of magnetic ceramic products will be further expanded, providing more high-performance magnetic materials for various industries.

Application field and development of magnetic ceramics

The application field of magnetic ceramic materials continues to expand, from the initial electronics, automotive industry, and gradually extended to new energy, aerospace, medical equipment and other high-tech fields.

1. In the electronics industry, magnetic ceramics are widely used in key components such as high-frequency filters, transformers, motors, etc., and their high-performance magnetism and heat resistance meet the high requirements of electronic equipment for material performance.

2. In the automotive industry, magnetic ceramic materials are widely used in key components such as motors, generators, and sensors, which are of great significance for improving vehicle performance and saving energy and reducing consumption. With the rise of new energy vehicles, the application prospect of magnetic ceramics in the automotive field is broader.

3. In the field of new energy, the application of magnetic ceramics in wind power generation and solar power generation is increasing. For example, in wind generator, magnetic ceramic materials are used to make high-performance permanent magnet motors, improving power generation efficiency and reliability. In the field of solar energy, magnetic ceramic materials are used in photovoltaic power generation systems to optimize the current path and improve the conversion efficiency of photovoltaic modules.

4. There is also a growing demand for magnetic ceramic materials in the aerospace sector. In aerospace vehicles, magnetic ceramic materials are used to manufacture key components such as engine parts and sensors due to their lightweight, high

strength, and high temperature resistance. In addition, in the field of medical devices, the application of magnetic ceramic materials is also gradually expanding, such as for the manufacture of magnetic components in cardiac pacemakers, to ensure the stability and reliability of the device (Soboyejo et al., 2015).

With the continuous progress of technology and the improvement of material properties, the application field of magnetic ceramics is expected to be further expanded.

Prospect and Development Trend of Advanced Functional Ceramic Materials

As a cornerstone of modern industry, advanced functional ceramic materials are continuously evolving, driven by technological breakthroughs and burgeoning market demands. Their future development is characterized by precision at the micro-level, integration of multiple functions, and expansion into new technological frontiers.

Technology Development Trends: From Micro-Nano Engineering to Intelligence

The pursuit of superior performance is pushing the boundaries of material design and processing towards greater precision and complexity.

Precision in Microstructure Control: The future lies in precisely engineering ceramics at the micro- and nano-scale. Technologies like spark plasma sintering (SPS) and field-assisted sintering techniques (FAST) enable the fabrication of ceramics with ultrafine or even nano-sized grains. This results in exceptional mechanical properties, such as superplasticity and ultra-high strength, by effectively inhibiting crack propagation. For instance, the development of nano-grained zirconia (ZrO_2) significantly enhances its fracture toughness, making it more reliable for demanding applications like artificial joints and cutting tools.

Advanced Composite Architectures: The trend is moving beyond single-phase materials towards sophisticated multi-material systems. Ceramic Matrix Composites (CMCs), particularly those reinforced with carbon or silicon carbide fibers, exemplify this direction. By combining the high-temperature resistance and wear resistance of ceramics with the toughness of fibers, CMCs overcome the inherent brittleness of monolithic ceramics. They are becoming indispensable in aerospace for components like engine turbine blades and thermal protection systems, where their high strength-to-weight ratio and thermal stability are critical. Similarly, Functionally Graded Materials (FGMs), which exhibit a gradual transition in composition and structure, are designed to manage extreme thermal stresses in applications like fusion reactor walls.

Integration of Intelligent Functions: The next frontier is embedding "intelligence" into ceramics. Piezoelectric ceramics (e.g., lead zirconate titanate, PZT, and its lead-free alternatives like potassium sodium niobate, KNN) are evolving from simple actuators and sensors into core components for energy harvesting and self-powered sensing systems (Lee et al., 2011). Furthermore, research into self-healing ceramics, materials that can autonomously repair micro-cracks, often through the oxidation of embedded phases like silicon carbide (SiC) at high temperatures-promises to dramatically extend the service life and reliability of components in extreme environments.

Expansion of Application Frontiers: Driven by Emerging Industries

The unique properties of advanced ceramics are catalyzing innovations across a spectrum of high-tech sectors.

Next-Generation Electronics and Communications: The rollout of 5G/6G networks, the Internet of Things (IoT), and artificial intelligence (AI) demands ceramic materials with higher frequency, better thermal management, and greater integration density. Silicon Nitride (Si_3N_4)substrates are gaining traction over aluminum oxide due to their high thermal conductivity and excellent mechanical strength, making them ideal for high-power semiconductor packaging. Low-Temperature Cofired Ceramics (LTCC) and glass-ceramic compositesare key for creating miniaturized, high-frequency modules and antennas. For future terahertz applications, ultra-low loss dielectric ceramics based on $ZrTiO_4$ or $MgTiO_3$ systems are under intensive research.

Sustainable Energy and Environmental Technologies:In alignment with global carbon neutrality goals, ceramics are pivotal in green technologies. In the hydrogen economy, dense zirconia (YSZ)membranes are essential for high-efficiency solid oxide electrolysis cells (SOEC) for hydrogen production, while porous silicon carbide (SiC)and alumina (Al_2O_3)ceramics serve as robust supports for catalysts in fuel reforming and exhaust gas purification. For energy storage, ceramic coatings and solid electrolytes (e.g., garnet-type $Li_7La_3Zr_2O_{12}$) are critical for improving the safety and energy density of next-generationall-solid-state batteries.

Advanced Biomedical and Aerospace Systems: The trend in biomedicine is towards personalization and bioactivity. Additive manufacturing (3D printing) of bioactive glasses and hydroxyapatite (HAp)scaffolds allows for patient-specific implants that promote bone regeneration. In aerospace, the application of Ultra-High Temperature Ceramics (UHTCs), such as ZrB_2 and HfC-based composites, is being matured for hypersonic vehicle leading edges and propulsion components, where they must withstand temperatures exceeding 2000 °C (Moslim et al., 2017).

In summary, the future of advanced functional ceramic materials is one of convergence-where materials science, nanotechnology, and digital manufacturing intersect. The focus will shift from “material manufacturing” to “material systems engineering”, creating intelligent, multi-functional, and highly reliable components that are fundamental to the technological advancements of the coming decades.

REFERENCES

Alallaq, A. A., & Kashan, J. S. (2023). A review: In vivo studies of bioceramics as bone substitute materials. *Nano Select*, *4*(2), 123–144. https://doi.org/. DOI: 10.1002/nano.202200222

Best, S. M., Porter, A. E., Thian, E. S., & Huang, J. (2008). Bioceramics: Past, present and for the future. *Journal of the European Ceramic Society*, *28*(7), 1319–1327. DOI: 10.1016/j.jeurceramsoc.2007.12.001

Bhadrakumari, S., & Predeep, P. (2011). Performance of high temperature superconducting ceramic/thermoplastic polymer composites. *Journal of Polymer Engineering*, *31*(2-3), 205–207. DOI: 10.1515/polyeng.2011.043

Bochenek, D. (2022). Advanced ceramic materials with functional properties. *Materials (Basel)*, *15*(18), 6439–6440. DOI: 10.3390/ma15186439 PMID: 36143751

Cavelier, S., Mirmohammadi, S. A., & Barthelat, F. (2021). Titanium mesh-reinforced calcium sulfate for structural bone grafts. *Journal of the Mechanical Behavior of Biomedical Materials*, *118*, 10446–10450. DOI: 10.1016/j.jmbbm.2021.104461 PMID: 33756418

Chen, Z., Li, Z., Li, J., Liu, C., Lao, C., Fu, Y., & He, Y. (2019). 3D printing of ceramics: A review. *Journal of the European Ceramic Society*, *39*(4), 661–687. https://doi.org/. DOI: 10.1016/j.jeurceramsoc.2018.11.013

Cui, W., Cao, L., & Zhang, X. (2022). Bioceramics: Materials, properties, and applications. *Ceramic Science and Engineering*, *65*, 110. DOI: 10.1016/B978-0-323-89956-7.00017-6

Dorozhkin, S. V. (2010). Calcium orthophosphates as bioceramics: State of the art. *Journal of Functional Biomaterials*, *1*(1), 22–107. DOI: 10.3390/jfb1010022 PMID: 24955932

Engstrand, T., Kihlström, L., Lundgren, K., Trobos, M., Engqvist, H., & Thomsen, P. (2015). Bioceramic implant induces bone healing of cranial defects. *Plastic and Reconstructive Surgery. Global Open*, *3*(8), e491. DOI: 10.1097/GOX.0000000000000467 PMID: 26495204

Guillon, O. (2021). Ceramic materials for energy conversion and storage: A perspective. *International Journal of Ceramic Engineering & Science*, *3*(3), 100–104. DOI: 10.1002/ces2.10086

Han, Y., Gao, R., Bajpai, I., Kim, B., Yoshida, H., Nieto, A., Son, H.-W., Yun, J., Jang, B.-K., Jhung, S., Jingming, Z., Hwang, K.-H., Chen, F., Shackelford, J. F., & Kim, S. (2020). Spark plasma sintered bioceramics-from transparent hydroxyapatite to graphene nanocomposites: A review. *Advances in Applied Ceramics*, *119*(2), 57–74. DOI: 10.1080/17436753.2019.1691871

Hench, L. (1991). Bioceramics: From concept to clinic. *Journal of the American Ceramic Society*, *74*(7), 1487–1510. DOI: 10.1111/j.1151-2916.1991.tb07132.x

Jayaswal, G. P., Dange, S. P., & Khalikar, A. N. (2010). Bioceramic in dental implants: A review. *Journal of Indian Prosthodontic Society*, *10*(1), 8–12. DOI: 10.1007/s13191-010-0002-4 PMID: 23204715

Kempa, M., Bovtun, V., & Repek, D. (2024). Microwave magnetic excitations in U-type hexaferrite $Sr_4CoZnFe_{36}O_{60}$ ceramics. *Journal of Applied Physics*, *136*(7), 15. DOI: 10.1063/5.0222910

Kim, I. D. (2014). Editorial: Advances in functional ceramic materials. *Journal of Electroceramics*, *33*(1), 11–15. DOI: 10.1007/s10832-014-9970-9

Lee, S. G., Seo, J. Y., Lee, J. W., Park, W. B., Sohn, K. S., & Pyo, M. (2021). Composition-tuned lithium aluminosilicate as a new humidity-sensing ceramic material with high sensitivity. *Sensors and Actuators. B, Chemical*, *339*, 129928. DOI: 10.1016/j.snb.2021.129928

Li, M., Wu, D., Zhu, Y., Guan, L., Huang, Y., & Song, X. (2023). Low-temperature preparation of $La_{0.9}Ca_{0.1}CrO_3$ semiconductor ceramics by flash sintering. *Ceramics International*, *49*(17), 29364–29369. DOI: 10.1016/j.ceramint.2023.06.203

Li, Q., Chen, Y., Yu, C., Young, L., Spector, J., & Harris, V. G. (2022). Emerging magnetodielectric materials for 5G communications: 18H hexaferrites. *Acta Materialia*, *231*, 117854. DOI: 10.1016/j.actamat.2022.117854

Moslim, N., Beh, C., Kasim, S., & Ramakrishnan, S. (2017), Effect of composition and temperature to the HA/beta-TCP composite, in: *Regional Conference on Materials and Asean Microscopy Conference*. 1082, 12023-12027. https://doi.org/DOI: 10.1088/1742-6596/1082/1/012023

Mu, Y. G. (2010). Applications of bioceramics and its future. *Yaowu Fenxi Zazhi*, *30*(7), 1343–1347. DOI: 10.16155/j.0254-1793.2010.07.058

Mustafa, B. M. (2026). Crystal symmetry effects on oxygen doping distribution, SC, and phase diagram in $YBa_2Cu_3O_7$. Journal of Superconductivity and Novel Magnetism. *Journal of Superconductivity and Novel Magnetism*, *39*(55), 152. DOI: 10.1007/s10948-026-07136-5

Niska, J., & Loberg, B. (1994). Hot isostatic pressing of high temperature superconducting ceramics. *Materials Technology*, *9*(9-10), 211–215. DOI: 10.1080/10667857.1994.11785072

Pattipaka, S., Lim, Y., Son, Y. H., Bae, Y. M., Peddigari, M., & Hwang, G. T. (2024). Ceramic-based dielectric materials for energy storage capacitor applications. *Materials (Basel)*, *17*(10), 2277. DOI: 10.3390/ma17102277 PMID: 38793340

Punj, S., Singh, J., & Singh, K. (2021). Ceramic biomaterials: Properties, state of the art and future prospectives. *Ceramics International*, *47*(20), 28059–28074. DOI: 10.1016/j.ceramint.2021.06.238

Randhawa, K. S. (2024). Advanced ceramics in energy storage applications: Batteries to hydrogen energy. *Journal of Energy Storage*, *98*, 113122. DOI: 10.1016/j.est.2024.113122

Reclaru, L., Ardelean, L., Miu, C., & Grecu, A. (2020). Are zirconia bioceramics and ceramics intended to come in contact with skin inert. *Materials (Basel)*, *13*(7), 1697–1672. DOI: 10.3390/ma13071697 PMID: 32260495

Salimi, E. (2020). Functionally graded calcium phosphate bioceramics: An overview of preparation and properties. *Ceramics International*, *46*(12), 19664–19668. DOI: 10.1016/j.ceramint.2020.05.077

Sanjabi, S., & Salehirad, A. (2025). $MgAl_2O_4/CoFe_2O_4$/Zeolite a magnetic ceramic nanocomposite for efficacious elimination of Methylene blue from aqueous solutions: Composition influence. *Research on Chemical Intermediates*, *51*(9), 4907–4935. DOI: 10.1007/s11164-025-05679-1

Selicani, G. V., Mobin, M., Esposito, V., Insinga, A. R., & Haugen, A. B. (2025). Shape engineering and manufacturing of piezoceramics for energy conversion-a review. *Journal of Physics: Energy*, *7*(2), 022004. DOI: 10.1088/2515-7655/adb0ee

. Shekhawat D., A. Singh, M. Banerjee, T. Singh, A (2021). Patnaik, Bioceramic composites for orthopaedic applications: a comprehensive review of mechanical, biological, and microstructural properties, Ceramics International, 47, 3013e3030. https://doi.org/.DOI: 10.1016/j.ceramint.2020.09.214

Soboyejo, W., Obayemi, J., Annan, E., Ampaw, E. K., Daniels, L., & Rahbar, N. (2015). Review of high temperature ceramics for aerospace applications. *Advanced Materials Research*, *1132*, 385–407. . DOI: 10.4028/www.scientific.net/AMR.1132 .385

Surana, P., Singh Dhull, K., Arya, A., Samreen, S., Rajan, M., & Parihar, A. S. (2024). Bio-ceramics application in dentistry. *Bioinformation, 20*(2), 136–139. DOI: 10.6026/973206300200136 PMID: 38497078

Tavoosi, M. (2023). The precipitation of $PbFe_{12}O_{19}$ hard magnetic phase in PbO-SiO_2-B_2O_3-Fe_2O_3 glass-ceramics. Journal of Non-Crystalline Solids: A Journal Devoted to Oxide, Halide, Chalcogenide and Metallic Glasses, Amorphous Semi-conductors, Non-Crystalline Films. *Journal of Non-Crystalline Solids, 615*, 122402. DOI: 10.1016/j.jnoncrysol.2023.122402

Vaiani, L., Boccaccio, A., Uva, A. E., Palumbo, G., Piccininni, A., Guglielmi, P., & Ballini, A. (2023). Ceramic materials for biomedical applications: An overview on properties and fabrication processes. *Journal of Functional Biomaterials, 14*(3), 146–152. DOI: 10.3390/jfb14030146 PMID: 36976070

. ValletRegí M., Ruiz-Hernandez, E(2011). Bioceramics: from bone regeneration to cancer nanomedicine, Advanced Materials, 23, 5177e5218. https://doi.org/.DOI: 10.1515/pac-2018-0505

Vassilios, Z. (2004). Influence of compaction and granulation processes on the properties of soft magnetic ceramics. *Journal of the American Ceramic Society, 87*(5), 824–829. DOI: 10.1111/j.1551-2916.2004.00824.x

Wan, B., Insinga, A. R., Dash, A., & Grivel, J. C. (2024). Application of spark plasma sintering to the manufacture of $La_{2-x}Sr_xCuO_4$ superconducting ceramics. *Ceramics International, 50*(23), 51421–51428. DOI: 10.1016/j.ceramint.2024.10.058

Wen, Y., Zhang, H., Li, K., Wang, H., & Li, Z. (2024). Bi-modified samarium cuprate ($Bi_xSm_{2-x}CuO_4$ ($0 \leq x \leq 0.1$)) thermosensitive ceramics with low sintering temperature. *Journal of Materials Science Materials in Electronics, 35*(4), 266–272. DOI: 10.1007/s10854-024-12019-3

Chapter 2
Advanced Zeolitic Catalysis for Sustainable Hydrogen Production

Imen Amri
http://orcid.org/0009-0004-1090-932X
University of Évora, Portugal

Asma Nouira
http://orcid.org/0000-0001-5982-060X
University of Évora, Portugal

Noureddine Hamdi
University of Gabes, Tunisia

Diogo Canavarro
http://orcid.org/0000-0001-9412-3252
University of Évora, Portugal

ABSTRACT

Zeolites have proven to be highly effective catalysts in various reactions, including acid, base, and redox catalysis. This chapter discusses the use of zeolitic materials as an affordable and efficient approach to producing green hydrogen. It focuses on three main goals: first, providing a general overview of zeolites and their diverse applications across multiple fields; second, examining the role of zeolitic materials as catalysts in hydrogen production processes; and finally, presenting case studies that highlight the mechanisms behind catalytic reactions during hydrogen generation with zeolites.

DOI: 10.4018/979-8-3373-4402-7.ch002

1. INTRODUCTION

Reducing CO_2 and other greenhouse gas emissions is crucial in combating global warming (Jeje et al., 2024). The Intergovernmental Panel on Climate Change (IPCC) and other scientific bodies have repeatedly warned that inadequate emission reductions could lead to dire environmental outcomes, including rising sea levels, erratic weather patterns, and extensive ecological harm. In response, governments, industries, and research institutions globally are intensifying efforts to create sustainable alternatives to fossil fuels, particularly in energy-intensive sectors like transportation, power generation, and heavy industry. This urgency has driven a worldwide shift toward reducing reliance on hydrocarbon-based energy and transitioning to cleaner, renewable sources.

International organizations such as the European Commission and the International Renewable Energy Agency (IRENA), alongside national energy transition initiatives, are accelerating innovation and investment in decarbonization technologies. Key areas of focus include scaling carbon capture, utilization, and storage (CCUS), promoting electrification through renewable energy, and advancing alternative fuels.

One of the most effective strategies for achieving a net-zero emissions future is the adoption of green hydrogen, produced through renewable energy sources (Owusu & Asumadu-Sarkodie, 2016; Kim et al., 2021).

Green hydrogen is widely regarded as a leading alternative to fossil fuels due to its eco-friendliness, efficiency, abundance, high energy density, and renewable nature. Unlike blue or grey hydrogen, which are derived from fossil fuels and generate significant CO_2 emissions, green hydrogen is produced by water electrolysis powered by renewable electricity. This carbon-neutral process makes it ideal for integration into zero-emission energy systems (Sun et al., 2019).

Hydrogen production involves various techniques like pyrolysis, methanation, gasification, and electrolysis, often paired with renewable energy sources like solar and wind to promote sustainability and reduce environmental impact (Benghanem et al., 2023). Each method differs in cost, efficiency, purity, and energy requirements. For instance, electrolysis provides significant environmental advantages when powered by renewables but is less cost-effective compared to steam methane reforming (SMR). While SMR is more economical and established, it produces substantial CO_2 unless coupled with CCUS (Wheelock Gutierrez, 2023; Aliyu & AlZahrani, 2024). On the other hand, gasification and pyrolysis can use biomass or waste as feedstocks, supporting circular economy models and waste valorization, though they face challenges like reaction selectivity, carbon residue formation, and catalyst degradation (Belluati et al., 2024).

To address these issues, extensive research aims to improve hydrogen production methods, focusing on increasing yield and purity. Many improvements rely on developing efficient purification and catalytic systems that can operate under varied thermal and chemical conditions. However, these advanced systems often come with high economic costs due to expensive materials, complex multistage separation processes, and the limited durability of conventional catalysts under extreme conditions like high temperatures, steam, or corrosive gases.

To tackle these challenges, researchers are increasingly utilizing low-cost, high-performance catalysts and supports to accelerate hydrogen production reactions and improve process efficiency. Zeolites, with their remarkable physicochemical properties, have garnered significant attention. These microporous crystalline aluminosilicates are distinguished by their high surface area, ion-exchange capacity, acidity, and thermal stability, key attributes for achieving optimal catalytic performance (Chica, 2013). The application of zeolites in hydrogen production processes shows potential in enhancing reaction selectivity, minimizing coke formation, and maintaining active catalytic sites over prolonged operational periods. Additionally, the inherent flexibility of zeolite frameworks enables the fine-tuning of their acidity, pore size, and metal dispersion properties, making them versatile for various catalytic environments(Chica, 2013). Zeolites can function either as active catalysts themselves, utilizing acidic sites or embedded metal species to mediate specific reactions, or as catalyst supports that stabilize reactive metals such as nickel, cobalt, or platinum under extreme conditions (Sun et al., 2021).

Zeolites may be employed in their natural, synthetic, or modified forms, depending on the desired quantity and quality of the final hydrogen product. Naturally occurring zeolites, like clinoptilolite and mordenite, offer a cost-effective solution for large-scale applications but are often constrained by lower purity and inconsistent pore structures (Pérez-Botella et al., 2022). Conversely, synthetic zeolites, manufactured under controlled conditions, provide tailored properties suited to specific applications, including higher surface areas, uniform pore channels, and improved thermal or hydrothermal stability (Chen et al., 2020).

The modification of zeolites through processes such as ion exchange, metal impregnation, or post-synthetic treatments significantly enhances their functional properties, enabling the selective activation of feedstocks like methane, ethanol, glycerol, or biomass-derived compounds (Sánchez-López et al., 2021).

Although around 40 types of zeolite minerals occur naturally, recent studies highlight a growing trend in synthesizing zeolite particles, with over 140 new types now available. Various synthesis methods have been developed to meet the increasing demand for application-specific zeolite structures (Chen et al., 2020). The diverse compositions and architectures of zeolites enable the development of highly selective and efficient catalytic systems for processes like steam reforming, partial oxidation, pyrolysis, and autothermal reforming. For instance, zeolite-supported Ni or Co catalysts have shown enhanced hydrogen yields and reduced carbon deposition in biomass pyrolysis (Dutta et al., 2015).

In this chapter offer a concise overview of zeolites, covering their features, synthesis methods and applications in science and industry. We explore their role in hydrogen production and how they improve efficiency, selectivity and stability in water-splitting and reforming. We also cover innovations in catalyst design and case studies examining catalytic mechanisms and zeolite performance. These illustrate how zeolites can lower activation energy barriers and enhance thermal stability. Finally, we highlight zeolite importance in hydrogen production systems, both large-scale industrial and small distributed reactors.

2. GENERALITY ABOUT ZEOLITES

Zeolites are microporous, crystalline aluminosilicates first identified by Swedish mineralogist A. F. Cronstedt in 1756 (Guisnet & Ribeiro, 2006). They are typically represented by the empirical formula $M_{x/n}(AlO_2)_x(SiO_2)_y.zH_2O$, where M denotes a charge-balancing cation with valence n, and z represents the number of water molecules per unit cell. This composition highlights silica and alumina as the primary components, with water molecules residing within the framework's pores. Structurally, zeolites comprise a three-dimensional framework of corner-sharing $[SiO_4]^{4-}$ and $[AlO_4]^{5-}$ tetrahedra. These tetrahedra, known as primary building units (PBUs), assemble into more intricate structures called secondary building units (SBUs). The SBUs are formed through systematic oxygen bridge linkages, creating a periodic long-range framework (Petrov, 2012; Mozgawa & Król, 2011).

The organization of these SBUs is fundamental in zeolite science, serving as the basis for classification and property prediction. Each zeolite framework is assigned a unique three-letter code standardized by the International Zeolite Association (IZA). These codes describe the framework's crystallographic symmetry and connectivity, aiding its topological classification and enhancing scientific communication (Figure 1) (Baerlocher & Ch. D. H. O., 2007). This classification system is critical for understanding structure-function relationships and enabling the rational design of zeolite materials for applications such as heterogeneous catalysis, ion exchange, and molecular sieving.

Figure 1. Examples for crystalline structure of zeolites; (a) Mobil Five: MFI (ZSM 5) (b) Faujasite: FAU (X), (c) Linde Type A: LTA, (d) Gismondine: GIS (Na-P1)

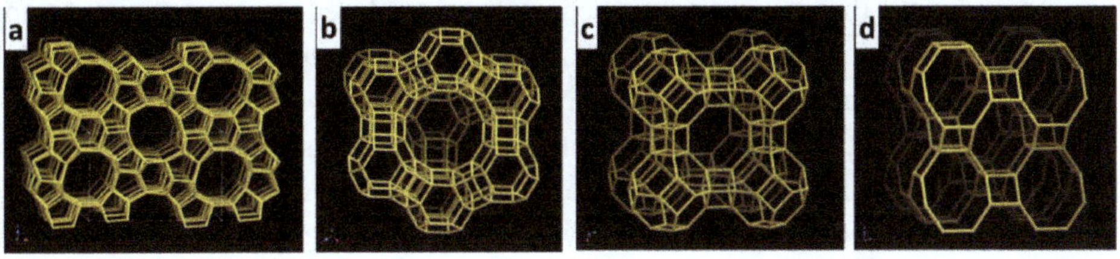

Natural zeolites are formed through various geological processes, classifying them as igneous, metamorphic, or sedimentary rocks (Clifton, 1987; Demirel & Ecenur, 2022). Their formation environment, mineral composition, and geochemical conditions contribute to the diversity of crystal morphology and chemical properties. With a large specific surface area, high ion exchange capacity, and excellent chemical and mechanical stability (Demirel & Ecenur, 2022), natural zeolites are widely used in applications such as water purification, gas separation, agriculture, and environmental remediation. Additionally, recent advancements in green energy technologies have spurred interest in utilizing natural zeolites for hydrogen storage and production (Benghanem et al., 2023; Kalló, 2001; Misaelides, 2011; Eroglu & Athanassiou, 2017).

Despite their advantageous properties, the industrial-scale use of natural zeolites is limited. Studies have shown that impurities and non-homogeneous crystal structures from their formation processes can lead to unpredictable performance. These imperfections impact pore uniformity, reducing ion exchange capacity and catalytic efficiency (Król, 2020). As a result, their behavior can be inconsistent, particularly in critical applications requiring uniformity, purity, and reliability. These challenges have driven interest in synthetic zeolites, which can be tailored to exhibit specific properties and controlled compositions, ensuring improved performance and consistency.

The first successful synthesis of zeolites using pure chemical compounds was achieved in the 1940s by the research groups of Barrer and Milton (Ozdemir & Pişkin, 2013). These pioneering efforts produced zeolite materials with well-defined crystal structures, uniform particle sizes, high surface areas, and remarkable cation exchange capacities. Synthetic zeolites were structurally purer and exhibited more consistent physicochemical properties than their natural counterparts. However, the synthesis process depends on high-purity reagents like sodium silicate and aluminium hydroxide, which pose significant environmental and economic challenges. The chemical reactions involved can generate harmful by-products that contribute to environmental pollution if not properly managed (Garcia et al., 2018). Additionally, the high cost of starting materials and the energy-intensive synthesis process render large-scale production of synthetic zeolites economically impractical.

To overcome these issues, researchers have explored more sustainable and cost-effective alternatives. Industrial waste and naturally occurring materials rich in silica and alumina have emerged as promising precursors for zeolite synthesis. These materials not only provide suitable raw materials but also reduce waste and support environmental protection, aligning with the principles of the circular economy (Yoldi et al., 2019). Utilizing waste-derived raw materials helps meet environmental regulations, lower synthesis costs, and foster sustainable industrial practices.

Ariela M. Cardoso et al. (2015) conducted a study on the use of Brazilian coal fly ash as a source material for synthesizing Gismondine-type zeolite (Na-P1). The fly ash, collected from electrostatic precipitators at a thermoelectric power plant, yielded an impressive 77% under optimized hydrothermal conditions, demonstrating both the efficiency and viability of this method (Cardoso et al., 2015). Similarly, another study utilized oil shale ash to synthesize Na-A and Na-X type zeolites, which are highly valued for applications in water softening, gas separation, and catalysis. Reported yields were 20% for Na-A and 71% for Na-X, showing that low-grade waste materials can serve as valuable feedstocks for zeolite production (Fernandes Machado & Miotto, 2005).

Expanding the exploration of alternative raw materials, Raquel Vigil de la Villa Mencía et al. (2019) investigated various industrial waste products, including recycled glass, silica fume, siliceous concrete waste aggregates, sterile coal, and foundry sand from the steel industry. Utilizing a hydrothermal synthesis technique, the researchers successfully transformed these diverse waste materials into ANA zeolite particles characterized by high crystallinity and a well-developed porous structure. This study underscores the technical feasibility of employing heterogeneous industrial waste streams for zeolite synthesis, helping to mitigate the environmental impact of waste disposal (Vigil et al., 2019).

3. DIFFERENT ZEOLITES SYNTHESIS PROCESS

The synthesis of zeolites generally involves a series of precise chemical and thermal steps. It begins with the preparation of a hydrogel composed of silica and aluminium sources, dissolved or dispersed in an alkaline aqueous medium, typically sodium hydroxide (NaOH) or potassium hydroxide (KOH). Water is essential in facilitating the chemical interactions among these components. Once the gel is formed, the mixture undergoes thermal activation, a critical step to initiate the zeolitization process (table 1).

After thermal treatment, the process enters the 'ageing' stage, where the silica and alumina sources partially dissolve to create a supersaturated solution, ideal for the nucleation phase. During nucleation, small clusters of atoms or molecules organize into patterns that serve as nuclei for zeolite crystals. In the subsequent growth phase, these nuclei develop into mature zeolite crystals under carefully controlled temperature and reaction time conditions (Feijen et al., 1994; Guth & Kessler, 1999).

Zeolite synthesis is a complex process influenced by numerous variables. Researchers have implemented diverse zeolitization strategies, using natural sources and biowastes to improve efficiency and sustainability.

Successful formation of zeolite structures depends on meeting specific reaction conditions, with key factors including temperature, crystallization duration, and the silicon-to-aluminium (Si/Al) molar ratio. The Si/Al ratio, in particular, plays a dominant role in determining the type of zeolite formed, significantly influencing its properties and applications.

The Si/Al ratio is used to classify zeolites into three primary categories (Guth & Kessler, 1999):

- Low silica zeolites (Si/Al <5): These are characterized by high cation exchange capacity and strong hydrophilicity.
- Medium silica zeolites ($5 < Si/Al < 10$): These provide a balance between hydrophilicity and hydrophobicity, making them suitable for applications in catalysis and separation processes.
- High silica zeolites (Si/Al >10): Known for their high thermal stability and hydrophobic nature, these are commonly used in organic transformations and petrochemical applications.

Among synthesis techniques, the alkaline fusion method is notable for its effectiveness, reproducibility, and cost-efficiency in producing various zeolites, including sodalite (SOD), gismondine (Na-P1), faujasite-X (FAU-X), and linde type A (LTA). It involves fusing silica- and alumina-rich precursors with an alkaline reagent at high temperatures, breaking their crystalline structure and converting silicon and aluminum into reactive forms that promote zeolite framework formation (Tahmasebi, 2021).

Clay minerals are preferred raw materials for their abundance and low cost. When treated with NaOH or KOH and heated, their lattice breaks down, activating silicate and aluminate phases that dissolve to form a reactive gel for zeolite formation. Amorphous silica, with higher Gibbs free energy, is especially reactive and beneficial for geopolymerization and zeolite synthesis (Gunasekara et al., 2020).

After fusion, the calcined product is cooled, mixed with distilled water, and stirred to ensure uniformity and partial dissolution. The slurry is then incubated under conditions promoting zeolite crystallization. Once the reaction ends, the product is filtered, washed to remove residual alkali, and dried to obtain the final crystalline zeolite (Amri et al., 2022) (Figure 2).

Figure 2. Synthesis of zeolites using alkaline fusion process

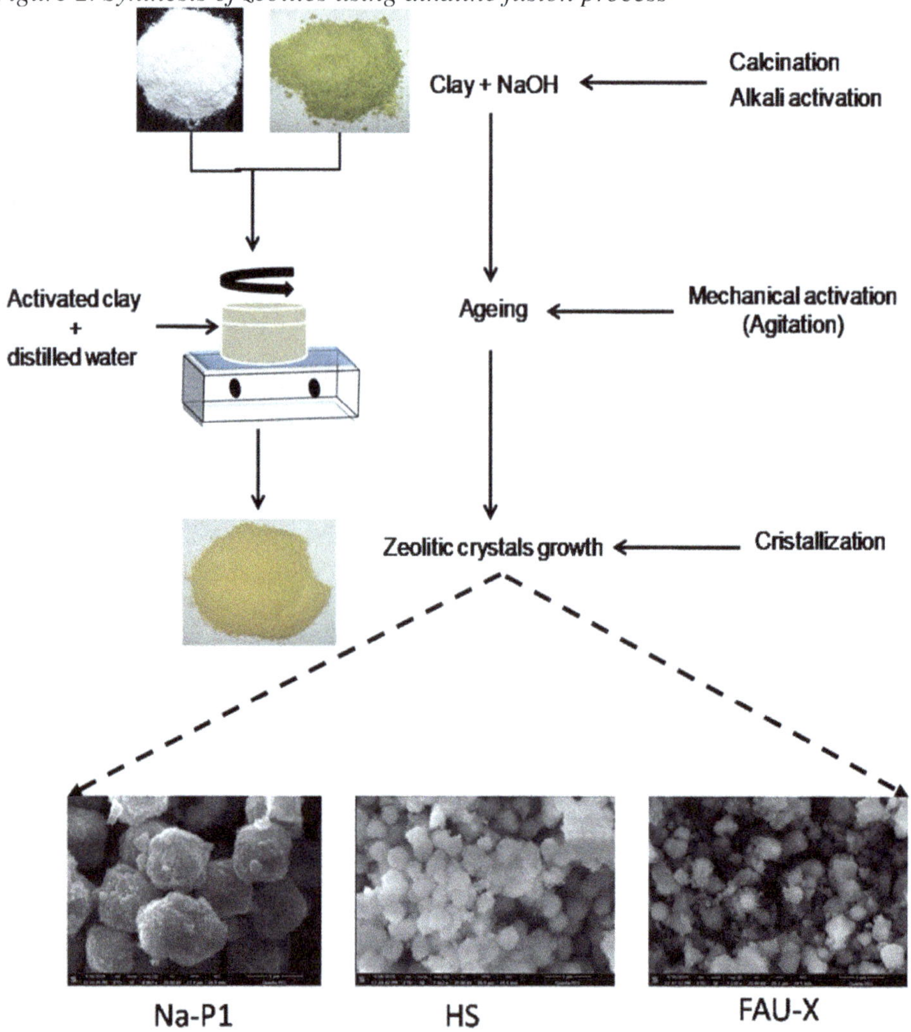

The alkaline fusion technique serves not only as an independent method but also as a crucial preparatory step for other synthesis approaches. Notably, it has been shown to improve the efficiency of subsequent methods, such as hydrothermal synthesis and microwave irradiation techniques. This underscores its significance in obtaining pure, well-crystallized, nanosized zeolitic materials (Novembre et al., 2011; Kazemian et al., 2010).

Among the earliest and most widely used techniques for zeolite synthesis is the hydrothermal method, as detailed in Barrer's 1948 study. In the hydrothermal process, the gel or solution is placed in a Teflon-lined stainless-steel autoclave and heated to a defined temperature, generally between 80°C and 200°C, depending on the type of zeolite desired. The crystallization process can be protracted, with a duration ranging from several hours to several days. This method is distinguished by its capacity to yield zeolites with high crystallinity, uniform pore structures and tunable properties (Cundy & Cox, 2003; Abdullahi et al., 2017).

In recent years, microwave-assisted synthesis has emerged as a faster and more energy-efficient alternative to traditional hydrothermal methods. As posited by Geoffrey A. Tompsett et al. (2006), microwave synthesis has been demonstrated to significantly reduce reaction time whilst concomitantly improving the homogeneity of crystal size distribution (Tompsett et al., 2006). The uniform energy dispersion that is afforded by microwave irradiation serves to minimise temperature gradients, a common issue in conventional heating methods, thereby ensuring consistent product quality (Jawor et al., 2009).

Table 1. Synthesis of different zeolite types using various processes

Zeolite type	Synthesis process	Ref
LTA	Hydrothermal treatment	(Srilai et al., 2020)
LTA	Hydrothermal treatment	(Cecilia et al., 2022)
LTA	Hydrothermal treatment	(Ríos et al., 2007)
Na-X, Na-P1	Hydrothermal treatment	(Gualtieri & Modena, 2001)
ZSM-5	vapor-induced transformation (VIT)	(Li et al., 2017)
SOD	Microwave irradiation	(Zeng et al., 2016)
FAU-X	Alkaline fusion	(Mezni et al., 2011)
SOD, X	Alkaline fusion	(Musyoka et al., 2014)
LTA	Alkaline fusion	(Amri et al., 2022)
FAU-X	Alkaline fusion	(Joseph et al., 2021)
FAU-X, LTA	Alkaline fusion	(Belviso et al., 2013)
FAU-X	Alkaline fusion prior to hydrothermal treatment	(Moneim & Ahmed, 2015)
FAU-X	Alkaline fusion	(Fourmentin & Benzina, 2015)
SOD	Alkaline fusion	(El Gaidoumi et al., 2017)

continued on following page

Table 1. Continued

Zeolite type	Synthesis process	Ref
LTA	Alkaline fusion prior to hydrothermal treatment	(Pereira et al., 2018)
Na-Y	Hydrothermal treatment	(Warzybok, 2018)
MFI	Hydrothermal treatment	(Ruiz-Hitzky & Camblor, 2014)

4. CHARACTERIZATION OF SYNTHESIZED ZEOLITES FROM KAOLINITIC CLAYS

Various analytical techniques have been developed to study the structural and morphological properties of clay minerals and nanoparticles. In mineralogy, several methods are used to identify and quantify mineral phases. X-ray diffraction (XRD) analyzes mineralogical composition, while Fourier transform infrared spectroscopy (FT-IR) provides insights into crystal chemistry. Thermogravimetric analysis (TGA) evaluates thermal behavior, compositional stability, and moisture content, and scanning electron microscopy (SEM) determines the morphology and particle size of clays and zeolites X-ray diffraction analysis.

4.1. X-Ray Diffraction Analysis

Clay minerals and zeolites are characterized by diffraction peaks indexed with basal spacing values, also known as d-spacing (d_{hkl}), which reflect the distance between equivalent crystal planes. This parameter is crucial for identifying minerals and is derived from the angle at which X-rays of a specific wavelength are diffracted by the sample's crystal structure. Based on Bragg's Law, diffraction occurs when the condition $n_{\lambda} = 2d \sin \theta$ is met, where n represents the order of diffraction, λ is the X-ray wavelength, d is the interplanar distance, and θ is the diffraction angle. The d-spacing, expressed in angstroms (Å), provides an accurate crystallographic fingerprint of the analyzed material (Gualtieri & Modena, 2001;Warzybok, 2018).

The diffraction pattern of a sample reveals a series of peaks corresponding to various crystallographic planes. Among these, the first-order diffraction peak, known as d_{001}, is particularly significant. This peak, typically the most intense, is often sufficient to identify the clay mineral type. For example, kaolinite exhibits a prominent d_{001}= 7 Å, illite d_{001}= 10 Å, and chlorite d_{001}= 14 Å (Barton, 2000; Manea et al., 2006).

Non-clay minerals, often present as impurities or secondary phases, are readily identifiable by their distinct diffraction peaks. Common examples, such as quartz, calcite, and dolomite, display unique basal spacing values, with quartz indexed at 3.34 Å, calcite at 3.03 Å, and dolomite at 2.88 Å (Li et al., 2017; Akolekar et al., 1997).

Extensive research in the literature has identified kaolinitic clays as a promising starting material for zeolite synthesis. The abundance of kaolinite, its high aluminosilicate content, and favorable crystallinity make it an ideal precursor for low-silica zeolites. These include widely used types such as FAU-X (Faujasite-X) and Na-P1, which are in high demand for applications in catalysis, water treatment, gas separation, and agriculture (Jha & Singh, 2011; Choi & Hong, 2022). The transformation of kaolinite into zeolites generally follows a two-step process. First, calcination at moderate to high temperatures (500–800°C) converts kaolinite into metakaolinite, a disordered amorphous phase with enhanced reactivity. Second, hydrothermal treatment or alkaline fusion is employed to facilitate the nucleation and growth of zeolite crystals under controlled conditions.

In some regions, pure kaolinite deposits are scarce or economically impractical for large-scale exploitation. Consequently, alternative raw materials are utilized, with stratified or mixed-layer clays, especially those rich in kaolinite, proving particularly effective. A notable example is illito-kaolinitic clays, which combine kaolinite and illite. These clays strike a practical balance between availability and reactivity and have been widely used in zeolite synthesis studies (Gunasekara et al., 2020). Their structural complexity necessitates careful pretreatment to enhance reactivity, often involving milling, acid leaching, or chemical activation before calcination.

Numerous studies successfully synthesized FAU-X zeolite (d_{001} = 14 Å) from Tunisian illito-kaolinitic clay using the alkaline fusion method (Olson et al., 1970; Fourmentin & Benzina, 2015). This method was further developed in a recent comparative study that introduced two types of illito-kaolinitic clays, designated as C (red clay) and C2 (green clay) (Figures 3 and 4). These clays served as precursor materials for synthesizing FAU-X and Na-P1 zeolites. XRD analysis of the synthesized materials revealed that the transformation process led to the complete replacement of the original clay peaks with new peaks corresponding to the targeted zeolite phases (Amri et al., 2022). Notably, the d_{001} reflection of Na-P1 zeolite, observed at 7 Å, appeared in the final diffraction patterns, confirming successful crystallization and phase transformation (Baerlocher & Ch. D. H. O., 2007).

The transformation is visually depicted in Figures 3 and 4 from the original study. In the initial diffractograms, strong peaks representing clay minerals, such as the 7 Å reflection for kaolinite and 10 Å for illite, are prominently visible. After synthesis, these peaks are significantly reduced or absent, while new peaks corresponding to the zeolite frameworks (FAU-X and Na-P1) emerge with greater intensity and sharpness. This evolution in the XRD pattern confirms phase transformation and crystal growth, providing robust evidence of successful zeolite synthesis.

Figure 3. XR patterns of, C1: raw red clay, a and b: Na-P1 zeolite (Amri et al., 2022)

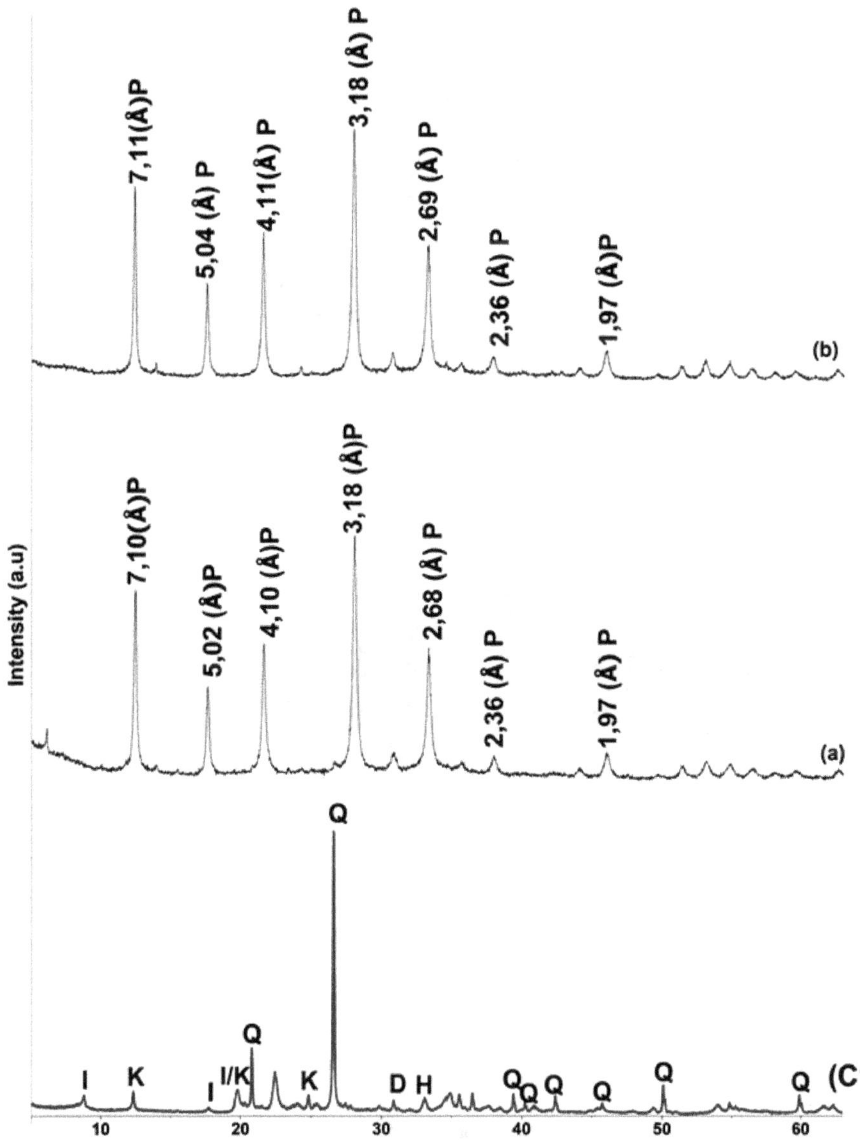

Figure 4. XR patterns of, C2: raw green clay, a and b: FAU-X zeolite (Amri et al., 2022)

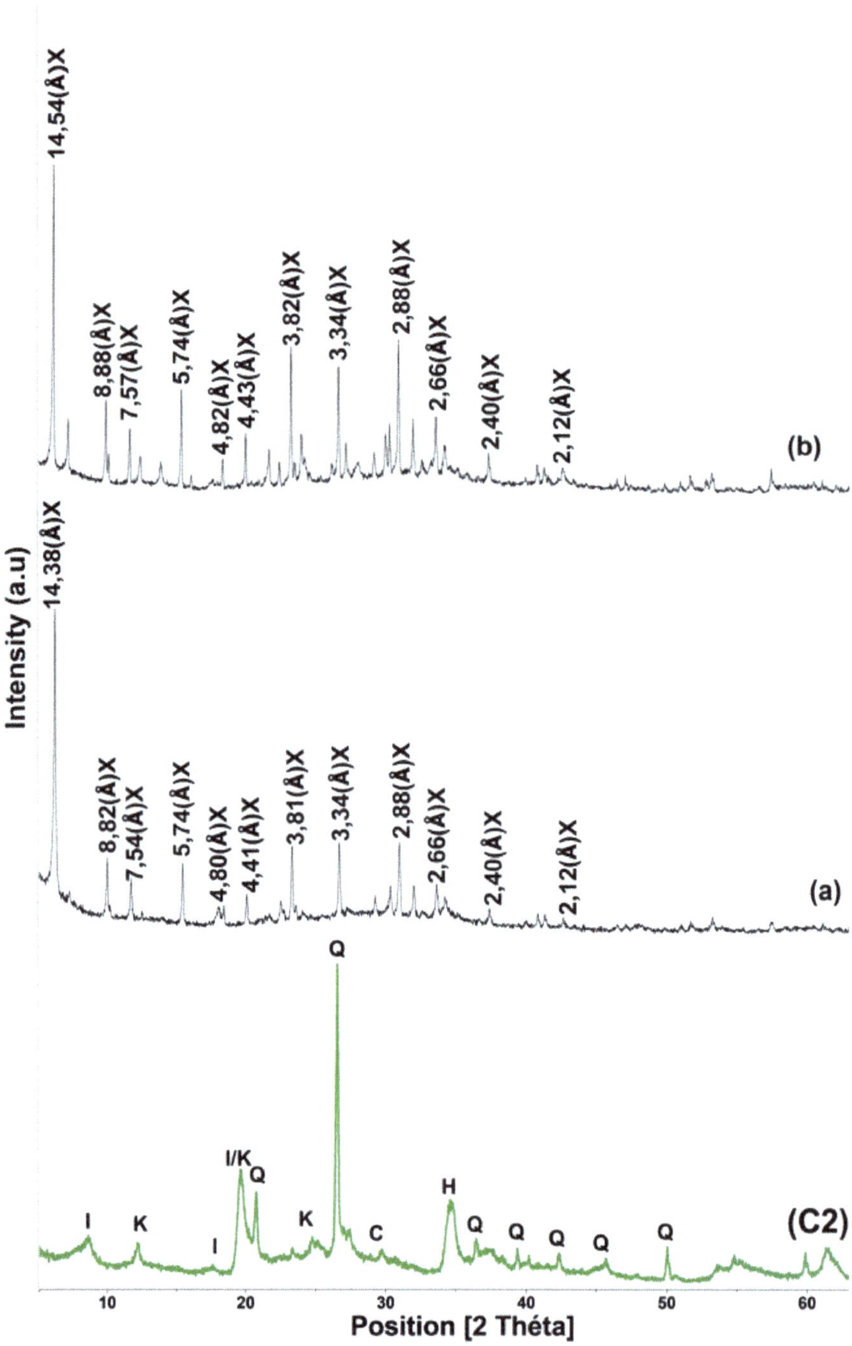

In the study undertaken by Daniela Novembre et al. in 2021, a kaolinitic rock was utilised for the synthesis of Na-P1 zeolites. The formation of Na-P1 zeolite was achieved through synthesis in an alkaline environment, followed by hydrothermal treatment. The final zeolitic product in this study demonstrated a purity of up to 92% after 54 hours of reaction time (Novembre et al., 2021).

In another study, Meyga Evi Ferama Sari et al. (2018) used natural kaolin as a precursor material for SOD synthesis. It was reported that zeolitic particles crystallized within 24 hours at a temperature of 100°C (Sari et al., 2018).

Carlos A. Ríos et al., in their 2012 study, synthesized FAU-X and LTA zeolites by subjecting natural kaolin to alkaline fusion treatment (Ríos et al., 2012).

4.2. Fourier Transforms Infra-Red Analysis (ATR/FT-IR)

Fourier-transform infrared (FT-IR) spectroscopy is a highly effective method for characterizing clay minerals, as these minerals exhibit distinct and identifiable vibrational features associated with hydroxyl groups (–OH), silicate frameworks (Si–O, Al–O–Si), and interlayer water molecules. Hydroxyl groups, in particular, serve as significant markers for identifying various types of clays. FT-IR analysis of different clay types has shown that hydroxyl groups are often bonded to structural aluminum or magnesium and are also associated with adsorbed or interlayer water, contributing to characteristic spectral features. For example, kaolinite shows a distinct –OH stretching band at 3696 cm^{-1}, illite displays a peak at 3600 cm^{-1}, and chlorite presents a maximum at 3670 cm^{-1} (Djomgoue & Njopwouo, 2013). These bands are linked to inner-surface hydroxyls or structural hydroxyls bonded within the tetrahedral-octahedral-tetrahedral (TOT) layers of clay minerals.

Another critical region of the FT-IR spectrum is the Si–O stretching vibration, which appears at lower wavenumbers and provides essential insights into the silicate framework of phyllosilicates and zeolites. The Si–O bond in clay minerals exhibits asymmetric stretching vibrations in the range of 460–570 cm^{-1}, depending on the specific mineral and its bonding environment. For instance, kaolinite displays a distinctive band at 470.88 cm^{-1}, bentonite shows a peak at 470.79 cm^{-1}, and illite exhibits a more pronounced peak at 560 cm^{-1}, reflecting the structural arrangement of silicon-oxygen tetrahedra (Essaidi et al., 2014; Kumar & Lingfa, 2019). For instance, the transformation of kaolinite into zeolite includes intermediate phases like metakaolin, characterized by a loss of hydroxyl groups and shifts in the Si–O–Al vibration regions. The disappearance of characteristic bands associated with the parent clay, alongside the emergence of new bands corresponding to zeolitic frameworks, directly confirms successful phase transformation.

The importance of FT-IR in this domain was aptly demonstrated by Włodzimierz Mozgawa et al. (2011), who conducted a detailed vibrational study of natural zeolites. Their research emphasized FT-IR's ability to identify structural building units (SBUs) within zeolitic frameworks, particularly single and double four- and six-membered rings (4MR and 6MR), which form the foundation of zeolite topologies and correspond to specific vibrational bands in the FT-IR spectrum (Mozgawa & Król, 2011). Further evidence of FT-IR's utility in zeolite identification comes from Cristian Covarrubias et al. (2006), who synthesized FAU-X and Na-P1 zeolites from kaolinite and mordenite, using FT-IR to detect and characterize the transformation into desired zeolitic phases. The characteristic bands associated with kaolinite, as identified by H.J. Percival et al (1974), gradually dissipated during the synthesis process, indicating successful dehydroxylation and structural rearrangement (H. J. P. & Duncan, 1974). Transformation was marked by the emergence of new FT-IR bands unique to zeolite formation. For the FAU-X zeolite, indicator bands appeared at 462, 562, 748, and 994 cm^{-1}, attributed to framework vibrations of Si–O–Al linkages and characteristic of the faujasite structure. The 560 cm^{-1} band, specifically, corresponds to the vibration of double six-membered rings (D6R), a common motif in many zeolites (Ozdemir & Pişkin, 2013). This band serves as a spectral signature for FAU-type zeolites, being sensitive to framework aluminum content and symmetry. For Na-P1 zeolite, the same D6R-related band shifted slightly to 599 cm^{-1}, reflecting the structural differences in the framework and allowing differentiation through subtle spectral variations.

As demonstrated in multiple studies, these FT-IR transformations are effectively illustrated using overlaid or comparative spectra. Figure 5, from J.A. Cecilia et al. (2022), provides a well-resolved FT-IR analysis of the transformation from natural kaolin to LTA zeolite (Cecilia et al., 2022). This study employed an alkaline fusion method followed by hydrothermal treatment, a process that enhances the reactivity of aluminosilicate precursors and promotes the nucleation of highly crystalline zeolite structures. The spectra revealed the progressive disappearance of kaolinite's hydroxyl bands, particularly the 3696 cm^{-1} signal, during metakaolinization. This critical step involves the thermal disruption of the layered kaolinite structure and the loss of interlayer water.

Following hydrothermal treatment, the FT-IR spectra of the resultant product were compared with those of commercially available LTA zeolite, revealing a striking similarity in band position, intensity, and profile. This alignment confirms the successful synthesis of high-purity LTA zeolite and underscores the effectiveness of FT-IR in verifying both the presence and quality of zeolitic phases. Strong bands at 985 cm-1 (Si-O stretching), 660-680 cm-1 (T-O-T bending, where T = Si or Al), and shifts in hydroxyl bands further validate the structural correspondence with the reference material.

Figure 5. FT-IR spectra of raw kaolin, metakaolin, synthesized zeolite and commercial zeolite

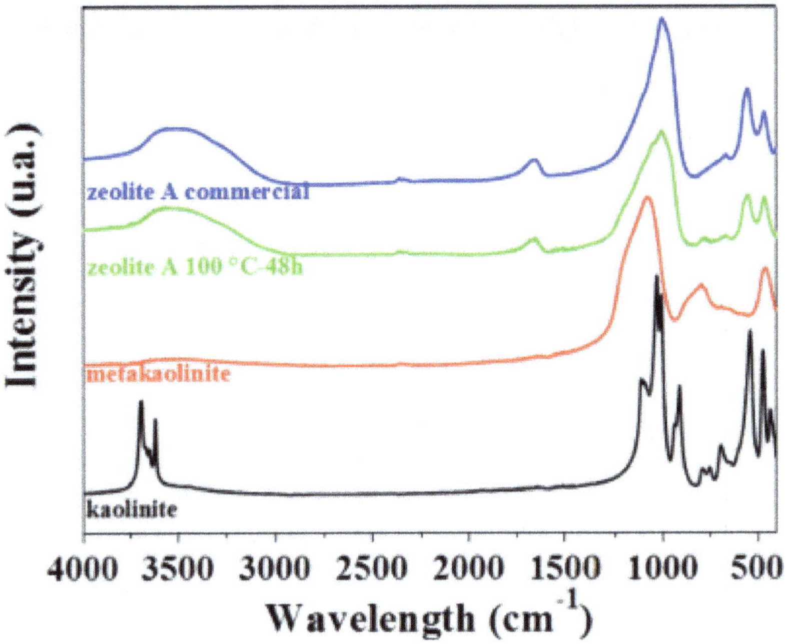

4.3. Thermogravimetric Analysis (TGA)

TGA plays a significant role in studying the thermal transformations occurring during the calcination and synthesis of zeolites from clay precursors. It is vital for optimizing processing temperatures, ensuring phase purity, and tailoring material properties for industrial applications, including adsorption, catalysis, and ion exchange.

A notable study by Adane Adugna Ayalew (2020) analyzed the thermal behavior of kaolin clay using TGA (Ayalew, 2020). The analysis revealed a multi-stage thermal decomposition pattern characteristic of hydrated clay minerals. The initial significant weight loss occurred between 75°C and 200°C, attributed to the evaporation of physically adsorbed surface moisture. This loss is typical of layered silicate materials due to the presence of weakly bound water in interparticle and interlayer voids. Following this, a secondary, more gradual weight loss was observed from 300°C to 430°C, attributed to the elimination of volatile organic residues and the removal of chemically bonded water, particularly hydroxyl groups (–OH) in the octahedral sheets of kaolinite (figure 6).

Figure 6. TGA curves of fly ash-based zeolite X and commercial Kaolin

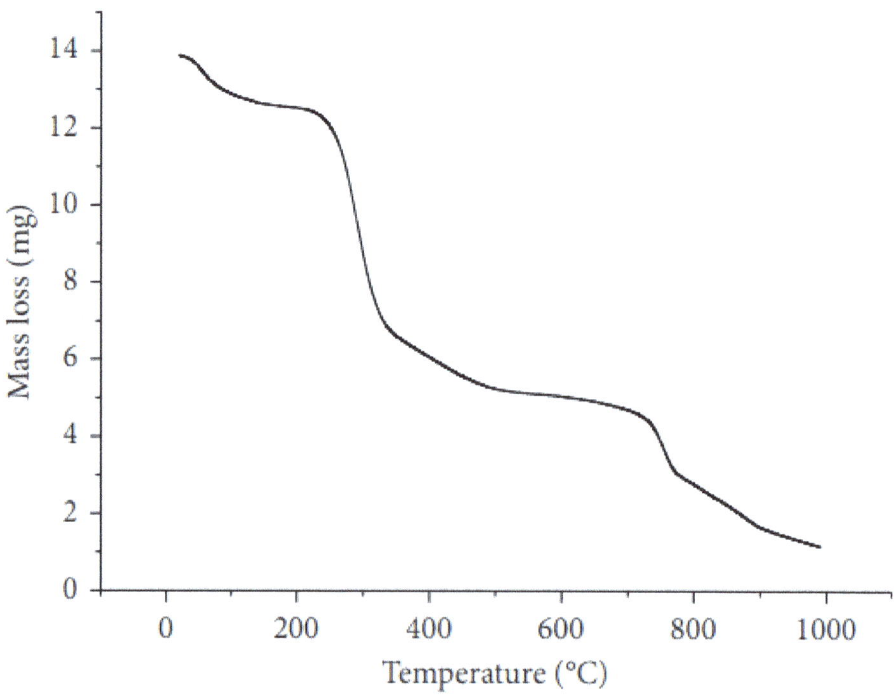

As noted by Musyoka et al., both zeolites maintained their structural integrity up to approximately 800°C, demonstrating a high degree of thermal stability (Musyoka et al., 2015) (Figures 7 and 8). The absence of significant weight loss or derivative weight change beyond 200°C suggests that no decomposition of the aluminosilicate framework or structural collapse occurred under these conditions. This thermal resilience is a defining feature of many zeolite types and is linked to the three-dimensional tetrahedral framework formed by SiO4 and AlO4 units, which are covalently bonded through shared oxygen atoms to form a robust, thermally resistant network.

Moreover, the absence of further weight reduction at elevated temperatures suggests that the zeolite samples examined were mostly free of organic impurities or synthesis by-products. This confirms their suitability for high-temperature reactions, adsorption cycles and other thermally demanding processes.

Figure 7. TGA curves of fly ash-based zeolite X and commercial zeolite X

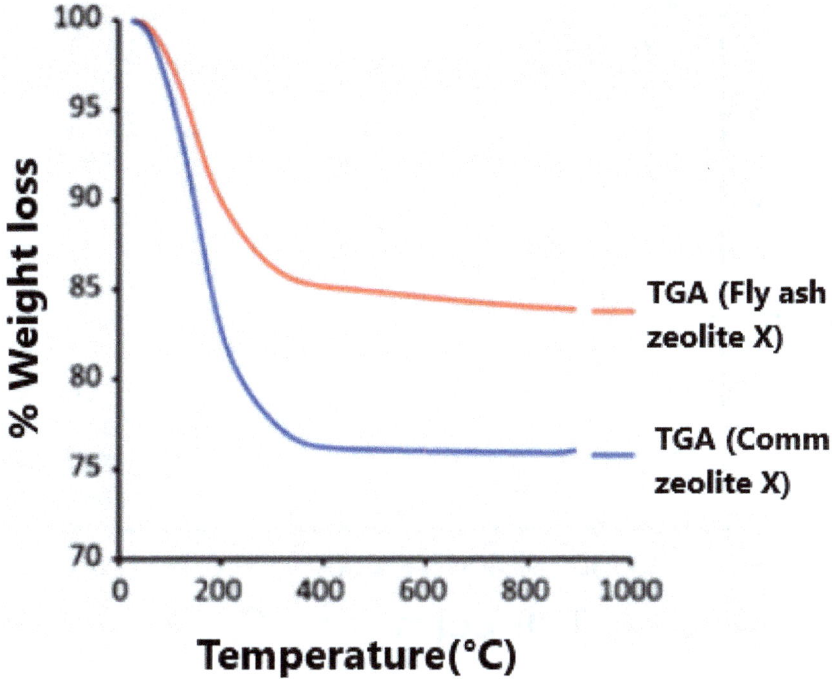

Figure 8. TGA curves of fly ash-based zeolite A and commercial zeolite A

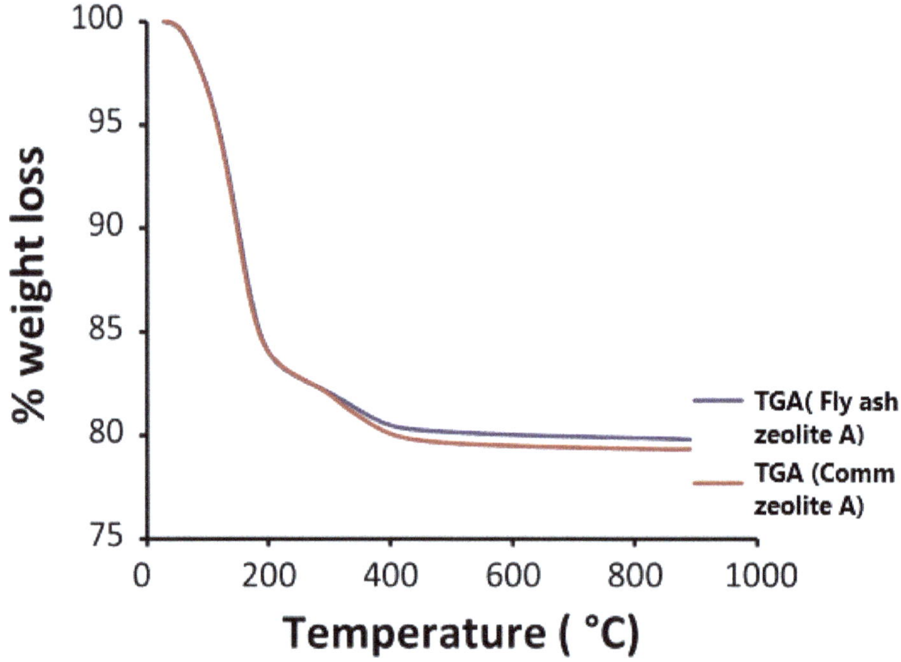

4.4. Scanning Electron Microscopy Analysis

In zeolite synthesis, SEM is crucial for assessing crystallinity, homogeneity, and growth, revealing how parameters like temperature, precursor type, and Si/Al ratio affect morphology. Pure zeolites from kaolin, bentonite, or illito-kaolinitic clays display uniform geometries, sharp edges, and consistent particle sizes. The Si/Al ratio strongly influences structure, porosity, cation exchange capacity, and function. Pankaj Sharma et al. (2016) found that increasing Si/Al (5–11) in Na-P1 synthesis shifted crystals from compact to irregular, likely due to altered gel dynamics (figure 9) (Sharma et al., 2016). Low ratios favor low-silica zeolites (FAU-X, Na-P1, LTA) with high hydrophilicity for water treatment and gas adsorption, while high ratios yield hydrophobic, thermally stable zeolites (ZSM-5, BEA, ZSM-11) for catalysis. Choi & Hong (2022) and Choi et al. (2020) showed low Si/Al (<2) produces ordered Na-P1 with enhanced reactivity (Manuscript, 2020). Johnson & Arshad (2014) linked low ratios to FAU-X, LTA, SOD, and high ratios to BEA, ZSM-5, ZSM-11 (Johnson & Arshad, 2014).

Figure 9. SEM images for synthesized Na-P1 zeolites using different Si/Al ratio (Sharma et al., 2016)

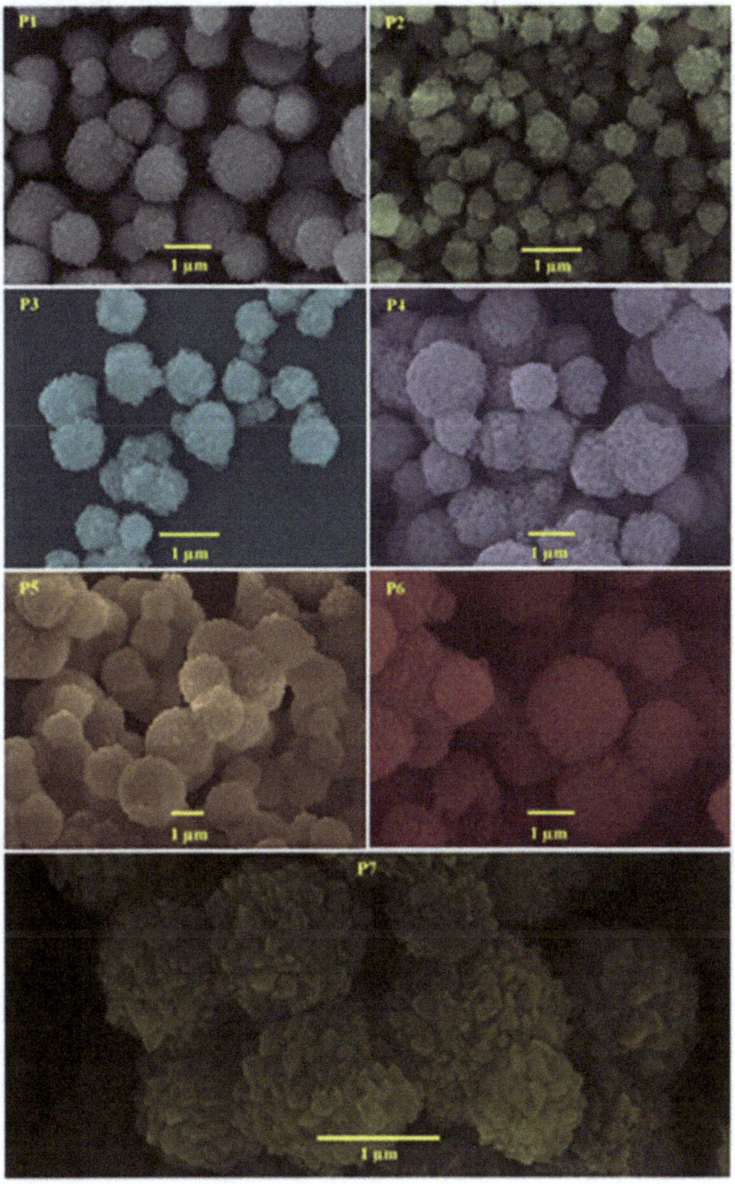

5. DIFFERENT APPLICATION OF ZEOLITES

In contrast to natural zeolites, synthetic and modified zeolites have superior physicochemical properties, making them better suited to a wide range of high-performance applications. They offer higher purity, uniform crystal morphology, and well-developed porosity, which enhance adsorption capacity, catalytic activity, and thermal stability. These attributes have expanded the potential uses of synthetic zeolites, including environmental remediation, catalysis, and advanced energy systems (Figure 10).

Figure 10. Different industrial applications for synthetic zeolites (Michalkiewicz & Koren, 2015)

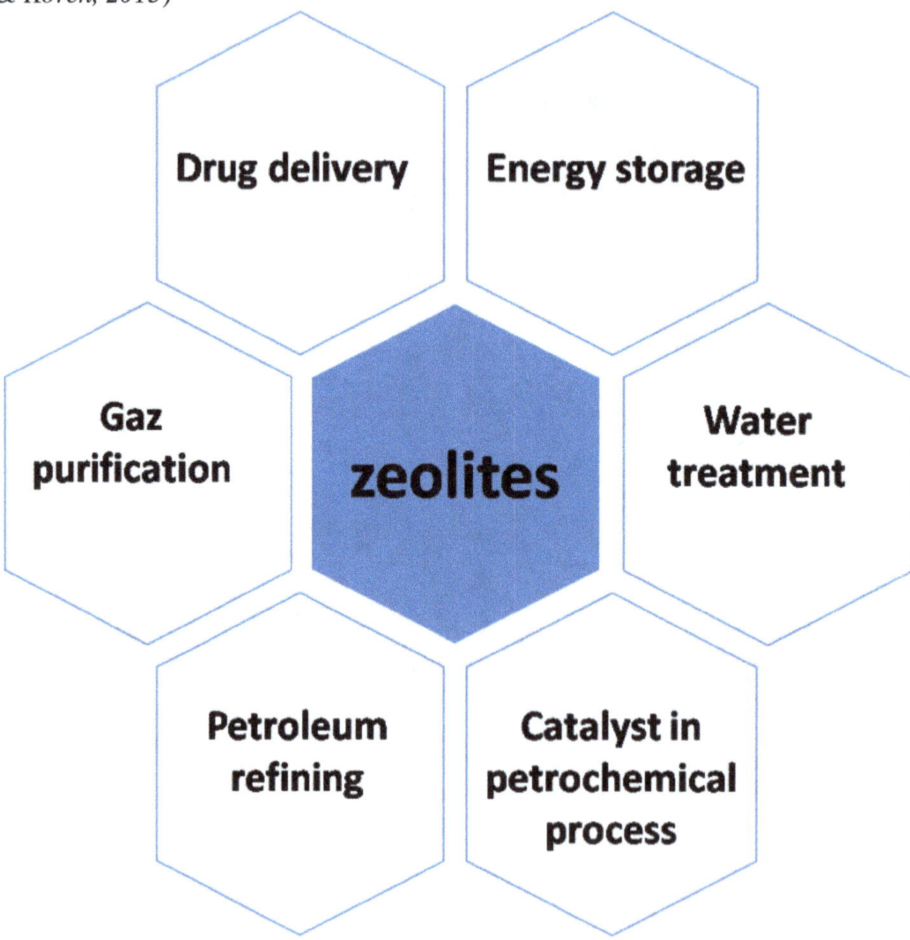

Ximena Jaramillo-Fierro et al. (2021) studied the adsorption performance of three synthetic zeolites, FAU-X, Na1, and LTA, for removing methylene blue dye from aqueous solutions. FAU-X and Na-P1 demonstrated strong affinity with adsorption capacities up to 46.36 mg/g, while LTA showed slightly lower performance at 36.24 mg/g (Naber & De Jong, 1994). This underscores how variations in crystal structure and surface chemistry influence pollutant removal efficiency.

In another study focused on heavy metal removal, Zhao (2016) used SOD-type zeolite to adsorb Cr^{3+} ions from water. The research showed that SOD zeolite achieved an adsorption capacity of 34.7 mg/g at an initial metal concentration of 500 mg/L, highlighting its potential for toxic metal remediation (Marcilly, 2001). These findings emphasize the strong cation-exchange capacity and selectivity of synthetic zeolites for various contaminants.

Beyond aqueous applications, synthetic zeolites have recently been used for air purification, particularly for capturing CO_2 and adsorbing volatile organic compounds (VOCs). Significant research, such as that noted in reference (Mravec et al., 2005), has focused on their use in atmospheric CO_2 sorption, attributed to their tunable pore sizes and strong affinity for polar molecules. Magdalena Warzybok and Jolanta Warchoł (2018) synthesized Na-Y zeolite from natural kaolinite and assessed its efficiency in removing VOCs like acetone and methyl ethyl ketone from indoor air. Despite its relatively low specific surface area, Na-Y showed a strong adsorption capacity, outperforming several commercial adsorbents due to its favorable surface chemistry and framework structure (Ronchi et al., 2018).

In petrochemical engineering, synthetic and modified zeolites are indispensable catalytic materials. Their integration into gasoline production has enhanced fuel quality and reduced coke formation in fuel reservoirs (Yao et al., 2018). The introduction of Y-type zeolites into fluid catalytic cracking (FCC) units in the 1960s marked a pivotal moment in petroleum refining. As highlighted by Christian Marcilly (1) and Dušan Mravec et al. (2005), zeolites possess strong Brønsted acidity, shape selectivity, and thermal stability, making them highly effective catalysts for processes such as FCC, hydrocracking, isomerization, dewaxing, and hydroisomerization (Salahudeen, 2022; Zhang et al., 2012).

Compared to corrosive and less selective catalysts like $AlCl_3$, zeolites facilitate the selective breakdown of heavy hydrocarbons into lighter fractions with high octane ratings and low sulfur content, meeting modern environmental standards. Their recyclability and ability to function under high-pressure and high-temperature conditions further enhance their value in continuous industrial operations.

Building on this, Yuansheng Zhao et al. (2013) conducted an in-depth study on the catalytic roles of high-silica Y-type (faujasite) and small-crystal Na-Y zeolites in hydrocracking, with a particular focus on converting vacuum gas oil (VGO). These zeolites function as structured, microporous frameworks that provide strong acidity and selective diffusion pathways. These pathways minimize coke formation, enhance catalyst stability, and promote the production of middle distillates, such as kerosene and diesel. Their effectiveness is influenced by synthesis parameters, including the SiO_2/Al_2O_3 ratio, crystal size, ageing time, and Na_2O content, which affect active site availability and mass transfer efficiency (Mohd et al., 2021).

As well as their well-established roles in catalysis and environmental remediation, synthetic zeolites have also shown significant promise in energy storage, particularly for mechanical energy applications. Laura Ronchi et al. (2017) explored extra-large-pore pure silica zeolites with CFI and DON topologies and emphasized their exceptional mechanical flexibility, attributed to their distinctive crystalline framework structures. Unlike conventional rigid zeolites, these engineered zeosils exhibit reversible structural deformation under mechanical pressure, allowing them to absorb, store, and release mechanical energy. This elasticity arises from their large-pore architectures, which permit framework shifts without structural collapse. These findings point to their potential use in innovative nanoscale energy storage systems where compact, efficient, and reversible mechanical energy solutions are vital. Additionally, possible applications include advanced shock-absorbing systems and highly sensitive pressure sensors, expanding the functional scope of zeolites beyond traditional applications (Musyoka et al., 2018).

An emerging application of synthetic zeolites is in waste-to-energy (WTE) conversion technologies, addressing the dual challenges of managing plastic waste and producing hydrogen sustainably. Yao et al. (2018) explored the use of synthetic zeolites as nickel catalyst supports in a two-stage pyrolysis–catalytic steam reforming system for converting plastic waste into hydrogen and synthesis gas (syngas). In this process, plastic undergoes thermal decomposition via pyrolysis, followed by catalytic reforming into gaseous fuels. Zeolites such as ZSM-5, β-zeolite, and Y-zeolite were evaluated for their ability to disperse and stabilize nickel nanoparticles, enhance catalytic activity, and resist coke formation. ZSM-5 with a Si/Al ratio of 30 demonstrated superior performance due to its well-defined microporous structure and optimal acidity. This enabled higher gas yields, better metal dispersion, and extended catalyst life. The unique channel system of ZSM-5 facilitated efficient diffusion and minimized the formation of coke-producing by-products, while its moderate Brønsted acidity supported hydrocarbon cracking and reforming without significant deactivation. These features establish synthetic ZSM-5 as an effective support for WTE processes, enabling cleaner hydrogen production and advancing circular economy solutions by converting non-recyclable plastic waste into valuable energy carriers (Dragomirova & Wohlrab, 2015).

6. HYDROGEN PRODUCTION PROCESS ASSISTED BY ZEOLITIC CATALYST

Zeolites, spanning from naturally occurring minerals to highly engineered synthetic variants, have become essential and versatile materials in catalysis, membrane separations, and hydrogen storage technologies. Their crystalline, microporous structures, high surface areas, thermal stability, and adjustable acidity make them ideal for a broad range of industrial and environmental applications. A particularly intriguing feature of zeolites is their dual role as catalysts and supports for active catalytic species, especially in gas-phase reactions where selectivity, thermal stability, and molecular sieving are critical. Beyond traditional catalysis, zeolites excel in facilitating complex reaction networks such as the water–gas shift (WGS) reaction, steam reforming, and biomass conversion. This is especially significant in scenarios where conventional materials may fail due to thermal degradation or coke formation.

In this context, a pivotal study was carried out by Yuting Zhang et al. (2012) exploring the performance of a small MFI zeolite membrane reactor (MR) modified to operate in conjunction with a commercial $CuO/ZnO/Al_2O_3$ catalyst in the low-temperature WGS reaction (Figure 11). The reactor leveraged the permselectivity of zeolite membranes. A CO/N_2 gas stream containing steam was introduced to the membrane side, while nitrogen flowed counter-currently on the permeate side. The primary role of the MFI zeolite membrane was to selectively permit hydrogen to pass through, shifting the WGS reaction equilibrium and driving the conversion of CO to CO_2 and H_2 to completion. Figure 11 illustrates this mechanism: the left panel depicts the modified MFI zeolite structure with hydrogen-permeable channels, and the right panel shows the selective hydrogen extraction as it diffuses into the sweep gas. Continuous hydrogen removal enhanced CO conversion and enabled high-purity hydrogen recovery, critical for downstream fuel cell applications. Notably, the system achieved a CO conversion rate of 95.4%, significantly surpassing the 70% achieved by a conventional packed bed reactor (PBR). This improvement resulted from the zeolite membrane's ability to continuously extract hydrogen, minimizing the backward reaction and maximizing forward progression. Additionally, the compact design and high selectivity of the membrane reactor make it ideal for industrial hydrogen generation processes, especially those requiring high-purity hydrogen under mild reaction conditions (Ali et al., 2022).

Figure 11. Hydrogen Production via Low-Temperature Water-Gas Shift Reaction Using Modified MFI Zeolite Membrane (Ali et al., 2022)

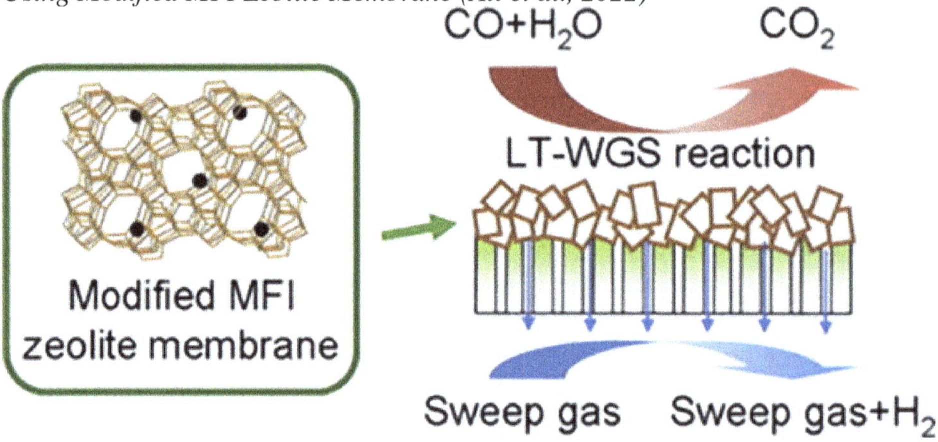

As well as enhancing reactions, zeolites are also being used to produce advanced porous carbon materials through templating methods. Nicholas M. Musyoka et al. (2018) emphasized the use of commercial zeolite 13X as a structural template for synthesizing Zeolite-Templated Carbons (ZTCs) for hydrogen storage. Their innovative approach employs waste tyre pyrolysis oil vapor as a carbon precursor, integrated into the porous zeolite framework via one-step and two-step synthesis methods. Remarkably, the two-step method produced carbon materials with higher specific surface areas and superior hydrogen storage capacities. These findings are noteworthy for demonstrating dual benefits: valorizing waste materials and creating high-performance hydrogen storage media. The zeolite's structured pore network directs the nanoscale carbon architecture, imparting ordered porosity, surface functionality, and structural rigidity to the ZTCs. This study underscores the critical role of zeolites as functional materials and sacrificial templates in the development of next-generation energy materials (Musyoka et al., 2018).

A more comprehensive perspective on zeolite's role in membrane-based catalysis is provided by Radostina Dragomirova and Sebastian Wohlrab (2015), who conducted a thorough review of zeolite membrane evolution, highlighting their transition from passive separation barriers to active catalytic components in integrated membrane reactor systems (Figure 12). The authors classify zeolite membrane configurations into three categories: catalytic membrane reactors, zeolite-coated catalyst particles, and spatially segregated membrane modules. These designs leverage the permselectivity and molecular sieving properties of zeolite membranes to enhance catalytic reactions. Figure 12 illustrates a sophisticated setup employing modified hollow fibre MFI zeolite membranes. In this system, feed gas containing CO and steam enters the reactor, where the WGS reaction occurs over a high-temperature WGS catalyst. The zeolite membrane, coated around the catalyst bed, selectively removes hydrogen as it is produced, assisted by a counter-diffusing steam sweep gas. The inset images depict the membrane's structure and function: the SEM image reveals the hollow fibre membrane's morphology, while the MFI framework schematic highlights the nanochannels facilitating H_2 diffusion. This arrangement supports pure hydrogen recovery and sustains the WGS reaction by continuously shifting the equilibrium. The study emphasizes the importance of optimizing process parameters, such as steam sweep rate, feed pressure, and the H_2O/CO molar ratio, to maximize efficiency. It underscores the role of zeolite membranes in process intensification, offering benefits like energy efficiency, reduced reactor footprint, and integration of reaction and separation in a single step (Dragomirova & Wohlrab, 2015)

Figure 12. Selective Hydrogen Production Using Modified MFI Zeolite Membrane in the Water-Gas Shift Reaction (Dragomirova & Wohlrab, 2015)

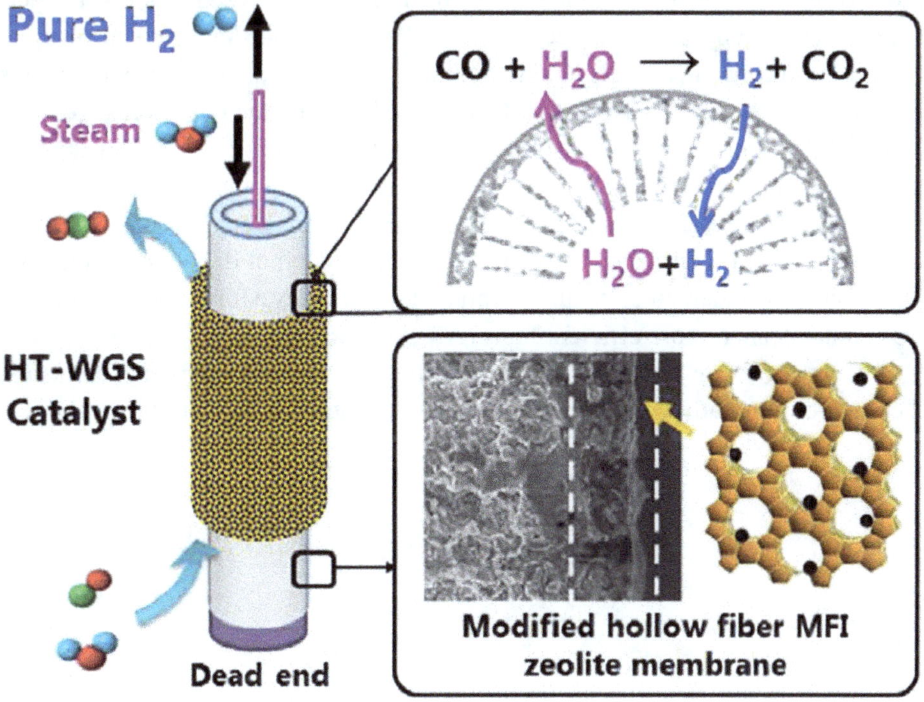

The efficiency of steam methane reforming (SMR) and WGS processes is increased by zeolite membranes, particularly MFI-type materials such as ZSM-5, as Beata Michalkiewicz and Zvi C. Koren (2015) discussed. This further reinforces the case for zeolite-based membrane reactors in hydrogen production. Their study highlights the dual role of zeolites as selective molecular sieves and vital components of catalytic systems. With a uniform microporous structure, high chemical and thermal stability, and low permeability to contaminants like H_2S, zeolite membranes effectively separate hydrogen under demanding conditions. In membrane reactor configurations, zeolites selectively extract hydrogen during the reaction, shifting the equilibrium to enhance conversion and yield. Adjustments to pore size, such as through catalytic silane cracking, have further improved H_2/CO_2 selectivity, a key factor in syngas purification. Compared to conventional packed bed reactors, these systems achieve higher conversion rates and hydrogen recovery, allow for compact reactor designs, promote faster reaction kinetics, and operate at reduced pressures and temperatures (Michalkiewicz & Koren, 2015).

Zeolites play a crucial role in reaction engineering and gas purification and are increasingly recognized for their catalytic efficiency in biomass conversion. Ghulam Ali et al. (2022) explored the thermo-catalytic pyrolysis of lignin using hydrogen-loaded zeolite Y as a catalyst. Conducted at 390°C with a 3% catalyst loading, this process demonstrated significant improvements over uncatalyzed systems. The activation energy for pyrolysis dropped from 157.96 to 141.33 kJ/mol, and the resulting bio-oil showed elevated levels of light hydrocarbons and aromatic fuel components, including benzene derivatives and phenols. Furthermore, the reduced oxygen content of the bio-oil enhanced its calorific value and compatibility with existing fuel infrastructure. The microporous structure of zeolite Y facilitates the breakdown of high-molecular-weight lignin into smaller, valuable molecules while promoting the production of hydrogen-rich gases. This combination of depolymerization and deoxygenation underscores zeolite Y's potential as a catalyst for converting biomass waste into cleaner, more efficient fuels (Ali et al., 2022).

To provide a broader understanding of zeolite-assisted hydrogen generation, table 2 compares key studies on hydrogen production using various zeolitic catalysts.

Table 2. Comparison of selected studies on hydrogen production using zeolitic catalysts

Catalyst type	Feedstock / Process	H_2 yield or conversion (%)	Operating conditions	Key findings	Ref
MFI zeolite membrane + $CuO/ZnO/Al_2O_3$	Water-gas shift (WGS) reaction	95.4	250 °C, 1 bar	Zeolite membrane selectively removed H_2, boosting CO conversion efficiency.	(Zhang et al., 2012)
H–Y zeolite	Lignin pyrolysis	High H_2-rich gas	390 °C	Zeolite Y reduced activation energy and increased aromatic product yield.	(Ghulam Ali et al., 2022)
Ni/ZSM-5	Plastic waste pyrolysis and steam reforming	80-90	600 °C	Enhanced H_2 production and reduced coke formation due to strong acidity and dispersion.	(Yao et al., 2021)
Metal clusters on zeolite	Photocatalytic water splitting	****	Visible light	Synergetic metal–zeolite interface improved hydrogen evolution activity.	(Sun et al., 2019)

7. CONCLUSION AND OUTLOOK

The transition to a hydrogen-based energy economy relies on the development of efficient, sustainable, and cost-effective hydrogen production technologies. This chapter has highlighted zeolitic materials as among the most promising candidates for advancing green hydrogen technologies, owing to their distinctive crystalline morphology, well-developed porous structure, and exceptional thermal and chemical stability. Zeolites have showcased their versatility in catalysis, membrane-based separations, and hydrogen storage, spanning from natural minerals to highly engineered synthetic variants. Their capacity to support and stabilize active metal species, enable shape-selective reactions, and withstand complex operating environments makes them indispensable in processes such as water, gas shift reactions, steam reforming, and biomass conversion.

Recent advancements in zeolite modification techniques, including ion exchange, doping, structural templating, and hybridization with metal nanoparticles, have significantly enhanced their catalytic activity and selectivity. Additionally, the integration of zeolite membranes within catalytic reactors has opened new avenues for process intensification, resulting in higher hydrogen purity and conversion rates. As research continues to explore innovative synthesis strategies, waste-derived feedstocks, and multifunctional reactor designs, zeolite-based catalysis is expected to play a pivotal role in achieving sustainable and scalable hydrogen production.

ACKNOWLEDGMENT

The authors gratefully acknowledge the support received from the FUTERTECH_H2 project. Fomento y desarrollo tecnológico de la industria del hidrógeno verde (ref. 0108_futuretech_h2_5_e). We also extend our gratitude to AIHRE project: Análisis e Impulso del H2 REnovable en la región POCTEP (REF. 0093_AIHRE_6_E).

Competing Interests: The authors declare no competing interests.

Funding: Not applicable

Ethic approval: Not applicable

Informed consent: Not applicable

REFERENCES

Abdullahi, T., Harun, Z., Hafiz, M., & Othman, D. (2017). A review on sustainable synthesis of zeolite from kaolinite resources via hydrothermal process. *Advanced Powder Technology*, *28*(8), 1827–1840. DOI: 10.1016/j.apt.2017.04.028

Akolekar, D., Chaffee, A., & Howe, R. F. (1997). The transformation of kaolin to low-silica X zeolite. *Zeolites*, *19*(5-6), 359–365. DOI: 10.1016/S0144-2449(97)00132-2

Ali, G., Afraz, M., Muhammad, F., Nisar, J., Shah, A., Munir, S., & Tasleem Hussain, S. (2022). Production of fuel range hydrocarbons from pyrolysis of lignin over zeolite Y, hydrogen. *Energies*, *16*(1), 215. DOI: 10.3390/en16010215

Aliyu, M., & AlZahrani, A. A. (2024). Integrated SMR-SOFC system for efficient hydrogen and power production. *International Journal of Hydrogen Energy*. Advance online publication. DOI: 10.1016/j.ijhydene.2024.12.354

Amri, I., Ouakouak, A., Hamdi, W., Srasra, E., & Hamdi, N. (2022). Removal of non-steroidal drug from waste water using synthetic zeolites from illito-kaolinitic clays. *Water, Air, and Soil Pollution*, *233*(9), 369. DOI: 10.1007/s11270-022-05846-9

Ayalew, A. A. (2020). Development of kaolin clay as a cost-effective technology for defluoridation of groundwater. *International Journal of Chemical Engineering*, *2020*(1), 8820727. DOI: 10.1155/2020/8820727

Baerlocher, L. B. M., & Ch, D. H. O. (2007). *Atlas of zeolite framework types.*

Barton, A. D. K. C. D. (2000). Clay minerals. In *Encyclopedia of Soil Science*. https://doi.org/DOI: 10.2118/59292-MS

Belluati, M., Tabasso, S., Gaudino, E. C., Cravotto, G., & Manzoli, M. (2024). Biomass-derived carbon-based catalysts for lignocellulosic biomass and waste valorisation: A circular approach. *Green Chemistry*, *26*(15), 8642–8668. DOI: 10.1039/D4GC00606B

Belviso, C., Cavalcante, F., Lettino, A., & Fiore, S. (2013). A and X-type zeolites synthesised from kaolinite at low temperature. *Applied Clay Science*, *80–81*, 162–168. DOI: 10.1016/j.clay.2013.02.003

Benghanem, M., Mellit, A., Almohamadi, H., Haddad, S., Chettibi, N., Alanazi, A. M., Dasalla, D., & Alzahrani, A. (2023). Hydrogen production methods based on solar and wind energy: A review. *Energies*, *16*(2), 757. DOI: 10.3390/en16020757

Cardoso, A. M., Paprocki, A., Ferret, L. S., Azevedo, C. M. N., & Pires, M. (2015). Synthesis of zeolite Na-P1 under mild conditions using Brazilian coal fly ash and its application in wastewater treatment. *Fuel*, *139*, 59–67. DOI: 10.1016/j.fuel.2014.08.016

Cecilia, J. A., Vilarrasa-García, E., Morales-Ospino, R., Finocchio, E., Busca, G., Sapag, K., Villarroel-Rocha, J., Bastos-Neto, M., Azevedo, D. C. S., & Rodríguez-Castellón, E. (2022). Kaolinite-based zeolites synthesis and their application in CO_2 capture processes. *Fuel*, *320*, 123953. DOI: 10.1016/j.fuel.2022.123953

Chen, L. H., Sun, M. H., Wang, Z., Yang, W., Xie, Z., & Su, B. L. (2020). Hierarchically structured zeolites: From design to application. *Chemical Reviews*, *120*(20), 11194–11294. DOI: 10.1021/acs.chemrev.0c00016

Chica, A. (2013). Zeolites: Promised materials for the sustainable production of hydrogen. *International Scholarly Research Notices*, *2013*(1), 907425. DOI: 10.1155/2013/907425

Choi, H. J., & Hong, S. B. (2022). Effect of framework Si/Al ratio on the mechanism of CO_2 adsorption on the small-pore zeolite gismondine. *Chemical Engineering Journal*, *433*(P3), 133800. DOI: 10.1016/j.cej.2021.133800

Clifton, R. A. (1987). *Natural and synthetic zeolites*. U.S. Department of the Interior, Bureau of Mines.

Cundy, C. S., & Cox, P. A. (2003). The hydrothermal synthesis of zeolites: History and development from the earliest days to the present time. *Chemical Reviews*, *103*(3), 663–702. Advance online publication. DOI: 10.1021/cr020060i

. Demirel, S., & Ecenur, İ. (2022). Synthesis and characterization of Ppy and Ppy/zeolite and their use as adsorbents in removal of diclofenac sodium.

Djomgoue, P., & Njopwouo, D. (2013). FT-IR spectroscopy applied for surface clays. *Journal of Surface Engineered Materials and Advanced Technology*, *3*(4), 275–282. DOI: 10.4236/jsemat.2013.34037

Dragomirova, R., & Wohlrab, S. (2015). Zeolite membranes in catalysis: From separate units to particle coatings. *Catalysts*, *5*(4), 2161–2222. DOI: 10.3390/catal5042161

Dutta, A., Sahir, A. H., Tan, E., Humbird, D., Snowden-Swan, L. J., Meyer, P. A., . . . Lukas, J. (2015). *Process design and economics for the conversion of lignocellulosic biomass to hydrocarbon fuels: Thermochemical research pathways with in situ and ex situ upgrading of fast pyrolysis vapors* (No. PNNL-23823). Pacific Northwest National Laboratory.

El Gaidoumi, A., Benabdallah, A. C., & El Bali, B. (2017). Synthesis and characterization of zeolite HS using natural pyrophyllite as new clay source. *Arabian Journal for Science and Engineering*. Advance online publication. DOI: 10.1007/s13369-017-2768-8

Eroglu, N., & Athanassiou, C. G. (2017). Applications of natural zeolites on agriculture and food production. *Journal of the Science of Food and Agriculture*, *97*(11), 3487–3499. Advance online publication. DOI: 10.1002/jsfa.8312

Essaidi, N., Samet, B., Baklouti, S., & Rossignol, S. (2014). Science of producing geopolymers from two different Tunisian clays before and after calcination at various temperatures. *Applied Clay Science*, *88–89*, 221–227. DOI: 10.1016/j.clay.2013.12.006

Feijen, E. J. P., Martens, J. A., & Jacobs, P. A. (1994). Zeolites and their mechanism of synthesis. *Studies in Surface Science and Catalysis*, *84*, 3–21. DOI: 10.1016/S0167-2991(08)64074-4

Fernandes Machado, N. R. C., & Miotto, D. M. M. (2005). Synthesis of Na-A and -X zeolites from oil shale ash. *Fuel*, *84*(18), 2289–2294. DOI: 10.1016/j.fuel.2005.05.003

Fourmentin, S., & Benzina, M. (2015). Synthesis of faujasite type zeolite from low grade Tunisian clay for the removal of heavy metals from aqueous waste by batch process: Kinetic and equilibrium study. *Comptes Rendus. Chimie*, *18*(10), 1123–1133. DOI: 10.1016/j.crci.2015.03.013

Fourmentin, S., & Benzina, M. (2015). Synthesis of faujasite type zeolite from low grade Tunisian clay for the removal of heavy metals from aqueous waste by batch process: Kinetic and equilibrium study. *Comptes Rendus. Chimie*, *18*(10), 1123–1133. DOI: 10.1016/j.crci.2015.03.013

Garcia, G., Cabrera, S., Hedlund, J., & Mouzon, J. (2018). Selective synthesis of FAU-type zeolites. *Journal of Crystal Growth, 489*, 36–41. DOI: 10.1016/j.jcrysgro.2018.02.022

Ghulam Ali, S., Khan, R., Rehman, F., Bilal, M., & Ullah, S. (2022). Thermo-catalytic pyrolysis of lignin using hydrogen-loaded zeolite Y as catalyst for fuel range hydrocarbons production. *Energies, 15*(18), 6672. DOI: 10.3390/en15186672

Gualtieri, A. F., & Modena, Á. (2001). Synthesis of sodium zeolites from a natural halloysite. *Physics and Chemistry of Minerals, 28*(2), 719–728. DOI: 10.1007/s002690100197

Guisnet, M., & Ribeiro, F. R. (2006). *Les zéolithes, un nanomonde au service de la catalyse*. EDP Sciences.

Gunasekara, C., Li, J., Tahmasebi, M., Law, D., Brkljac, R., & Patnaikuni, I. (2020). Geopolymer synthesis using low-grade clays. *Construction & Building Materials, 121066*. Advance online publication. DOI: 10.1016/j.conbuildmat.2020.121066

Guth, J., & Kessler, H. (1999). Synthesis of aluminosilicate zeolites and related silica-based materials. In *Catalysis and zeolites* (pp. 1–52). DOI: 10.1007/978-3-662-03764-5_1

Jaramillo-Fierro, X., González, S., Montesdeoca-Mendoza, F., & Medina, F. (2021). Structuring of $ZnTiO_3/TiO_2$ adsorbents for the removal of methylene blue, using zeolite precursor clays as natural additives. *Nanomaterials (Basel, Switzerland), 11*(4), 898. DOI: 10.3390/nano11040898

Jawor, A., Jeong, B. H., & Hoek, E. M. (2009). Synthesis, characterization, and ion-exchange properties of colloidal zeolite nanocrystals. *Journal of Nanoparticle Research, 11*(7), 1795–1803. DOI: 10.1007/s11051-009-9688-9

Jeje, S. O., Marazani, T., Obiko, J. O., & Shongwe, M. B. (2024). Advancing the hydrogen production economy: A comprehensive review of technologies, sustainability, and future prospects. *International Journal of Hydrogen Energy, 78*, 642–661. DOI: 10.1016/j.ijhydene.2024.06.344

Jha, B., & Singh, D. N. (2011). A review on synthesis, characterization and industrial applications of fly ash zeolites. *The Journal of Materials Education, 33*(1–2), 65–132.

Johnson, E. B. G., & Arshad, S. E. (2014). Hydrothermally synthesized zeolites based on kaolinite: A review. *Applied Clay Science, 97–98*, 215–221. DOI: 10.1016/j.clay.2014.06.005

Joseph, I. V., Tosheva, L., Miller, G., & Doyle, A. M. (2021). FAU-type zeolite synthesis from clays and its use for the simultaneous adsorption of five divalent metals from aqueous solutions. *Materials (Basel), 14*(13), 3738. DOI: 10.3390/ma14133738

Kalló, D. (2001). Applications of natural zeolites in water and wastewater treatment. *Reviews in Mineralogy and Geochemistry, 45*(1), 1–67. DOI: 10.2138/rmg.2001.45.15

Kazemian, H., Naghdali, Z., Ghaffari Kashani, T., & Farhadi, F. (2010). Conversion of high silicon fly ash to Na-P1 zeolite: Alkaline fusion followed by hydrothermal crystallization. *Advanced Powder Technology, 21*(3), 279–283. DOI: 10.1016/j.apt.2009.12.005

Kim, I. T., Ahn, K. H., Jung, J., Jeong, Y., Shin, D. C., & Lee, Y. E. (2021). Removal of tar contents derived from lignocellulosic biomass gasification facilities using MgAl-LDH@clinoptilolite. *Catalysts, 11*(9), 1111. DOI: 10.3390/catal11091111

Król, M. (2020). Natural vs. synthetic zeolites. *Crystals, 10*(7), 1–8. DOI: 10.3390/cryst10070622

Kumar, A., & Lingfa, P. (2019). Sodium bentonite and kaolin clays: Comparative study on their FT-IR. *Materials Today: Proceedings.* Advance online publication. DOI: 10.1016/j.matpr.2019.10.037

Li, X., Jiang, Y., Liu, X., Shi, L., Zhang, D., & Sun, L. (2017). Direct synthesis of zeolites from a natural clay, attapulgite. *ACS Sustainable Chemistry & Engineering, 5*(7), 6124–6130. Advance online publication. DOI: 10.1021/acssuschemeng.7b01001

Manea, F., Troquet, M., & Avram, L. (2006). Analyse de certains minéraux argileux. *Scientific Study & Research, 7*(3), 689–696.

Manuscript, A. (2020). Framework flexibility-driven CO_2 adsorption on a zeolite. *Materials Horizons, 7*(6), 1528–1532. DOI: 10.1039/D0MH00307G

Marcilly, C. (2001). Evolution of refining and petrochemicals: What is the place of zeolites. In *Studies in Surface Science and Catalysis* (Vol. 135, pp. 37–60). Elsevier., DOI: 10.1016/S0167-2991(01)81185-X

Mezni, M., Hamzaoui, A., Hamdi, N., & Srasra, E. (2011). Synthesis of zeolites from the low-grade Tunisian natural illite by two different methods. *Applied Clay Science, 52*(3), 209–218. DOI: 10.1016/j.clay.2011.02.017

Michalkiewicz, B., & Koren, Z. C. (2015). Zeolite membranes for hydrogen production from natural gas: State of the art. *Journal of Porous Materials, 22*(3), 635–646. DOI: 10.1007/s10934-015-9936-6

Misaelides, P. (2011). Application of natural zeolites in environmental remediation: A short review. *Microporous and Mesoporous Materials, 144*(1–3), 15–18. DOI: 10.1016/j.micromeso.2011.03.024

Mohd, R., Desforges, A., Rahman, A., & Vigolo, B. (2021). Progress in adsorption capacity of nanomaterials for carbon dioxide capture: A comparative study. *Journal of Cleaner Production, 328*, 129553. DOI: 10.1016/j.jclepro.2021.129553

Moneim, M. A., & Ahmed, E. A. (2015). Synthesis of faujasite from Egyptian clays: Characterizations and removal of heavy metals. *Geomaterials, 5*(2), 68–76. DOI: 10.4236/gm.2015.52007

Mozgawa, W., & Król, M. (2011). FT-IR studies of zeolites from different structural groups. *Chemik, 65*(7), 671–674.

Mravec, D., Hudec, J., & Janotka, I. (2005). Some possibilities of catalytic and noncatalytic utilization of zeolites. *Chemical Papers, 59*(1), 62–66.

Musyoka, N. M., Missengue, R., Kusisakana, M., & Petrik, L. F. (2014). Conversion of South African clays into high quality zeolites. *Applied Clay Science, 97–98*, 182–186. DOI: 10.1016/j.clay.2014.05.026

Musyoka, N. M., Petrik, L. F., Hums, E., Kuhnt, A., & Schwieger, W. (2015). Thermal stability studies of zeolites A and X synthesized from South African coal fly ash. *Research on Chemical Intermediates, 41*(2), 575–582. DOI: 10.1007/s11164-013-1211-3

Musyoka, N. M., Rambau, K. M., Manyala, N., Ren, J., Langmi, H. W., & Mathe, M. K. (2018). Utilization of waste tyres pyrolysis oil vapour in the synthesis of zeolite templated carbons (ZTCs) for hydrogen storage application. *Journal of Environmental Science and Health. Part A, Toxic/Hazardous Substances & Environmental Engineering, 53*(11), 1022–1028. DOI: 10.1080/10934529.2018.1471099

Naber, J. E., & De Jong, K. P. (1994). Industrial applications of zeolite catalysis. *Studies in Surface Science and Catalysis, 84*, 219–294.

Novembre, D., Gimeno, D., & Del Vecchio, A. (2021). Synthesis and characterization of Na-P1 (GIS) zeolite using a kaolinitic rock. *Scientific Reports, 1*(1), 1–11. DOI: 10.1038/s41598-021-84383-7

Novembre, D., Sabatino, B. D. I., Gimeno, D., & Annunzio, S. G. D. (2011). Synthesis and characterization of Na-X, Na-A and Na-P zeolites and hydroxysodalite from metakaolinite. *Clay Minerals, 46*(3), 339–354. DOI: 10.1180/claymin.2011.046.3.339

Olson, D. H.. (1970). A reinvestigation of the crystal structure of the zeolite hydrated NaX. *Journal of Physical Chemistry*, *74*(14), 2758–2764. DOI: 10.1021/j100708a006

Owusu, P. A., & Asumadu-Sarkodie, S. (2016). A review of renewable energy sources, sustainability issues and climate change mitigation. *Cogent Engineering*, *3*(1), 1167990. DOI: 10.1080/23311916.2016.1167990

Ozdemir, O. D., & Pişkin, S. (2013). Zeolite X synthesis with different sources. *International Journal of Chemical* [IJCEBS]. *Environmental & Biological Sciences*, *1*(2), 229–232.

Percival, H. J., Duncan, J. F., & Foster, P. K.H. J. P. (1974). Interpretation of the kaolinite-mullite reaction sequence. *Journal of the American Ceramic Society*, *57–61*(2), 57–61. Advance online publication. DOI: 10.1111/j.1151-2916.1974.tb10813.x

Pereira, P. M., Ferreira, B., Oliveira, N., Nassar, E., Ciuffi, K., Vicente, M., Trujillano, R., Rives, V., Gil, A., Korili, S., & De Faria, E. (2018). Synthesis of zeolite A from metakaolin and its application in the adsorption of cationic dyes. *Applied Sciences (Basel, Switzerland)*, *8*(4), 608. DOI: 10.3390/app8040608

Pérez-Botella, E., Valencia, S., & Rey, F. (2022). Zeolites in adsorption processes: State of the art and future prospects. *Chemical Reviews*, *122*(24), 17647–17695. DOI: 10.1021/acs.chemrev.2c00140

Petrov, T. M. I. (2012). *Synthesis of zeolite A: A review*. https://conf.uni-ruse.bg/bg/docs/cp12/9.1/9.1-5.pdf

Ríos, C., Williams, C., & Maple, M. (2007). Synthesis of zeolites and zeotypyes by hydrothermal transformation of kaolinite and metakaolinite. *Bistua*, *5*(1), 15–26.

. Ríos, C. A., Williams, C. D., & Castellanos, O. M. (2012). Crystallization of low silica Na-A and Na-X zeolites from transformation of kaolin and obsidian by alkaline fusion. *Ingeniería y competitividad*, *14*(2), 125-137.

Ronchi, L., Ryzhikov, A., Nouali, H., Daou, T. J., Albrecht, S., & Patarin, J. (2018). Extra large pore opening CFI and DON-type zeosils for mechanical energy storage. *Microporous and Mesoporous Materials*, *255*, 211–219. DOI: 10.1016/j.micromeso.2017.07.039

. Ruiz-Hitzky, P. A. E., & Camblor, M. A. (2014). Zeolite–sepiolite nanoheterostructures. https://doi.org/DOI: 10.1007/s40097-014-0090-5

Salahudeen, N. (2022). *A review on zeolite: Application, synthesis and effect of synthesis parameters on product properties*. Chemistry Africa., DOI: 10.1007/s42250-022-00471-9

Sánchez-López, P., Kotolevich, Y., Yocupicio-Gaxiola, R. I., Antúnez-García, J., Chowdari, R. K., Petranovskii, V., & Fuentes-Moyado, S. (2021). Recent advances in catalysis based on transition metals supported on zeolites. *Frontiers in Chemistry*, *9*, 716745. DOI: 10.3389/fchem.2021.716745

Sari, M. E. F., Suprapto, S., & Prasetyoko, D. (2018). Direct synthesis of sodalite from kaolin: The influence of alkalinity. *Indonesian Journal of Chemistry*, *18*(4), 607–613. DOI: 10.22146/ijc.25191

Sharma, P., Song, J. S., Han, M. H., & Cho, C. H. (2016). GIS-NaP1 zeolite microspheres as potential water adsorption material: Influence of initial silica concentration on adsorptive and physical/topological properties. *Scientific Reports*, *6*(1), 1–26. DOI: 10.1038/srep22734

Srilai, S., Tanwongwan, W., Onpecth, K., Wongkitikun, T., & Panpiemrasda, K. (2020). Synthesis of zeolite A from bentonite via hydrothermal method: The case of different base solution. In *The Second Materials Research Society of Thailand International Conference* (Vol. 060006). https://doi.org/DOI: 10.1063/5.0025043

Sun, Q., Wang, N., Bai, R., Hui, Y., Zhang, T., Do, D. A., Zhang, P., Song, L., Miao, S., & Yu, J. (2019). Synergetic effect of ultrasmall metal clusters and zeolites promoting hydrogen generation. *Advanced Science (Weinheim, Baden-Württemberg, Germany)*, *6*(10), 1802350. DOI: 10.1002/advs.201802350

Sun, Q., Wang, N., & Yu, J. (2021). Advances in catalytic applications of zeolite-supported metal catalysts. *Advanced Materials*, *33*(51), 2104442. DOI: 10.1002/adma.202104442

Tahmasebi, Y. M. (2021). Alkali activation of mechanically activated low-grade clay. *Journal of Sustainable Cement-Based Materials*, *5*(10), 272–288. DOI: 10.1080/21650373.2020.1838355

Tompsett, G. A., Conner, C., & Yngvesson, K. S. (2006). Microwave synthesis of nanoporous materials. *Chemphyschem : A European Journal of Chemical Physics and Physical Chemistry*, *7*(2), 296–319. DOI: 10.1002/cphc.200500449

Vigil, R., Villa, D., Goiti, E., Ocejo, M., & Gim, R. G. (2019). Synthesis of zeolite type analcime from industrial wastes. *Microporous and Mesoporous Materials*, *293*, 109817. DOI: 10.1016/j.micromeso.2019.109817

Warzybok, M. (2018). Synthesis of kaolin-based zeolite Y and its application for adsorption of two carbonyl compound gases. *Czasopismo Inżynierii Lądowej, Środowiska i Architektury, 65*(3), 13–26. DOI: 10.7862/rb.2018.2

Warzybok, M., & Warchoł, J. (2018). Synthesis of kaolin-based zeolite Y and its application for adsorption of two carbonyl compound gases. *Czasopismo Inżynierii Lądowej, Środowiska i Architektury. Czasopismo Inżynierii Lądowej, Środowiska i Architektury*. Advance online publication. DOI: 10.7862/rb.2018.2

. Wheelock Gutierrez, F. E. (2023). *A comparative study of low-emissions hydrogen production processes: Technical limitations and future trends.*

Yao, D., Yang, H., Chen, H., & Williams, P. T. (2018). Investigation of nickel-impregnated zeolite catalysts for hydrogen/syngas production from the catalytic reforming of waste polyethylene. *Applied Catalysis B: Environment and Energy, 227*, 477–487. DOI: 10.1016/j.apcatb.2018.01.050

Yao, D., Yang, H., Chen, H., & Williams, P. T. (2021). Hydrogen production by pyrolysis–nonthermal plasma/catalytic reforming of waste plastic over different catalyst support materials. *Energy & Fuels, 35*(10), 8280–8293. DOI: 10.1021/acs.energyfuels.1c04186

Yoldi, M., Korili, S. A., & Gil, A. (2019). Zeolite synthesis from industrial wastes. *Microporous and Mesoporous Materials, 287*, 183–191. DOI: 10.1016/j.micromeso.2019.06.009

Zeng, S., Wang, R., Zhang, Z., & Qiu, S. (2016). Solventless green synthesis of sodalite zeolite using diatomite as silica source by a microwave heating technique. *Inorganic Chemistry Communications, 70*, 168–171. DOI: 10.1016/j.inoche.2016.06.013

Zhang, Y., Wu, Z., Hong, Z., Gu, X., & Xu, N. (2012). Hydrogen-selective zeolite membrane reactor for low temperature water gas shift reaction. *Chemical Engineering Journal, 197*, 314–321. DOI: 10.1016/j.cej.2012.05.064

Chapter 3
Research on Preparation and Application Technology of Superhard Materials

Lei Guo
http://orcid.org/0000-0001-7849-9583
Tongren University, China

Yan Tan
http://orcid.org/0009-0003-9778-5851
Tongren University, China

Hassane Lgaz
Hanyang University, South Korea

Bochuan Tan
Chongqing University of Science and Technology, China

ABSTRACT

Owing to their exceptional hardness, wear resistance, and thermal stability, superhard materials have become a cornerstone of modern industrial innovation. This chapter provides a systematic review of the synthesis techniques, applications, and current research progress in this field. Following the dual themes of "preparation technology" and "application research," it analyzes the complete knowledge system of superhard materials, from fundamental theory to industrial implementation. By examining classic cases, such as the commercialization of CVD diamond for semiconductor heat dissipation, and recent advances like machine learning-driven

DOI: 10.4018/979-8-3373-4402-7.ch003

material design, this chapter aims to equip readers with a holistic perspective of the structure-property-process-application relationship and to foster the innovative thinking necessary to solve complex engineering problems.

INTRODUCTION

As the "pinnacle of materials science", superhard materials have become one of the core driving forces of modern industrial technological innovation due to their excellent hardness, wear resistance and thermal stability(Jazmina et al., 2022; Perevislov., 2022). From the shock of diamond cutting tools in precision machining, to the stable performance of cubic boron nitride (cBN) in high-temperature cutting, to the exploration of nano-polycrystalline materials to break through the traditional limits, superhard materials have always played a key role in breaking through the physical boundaries of mankind. With the rapid development of high-end manufacturing, energy exploration, semiconductor and quantum technology, the preparation and application of superhard materials is not only an academic subject in materials science, but also an important reflection of the competitiveness of national strategic industries(Anderson et al., 2024; Li et al., 2024).

Since the industrialization of artificial diamond by high temperature and high pressure (HPHT) technology in the middle of the 20th century, the development of superhard materials has undergone a metamorphosis from laboratory breakthroughs to large-scale applications(Fernando et al., 2023). Nowadays, the maturity of advanced preparation technologies such as Chemical Vapor Deposition (CVD) and Physical Vapor Deposition (PVD) has not only drastically reduced the production cost, but also given rise to new material forms such as diamond films and superhard coatings, which provide revolutionary solutions for cutting-edge fields such as aerospace and precision optics(Wang et al., 2025). At the same time, the boundaries of applications of superhard materials continue to expand - from cutting tools for traditional machining, to radiation-resistant optical windows in extreme environments, to sensing units based on diamond color centers in quantum computers - and their interdisciplinary integration is driving the in-depth crossover of materials science, physics, chemistry and engineering(Emanuele et al., 2023; Tang et al., 2012).

However, the research and development of superhard materials still faces many challenges: how to realize the efficient preparation of large-size single crystals? How to optimize the bonding strength of composite materials through interface engineering? How to maintain the long-life stability of materials under extreme working conditions? These issues involve both theoretical breakthroughs in basic science and the need for continuous innovation in engineering technology. In addition, the global consensus on green manufacturing, resource recycling and carbon-neutral targets further requires the transformation of superhard material preparation technologies in the direction of low energy consumption and sustainability.

Based on the main lines of "preparation technology" and "application research", this chapter systematically analyzes the whole chain of knowledge system of superhard materials from basic theory to industrial realization. By analyzing classic cases (e.g., the commercialization path of CVD diamond in semiconductor heat dissipation) and cutting-edge advances (e.g., machine learning-driven material design), this chapter aims to help readers build up a global perspective of "structure-property-process-application", and cultivate the innovative ability to solve complex engineering problems. Whether it is the "hard core" performance of diamond in deep earth drilling, the "invisible armor" given to cutting tools by superhard coatings, or the subversive challenges to traditional theories posed by nanopolycrystalline materials, this chapter will lead the readers into this fascinating and opportunistic field by means of rigorous scientific logic and vivid engineering practices.

1 OVERVIEW OF SUPERHARD MATERIALS

1.1 Definition and Classification of Superhard Materials

1.1.1 Scientific Definition and Physical Nature

The core characteristic of superhard materials is their extremely high hardness, which is usually defined in two ways(Hou et al., 2022; Zhao et al., 2016):

Mohs hardness $\geq$ 9 (10 on the Mohs scale for diamond; 9.5 for cubic boron nitride cBN).

Vickers hardness $\geq$ 40 GPa (up to 70-100 GPa for monocrystalline diamonds, much higher than 15-20 GPa for conventional cemented carbides)

The physical nature of its high hardness stems from the atomic bonding form and crystal structure(Zulkharnay & May, 2024):

Diamond: A tetrahedral structure formed by carbon atoms through sp^3 hybridization, with a network of strong covalent bonds making it the hardest substance in nature;

Cubic Boron Nitride (cBN): Boron and nitrogen atoms are arranged in a diamond-like structure with sp^3 bonds coexisting with some ionic bonds, combining high hardness and chemical stability.

Classification System and Typical Materials

Superhard materials can be divided into the following categories according to composition, structure and application scenarios:

1.1.2 Monolithic superhard materials

Diamond (natural/man-made)(Qian et al., 2012; Xiao et al., 2025):

Monocrystalline diamond: used for precision tools, optical windows;

Polycrystalline Diamond (PCD): made from micron-sized diamond particles sintered with a metal bond, used in oil drilling bits;

Chemical vapor deposition (CVD) diamond films: for semiconductor heat sinks, high power laser windows.

Cubic boron nitride (cBN)(Wang et al., 2024):

Monocrystalline cBN: High-temperature cutting tools (e.g. for machining hardened steel);

Polycrystalline Cubic Boron Nitride (PCBN): Composite with ceramic bond for automotive engine parts machining.

1.1.3 Compound-type superhard materials

Boron Carbon Nitride (B-C-N system): Synthesized by modulating the ratio of B, C, and N atoms, such as BC_2N, the theoretical hardness is close to that of diamond, and the oxidation resistance is superior;

Transition metal borides/carbides: Such as ReB_2 (rhenium diboride) and WB_4 (tetraboride), superhard properties are achieved through high-density covalent bonding.

1.1.4 Novel Nano and Composite Superhard Materials

Nanopolycrystalline Diamond (NPD): Formed from nano-sized diamond particles by high pressure sintering, it combines high hardness (>100 GPa) and toughness (5 times higher fracture resistance than single crystal diamond);

Diamond-metal composites: For example, diamond-copper composites, which are used for heat dissipation in high-power electronic devices, have a thermal conductivity of up to 800 W/m-K.

1.2 Historical development and technological milestones

Early Exploration and Application of Natural Materials:

Before the 19th century: natural diamonds were only used for jewelry and carving tools due to their scarcity, and there were records of diamonds being used to polish jade during the Han Dynasty in China; During the industrial revolution: the limitations of HSS tools (lack of hardness, wear) created a need for superhard materials.

The birth and industrialization of artificial superhard materials:

1954: Tracy Hall, a scientist at General Electric (GE), synthesized synthetic diamond for the first time through the high-temperature, high-pressure method (HPHT), with a pressure of 5.5 GPa and a temperature of 1,400°C, opening the era of the industrialization of superhard materials;

1957: The USSR Academy of Sciences successfully synthesizes cBN to make up for the lack of diamond in the machining of iron-based materials (diamond reacts chemically with iron);

1963: China Zhengzhou Abrasives Grinding Research Institute broke through the HPHT technology, realized the mass production of domestic artificial diamond, and got rid of the dependence on imports;

1990s: China's annual diamond production reached 1.2 billion carats, accounting for 80% of the world's output, but high-end products (e.g., CVD diamond films) were still dependent on imports.

Technological innovation and multidisciplinary penetration:

1980s-2000s: CVD/PVD technology breakthrough: preparation of diamond thin films under low pressure (NIRIM Labs, Japan), promoting its application in semiconductor heat dissipation and optical coatings; Nanotechnology convergence: nanodiamond powders (particle size <10 nm) for biomedical imaging and drug carriers;

21st Century: Extreme environment applications: Diamond windows are used in NASA's Trail Rover Laser Spectrometer to withstand extreme temperatures ranging from -120°C to +80°C; Extreme environment applications: Diamond windows are used in NASA's Trail Rover Laser Spectrometer to withstand extreme temperatures ranging from -120°C to +80°C;

Industrial importance is reflected in:

High-end manufacturing: superhard cutting tools increase the machining efficiency of aerospace titanium alloys by more than three times610;

Energy exploration: diamond drill bits have 10 times the life of traditional cemented carbide in shale gas extraction11;

Strategic position: China accounts for 80% of global diamond production, but the market for high-end products is still dominated by Europe and the US (e.g. Element Six accounts for 30% of the global diamond tool market)

1.3 Basic Properties and Characterization Methods

Core Performance Indicators:

1.Hardness: Diamond hardness up to 100 GPa2 as quantified by Vickers hardness tester or nanoindentation.

2. Wear resistance: Evaluated using a friction wear tester, the wear rate of cBN tools when cutting hardened steel is only 1/510th of that of carbide.
3. Thermal stability: Thermogravimetric analysis (TGA) shows that diamond can withstand 800°C in an inert atmosphere, while cBN is stable up to 1400°C in an oxidizing environment.
4. Chemically inert: X-ray photoelectron spectroscopy (XPS) confirms that diamond is virtually non-corrosive in strong acids, making it suitable for coating chemical reactors.

Table 1. Performance comparison (with conventional materials)

Materials	Hardness(GPa)	Heat conductivity (W/m·K)	maximum service temperature(°C)
Diamond	70-100	2000	800(inert atmosphere)
cBN	40-50	750	1400
Hard alloy (WC)	15-20	100	1000
High speed steel(HSS)	8-10	30	600

2 PREPARATION AND TECHNOLOGY OF SUPERHARD MATERIALS

2.1 High Temperature, High Pressure Synthesis (HPHT)

High Pressure High Temperature (HPHT) is a core technology for synthesizing superhard materials (e.g., diamond, cubic boron nitride) by simulating the extreme environmental conditions (high temperature and high pressure) in the deep Earth(Chen et al., 2024; Hou et al., 2025; Huang et al., 2024). The core principle is to utilize the law of thermodynamic phase transition to drive the structural reorganization of raw materials through externally applied high pressure and high temperature to form sub-stable superhard phase materials(Xiaoran et al., 2024). HPHT mainly relies on the synergistic effect of thermodynamic stability and kinetic conditions to synthesize superhard materials. Taking diamond as an example, graphite is a thermodynamically stable phase of carbon at room temperature and pressure, while diamond becomes stable only at high pressure (above about 5-6 GPa).According to the phase diagram of carbon, the Gibbs free energy of graphite is higher than that of diamond at high pressures (5-10 GPa) and high temperatures (1200-2000°C), resulting in a spontaneous transition to the diamond phase. However, relying solely on thermodynamic conditions, the phase transition rate is extremely low, and metal catalysts (e.g., Fe, Ni, Co alloys) need to be introduced to reduce the reaction activation energy and accelerate the atomic rearrangement process.

2.1.1 Phase Change Mechanisms and Catalyst Role

(1) Catalytic melting mechanism for diamond synthesis

Metal catalysts (e.g. Ni, Co, Fe or their alloys) melt to liquid state at high temperatures and are highly soluble in graphite. It's action can be divided into three stages:

Dissolution: the dissolution of graphite carbon atoms in a liquid metal;

Diffusion: diffusion of carbon atoms in the metal melt to low-temperature, low-pressure regions (e.g., the surface of diamond crystal species);

Precipitation: Carbon atoms precipitate as sp^3 hybrids on the surface of diamond seeds to form diamond crystals. This process can be described by the VLS (Vapor-Liquid-Solid) model, in which the liquid metal serves as the carbon transport medium and the diamond seed provides the nucleation sites.

(2) Solid-state phase transition mechanism for cubic boron nitride (cBN) synthesis

When hexagonal boron nitride (hBN) is used as the raw material, the HPHT method directly induces the collapse of the hBN layered structure by high pressure, and the B-N bonds undergo angular distortion to form a diamond-like cubic phase. The catalysts (e.g., Li_3N, Mg_3N_2) act as embedded in the hBN interlayers, weakening the interlayer van der Waals forces, and facilitating the fracture and reconstruction of the B-N bonds. For example, Li_3N reacts with hBN under high pressure to produce the intermediate $LiBN_2$, which reduces the pressure required for the phase transition to 5-7 GPa (more than 10 GPa is required in the absence of catalyst).

Figure 1. (a) Cubic high-pressure apparatus. (b) Schematic diagrams of the high-pressure anvil. HPHT assembly: 1: steel cap; 2: copper; 3: dolomite; 4: NaCl+ZrO2; 5: carbon source; 6: metal alloy catalyst; 7: crystal seed; 8: graphite heater; 9: MgO insulating medium; 10: graphite plate; and 11: pyrophyllite.

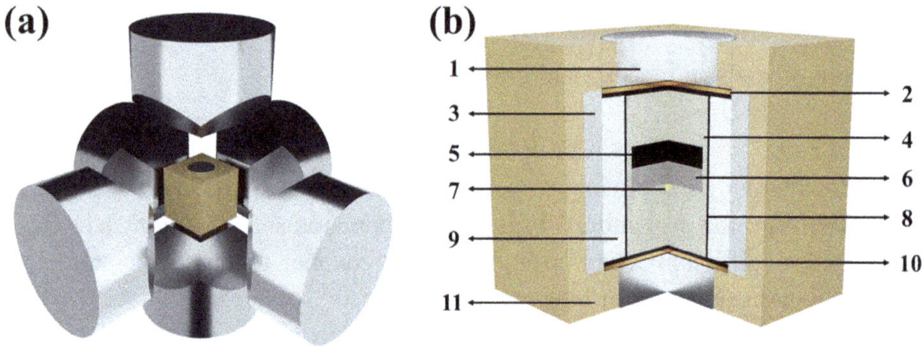

DOI: 10.1016/J.DIAMOND.2024.111073

2.1.2 Processes and equipment

(1) Typical process flow

Ingredients Preparation:

Diamond synthesis: high purity graphite (carbon source) + metal catalyst (Ni-Co-Fe alloy) + diamond crystal species;

cBN synthesis: hBN powder + nitride catalyst (e.g. Li_3N).

Assembly of the synthesis chamber: Encapsulation of the raw material with the pressure transfer medium (chlorite, sodium chloride) in a cylindrical or cubic chamber.

Apply pressure and heat: Apply 5-7 GPa of pressure using a six-sided top press or a multi-sided top press; Heating by resistance heating or high-frequency induction to 1300-2000°C (1400-1600°C for diamond, 1500-2000°C for cBN).

Holding pressure: Maintain high pressure and high temperature conditions for 10 minutes to several hours (adjusted according to crystal size requirements).

Buckling cooling and post-processing: Slow depressurization to avoid crystal rupture; Acid washing (e.g. nitric acid, aqua regia) to remove metal catalysts and obtain pure superhard materials.

(2) Equipment structure

Six-sided top press: It consists of six carbide top hammers, synchronized pressure application by hydraulic system, pressure uniformity up to ±1%;

Pressure-transferring medium: Leaf wax stone or salt (NaCl) is used to transfer pressure and insulate;

Heating element: graphite tube or tantalum sheet as resistance heat generator, temperature control accuracy ±10°C.

High-temperature and high-pressure synthesis method realizes the efficient preparation of superhard materials by precisely regulating pressure, temperature and catalyst system. Despite the challenges of cost and crystal quality, its irreplaceability in industrial tools, optical devices and quantum materials continues to drive the development of equipment intelligence (e.g. AI pressure feedback control) and process decarbonization. In the future, HPHT technology is expected to integrate with CVD, additive manufacturing and other technologies to expand the boundaries of superhard materials applications in extreme environments.

Figure 2. A schematic diagram of the synthetic assembly process: A1–A4 experimental design ideas.

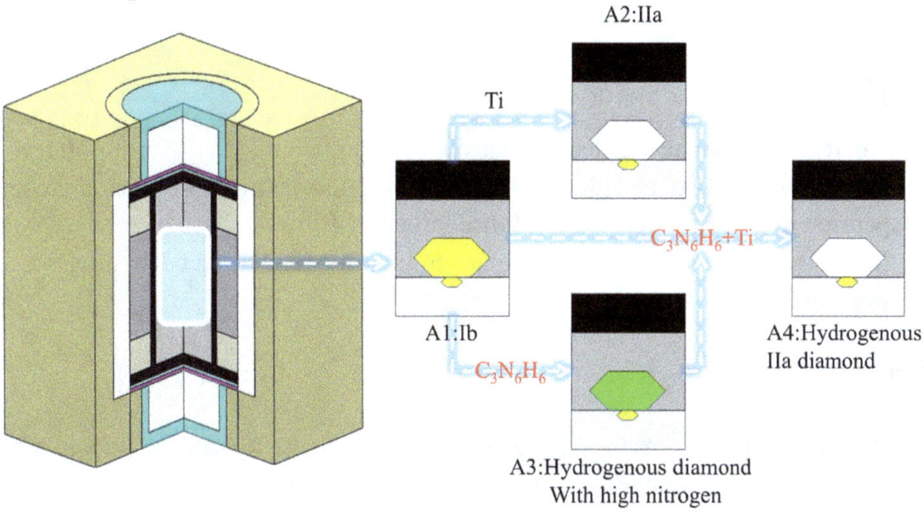

DOI: 10.1088/1674-1056/AD2A6D

2.2 Chemical Vapor Deposition

Chemical Vapor Deposition (CVD) is a technique for depositing solid materials on the surface of a substrate by means of a gas-phase chemical reaction, and is widely used in the preparation of diamond films, cubic boron nitride (cBN) coatings, and other superhard materials(Feng et al., 2025). The core principle is to generate thin films or bulk materials with specific crystal structures and functions on a substrate by controlling the chemical reaction of a gaseous precursor(Fang et al., 2025).

Nucleation and growth mechanisms of CVD diamond films

The chemical vapor deposition (CVD) technique deposits diamond films on the substrate surface by means of a gas-phase chemical reaction and is characterized by low temperatures (800-1000°C) and low pressures (<1 atm).The process is described below:

Precursor gas activation: A gas mixture of methane (CH_4) and hydrogen (H_2) decomposes under microwave plasma or hot filament excitation, producing carbon-containing reactive groups (e.g., CH_3, C_2H_2) and a large number of hydrogen atoms.

Hydrogen dissociates into active hydrogen atoms (H^+) in a high-energy environment:

$$H_2 \rightarrow 2H$$

The methane molecule is attacked by H^+ and progressively dehydrogenated to produce methyl radicals (CH_3^+):

$$CH_4 + H^{\cdot} \rightarrow CH_3^{\cdot} + H_2$$

Nucleation control: Nucleation density can be increased to 10^{10} cm-^{2} after the substrate surface is pre-treated with diamond micropowder milling or nanodiamond suspension to avoid the formation of discontinuous films.

Methyl radicals are adsorbed on the substrate surface and further dehydrogenated to form carbon atoms:

$$CH_3^{\cdot} \rightarrow C + 3H^{\cdot}$$

Reactive carbon atoms form a diamond structure by sp^3 hybridization, while sp^2 hybridized carbon (graphite) is removed by H^+ etching for selective diamond growth

Crystal growth and film formation: After hydrogen forms diamond nuclei on the substrate surface, the nuclei gradually expand and form a continuous film by continuous supply of precursors. The synergistic control of substrate temperature, gas flow and plasma power determine the grain size (nanometer to micron scale) and orientation.

Figure 3. Schematics of the low pressure CVD set-up.

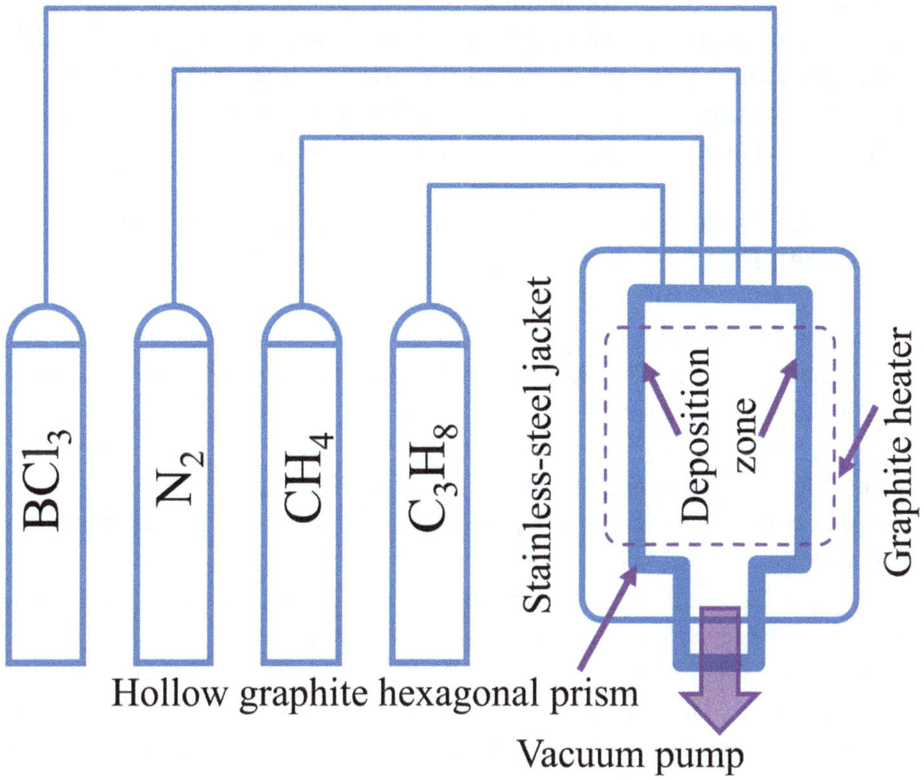

DOI: 10.1016/J.MSEB.2024.117491

2.3 Physical vapor deposition

Physical Vapor Deposition (PVD) is a technology that physically converts a solid material into a gaseous state and then condenses it on the surface of a substrate to form a thin film. Its core feature is that it does not rely on chemical reaction, but only through the energy input (such as heating, sputtering) to realize the phase change and deposition of materials(Ihara et al., 2025; Lavrinenko, 2025). PVD technology is widely used in the preparation of super-hard coatings (e.g. titanium nitride, diamond-like carbon film), optical thin films and functional devices(Ipatov & Kharanzhevskiy, 2019).

The physical process of PVD follows the three stages of "vaporization - transport - deposition":

Vaporization: The removal of atoms or molecules from the surface of a solid target into the gas phase by high energy input (e.g. thermal evaporation, ion bombardment).

Energy forms: resistance heating, electron beam, laser, magnetron sputtering, etc.

Transport: Gaseous particles migrate toward the substrate in a vacuum or low-pressure environment and may collide or ionize with reactive gases (e.g., N_2, CH_4).

Deposition: Gas-phase particles condense on the substrate surface to form a film, and the film structure (e.g., crystalline, amorphous) is regulated by controlling the substrate temperature, bias pressure, and other parameters.

2.3.1 PVD technologies can be classified into the following three categories based on the gasification mechanism:

Thermal Evaporation:

Principle: melting and evaporation of the target material by resistive heating or electron beam bombardment.

Characteristics: High deposition rate (1-10 μm/h), but weak film bonding, suitable for low melting point materials (e.g. Al, Au).

Applications: Aluminum plating for optical mirrors, solar cell electrodes.

Sputtering:

Principle: Bombardment of the target with energetic ions (e.g. Ar^+), which sputter atoms of the target into the gas phase by momentum transfer.

Magnetron Sputtering: A magnetic field is applied to the target surface to confine the plasma and increase the sputtering efficiency (deposition rate 0.1-1 μm/h).

Reactive Sputtering: Reactive gases (e.g. N_2, C_2H_2) are introduced to generate compound films (e.g. TiN, DLC).

Characteristics: Dense film, strong adhesion, suitable for high melting point materials (e.g. W, Ta).

Ion Plating:

Principle: Combining evaporation and plasma technology, a negative bias voltage (-100~-1000 V) is applied to the substrate during the deposition process, which attracts ions to bombard the film layer and enhance the densification.

Characteristics: Excellent film base bonding (>50 N), thick films (>10 μm) can be prepared.

Applications: TiN coatings for cutting tools, protective films for aero-engine blades.

PVD technology realizes the efficient preparation of functional films such as super-hard, wear-resistant and optical films through physical phase change and precise process control. Its wide application in precision manufacturing, new energy, biomedicine and other fields has pushed the limit breakthrough of coating performance (e.g., ultra-low friction DLC, 1500°C resistant AlCrSiN coatings).In the future, with the integration of high-energy beam technology (e.g., laser-assisted sputtering) and nanostructure design, PVD will further expand its application scenarios in cutting-edge fields such as microelectronics and quantum devices.

Table 2. Comparison of key properties of diamond synthesized by HPHT, CVD and PVD Table

Performance indicators	HPHT	CVD	PVD
Crystal Type	Monocrystalline/ Polycrystalline	Monocrystalline/ Polycrystalline	Amorphous/ Nanocrystalline
Hardness (GPa)	80-100 (monocrystalline)	70-95 (polycrystalline)	10-80
Thermal conductivity (W/m·K)	1800-2200	1200-2000	1-10
maximum size	Single crystal: 10 carats (~1 cm^3)	Wafer level: 200 mm diameter (polycrystalline thin film)	Film thickness: 0.1-10 μm
growth rate	0.1-1 mm/day (single crystal)	1-50 μm/h (polycrystalline)	0.1-5 μm/h
Impurity control	Metal catalyst residues (Fe, Ni)	Hydrogen atom doping (0.1-1 at%)	Gas impurities (Ar, N, etc.)
Temperature/ pressure conditions	1300-1600°C, 5-6 GPa	700-1000°C, 1-30 kPa	200-500°C, 10^{-4}-10^{-1} Pa
Surface roughness (Ra)	After single crystal polishing: <1 nm	Polycrystalline films: 10-100 nm	Coating: 5-50 nm
typical application	Industrial tools, jewelry, quantum sensors	Optical window, heat sink, wear-resistant coating	Tool coatings, biomedical devices, optical films
Technical constraints	Difficult to control defects in large single crystals, high equipment cost	Thick film growth is slow and hydrogen impurities affect thermal conductivity	Poor platability, substrate bonding dependent on pretreatment

Figure 4. The schematic representation of the MW-PECVD setup with the gas heating arrangement

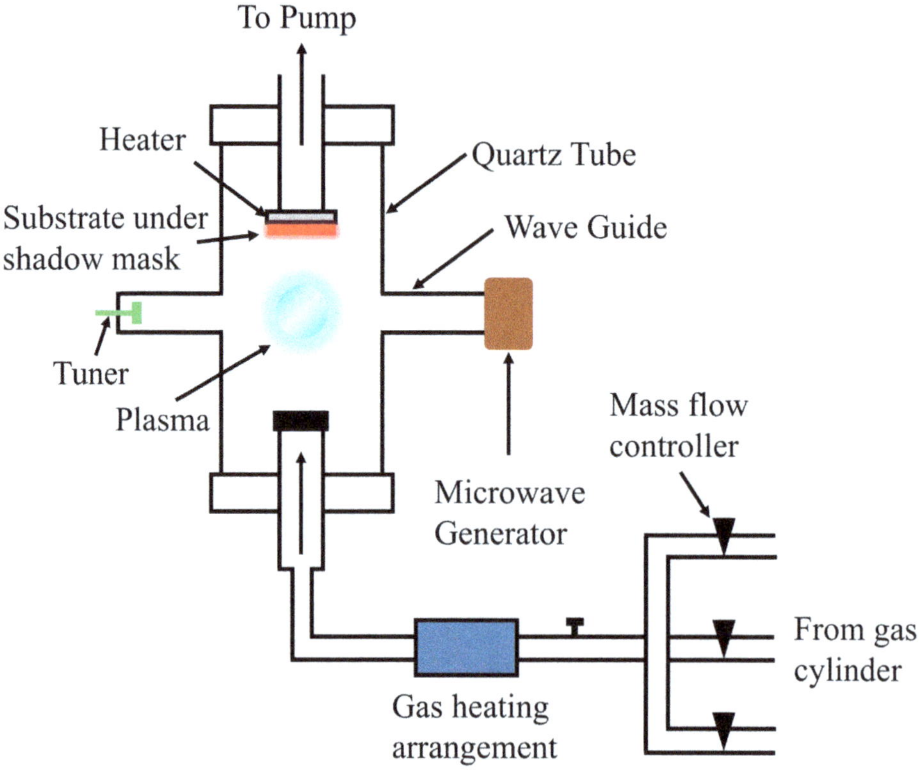

DOI: 10.1016/J.JPCS.2025.112554

2.4 Static High-Pressure Catalyst Method (for cBN)

As a superhard material second only to diamond, cubic boron nitride (cBN) shows irreplaceable advantages in the field of high-temperature cutting due to its excellent resistance to high-temperature oxidation and chemical stability(Ipatov & Kharanzhevskiy, 2019; Tang et al., 2012). High-Pressure High-Temperature (HPHT) is the mainstream technology to synthesize cBN in industry. The core of HPHT is to transform hexagonal boron nitride (hBN) into cubic phase structure by simulating the high-pressure, high-temperature environment in the deep crust of the earth, combined with the action of catalyst(Li et al., 2024). This process not only breaks through the limiting conditions of material synthesis, but also lays the foundation for the manufacture of high-performance cutting tools.

Principle and Synthesis Mechanism

The synthesis of cubic boron nitride is based on the phase transition properties of hBN. Hexagonal boron nitride (hBN) has a graphite-like layered structure, consisting of B-N six-membered rings connected by strong covalent bonds within the layers and weak van der Waals forces between the layers. Under high hydrostatic pressure (5-7 GPa) and high temperature (1500-2000°C), the layered structure of hBN is disrupted by external forces, and the B-N bonds are rearranged, resulting in the formation of a three-dimensional cubic lattice structure similar to that of diamond. The introduction of catalysts (e.g., magnesium, lithium, or lithium nitride) can significantly reduce the activation energy required for the phase transition: for example, lithium-based catalysts can form the intermediate compound Li_3BN_2 with the boron atoms in hBN, which facilitates the breaking and rearranging of the B-N bonds, and reduces the pressure of the phase transition from more than 10 GPa in the absence of a catalyst to 5-7 GPa, which greatly enhances the synthesis efficiency.

Processes and key technologies

Typical HPHT synthesis units utilize either a multi-face head press or a hinged six-face head press with a carbide head hammer to apply pressure to the reaction chamber. The synthesis process is divided into three stages:

Raw material pretreatment: hBN powder is mixed with catalyst (e.g., Mg_3N_2) in a proportional manner and pressed into a blank;

High-pressure & high-temperature reaction: In the sealed foliated paraffin chamber, the temperature is raised to more than 1500°C in 10-20 minutes by resistance heating or high-frequency induction heating, while maintaining a pressure of more than 5 GPa and continuing the reaction for 5-30 minutes;

Post-treatment and purification: Remove residual catalyst and unreacted hBN by acid washing (e.g., hydrochloric acid) to obtain micron-sized cBN particles.

The control of the pressure-temperature gradient in the reaction chamber is critical in determining the cBN purity and grain size.

Advantages of cBN in high-temperature cutting tools

The oxidation resistance (stable in air up to 1400°C) and chemical inertness of cBN make it far superior to diamond when machining iron-based materials. While diamond tools tend to carbonize with iron at high temperatures ($Fe + C \rightarrow Fe_3C$), resulting in rapid edge wear, cBN is virtually non-reactive with iron below 1200°C. For example, in the automotive engine block hardened steel (HRC 60-65) milling, cBN tool cutting speed up to 300-500 m/min, life is more than 50 times that of the carbide tool, and machining surface roughness (Ra) can be controlled within 0.4 μm.

2.5 Novel Preparation Technology Frontiers

Discharge plasma sintering (SPS)

Spark Plasma Sintering (SPS) is a rapid densification technique especially suitable for the preparation of nanostructured superhard materials(Daochun et al., 2023; Jin et al., 2013):

Principle: Fill the mold with diamond micropowder and metal binder (e.g. Co, Ni), apply pulsed DC current (1000-5000 A) to generate partial discharge and Joule heat between particles to achieve grain boundary diffusion and densification.

Advantages: fast heating rate (>100°C/min) and short sintering time (5-10 min), which can inhibit grain growth. For example, nanocrystalline diamond-tungsten carbide composites prepared with SPS have a hardness of 40 GPa and 20% higher fracture toughness.

Supercritical Fluid Synthesis Technology

Supercritical fluids (e.g., CO_2) combine gas diffusivity with liquid solubility, providing new avenues for nanodiamond synthesis:

Process steps: Organic precursors (e.g., polycarbosilanes) are dissolved in supercritical CO_2 (temperature >31°C, pressure >7.38 MPa), and the oriented arrangement of carbon atoms into diamond structure is induced by rapid pressure reduction.

Application prospects: nanodiamonds (particle size <10 nm) synthesized by this technique have low defect density and are suitable for quantum sensing. For example, nanodiamonds containing nitrogen-vacancy (NV) color centers can be used as highly sensitive magnetic sensors for biomedical imaging.

2.6 Key Challenges in the Preparation Process

Grain size control and interface optimization

The properties of superhard materials are highly dependent on the microstructure:

HPHT and CVD: The addition of boron (B) or nitrogen (N) can inhibit abnormal grain growth; in diamond-metal composites, the introduction of titanium (Ti) as an interfacial active agent reduces interfacial porosity and enhances the bond strength.

Figure 5. Experimental roadmap. The schematic shows the process of HPHT welding of CVD polycrystalline diamond: graphite is added to the laser-treated CVD diamond and encapsulated, after which it is HPHT welded in a cubic press.

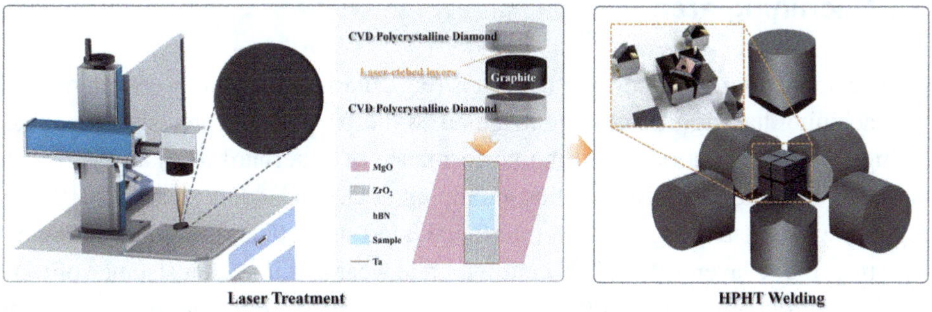

DOI: 10.1016/J.JEURCERAMSOC.2025.117246

SPS technology: By controlling the pulse current frequency and pressure distribution, the grain size gradient design is realized to balance the hardness and toughness.

Containment Strategies for Impurities and Defects

Defect control is the key to improving material performance:

HPHT process: purification of graphite raw material by zone melting technology, which makes the purity of metal catalyst more than 99.99%; gradient boosting (0.1 GPa/min) reduces the dislocation density in diamond crystals.

CVD process: the use of high-purity hydrogen (purity >99.999%) and microwave plasma (MPCVD) reduces the introduction of impurities; in-situ laser interferometry allows real-time monitoring of film defects and adjustment of deposition parameters.

2.7 Application Cases of Superhard Materials in Energy Storage

Field Diamond Heat Sinks for High Power Battery Thermal Management With the demand for high power densities in electric vehicles and energy storage systems, battery thermal management has become a key challenge. cvd diamond films are used to make highly efficient heat sinks due to their ultra-high thermal conductivity (>2000 W/m-K).For example, integrating a diamond heat sink layer in a lithium-ion battery module can reduce the battery operating temperature by 15-20°C and significantly extend cycle life. Ultra-hard coating to enhance the durability of energy storage devices TiAlN coating prepared by PVD technology can be used on the surface of supercapacitor collector to reduce electrolyte corrosion and lower contact resistance. Experiments show that the cycling stability of the collector after coating is improved by more than 50%, and the power density is increased by 30%.

3 APPLICATION TECHNOLOGY RESEARCH OF SUPERHARD MATERIALS

3.1 Machining area

The application of superhard materials (e.g., diamond and cubic boron nitride) in machining has significantly contributed to the progress of precision manufacturing. The optimization of cutting performance of superhard tools and abrasives is one of the core research directions(Caoping et al., 2023; János et al., 2019). By regulating the material grain size, interfacial bond strength and surface coating design (e.g. multi-layer TiN/AlCrN coatings), the wear and impact resistance of tools can be significantly improved(Zulkharnay & May, 2024). For example, the cutting life of nanocrystalline diamond tools can be extended to more than five times that of conventional carbide tools when machining high-silicon aluminum alloys(M., 2024). In addition, surface integrity control in precision machining relies on the high hardness and low coefficient of friction properties of super abrasives. Studies have shown that the use of chemical mechanical polishing (CMP) in combination with super abrasives can achieve sub-nanometer surface roughness while suppressing microcracks caused by thermal stresses during machining, thus ensuring high reliability of aerospace components(Koroviaka et al., 2023).

3.2 Energy and Mining

In the field of energy exploration, the application of diamond drill bits has greatly improved the drilling efficiency of oil and gas wells. Polycrystalline diamond composite (PDC) bits can maintain excellent anti-wear performance in high-temperature and high-pressure formations by combining diamond micro powder with cemented carbide matrix(Demidenko et al., 2024). For example, in shale gas horizontal well drilling, the average feed rate of PDC bits is more than 30% higher than that of conventional bits. In addition, the wear-resistant design of superhard materials in shale gas extraction has become a research hotspot. By introducing gradient-structured diamond coatings or silicon carbide-diamond composites, the problem of erosion and wear of hydraulic fracturing tools in high-pressure sandy fluids can be effectively mitigated, and the service life of key components can be extended(Loginov et al., 2020).

3.3 Electronics and Optics

The application of superhard materials in electronics and optics is gradually expanding to extreme environments(Jin et al., 2013). The development of CVD diamond heat sinks is a breakthrough in thermal management for electronics. Thanks to diamond's thermal conductivity of up to 2000 W/(m-K), its heat sinks can reduce the operating temperature of high-power laser diodes by more than 40%, significantly improving device stability. On the other hand, the application of ultra-hard window materials (e.g., single-crystal diamond) in extreme environments has attracted much attention. For example, diamond windows are used in high-energy laser systems, where their excellent light transmission (from the ultraviolet to the far-infrared bands) and radiation resistance make them indispensable for space exploration and nuclear fusion devices. High Power Laser Windows and Optics Diamond is the only known broad-spectrum light-transmitting material that can simultaneously cover the ultraviolet (UV), visible (VIS), infrared (IR) to terahertz bands. Its transmittance is over 90% in the range of 0.22 μm (UV) to 25 μm (FIR), and its thermal conductivity is as high as 2000 W/(m-K), which is five times higher than that of copper. This property makes it an ideal window material for high-power laser systems. For example, in industrial-grade CO_2 laser cutters (10.6 μm wavelength), diamond windows can withstand laser power densities in excess of 10 kW/cm^2, whereas conventional zinc selenide (ZnSe) windows are susceptible to rupture due to heat buildup at this power. In addition, diamond is used in exit mirrors for free electron lasers (FELs), where its radiation-resistant properties withstand bombardment by high-energy electron beams, significantly extending equipment life.

3.4 New Energy and Environmental Protection

3.4.1 hydrogen technology

(1) Electrolyzed water to hydrogen electrode

The anodes of traditional proton exchange membrane (PEM) electrolyzers mostly use precious metal catalysts such as iridium (Ir) and ruthenium (Ru), but their corrosion problems in acidic environments lead to limited lifetimes (usually less than 10,000 hours) and high costs. The introduction of boron doped diamond (BDD) electrodes has changed this situation:

Corrosion resistance: The chemical inertness of diamond makes it virtually lossless in strong acids (e.g. 1 M H_2SO_4) and strongly oxidizing environments.

Catalytic efficiency: By surface nanostructuring (e.g., preparation of porous diamond arrays), the oxygen precipitation reaction (OER) overpotential can be reduced to below 300 mV, which is comparable to that of iridium-based catalysts.

(2) Catalyst carrier for fuel cells

In proton exchange membrane fuel cells (PEMFC), platinum (Pt) catalysts are susceptible to agglomeration deactivation due to carbon carrier corrosion, leading to cell performance degradation. Nanodiamond (ND), as a novel carrier, enhances the durability by the following mechanism:

High specific surface area and stability: The abundant functional groups (e.g. -OH, -COOH) on the surface of nanodiamond can anchor Pt particles and inhibit their migration.

Synergistic catalytic effect: sp^3-hybridized carbon on the diamond surface can form strong electronic interactions with platinum to optimize the oxygen reduction reaction (ORR) pathway.

3.4.2 In the field of nuclear energy

(1) Radiation detectors

With its wide forbidden bandwidth (5.5 eV) and high displacement threshold (43 eV), the diamond detector can maintain stable operation in strong radiation fields, and its performance far exceeds that of silicon-based detectors:

Fast response: the carrier mobility in diamond is up to 4500 $cm^2/(V\text{-}s)$, the response time to γ-rays and neutrons is less than 1 ns, which can be used for transient radiation monitoring of nuclear reactors.

Resistant to high-temperature radiation: Under fast neutron irradiation (injection rate 10^{14} n/cm^2), the signal drift rate of diamond detectors is two orders of magnitude lower than that of silicon detectors.

(2) Nuclear waste disposal containers

The long-term safety of nuclear waste storage tanks depends on the material's resistance to radiation attack, and conventional stainless steel vessels are susceptible to lattice damage from alpha particle bombardment. The application of diamond coatings significantly improves the protective performance:

Resistant to irradiation damage: Diamond has an atomic displacement threshold 4 times that of steel and maintains structural integrity at a cumulative dose of 1000 dpa (atomic displacement/atom).

Chemical corrosion resistance: diamond has almost no reaction in strong corrosive media such as concentrated nitric acid, aqua regia, etc. It is suitable for acid recovery equipment in spent fuel reprocessing plants.

3.5 Frontier technology and quantum field

Superhard materials show disruptive potential in the fields of quantum technology and extreme environment detection, and their unique physical properties provide revolutionary solutions for high-precision quantum sensing, scalable quantum computation, and deep-space and deep-sea exploration equipment. In the following, we will discuss the two aspects of quantum technology and extreme environment detection.

3.5.1. Quantum technology

(1) NV color center quantum sensor: "microscopic probe" with atomic-level precision Nitrogen-vacancy (NV color center) in diamond is a defective structure formed by a nitrogen atom replacing a carbon atom and accompanied by adjacent vacancies, and its electronic spin state can be manipulated by laser and microwave synergy to realize the detection of magnetic field, temperature, electric field and other physical quantities with nanoscale resolution.

Magnetic field sensing: The magnetic field sensitivity of the NV color center is up to 1 $pT/\sqrt{Hz}$ (picoTesla/Hertz ½), which far exceeds that of conventional superconducting quantum interferometers (SQUID).

Temperature measurement: Zero-Field Splitting of NV color centers is highly sensitive to temperature changes with an accuracy of 1 mK.

3.6 Challenges and solutions

Despite the wide range of applications of superhard materials, they still face serious challenges in actual working conditions. The analysis of material failure mechanisms under complex working conditions is the key to solving the problem. For example, diamond tools are susceptible to thermo-chemical wear during intermittent cutting of titanium alloys, which needs to be combined with molecular dynamics simulation and in-situ characterization techniques to reveal its atomic-level damage mechanism. Meanwhile, the synergistic effect study of composite coating technology provides new ideas for performance enhancement. By designing a diamond-silicon carbide multilayer heterogeneous structure, the hardness and toughness of the coating can be optimized synchronously to solve the problem of brittle fracture of a single material under alternating loads. In the future, artificial intelligence-driven material design and in-situ performance monitoring technology will further promote the engineering application of superhard materials.

4 CUTTING-EDGE RESEARCH AND TECHNOLOGICAL BREAKTHROUGHS

4.1 New superhard material design

With the continuous improvement of material performance requirements in modern industry, traditional superhard materials (e.g., diamond and cubic boron nitride) have been difficult to meet the diverse needs in extreme environments. In this context, the design of new superhard materials has become a hot research topic in the field of materials science. In recent years, the combination of theoretical calculations and experimental means has significantly accelerated the discovery and optimization of materials, while the breakthrough of multiphase composite technology has opened up a brand-new path for the performance enhancement of superhard materials.

4.1.1 Material prediction guided by theoretical calculations

Traditional materials research and development relies on a “trial and error” approach, which is long and costly. In recent years, the combination of high-throughput computation based on Density Functional Theory (DFT) and machine learning techniques has revolutionized material design(Tang et al., 2012; Zulkharnay & May, 2024). For example, by building material databases (e.g., Materials Project), researchers are able to quickly screen candidate materials with high stiffness, high thermal stability, and chemical inertness. Machine learning algorithms (e.g., neural networks, random forests) can identify complex correlations between a material's crystal structure, chemical bonding characteristics, and mechanical properties to predict potential superhard phases(Caoping et al., 2023).

Taking the boron carbon nitride (BCN) system as an example, it was found through theoretical simulations that certain ternary compounds (e.g., B_3C_2N) can have hardnesses close to that of diamond, while possessing higher oxidation resistance. Similarly, transition metal borides (e.g., ReB_2) were computationally simulated to verify their superhardness and subsequently successfully synthesized experimentally. This “computation-first” model significantly shortens the cycle time from theory to application of new materials.

4.1.2 Development of multi-phase composite superhard materials

Single superhard materials often have performance shortcomings (e.g. poor high-temperature oxidizability of diamond, insufficient toughness of cubic boron nitride). Multi-phase composite technology has become an important research direction by realizing complementary performance through microstructure design. For example, diamond-silicon carbide (Diamond-SiC) composites are prepared by chemical vapor deposition (CVD) with high interfacial bonding strength, combining the hardness of diamond and the impact resistance of silicon carbide, and are suitable for high-performance cutting tools in the aerospace field.

Another breakthrough is the design of nanocomposite superhard materials. For example, the introduction of nanoscale cubic boron nitride particles into the diamond matrix inhibits crack extension through the grain boundary pinning effect, increasing the fracture toughness of the material by more than 30%.In addition, gradient composite structures (e.g., diamond on the surface and cemented carbide in the inner layer) show excellent performance in the field of tool coating, which can significantly extend tool life.

4.2 Innovations in preparation technology

The performance of superhard materials not only depends on the composition design, but is also closely related to the preparation process. In recent years, breakthroughs in technologies such as ultrafast laser processing and atomic layer deposition (ALD) have provided brand-new solutions for the precision fabrication and nanostructure modulation of superhard materials.

4.2.1 Ultra-fast laser-assisted processing technology

While conventional mechanical processing of ultra-hard materials is prone to chipping and thermal damage, ultra-fast lasers (femto- to picosecond pulses) can achieve a "cold processing" effect thanks to their extremely short energy action time. The principle lies in the fact that the ultra-short pulses plasmaize the material instantaneously and the heat-affected zone is extremely small, thus avoiding micro-cracks and phase transitions. For example, the use of femtosecond lasers to process micro-nanostructures (e.g., gratings, microholes) on the surface of diamond can reach submicron accuracy, laying the foundation for the fabrication of optical and quantum devices.

In addition, ultrafast lasers can be used for the welding of superhard materials. Through precise control of laser parameters, heterogeneous joining of diamond and metal (e.g., titanium alloys) can be realized with an interfacial bond strength of more than 200 MPa, which provides key technical support for heat dissipation modules for high-power electronic devices.

4.2.2 Atomic Layer Deposition (ALD) in Nanostructures

The ALD technique allows precise control of film growth at the atomic scale by alternately energizing precursor gases, and is particularly suitable for the preparation of complex nanostructures. For example, the deposition of alumina (Al_2O_3) or titanium nitride (TiN) nanolayers on diamond surfaces significantly improves the chemical inertness of their surfaces, allowing them to remain stable in corrosive environments.

In the field of nanocomposite coatings, ALD technology shows unique advantages. Taking diamond-tungsten carbide (WC) composite coatings as an example, by alternately depositing diamond and WC nanolayers to form a "sandwich" structure, its hardness reaches 50 GPa, while the coefficient of friction is reduced to less than 0.1, which is suitable for high-end bearings and precision molds. In addition, ALD can prepare ultra-thin (<10 nm) diamond films for thermal management of next-generation semiconductor devices.

4.3 Interdisciplinary integration of applications

The application fields of superhard materials are expanding from traditional mechanical processing to emerging fields such as biomedicine and quantum information, showing great interdisciplinary potential.

4.3.1 The potential of superhard materials in biomedical devices

Biomedical devices require extremely high biocompatibility, wear resistance and antimicrobial properties of materials, and superabrasives are ideal candidates due to their unique properties. For example, nitrogen-doped diamond coatings have been used for artificial joint surfaces, where the coefficient of friction is close to that of natural cartilage (0.01-0.05), and nitrogen doping inhibits bacterial biofilm formation and reduces the risk of postoperative infection.

In the field of minimally invasive surgical instruments, the technology of miniaturized processing of superhard materials is particularly important. Nanoscale blades processed by femtosecond lasers at the tip of diamond needles can realize cutting operations with cellular-level precision for applications in ophthalmic surgery and neurosurgery. In addition, porous diamond scaffolds are being explored as drug slow-release carriers for targeted therapy due to their high specific surface area and chemical stability.

4.3 2 Study of diamond nitrogen-vacancy color centers in quantum devices

Nitrogen-vacancy (NV) color centers in diamond are atomic-scale defects whose electronic spin states can be manipulated at room temperature by light-detectable magnetic resonance (ODMR) techniques, making them a central platform for quantum sensing and quantum computation. NV color centers are extremely sensitive for detecting changes in the magnetic field of individual molecules, and have great potential for use in biomolecular imaging and early disease diagnosis.

In recent years, high-purity single-crystal diamond prepared by chemical vapor deposition (CVD) has enabled the controlled doping and arrayed integration of NV color centers. For example, etching microcavity structures on the diamond surface can improve the fluorescence collection efficiency of NV color centers by more than 10 times, laying the foundation for the development of high-precision quantum gyroscopes and gravimeters. In addition, the NV color-center-based quantum bit (Qubit) has a coherence time of more than 1 second at low temperature, which provides a new idea for building scalable quantum computers.

REFERENCES

Anderson, K. P., Wollmershauser, J. A., Ryou, H., Goswami, R., Rommel, S., Aindow, M., Gorzkowski, E. P., & Feigelson, B. N. (2024). Nanostructural effects beyond Hall-Petch: Towards superhard tungsten carbide. *Acta Materialia*, *275*, 120004. DOI: 10.1016/j.actamat.2024.120004

Caoping, N., Jing, Z., Wenming, X., Guo, C., Wei, L., Jie, Z., Rajeev, A., & Xianlong, W. (2023). Configuration stability and physical properties of new diamondene structure. *Materials Today. Communications*, *36*, 106465. DOI: 10.1016/j.mtcomm.2023.106465

Chen, Y., Li, J., Li, L., Han, M., & He, J. (2024). In Situ Synthesis and Tribological Characterization of TiC–Diamond Composites: Effect of the Counterface Material on Wear Rate and Mechanism. *Coatings*, *14*(6), 735–755. DOI: 10.3390/coatings14060735

Daochun, H., Lei, W., Minghe, C., Shaohui, F., Ning, W., & Ying, Z. (2023). Thermal conduction and strength of diamond-copper composite sandwich obtained by SPS diffusion bonding with Ti interlayer. *Journal of Materials Research and Technology*, *26*, 8806–8812. DOI: 10.1016/j.jmrt.2023.09.188

Demidenko, M., Adamchuk, D., Liubimau, A., Uglov, V., Ishchenko, A., Chekan, M., Khama, M., & Maksimenko, S. (2024). High temperature synthesis and material properties of boron-enriched balk pyrolytic carbon. *Materials Science and Engineering B*, *307*, 117491. DOI: 10.1016/j.mseb.2024.117491

Emanuele, G., Giovanni, B., Alessandro, B., & Emanuela, C. (2023). Diamond-Like Carbon (DLC) and AlCrN films onto Ti-6Al-4V substrates by Laser-Powder Bed Fusion (L-PBF): Effect of substrate heat treatment and surface finish. *Surface and Coatings Technology*, *475*, 130128. DOI: 10.1016/j.surfcoat.2023.130128

. Fang, K., Zhang, X., Deng, M., Gai, Z., Zhang, M., & Yang, Y. (2025). Effect of boron doping levels on the piezoresistive properties of boron-doped diamond films prepared by HFCVD. *Vacuum, 231*(PA), 113768. DOI: 10.1016/j.vacuum.2024.113768

Feng, X., Yuan, X., Zhou, J., An, K., Zhu, F., Wei, X., Huang, Y., Zhang, J., Chen, L., Liu, J., Li, C., & Wei, J. (2025). A review: CNT/diamond composites prepared via CVD and its potential applications. *Materials Science in Semiconductor Processing*, *186*, 109008. DOI: 10.1016/j.mssp.2024.109008

Fernando, R.-R., Victor, Z., & Fernando, G. (2023). Solid-state spark plasma sintering of super wear resistant B4C–SiC–TiB2 triplex-particulate composites. *Ceramics International*, *49*(3), 5532–5537. DOI: 10.1016/j.ceramint.2022.11.181

Hou, H. Z., Haikuo, W., Yao, T., Jiakun, W., Chao, W., Zhicai, Z., & Xiaoping, O. (2022). Research progress of spectra and properties of ultrahard carbon materials at high pressure and high temperature. *Functional Diamond, 2*(1), 245–257. DOI: 10.1080/26941112.2022.2163857

Hou, X., Zhao, Y., Shang, Y., Shen, F., Wu, B., Chen, D., Liu, Z., Yao, M., & Liu, B. (2025). Nanostructure engineering of superhard nano-polycrystalline diamond by compressing different fullerene precursors. *Carbon, 235*, 120078. DOI: 10.1016/j.carbon.2025.120078

Huang, G., Chen, L., & Fang, C. (2024). Synthesis and nitrogen content regulation of diamond in a high-pressure hydrogen-rich environment. *Chinese Physics B, 33*(6), 068102. DOI: 10.1088/1674-1056/ad2a6d

Ihara, L., Wu, G., Cavaleiro, A., Morina, A., & Yang, L. (2025). Environmental Impact of Physical Vapour Deposition and Plasma-Enhanced Chemical Vapour Deposition Technologies for Deposition of Diamond-like Carbon Coatings for Green Tribology. *Coatings, 15*(2), 218. DOI: 10.3390/coatings15020218

Ipatov, A. G., & Kharanzhevskiy, E. V. (2019). The Tribological Properties of Superhard and Functional Coatings Based on Carbide and Boron Nitride. *Journal of Friction and Wear, 40*(6), 588–592. DOI: 10.3103/S1068366619060114

. János, K., Vladimir, F., Ivan, P., & P., M. A. (2019). Improving the effectiveness of combined grinding processes for processing superhard materials. *Journal of Manufacturing Processes, 43*(PA), 270-275. DOI: 10.1016/j.jmapro.2019.05.004

Jazmina, N.-C., Tomas, S.-B., Lorena, L.-C., Steven, M., Patricia, A., & Nerea, G.-P. (2022). Hot isostatic pressing of WB4 and WB4-TaB2 based ultrahard materials. *International Journal of Refractory & Hard Metals, 109*, 105965. DOI: 10.1016/j.ijrmhm.2022.105965

Jin, Z., Shi, S., Lin, J., & Jin, L. (2013). Preparation and Performance of Dynamic Friction Polishing Plate for Diamond Film. *Materials and Manufacturing Processes, 29*(1), 20–26. DOI: 10.1080/10426914.2013.852219

Koroviaka, Y. A., Ihnatov, A. O., Pavlychenko, A. V., Valouch, K., Rastsvietaiev, V. O., Matyash, O. V., Mekshun, M. R., & Shypunov, S. O. (2023). Studying the Performance Features of Drilling Rock Destruction and Technological Tools. *Journal of Superhard Materials, 45*(6), 466–476. DOI: 10.3103/S1063457623060059

Lavrinenko, V. I. (2025). Recent Developments in Diamond Processing and Surface Modification Techniques for Tailoring Properties of Diamond Grains: A Review. *Journal of Superhard Materials, 46*(6), 462–481. DOI: 10.3103/S1063457624060054

Li, J., Niu, G., Mu, P., Yan, B., Liu, F., Zhao, S., Fang, L., & Gou, H. (2024). Amorphous diamond embedded in dense boron nitride with excellent mechanical properties. *Microstructures*, *4*(1), 2024010. DOI: 10.20517/microstructures.2023.54

Loginov, P. A., Zhassay, U. A., Bychkova, M. Y., Petrzhik, M. I., Mukanov, S. K., Sidorenko, D. A., Orekhov, A. S., Rupasov, S. I., & Levashov, E. A. (2020). Chromium-doped Fe-Co-Ni binders for diamond cutting tools: The features of the structure, mechanical properties, and adhesion to diamond. *International Journal of Refractory & Hard Metals*, *92*, 105289. DOI: 10.1016/j.ijrmhm.2020.105289

M., A. A. (2024). Heat Resistance of Tool Materials. *Russian Engineering Research, 44*(1), 61-69. DOI: 10.3103/S1068798X24010040

Perevislov, N. S. (2022). Superhard boron suboxide (B6O): Crystal structure, synthesis, properties, applications, and materials based thereon. *Mendeleev Communications, 32*(2), 152–163. DOI: 10.1016/j.mencom.2022.03.002

Qian, J., McMurray, C. E., Mukhopadhyay, D. K., Wiggins, J. K., Vail, M. A., & Bertagnolli, K. E. (2012). Polycrystalline diamond cutters sintered with magnesium carbonate in cubic anvil press. *International Journal of Refractory & Hard Metals, 31*, 71–75. DOI: 10.1016/j.ijrmhm.2011.09.008

Tang, M., He, D., Wang, W., Wang, H., Xu, C., Li, F., & Guan, J. (2012). Superhard solid solutions of diamond and cubic boron nitride. *Scripta Materialia*, *66*(10), 781–784. DOI: 10.1016/j.scriptamat.2012.02.006

Wang, X., Wu, X., Lu, K., & Ye, J. (2025). Influence of cemented carbide substrate microstructure on CVD diamond coating characteristics. *Diamond and Related Materials*, *152*, 111886. DOI: 10.1016/j.diamond.2024.111886

Wang, Z., Wang, Z., Liu, Y., Li, B., Zhao, H., Guo, Q., Ma, H., & Jia, X. (2024). Properties of boron-doped HPHT diamond single crystals grown in a Fe-Ti-B-C system. *Diamond and Related Materials*, *145*, 111073. DOI: 10.1016/j.diamond.2024.111073

Xiao, X., Zheng, L., Li, Q., Chen, C., He, D., Wei, X., Hu, Z., Yang, J., & Zhan, G. (2025). HPHT welding of CVD polycrystalline diamond films. *Journal of the European Ceramic Society*, *45*(7), 117246. DOI: 10.1016/j.jeurceramsoc.2025.117246

Xiaoran, Z., Jinyu, S., Xiangzhuo, X., Chutong, Z., Yonghao, S., & Xiaobing, L. (2024). Significant enhancement of electrical conductivity in boron-doped diamond through HPHT post-annealing treatment. *International Journal of Refractory & Hard Metals*, *119*, 106509. DOI: 10.1016/j.ijrmhm.2023.106509

Zhao, Z., Xu, B., & Tian, Y. (2016). Recent Advances in Superhard Materials. *Annual Review of Materials Research*, *46*(1), 383–406. DOI: 10.1146/annurev-matsci-070115-031649

Zulkharnay, R., & May, P. W. (2024). Applications of diamond films: A review. *Functional Diamond*, *4*(1), 2410160. DOI: 10.1080/26941112.2024.2410160

Chapter 4
Fundamentals and Technological Advances in Energy Storage Materials

Xiugang Li
http://orcid.org/0009-0003-1728-3545
Tongren University, China

Qing Zhang
Tongren University, China

Lingyu Zhang
Tongren University, China

Zhang-Hui Lu
http://orcid.org/0000-0001-6342-0879
Jiangxi Normal University, China

ABSTRACT

In the twenty-first century, human society is confronted with significant climate challenges, and the intermittent and variable nature of renewable energy sources limits their large-scale implementation. This situation underscores the importance of energy storage technology as a pivotal element in the ongoing energy revolution. This chapter systematically reviews the latest advancements in research on energy storage materials and devices, focusing on four primary areas: enhancing energy density, extending cycle life, improving safety, and reducing costs. The specific

DOI: 10.4018/979-8-3373-4402-7.ch004

content includes: (1) fundamental theories of energy storage materials; (2) the performance and challenges associated with typical energy storage systems; (3) ingenious material design strategies; (4) disruptive technological advances.

INTRODUCTION

In the twenty-first century, human society is confronted with unprecedented climate challenges. According to the report by the United Nations Intergovernmental Panel on Climate Change (IPCC), a global temperature increase surpassing the 1.5 °C threshold is likely to initiate a catastrophic ecological chain reaction (Armstrong McKay et al., 2022; McCulloch et al., 2024). In response to this situation, the Paris Agreement, endorsed by 178 countries worldwide, establishes carbon neutrality targets and advocates for a transition in the energy mix from fossil fuels to renewable energy sources, such as wind and solar power. Nonetheless, the intermittent and volatile nature of these clean energy sources, for instance, the diurnal fluctuations in photovoltaic power generation and the seasonal variations in wind power-significantly constrains their large-scale deployment (Chen & Ma, 2024). According to the International Energy Agency (IEA), the global economic loss resulting from the abandonment of renewable energy power in 2023 is projected to reach $12 billion. This situation has elevated energy storage technology to a central position in the ongoing energy revolution. Advances in energy storage performance are fundamentally driven by innovations in material systems, ranging from the intercalation of lithium ions in graphite anodes to the thermal phase change storage of molten salt materials. Presently, leading research laboratories worldwide are concentrating on four primary dimensions: enhancing energy density (e.g., silicon-based anode lithium), extending cycle life (e.g., monocrystalline ternary anode), improving safety (e.g., solid-state electrolytes), and reducing costs (e.g., sodium-ion batteries) (Li et al., 2025). Nonetheless, challenges such as the geopolitical risks associated with cobalt resources, the growth of lithium metal dendrites, and the corrosion failure of thermal storage materials remain significant concerns, akin to the proverbial Sword of Damocles. Addressing these challenges necessitates a comprehensive understanding of material behavior at the atomic scale, coupled with the integration of interdisciplinary methodologies, including artificial intelligence and high-throughput computing.

Figure 1. Historical Development of Energy Storage Materials

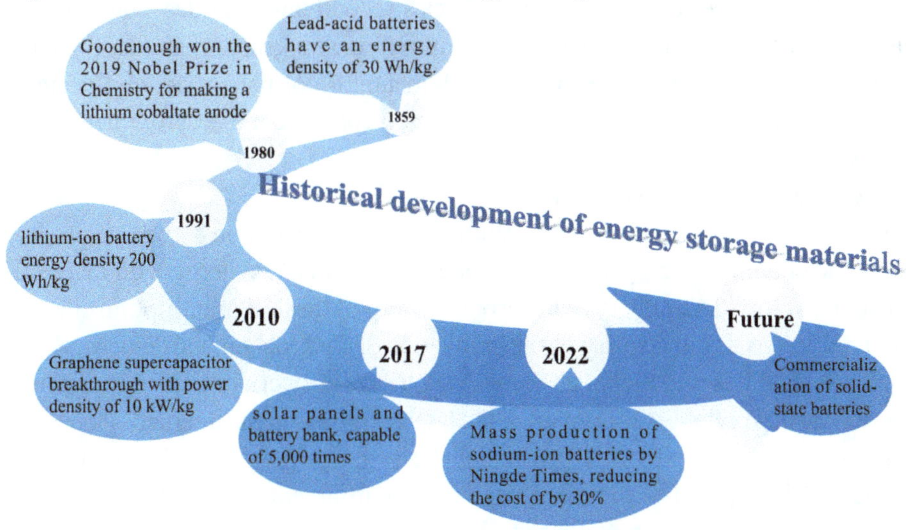

1. FUNDAMENTALS OF ENERGY STORAGE MATERIALS

1.1 Core Concepts of Materials Science

Beginning with the fundamental concepts of crystal structure and defects, this section integrates theoretical frameworks from thermodynamics and dynamics to establish a foundational understanding of electron and ion transport mechanisms in energy storage materials.

Crystal structure, defects and electron/ion transport mechanisms:

The crystal structure fundamentally influences the intrinsic properties of a material. For instance, the layered architecture of cathode materials in lithium-ion batteries, such as $LiCoO_2$, facilitates energy storage via the interlayer incorporation of lithium ions (Liu et al., 2025). Defects, including vacancies, dislocations, and grain boundaries, which are commonly observed in real crystals, play a dual role in influencing transport properties (Hong et al., 2021).

Thermodynamic and kinetic basis for energy storage:

Thermodynamics, specifically Gibbs free energy, dictates the feasibility of energy storage reactions. In the context of lithium-ion batteries, the interfacial reaction between the cathode material and the electrolyte must adhere to the condition $\Delta G < 0$. Conversely, the deposition and dissolution of lithium metal at the anode are governed by the electrode potential and activity (Bazant, 2013; Tang et al., 2022). The kinetics of these processes, described by the diffusion equation, determine the actual charge and discharge rates. For instance, Fick's second law provides a quantitative measure of the diffusion coefficient of lithium ions in a graphite anode, typically around $D \approx 10^{-10}$ cm^2/s. However, the implementation of nanostructured designs, such as porous carbon, can enhance the diffusion coefficient to approximately 10^{-8} cm^2/s (Gao et al., 2024).

Electrochemistry and Materials

Electrochemistry, a discipline that studies the interconversion between electrical energy and chemical energy, has established an inseparable synergy with materials science. This cross-disciplinary field not only underpins the design and optimization of functional materials but also drives the innovation of electrochemical devices, which are pivotal to addressing global challenges such as energy crisis and environmental pollution. In this review, we summarize the latest progress in the interaction between electrochemistry and materials, focusing on the design principles of electrochemical materials, their applications in key energy and environmental technologies, and the emerging trends shaping the future of this domain.

1.2 Energy Storage Material Performance Evaluation System

This section conducts a systematic analysis of the key performance indicators and failure mechanisms of energy storage materials, offering a quantitative foundation for material design and application.

Key Performance Indicators:

Energy Density: This parameter, defined as the energy stored per unit mass or volume of a material (expressed in Wh/kg or Wh/L), is fundamental for evaluating the performance of energy storage devices. For instance, lithium-ion battery cathode materials, such as high-nickel ternary (NMC811), can achieve an energy density of 250 Wh/kg. In contrast, silicon-carbon composite anodes can be enhanced to 450 Wh/kg through advanced nanostructure design.

Power Density: This metric, indicative of rapid charging and discharging capabilities, shows that supercapacitors can attain power densities of up to 10 kW/kg, owing to the double-layer adsorption mechanism, which significantly surpasses the 1-3 kW/kg range typical of lithium-ion batteries.

Cycle Life: This term denotes the number of charge-discharge cycles a battery can undergo before its capacity diminishes to 80% of its initial value. Commercial lithium-ion batteries generally exhibit a cycle life of 1,000-2,000 cycles, whereas liquid flow batteries, such as all-vanadium systems, can exceed 20,000 cycles due to the self-healing properties inherent in their liquid electrolytes.

Coulombic Efficiency: This parameter characterizes the reversibility of charge transfer processes. In lithium metal batteries, the initial coulombic efficiency is often below 90%. However, graphite anodes can achieve efficiencies as high as 99.9%, attributed to the stabilization provided by the solid-electrolyte interphase (SEI) film.

1.2.1 Reasons affecting the performance of energy storage materials

Dendrite Growth: During the electrochemical deposition process, such as in lithium or sodium metal anodes, the uneven distribution of ion flux results in the tip effect, where crystals preferentially grow in the direction of the electric field, forming tree-like dendrites (Wu et al., 2017; Wang et al., 2022).

Phase Transformation Stress: Changes in lattice volume during the charging and discharging processes, such as the significant expansion of silicon anodes by over 300%, induce localized stress concentrations (Mukhopadhyay & Sheldon, 2014). Furthermore, stress-induced cracking of the electrode facilitates accelerated electrolyte penetration, exacerbating the occurrence of side reactions at the electrode interface (Xu et al., 2020).

Interfacial Side Reactions: Interfacial side reactions can occur at three distinct interfaces: the positive electrode/electrolyte, the negative electrode/electrolyte, and the collector/electrolyte interfaces. These reactions significantly impact the energy conversion efficiency of batteries (Yu & Manthiram, 2018).

1.3 Physicochemical Principles of Energy Storage

1.3.1 Basic concepts and core differences

Faradaic Process

Definition: A redox reaction characterized by charge transfer, wherein the active substance undergoes a chemical transformation either at the electrode surface or within the bulk phase.

Fundamentals of Physical Chemistry:

Electron transfer: This process adheres to Faraday's law, with the reaction rate being directly proportional to the current.

Substance transformation: It involves the reversible conversion between oxidation and reduction states of reactive substances, such as the embedding and de-embedding of Lithium-ion or the valence changes in metal oxides.

Energy forms: The process facilitates the direct conversion of chemical and electrical energy, with the stored energy being contingent upon the redox capacity of the reactants.

Non-Faradaic Process

Definition: A physical process that does not involve charge transfer; instead, energy storage is accomplished through electrostatic adsorption or polarization phenomena.

Fundamentals of Physical Chemistry:

Bilayer theory: The electrode/electrolyte interface establishes charge-separated layers, known as the Helmholtz layer and the diffusion layer, which facilitate charge storage through ion adsorption.

Physical adsorption: This process depends on the high specific surface area of materials such as activated carbon and graphene to adsorb molecules, such as H_2, via van der Waals forces.

Energy form: Electrostatic energy is stored directly without the involvement of chemical bond breaking or reorganization.

1.3.2 Energy storage mechanisms in detail

Faraday Reaction Energy Storage Mechanism

(1) Redox reaction driven:

①**Lithium-ion batteries:** positive electrode (e.g., $LiCoO_2$) and negative electrode (e.g., graphite) realize energy storage by embedding/de-embedding of Li^+:

$$LiCoO_2 \leftrightarrow Li_{1-x}CoO_2 + xLi^+ + xe^-$$

②**Liquid flow batteries:** valence changes of active substances (e.g. vanadium ions) in the electrolyte:

$$VO^{2+} + H_2O \leftrightarrow VO^{2+} + 2H^+ + e^-$$

(2) Kinetics and Thermodynamics:

①The Nernst equation ($E = E^{\theta} - \frac{RT}{nF}\ln\frac{[\text{Reduced}]}{[\text{oxidized}]}$) determines the electrode potential.

②The reaction rate is described by the Butler-Volmer equation and is controlled by the activation energy and ion diffusion

Table 1. Summary of core physical chemistry principles

Parameters	Faraday reaction (physics)	Non-faraday reaction
The nature of energy storage	Chemical bond reorganization (redox reactions)	Physical adsorption/polarization (electrostatic energy storage)
Thermodynamic control	Nernst equation, reaction Gibbs free energy change (ΔG)	Double layer capacitance
Kinetic Limitations	Charge transfer rate, solid-state diffusion (Fick's law)	Ion migration rate (related to electrolyte conductivity)
Energy Density Sources	Redox capacity and potential difference of active substances	Redox capacity and potential difference of active substances
Failure mechanism	Material phase transition, electrolyte decomposition, lithium dendrite growth	Pore clogging, material pulverization

1.3.3 Synergies and optimization strategies

Hybrid energy storage systems

(1)Supercapacitor-battery hybrid system:

Batteries, operating on Faradaic principles, deliver high energy density, whereas supercapacitors, which function on non-Faradaic principles, provide immediate high-power essential for electric vehicle start-stop systems and regenerative braking.

(2)Pseudocapacitor material design:

The deposition of transition metal oxides, such as manganese dioxide (MnO_2), onto carbon-based materials enhances their capacity through surface redox reactions, which are characteristic of Faradaic processes.

Material and structure optimization

Nanostructure design:

The reduction of ion diffusion pathways, exemplified by the enhancement of the lithium-ion diffusion coefficient to 10^{-8} cm^2/s in titanium dioxide nanotubes, accelerates the kinetics of Faradaic reactions.

Porous material modification:

Modulation of MOFs pore size (e.g., UiO-66 pore size of 1.2 nm) to enhance the H_2 physisorption density balances the energy density limitations of non-Faraday processes.

Table 2. Comparison of typical application examples

Energy storage technology	Lead mechanism	specificities
Lithium-ion battery	Faraday reaction (physics)	High energy density (250 Wh/kg), medium charge/discharge rate (hourly), cycle life 500-2000 cycles.
Ultracapacitor	Non-Faraday reaction (EDLC) (EDLC)	Low energy density (5-10 Wh/kg), ultra-high power density (10 kW/kg), cycle life >100,000 cycles.
False capacitor	Hybridization mechanism	Combining bilayer (non-faradic) and surface redox (faradic), energy density is increased to 20-50 Wh/kg.
Hydrogen storage material	Non-Faraday (physisorption)	Rapid hydrogen absorption/discharge (minute level), low hydrogen storage capacity (<2 wt%), requires low temperature (77 K) or high-pressure conditions.
Metal hydride hydrogen storage	Faraday reaction (physics)	High hydrogen storage capacity (e.g., 7.6 wt% theoretical capacity of MgH_2), and has limited cyclic stability.

1.3.4 Hot Storage Mechanisms

Sensible Heat Storage: Thermal energy storage through changes in material temperature is exemplified by molten salt mixtures, specifically comprising 60% sodium nitrate ($NaNO_3$) and 40% potassium nitrate (KNO_3), which operate within a temperature range of 290-565°C and possess a specific heat capacity of 1.5 J/(g-K). However, these systems experience a heat loss rate of up to 30% per day (Zhang et al., 2020).

Latent Heat Storage: Phase change materials (PCMs) such as paraffin ($C_{18}H_{38}$) are capable of absorbing 200-300 kJ/kg of heat during the solid-liquid phase transition. To enhance their thermal conductivity from 0.2 to 2.5 W/(m-K), the incorporation of graphene at a concentration of 5 wt% is necessary (Aftab et al., 2021).

Thermochemical Storage: Energy storage systems based on reversible chemical reactions, such as the calcium oxide/water (CaO/H_2O) system, can achieve an energy storage density of up to 800 kJ/kg. However, these systems suffer from inadequate cycling stability, evidenced by a 40% capacity decay after 50 cycles (Ogura et al., 2003).

2. TYPICAL ENERGY STORAGE MATERIAL SYSTEMS

This chapter provides a comprehensive analysis of the fundamental components, technical characteristics, and recent advancements in current mainstream energy storage material systems. It emphasizes the examination of technological pathways and challenges associated with lithium-ion batteries, sodium/potassium ion batteries, liquid flow batteries, supercapacitors, and thermal storage materials, thereby offering a scientific framework for understanding the diversity of energy storage technologies.

2.1 Lithium-ion Battery Materials

Since their commercialization in 1991, Lithium-Ion Batteries (LIBs) have emerged as a fundamental energy technology for portable electronic devices, electric vehicles, and energy storage systems (Kim et al., 2019; Grey & Hall, 2020). The high energy density, extended cycle life, and lightweight properties of LIBs have significantly propelled advancements in contemporary energy technologies. Nonetheless, the increasing complexity of application scenarios, such as fast-charging requirements and extreme temperature conditions, has highlighted the issue of battery degradation. This section examines the synergistic optimization strategies for positive and negative electrode materials, as well as electrolytes.

Figure 2. Applications and problems of Li metal anode, and methods to solve problems

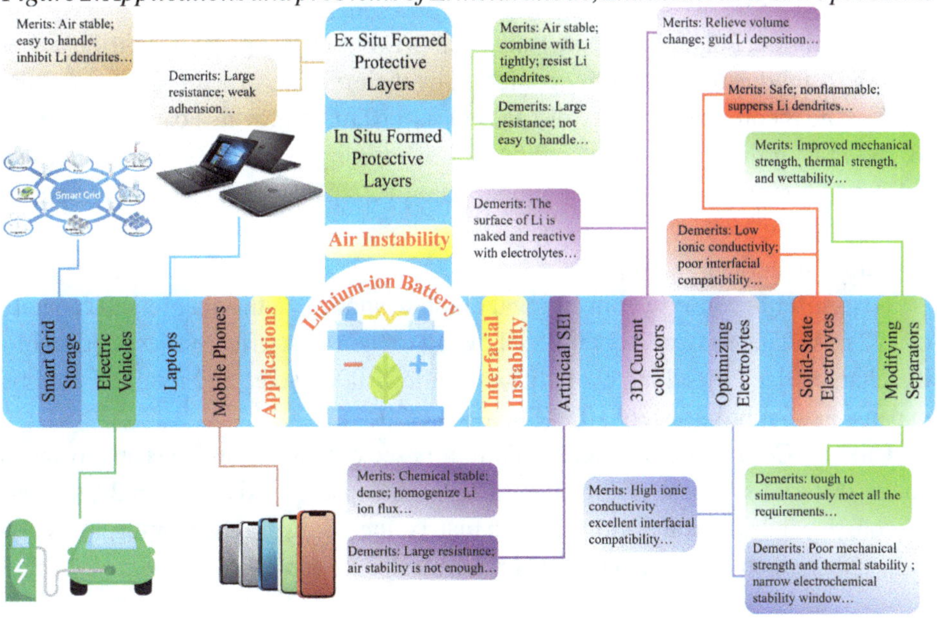

Anode materials: layered oxide and polyanionic

Layered Oxide: Nickel Manganese Cobalt Oxide (NMC) materials, exemplified by $LiNi_{1/3}Mn_{1/3}Co_{1/3}O_2$, can enhance energy density to as much as 250 Wh/kg through adjustments in nickel content (e.g., NMC811) (Li et al., 2004). However, reliance on cobalt resources and insufficient thermal stability remain significant challenges.

Research indicates that surface coating with Al_2O_3 can extend the cycle life to over 2000 cycles (Hu et al., 2021).

Figure 3. Schematic illustration of the coating process of Al_2O_3 film on the NCM811 sphere (Hu et al., 2021)

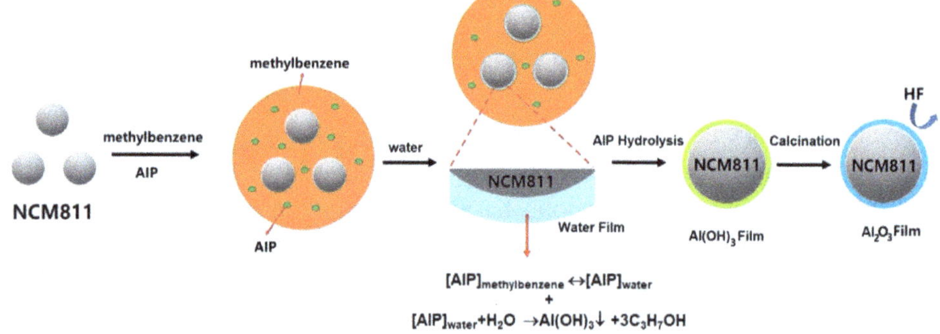

Polyanionic: Polyanionic materials, such as $LiFePO_4$, are prominent in the power battery sector due to the stability conferred by their olivine structure (Wang & Sun, 2015). The theoretical capacity of $LiFePO_4$ is 170 mAh/g, which can be enhanced to 160 mAh/g through carbon coating and nanosizing techniques. Additionally, it exhibits a high thermal runaway temperature of 270°C, which is notably superior to that of NMC36.

Anode materials

Graphite: In contrast, conventional graphite anodes, with a theoretical capacity of 372 mAh/g, are extensively utilized owing to their cost-effectiveness and stability, characterized by a long cycle life exceeding 1000 cycles and minimal volume expansion of approximately 10% (Shen et al., 2025).

Carbon Composite Anode: The carbon-based composite anode is an advanced anode material synthesized by integrating high-capacity active materials, such as silicon, tin, and antimony, with a carbon matrix, including graphite, amorphous carbon, and graphene, through physical or chemical methodologies (Jiang et al., 2022).

Figure 4. Overview of the various designs and respective solutions for structure, and functionalization of carbon materials in LMBs (Jiang et al., 2022)

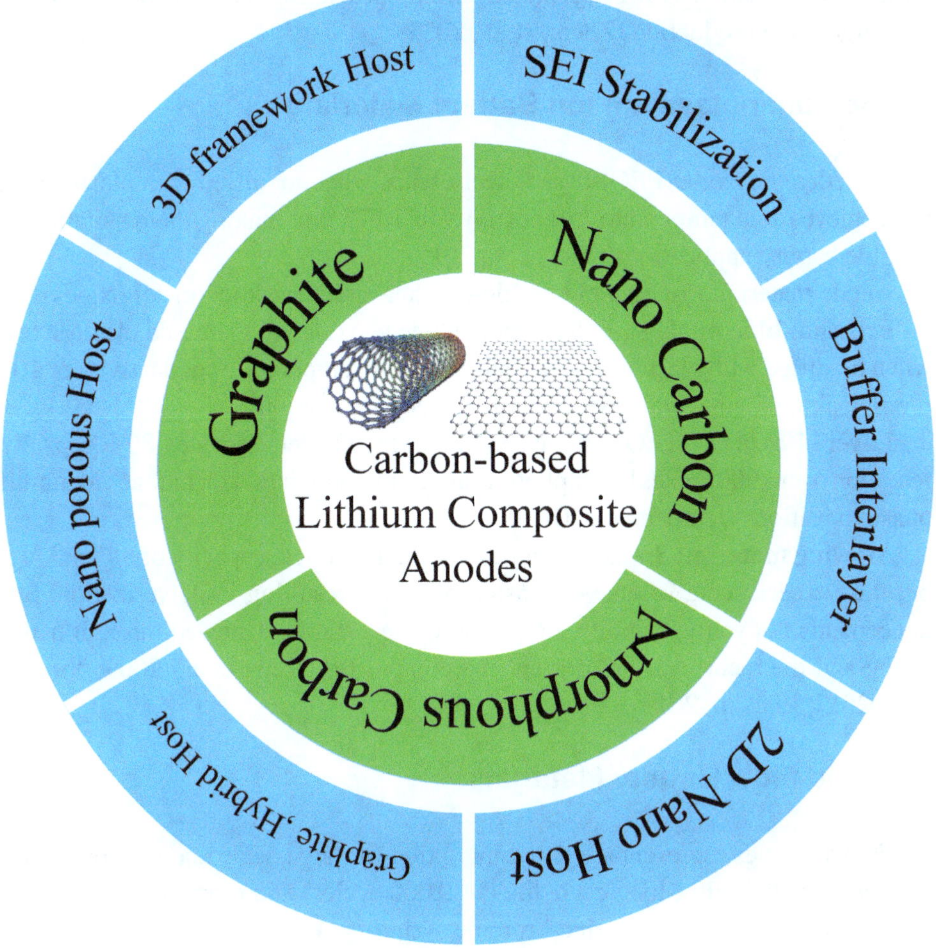

Pre-lithiation technology: The initial loss of lithium during the first week can be compensated by utilizing Li_5Si alloys or Li_2O additives, which have the potential to enhance the overall energy density of the battery by 8-10% (Maximov et al., 2020).

Electrolytes: liquid → polymer → solid state evolution

Liquid electrolyte: Despite its flammability limiting high-voltage applications, lithium hexafluorophosphate ($LiPF_6$)-based electrolyte remains prevalent due to its high ionic conductivity of 10^{-3} S/cm (Susarla & Ahmed, 2019).

Polymer electrolytes: Polyethylene oxide (PEO)-based electrolytes, when modified with plasticizers, can achieve room temperature conductivity of 10^{-4} S/cm, making them suitable for flexible battery applications (Cong et al., 2021).

Solid-state electrolytes: Sulfide ($Li_{10}GeP_2S_{12}$) and oxide (LLZO) electrolytes have the potential to enhance energy density to as much as 400 Wh/kg through interfacial optimization. However, challenges remain in terms of interfacial impedance and large-scale production (Zheng et al., 2023).

2.2 Sodium/potassium Ion Battery Materials

Polyethylene oxide (PEO)-based electrolytes, when modified with plasticizers, can achieve room temperature conductivity of 10^{-4} S/cm, making them suitable for flexible battery applications (Xue et al., 2015).

Anode material: Prussian blue analog with layered transition metal oxides

Prussian blue analogs (PBAs): polyethylene oxide (PEO)-based electrolytes, when modified with plasticizers, can achieve room temperature conductivity of 10^{-4} S/cm, making them suitable for flexible battery applications.

Layered oxides: the O_3-type $NaNi_{1/3}Fe_{1/3}Mn_{1/3}O_2$ demonstrates 80% capacity retention over 1000 cycles by employing structural modulation techniques to inhibit phase transitions (Jeong et al., 2019).

Alloying material: Sn-based anodes offer a theoretical capacity of 847 mAh/g, but they experience a volume expansion of 200%, resulting in pulverization. Enhancements in cycling stability, up to 500 cycles, can be achieved through nanosizing and the incorporation of graphene composites (Deng et al., 2009; Ying & Han, 2017; Li et al., 2017).

2.3 Fluid Flow Battery Materials

Flow batteries represent a significant option for large-scale energy storage, attributable to their ultra-long cycle life (>20,000 cycles) and modular design.

All-vanadium system vs. zinc-bromine system

All-vanadium liquid flow battery (VRFB): A 75% energy efficiency is attainable using VO^{2+}/VO^{2+} versus V^{3+}/V^{2+} oxidation pairs; however, the cost of vanadium constitutes 40% of the total system cost (Wu et al., 2011).

Zinc Bromine Flow Battery (ZBFB): The zinc anode deposition/dissolution mechanism offers a high theoretical capacity (820 mAh/g), but issues such as dendrite growth and bromine diffusion contribute to a high self-discharge rate. Efficiency can be enhanced to 70% through the implementation of a porous diaphragm design.

2.4 Supercapacitor Devices

Supercapacitors represent a class of energy storage devices that occupy a position between traditional capacitors and batteries, distinguished by their high-power density and extended cycle life. The ultra-high power density (10 kW/kg) and rapid charging and discharging capabilities of supercapacitors effectively bridge the performance gap between batteries and conventional capacitors (Myrthong et al., 2025).

Figure 5. Classification of supercapacitor (Myrthong et al., 2025)

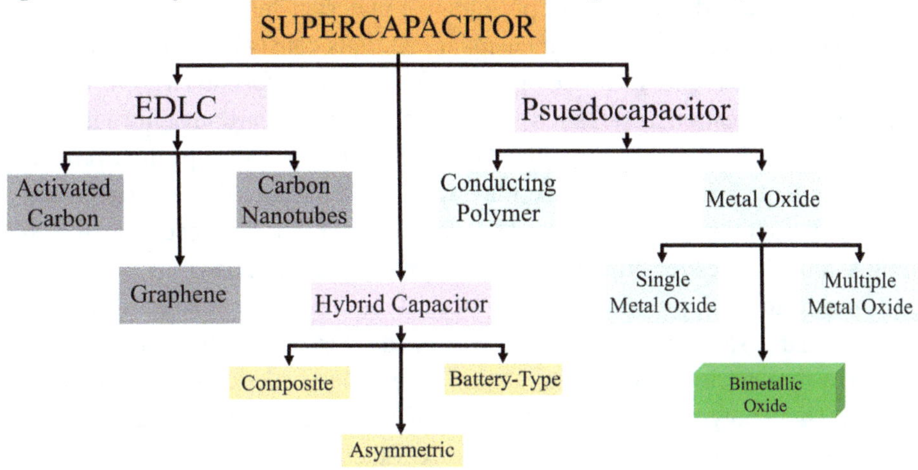

Electric Double Layer (EDL) materials:

Bilayer capacitors, a subset of supercapacitors, store energy through a physical adsorption/desorption process occurring between the electrode surface and the electrolyte ions, a mechanism that does not involve chemical reactions, thereby ensuring high cycling stability. Key materials include:

Activated Carbon (AC): Activated carbon is characterized by a hierarchical pore structure comprising micropores (<2 nm), mesopores (2-50 nm), and macropores (>50 nm), with a specific surface area ranging from 1,000 to 3,000 m^2/g. This extensive surface area facilitates numerous charge adsorption sites (Jiang et al., 2022).

Graphene: Graphene is a two-dimensional honeycomb lattice consisting of a single layer of carbon atoms, characterized by a theoretical specific surface area of 2630 m^2/g. It exhibits exceptional electrical conductivity (10^6 S/m) and mechanical strength. The unique structure of graphene materials facilitates short ion diffusion paths, enabling rapid charging and discharging processes (Shen et al., 2025).

Carbon Nanotubes (CNTs):

Carbon nanotubes (CNTs) are one-dimensional tubular structures, classified into single-walled (SWCNT) and multi-walled (MWCNT) varieties, which exhibit a combination of high electrical conductivity and mechanical robustness (Argirusis et al., 2024).

2.4.2 Pseudocapacitor materials

Pseudocapacitors store energy through rapid redox reactions occurring at or near the electrode surface, generally offering higher capacity than electric double-layer capacitors. However, their performance is constrained by reaction kinetics, resulting in a slightly reduced cycle life. Representative materials include:

MXenes: MXenes are two-dimensional materials consisting of transition metal carbides or nitrides, characterized by the general chemical formula $M_{n+1}X_nT_x$, where M represents transition metals such as Ti or Mo, X denotes carbon or nitrogen, and T refers to surface functional groups like -O and -OH. Their pseudocapacitive properties are attributed to surface redox reactions, which differ from bilayer capacitance, as the surface functional groups (-O, -OH) facilitate rapid redox reactions, resulting in bulk capacitances reaching up to 1500 F/cm^3 (Ali et al., 2024).

Conducting Polymers: The energy storage mechanism of conducting polymers involves charge storage through the doping and de-doping of polymer chains, exemplified by the transition from the oxidation state to the reduction state in polyaniline (PANI). These polymers offer the advantage of high theoretical specific capacitance, with PANI reaching up to 750 F/g (Okhay & Tkach, 2022).

2.5 Thermal Storage Materials

Heat storage materials are integral components of advanced energy storage systems, facilitating the absorption, storage, and release of thermal energy through either physical or chemical processes (Liu et al., 2012). These materials hold significant value in various applications, including solar energy utilization, industrial waste heat recovery, and building energy efficiency.

Thermal materials for heat storage: Thermal energy storage is achieved through the alteration of a material's temperature, with typical examples including water-based fluids, ceramic materials, and molten salt systems. Among these, molten nitrate salts, such as Solar Salt, have been extensively utilized in tower-type solar thermal power generation systems (Gabisa & Aman, 2016).

Latent heat storage materials: Phase change materials (PCMs) utilize the solid-liquid phase transition to achieve high-density thermal energy storage and are primarily categorized into organic (e.g., paraffin, fatty acids), inorganic (e.g., hydrated salts, metal alloys), and eutectic composite materials (Huang et al., 2021).

Thermochemical heat storage materials: Thermal energy storage via reversible chemical reactions offers distinct advantages, including high energy density (ranging from 800 to 1000 kJ/kg) and the capability for long-term storage without significant energy loss. Common systems employed for this purpose include metal oxides, such as the Co_3O_4/CoO system with an enthalpy change (ΔH) of 844 kJ/kg, hydroxides like the $Mg(OH)_2/MgO$ system with a ΔH of 1400 kJ/kg, and ammonia decomposition reaction systems (Liu et al., 2020).

Table 3. Key Performance Metrics of Advanced Energy Storage Devices

Device Type	Energy Density (Wh/kg)	Power Density (W/kg)	Cycle Life (Cycles)	Voltage Range (V)	Key Challenges	Cost (USD/kWh)	Ref
Lithium-Ion Batteries	250-300	200-500	1,000-2,000	3.0-4.2	Dendrite growth; Cobalt dependency	120-150	(Wang et al., 2023)
Solid-State Batteries	300-400	100-300	500-1,000	3.0-5.0	Interfacial resistance; Scalability	200-300 (est.)	(Yang et al., 2020)
Magnesium-Ion Batteries	150-200	50-150	300-500	1.0-2.5 vs. Mg/Mg^{2+}	Slow Mg^{2+} kinetics; Limited cathode options	100-200 (est.)	(Wei et al., 2022)
Aluminum-Ion Batteries	70-150	100-300	5,000-10,000	1.0-2.2 vs. Al/Al^{3+}	Low energy density; Corrosion issues	80-120 (est.)	(Nandi & Pumera, 2024)
Sodium-Ion Batteries	120-160	200-600	2,000-4,000	2.5-3.8	Lower energy density vs. LIBs	80-100	(Lin et al., 2025)
Supercapacitors	5-30	5,000-15,000	>100,000	2.5-3.0	Low energy density; High cost per Wh	3,000-10,000	(Ferdous et al., 2025)

3. RATIONAL DESIGN AND ADVANCED TECHNOLOGIES FOR MATERIALS

The energy density, cycle life, rate performance, and safety of energy storage devices are essentially determined by the structural characteristics, interfacial behavior, and dynamic evolution mechanisms of the core materials. Traditional materials research and development relies on the 'trial-and-error' method, which suffers from drawbacks such as long development cycles, high costs, and undirected performance optimization. Consequently, it struggles to meet the demands of next-generation energy storage technologies for precise materials control. This chapter focuses on the core theme of 'Rational Design and Advanced Technologies'.

3.1 Material Structure Design Strategies

In the rational design of materials, structure is fundamental to determining performance. Precise control over the structure of material at different scales—such as atomic, nano, and micro—is a core strategy for enhancing the performance of energy storage devices. These strategies primarily cover multiple aspects, from phase dimensions and surface/interfacial chemical environments to multifunctional synergy, among which nanostructure control, surface modification, and composite materials and interface optimization are the most critical directions in current research.

3.1.1 Nanostructure control

(1) Dimensional engineering

By precisely controlling the dimensional characteristics of materials, their energy storage behavior can be significantly optimized. Nanomaterials of different dimensions play distinct key roles in energy storage devices, such as lithium-ion batteries (LIBs), owing to their unique structures.

Zero-dimensional (0D) quantum dots (QDs): Particles with dimensions under 10 nm (e.g., carbon QDs, silicon QDs, black phosphorus QDs) exhibit significant quantum confinement effects. Their high specific surface area exposes more active sites while drastically shortening ion diffusion pathways. For example, when silicon QDs are used as anode materials for LIBs, their nanoscale dimensions can effectively mitigate the massive volume expansion (~300%) during charge/discharge cycles (Liu et al., 2025).

Two-dimensional (2D) nanosheets: Ultrathin layered structures (e.g., graphene, MoS_2, MXene, phosphorene, transition metal hydroxide nanosheets, covalent organic framework nanosheets) possess tunable interlayer spacing, which greatly facilitates the intercalation/deintercalation kinetics of ions (such as Li^+/Na^+). Graphene and MXene are typical conductive 2D materials.

Three-dimensional (3D) porous frameworks: Materials represented by Metal-Organic Frameworks (MOFs) and Covalent Organic Frameworks (COFs) (e.g., ZIF-8, ZIF-67, MOF-5, COF-1) feature high porosity and well-ordered channels, enabling rapid ion transport. Moreover, the porous structure provides buffer space, effectively mitigating the volume expansion of electrode materials during cycling (Nguyen et al., 2023).

(2) Surface modification

Surface modification aims to address key challenges in energy storage processes, such as interfacial side reactions, structural degradation, and sluggish ion/electron transport kinetics, through precise regulation of the physical and chemical properties of material surfaces and interfaces. Key technical approaches include atomic-scale coating, conductive layer construction, surface functionalization, and molecular interface engineering, each of which enhances interfacial stability and electrochemical performance from different dimensions.

Atomic-scale coating: This technique involves constructing an ultrathin, dense inert protective layer to physically suppress interfacial side reactions. Atomic layer deposition (ALD), as a representative method, enables sub-nanometer precision in thin-film growth through self-limiting surface reactions. For instance, Cao et al. constructed an ultrathin Al_2O_3 layer with a thickness of approximately 2 nm on the surface of hard carbon via atomic layer deposition (ALD), which can function as a solid electrolyte interphase (SEI) to significantly suppress electrolyte decomposition, thereby resulting in a thinner and more stable SEI (Wu et al., 2024). Similarly, coating high-nickel cathodes (e.g., NMC811) with a TiO_2 layer neutralizes surface residual lithium, suppresses phase transitions and microcrack propagation, and improves capacity retention by over 20% (Fan et al., 2020).

Conductive layer construction: This approach enhances both electron transport efficiency and interfacial stability by applying a continuous, highly conductive network onto active materials. Chemical vapor deposition (CVD), a typical method, enables *in-situ* growth of carbonaceous coatings such as graphene on material surfaces.

Surface functionalization: This strategy optimizes the surface chemical state and reaction pathways by introducing specific functional groups or defect structures. Plasma treatment, as an efficient dry modification technique, utilizes highly reactive particles to achieve precise control over surface composition. For instance, in the study of Plasma-induced ε-MnO_2 based aqueous zinc-ion batteries, Ar plasma treatment is applied to modify the surface of the ε-MnO_2 electrode (Zhang et al., 2022).

Molecular interface engineering: This technique employs molecular self-assembly to construct ordered molecular layers on electrode surfaces, enabling nanoscale regulation of interfacial properties (Kang et al., 2019).

3.1.2 Composite materials and interface optimization

Composite materials and interface engineering represent a critical pathway for overcoming the performance limitations of single-component materials and enhancing the overall performance of energy storage devices. Through the synergistic design of multi-component materials and precise modulation of interfacial behaviors, it is possible to optimize ion/electron transport, mechanical stability, and interfacial chemical environments.

(1) Core-shell structure design

Core-shell structures, composed of an active material core and a functional coating shell, aim to achieve synergistic optimization of volume buffering, interface stabilization, and electron conduction at the nanoscale. This 'core–shell' division of labor allows the advantages of each component to be fully utilized: the core is primarily responsible for energy storage, while the shell maintains structural stability, facilitates charge transport, and protects the core from electrolyte corrosion.

Constructing core–shell composite systems to achieve stress regulation and interface stabilization is an effective approach for simultaneously obtaining high energy density and long cycle life. For example, in sodium-ion battery cathode materials, a layered oxide heterostructure can be designed with an O_3-type crystal phase core (where 'O' indicates that sodium ions occupy octahedral sites) and a P_2-type crystal phase shell (where 'P' denotes that sodium ions occupy prismatic sites) (Zhu et al., 2024). This mechano-chemical coupling effect enables the cathode material to achieve a high energy density of 587.34 $Wh{\cdot}kg^{-1}$ along with excellent structural stability (Wei Seh et al., 2013).

(2) Artificial interphase engineering

The solid electrolyte interphase (SEI) is a naturally formed film resulting from reactions between the electrode and electrolyte. However, its 'inhomogeneous and fragile' nature often leads to lithium dendrite growth and capacity fading.

In practical applications, one common strategy involves *in-situ* construction via electrolyte additives. For example, introducing compounds such as fluoroethylene carbonate (FEC) or lithium nitrate ($LiNO_3$) into lithium-ion battery electrolytes. These additives possess specific electrochemical windows that allow them to undergo preferential reductive decomposition on the anode surface before conventional electrolyte solvents (Tran et al., 2025; Li et al., 2025).

Another strategy is *ex-situ* construction of artificial interphases, where protective layers are pre-formed through physical or chemical methods during electrode preparation. For instance, researchers have employed ALD to fabricate ultrathin Li_3PO_4 or Al_2O_3 layers on lithium metal anodes (Kim et al., 2024; Niu et al., 2021).

3.2 Advanced Characterization Techniques

Traditional offline characterization techniques can only acquire material information at the initial and final states, failing to capture dynamic evolution during charge/discharge processes (e.g., lithium dendrite growth, SEI layer restructuring, phase transition stress). This leads to unclear performance degradation mechanisms and blind design strategies. In recent years, breakthroughs in *in-situ* characterization techniques have enabled 'atomic-level, real-time, and multi-dimensional' observation, providing precise feedback for material design.

3.2.1 *In-situ* transmission electron microscopy (*in-situ* TEM)

In-situ TEM allows real-time observation of material structural changes at atomic resolution, serving as a core tool for revealing lithium dendrite growth and phase transition mechanisms. For instance, researchers utilized *in-situ* TEM combined with micro-heating chips to observe the charge/discharge processes of lithium-sulfur batteries at room temperature and elevated temperatures (100–300 °C) in real time (Wang et al., 2020).

In solid-state battery systems, *in-situ* TEM has been employed to track the penetration behavior of lithium dendrites within solid electrolytes. Studies show that in lithium aluminum titanium phosphate $Li_{1+x}Al_XTi_{2-x}(PO_4)_3$. (LATP) electrolytes, lithium dendrites initially nucleate at grain boundaries (t = 197.5 s), subsequently expand rapidly inducing particle volume expansion (t = 257 s), and ultimately cause electrolyte fracture and battery short-circuit (t = 399.5 s) (LaPotin et al., 2021). Further analysis using Electron Energy Loss Spectroscopy (EELS) revealed the reduction of Ti^{4+} to Ti^{3+} during lithiation, indicating irreversible phase transition and the formation of electronic conduction channels within the electrolyte (Li et al., 2021).

3.2.2 In-situ atomic force microscopy (*in-situ* AFM)

In-situ Atomic Force Microscopy (*in-situ* AFM) is a characterization technique capable of real-time monitoring of electrode surface morphological changes during electrochemical processes. This technique combines high-resolution surface topography imaging, mechanical property measurements (such as adhesion force and stiffness), and electrochemical signal monitoring.

In energy storage applications, the primary role of *in-situ* AFM is reflected in the *in-situ* characterization of electrochemical interfaces. For example, in lithium-ion battery research, it can monitor the formation and evolution of the SEI layer in real time, which is a critical factor for battery cycle stability (Tokranov et al., 2014). By observing SEI growth on silicon electrodes or metal anodes, *in-situ* AFM reveals deposition/stripping dynamics, aiding in the optimization of electrode design to reduce capacity fade and enhance cycle life (Demirocak & Bhushan, 2014).

The force-based measurement capability of *in-situ* AFM further extends its value in energy storage. It can not only image surfaces but also quantify local charge storage mechanisms, such as probing the mechanical response and charge transfer processes at the electrode-electrolyte interface in supercapacitors (Zhang et al., 2020; Wang et al., 2018).

3.2.3 Synchrotron radiation X-ray techniques

Synchrotron radiation sources provide intense continuous electromagnetic waves ranging from infrared to X-ray bands, offering unique advantages such as high intensity, high collimation, tunable energy, polarization, and pulsed time structure.

For instance, during the charge/discharge processes of lithium-ion batteries (e.g., with NCM or $LiCoO_2$ cathodes), synchrotron radiation XRD is used to track lattice expansion and phase transitions (e.g., from layered structure to rock-salt phase) induced by lithium ion intercalation/deintercalation in real time, aiding in the understanding of voltage fade and capacity loss mechanisms (Hua et al., 2019).

Synchrotron Radiation XAS: Synchrotron Radiation XAS, including X-ray Absorption Near Edge Structure (XANES) and Extended X-ray Absorption Fine Structure (EXAFS), is used to probe the electronic structure and local coordination environment of elements (Yin et al., 2025). Furthermore, researchers have used operando multi-phase EXAFS to study the structural evolution of the cathode material MnO_2 and the changes of Mn ions in the electrolyte during the cycling of aqueous Zn/MnO_2 batteries (Wu et al., 2022).

3.2.4 *In-situ* Raman spectroscopy and *in-situ* IR spectroscopy

As vibrational spectroscopy techniques, *in-situ* Raman and IR spectroscopy provide 'fingerprint' information about molecular structure, chemical bonding, and lattice vibrations. They are powerful tools for understanding complex chemical reactions at the electrode/electrolyte interface in energy storage devices. They play an irreplaceable role in revealing the solvation structure of electrolytes, the dynamic formation and evolution of SEI films, and the phase transition processes of active materials.

***In-situ* Raman spectroscopy:** With its high spatial resolution and sensitivity to vibrational modes, *in-situ* Raman spectroscopy is particularly suitable for monitoring the structural evolution of carbon-based materials, layered transition metal compounds, and interfacial SEI layers during energy storage processes. In lithium-ion batteries, Raman spectroscopy can track the intercalation/deintercalation process of graphite anodes in real time, revealing D/G band shifts and defect formation induced by Li^+ insertion, thereby clarifying capacity fade mechanisms (Cheng et al., 2021). Using a gap-enhanced Raman strategy, researchers observed charge transfer and ion adsorption processes at the multilayer graphene interface, confirming the critical role of pseudocapacitive contributions to total capacity enhancement (Yin et al., 2024).

***In-situ* infrared spectroscopy:** *In-situ* IR spectroscopy, with its unique sensitivity to functional groups and molecular adsorption, complements the shortcomings of Raman in energy storage monitoring, and is particularly suitable for analyzing electrolyte decomposition, gas generation, and surface adsorption processes. In lithium-ion batteries, *in-situ* IR can detect degradation products of carbonate solvents (such as Li_2CO_3 and $ROCO_2Li$) during SEI formation, revealing the chemical evolution at the electrolyte-electrode interface (Amaral et al., 2023). For example, monitoring changes in O-H and C=O stretching vibrations induced by ion adsorption on carbon nanotube electrodes confirms the synergistic effect of electric double-layer capacitance and pseudocapacitive mechanisms (Patra et al., 2021).

Table 4. Advantages, disadvantages and applications of advanced technologies

Technique	Advantages	Limitations	Typical Applications
***In-situ* XRD**	Tracks phase evolution in real-time. - Identifies crystallographic changes during cycling.	Limited sensitivity to amorphous phases. - Requires synchrotron sources for high resolution.	Monitoring structural transitions in Ni-rich cathodes.
***In-situ* XPS**	Provides chemical state analysis of surfaces/ interfaces- Detects SEI composition.	Ultra-high vacuum limits relevance to liquid electrolytes - Surface-sensitive only.	Analyzing solid electrolyte interphase (SEI) formation.
***In-situ* TEM**	Atomic-scale resolution; direct, real-time imaging of morphology and phase changes.	Requires high vacuum (or fragile liquid cell); electron beam can damage samples; small viewing area.	Visualizing Li dendrite growth; observing volume expansion in anodes (e.g., Si).
Cryo-electron microscopy (Cryo-EM)	Preserves native SEI nanostructure. - Atomic-scale resolution of beam-sensitive materials.	Complex sample preparation. - Limited to ex situ or post-mortem analysis.	Visualizing Li dendrites and SEI morphology.
***In-situ* AFM**	High-resolution surface topography; measures local mechanical properties (stiffness, adhesion).	Tip can damage soft samples (e.g., SEI); limited to surface analysis; slower scan speed.	Monitoring SEI morphological evolution; probing mechanical changes on electrode surfaces.
***In-situ* NMR**	Probes Li^+/Na^+ ion dynamics in electrodes.	Low spatial resolution. - Requires isotopic labeling for sensitivity.	Mapping ion diffusion in solid-state electrolytes.
***In-situ* XAS**	Element-specific; probes electronic structure and local coordination environment.	Requires synchrotron source; provides an averaged signal from the probed area.	Monitoring oxidation state changes in transition metal cathodes; analyzing local bond changes.

continued on following page

Table 4. Continued

Technique	Advantages	Limitations	Typical Applications
***In-situ* Raman Spectroscopy**	Detects local structural changes (e.g., bond breaking).	Weak signal for opaque samples. - Limited depth penetration.	Monitoring electrolyte decomposition in Li-S batteries.
***In-situ* IR Spectroscopy**	High sensitivity to functional groups and molecular bonds; identifies soluble species.	Signal can be weak for some components; cell design can be complex for liquid electrolytes.	Analyzing electrolyte decomposition pathways; identifying SEI layer chemical components.

3.3 Computational and Data-Driven Design

With the rapid development of computational power and continuous improvement of theoretical methods, computational and data-driven design approaches have become an indispensable part of the research and development of energy storage materials and devices. These methods can simulate the physicochemical behavior of materials at multiple scales (atomic, microscopic, and macroscopic), predict their electrochemical performance, and integrate with experimental data, greatly accelerating the innovation process of new energy storage technologies.

3.3.1 First-principles principles

First-principles calculations, particularly those based on Density Functional Theory (DFT), are computational methods derived from the fundamental principles of quantum mechanics, independent of any empirical parameters.

(1) DFT calculations

Electrochemical performance prediction: DFT calculations can predict the insertion and extraction voltage of lithium ions in electrode materials, which is a key parameter determining battery energy density. By calculating the total energy of the material at different lithium concentrations, voltage-capacity curves can be constructed, providing theoretical guidance for screening high-voltage cathode materials and low-voltage anode materials.

(2) High-throughput computational screening

High-throughput computational screening combines first-principles calculations with automated computational workflows and database technology to rapidly and systematically screen thousands of candidate materials theoretically.

(3) CALPHAD (Calculation of phase diagrams)

CALPHAD is a phase diagram calculation method based on thermodynamic data. By integrating thermodynamic data (such as compound formation enthalpy) obtained from DFT calculations with limited experimental data, CALPHAD can efficiently construct phase diagrams for multicomponent alloy or oxide systems.

3.3.2 Multiscale Modeling

Physical and chemical processes in energy storage systems occur across multiple time and spatial scales, from the atomic to the macroscopic level. Multiscale modeling aims to build a comprehensive model describing the behavior of the entire system by coupling computational methods at different scales.

3.3.3 Machine learning (ML)

Machine Learning, as a powerful data-driven approach, is bringing revolutionary changes to research in the energy storage field. By learning hidden patterns and 'material genes' from vast amounts of experimental and computational data, ML can efficiently predict material properties, optimize battery design, and manage battery operation. Key application directions include:

(1) Accelerated material screening and discovery: ML models can rapidly predict key electrochemical properties (e.g., voltage, capacity, ionic conductivity) based on the chemical composition and structural features of materials, thereby replacing some time-consuming high-throughput DFT calculations and screening potential new materials at a faster pace.

(2) Battery performance prediction and lifetime assessment: By analyzing data from early cycle charge/discharge curves, voltage, current, and temperature, ML models can accurately predict the Remaining Useful Life (RUL) and State of Health (SOH) of batteries. This is crucial for estimating the driving range of electric vehicles and maintaining large-scale energy storage stations, significantly reducing the cost and time associated with long-term testing.

(3) Optimizing battery management systems (BMS): ML algorithms can estimate the State of Charge (SOC) of batteries more accurately and optimize charge/discharge strategies based on real-time operating conditions, thereby maximizing battery performance, extending service life, and ensuring safety.

(4) Intelligent analysis of experimental data: In material characterization and failure analysis, machine learning (particularly computer vision) can be used to automatically identify and analyze microstructures (such as dendrites, cracks) in microscopic images, improving analysis efficiency and accuracy.

(5) Guiding experimental design: Based on strategies like active learning, ML models can suggest the next most promising material or condition for experimentation or computation, thereby finding the optimal solution with the fewest experiments, enabling 'intelligent experimentation'.

4. CUTTING-EDGE TECHNOLOGIES AND FUTURE CHALLENGES

With the accelerated global energy transition, innovation in energy storage technology has become a core driver of sustainable development. This chapter focuses on frontier technological breakthroughs, emerging application scenarios, and sustainability challenges in the battery field. It aims to review current research progress and industrialization bottlenecks, and explore potential pathways for future technological development.

4.1. All-Solid-State Batteries (ASSBs)

All-solid-state batteries represent a significant transformation in battery technology. Unlike traditional liquid lithium-ion batteries, ASSBs use solid electrolytes (e.g., oxide-, sulfide-, or polymer-based materials) to replace liquid electrolytes, thereby avoiding risks of leakage, combustion, and explosion.

(1) Breakthroughs in novel solid-state electrolyte materials

The ionic conductivity of the solid electrolyte is a core indicator determining the performance of all-solid-state batteries. Ideal SSEs should possess room-temperature ionic conductivity comparable to liquid electrolytes (>10^{-3} S cm^{-1}), a wide electrochemical stability window, and good mechanical properties.

Sulfide Electrolytes: Sulfide electrolytes (e.g., $Li_{10}GeP_2S_{12}$, Li_6PS_5Cl, Argyrodites) have attracted significant attention due to their extremely high room-temperature ionic conductivity (up to 10^{-2} S cm^{-1}) and good mechanical malleability (easy to cold-press into shape) (Kato et al., 2016). However, their high raw material costs (especially Li_2S) and sensitivity to air are major obstacles to commercialization. In this regard, researchers have made key breakthroughs in low-cost synthesis routes in recent years. For example, in 2024, a research team from the University of Science and Technology of China reported a novel $Li_7P_3S_{7.5}O_{3.5}$ (LPSO) solid electrolyte, the synthesis route of which does not require expensive Li_2S as a precursor (Li et al., 2024). This advancement significantly promotes the industrialization process of sulfide electrolytes.

Oxide and halide electrolytes: Oxide electrolytes (e.g., garnet-type $Li_7La_3Zr_2O_{12}$) offer advantages such as high chemical stability and a wide electrochemical window, but their intrinsic hardness leads to processing difficulties and poor solid-solid contact with electrode. For instance, researchers like Kota Motohashi from Osaka Metropolitan University reported in 2024 a novel solid electrolyte $Na_{2.25}TaCl_{4.75}O_{1.25}$, composed of nanoparticles embedded in an amorphous matrix, which exhibits high conductivity, good formability, and electrochemical stability in sodium-ion solid-state batteries (Wu et al., 2025).

(2) Engineering strategies for key interface issues

The performance bottleneck of ASSBs has shifted from the bulk conductivity of the electrolyte to the solid-solid interface between the electrode and the electrolyte. Poor interfacial contact leads to high interfacial impedance and induces lithium dendrite growth, causing battery short-circuit failure.

Anode/electrolyte interface: The poor "wettability" between the lithium metal anode and SSEs is a core challenge. To address this, the interface interlayer strategy is widely adopted. On one hand, introducing a flexible polymer or gel layer has proven effective. Studies show that introducing an *in-situ* polymerized succinonitrile gel electrolyte as an interlayer between a garnet-type ($Li_7La_3Zr_2O_{12}$) electrolyte and the lithium metal anode can significantly improve the physical contact and chemical stability of the interface, effectively suppress lithium dendrite growth, and enhance the battery's cycle performance (Wang et al., 2022). On the other hand, constructing a lithiophilic interface is also a mainstream strategy. For example, pre-depositing a lithiophilic metal (e.g., Al, Si) or alloy layer on the electrolyte surface can induce uniform nucleation and deposition of lithium metal, thereby fundamentally alleviating the dendrite problem and stabilizing the anode interface (Wang et al., 2025).

Cathode/electrolyte interface: In composite cathodes, active materials (e.g., LCO, NCM) and solid electrolyte powders struggle to form a tight ion/electron conductive network. Particularly under high areal capacity and charge-discharge cycling, the volume changes of active materials lead to interface separation. To improve the contact between garnet-type electrolytes and cathode active materials, researchers proposed using a deep eutectic electrolyte (DEE) as an interface wetting agent (Lin et al., 2025).

4.2. Lithium–Sulfur and Lithium–Air Batteries

Lithium–sulfur (Li–S) and lithium–air (Li–O_2) batteries represent cutting-edge 'post-lithium-ion' technological pathways aimed at overcoming the limitations of traditional intercalation-based electrochemistry and achieving ultra-high theoretical energy densities (reaching up to 2600 Wh/kg and 3500 Wh/kg, respectively). Both systems utilize high-capacity lithium metal as the anode and rely on multi-electron conversion reactions involving sulfur or oxygen for energy storage.

(1) Lithium–sulfur (Li–S) batteries: from 'Physical Confinement' to 'Catalytic Conversion'

Lithium–sulfur (Li–S) batteries operate based on the multi-electron conversion reaction between sulfur (S_8) and lithium (Li), offering a high theoretical energy density of 2600 Wh/kg. However, their commercialization has long been hindered by two key scientific issues. To address these challenges, the recent research paradigm has shifted from early strategies of 'physical confinement' (e.g., using porous carbon materials) towards more proactive approaches involving 'chemical adsorption and catalytic conversion' (Li et al., 2024). Furthermore, the nanostructured $CoTe_2$/Co-O-NC material designed by Wen Ming's team features Co-O-C bridges that chemically anchor LiPSs and catalyze their bidirectional conversion. This process, accelerating LiPSs reduction to Li_2S during discharge and Li_2S oxidation to sulfur during charging, effectively suppresses the shuttle effect, enhancing the cycle life and rate capability of battery (Yang et al., 2025).

(2) Lithium–air (Li–O_2) batteries: mitigating severe interfacial reactions

Lithium–air (Li–O_2) batteries, which operate via the reaction between lithium and oxygen to form lithium peroxide (Li_2O_2), exhibit an exceptionally high theoretical energy density of ~3500 Wh/kg but face even more daunting electrochemical challenges compared to Li–S batteries. The core difficulties stem from two main aspects: firstly, the reaction intermediate, the superoxide radical (O_2^-), is highly reactive and irreversibly attacks the electrolyte, causing severe parasitic reactions; secondly, the solid discharge product Li_2O_2 is insulating, and its electrochemical decomposition during charging (oxygen evolution reaction, OER) requires a high overpotential (Chen et al., 2013).

4.3. Other Emerging Electrochemical Energy Storage Systems

(1) Sodium-ion batteries

Sodium-ion batteries (SIBs) are a key emerging system addressing resource security and cost concerns. By utilizing sodium, which is highly abundant and widely distributed in the Earth's crust, SIBs fundamentally reduce reliance on scarce lithium resources. They demonstrate significant cost advantages for applications such as low-speed electric vehicles and large-scale grid energy storage.

(2) Batteries based on renewable materials

Simultaneously, batteries based on renewable materials (e.g., organic batteries) represent a highly sustainable technological pathway. They employ designable, renewable organic compounds (e.g., quinones, radical polymers) or biomass-derived substances as electrode active materials, aiming to circumvent the dependency on toxic and expensive heavy metals like cobalt and nickel. The frontier challenges in this field involve achieving multi-electron transfer through molecular structure design to enhance energy density, and resolving issues related to the dissolution of organic materials in electrolytes and ensuring long-term cycling stability.

4.4 Sustainability Challenges and Future Development Pathways

Energy storage technologies play a pivotal role in driving the energy transition. Despite significant progress in performance and cost, their large-scale application and long-term development face severe sustainability challenges. These challenges span the entire battery life cycle, from raw material extraction and manufacturing to final end-of-life disposal and recycling, constituting a complex set of systemic issues.

Firstly, the fragility of critical raw material supply chains and their environmental footprint are becoming increasingly prominent. Secondly, the energy consumption and carbon emissions from the battery manufacturing process cannot be ignored (Pehnt, 2006). Finally, the end-of-life disposal and recycling of retired batteries pose a major challenge. As the first generation of electric vehicles and large-scale energy storage power stations gradually reach their end-of-life, the number of retired batteries is growing exponentially.

In summary, the sustainable development of energy storage technology is a multi-dimensional systems engineering challenge. It requires not only continuous innovation at the material and chemistry levels but also urgent, profound transformations in manufacturing processes, business models, and global governance. This is essential to ensure that energy storage technologies, while driving the energy transition, themselves align with the long-term requirements of building an ecological civilization.

REFERENCES

Aftab, W., Usman, A., Shi, J., Yuan, K., Qin, M., & Zou, R. (2021). Phase change material-integrated latent heat storage systems for sustainable energy solutions. *Energy & Environmental Science, 14*(8), 4268–4291. DOI: 10.1039/D1EE00527H

Ali, S., Ahsan Farooq Qaisar, M., Sufyan Javed, M., Umer, K., Alarfaji, S. S., Mateen, M., Chhattal, M., Ali, S., Parkash, A., Lama Tamang, T., & Qi, J. (2024). Two-dimensional MXene based innovative electrode materials for supercapacitors: Recent advances and prospects. *Fuel, 377*, 132783. DOI: 10.1016/j.fuel.2024.132783

Amaral, M. M., Real, C. G., Yukuhiro, V. Y., Doubek, G., Fernandez, P. S., Singh, G., & Zanin, H. (2023). In situ and operando infrared spectroscopy of battery systems: Progress and opportunities. *Journal of Energy Chemistry, 81*, 472–491. DOI: 10.1016/j.jechem.2023.02.036

Argirusis, C., Angelara, C., Argirusis, N., Karantonis, A., Pandis, P. P., & Sourkouni, G. (2024). Preparation and characterization of supercapacitor cells using modified CNTs and bimetallic MOFs. *Processes (Basel, Switzerland), 12*(12), 2778. DOI: 10.3390/pr12122778

Armstrong McKay, D. I., Staal, A., Abrams, J. F., Winkelmann, R., Sakschewski, B., Loriani, S., Fetzer, I., Cornell, S. E., Rockström, J., & Lenton, T. M. (2022). Exceeding 1.5 °C global warming could trigger multiple climate tipping points. *Science, 377*(6611), eabn7950. DOI: 10.1126/science.abn7950 PMID: 36074831

Bazant, M. Z. (2013). Theory of chemical kinetics and charge transfer based on nonequilibrium thermodynamics. *Accounts of Chemical Research, 46*(5), 1144–1160. DOI: 10.1021/ar300145c PMID: 23520980

. Chen, L., & Ma, R. (2024). Clean energy synergy with electric vehicles: Insights into carbon footprint. Energy Strategy Reviews, 53, Article 101394.

Chen, Y., Freunberger, S. A., Peng, Z., Fontaine, O., & Bruce, P. G. (2013). Charging a Li–O2 battery using a redox mediator. *Nature Chemistry, 5*(6), 489–494. DOI: 10.1038/nchem.1646 PMID: 23695630

Cheng, X.-Q., Li, H., Zhao, Z.-X., Wang, Y.-Z., & Wang, X.-M. (2021). The use of in-situ Raman spectroscopy in investigating carbon materials as anodes of alkali metal-ion batteries. *New Carbon Materials, 36*(1), 93–105. DOI: 10.1016/S1872-5805(21)60007-0

Cong, X., Pierce, R., & Liu, X. L. (2021). Development of recycled polypropylene-based sustainable composites with recycled carbon fibre/Kenaf fibre hybrid reinforcements. *Journal of Physics: Conference Series*, *1765*(1), 012013. DOI: 10.1088/1742-6596/1765/1/012013

Demirocak, D. E., & Bhushan, B. (2014). In situ atomic force microscopy analysis of morphology and particle size changes in lithium iron phosphate cathode during discharge. *Journal of Colloid and Interface Science*, *423*, 151–157. DOI: 10.1016/j.jcis.2014.02.035 PMID: 24703680

Deng, D., Kim, M. G., Lee, J. Y., & Cho, J. (2009). Green energy storage materials: Nanostructured TiO2 and Sn-based anodes for lithium-ion batteries. *Energy & Environmental Science*, *2*(8), 818–837. DOI: 10.1039/b823474d

Fan, Q., Lin, K., Yang, S., Guan, S., Chen, J., Feng, S., Liu, J., Liu, L., Li, J., & Shi, Z. (2020). Constructing effective TiO2 nano-coating for high-voltage Ni-rich cathode materials for lithium ion batteries by precise kinetic control. *Journal of Power Sources*, *477*, 228745. DOI: 10.1016/j.jpowsour.2020.228745

. Ferdous, A. R., Shah, S. S., Hussain, A., Mirghni, A. A., Hardianto, Y. P., & Aziz, M. A. (2025). Review on current advancements in silica and their composites for high-performance batteries and supercapacitors. Sustainable Materials and Technologies, 43, Article e01296.

Gabisa, E. W., & Aman, A. (2016). Characterization and experimental investigation of NaNO3: KNO3 as solar thermal energy storage for potential cooking application. *Journal of Solar Energy*, *2016*(1), 2405094. DOI: 10.1155/2016/2405094

Gao, Y., Yan, L., Zhao, C., Chen, M., Yang, S., Shao, G., & Mao, J. J. (2024). Accurate calculation of solid-state Li+ diffusion coefficient and kinetic activation energies for an artificial graphite anode. *Journal of the Electrochemical Society*, *171*(2), 020558. DOI: 10.1149/1945-7111/ad27b4

Grey, C. P., & Hall, D. S. (2020). Prospects for lithium-ion batteries and beyond—A 2030 vision. *Nature Communications*, *11*(1), 6279. DOI: 10.1038/s41467-020-19991-4 PMID: 33293543

Hong, J., Kobayashi, S., Kuwabara, A., Ikuhara, Y. H., Fujiwara, Y., & Ikuhara, Y. J. M. (2021). Defect engineering and anisotropic modulation of ionic transport in perovskite solid electrolyte LixLa(1−x)/3NbO3. *ACS Nano*, *26*(12), 3559. PMID: 34200888

Hu, D., Du, F., Cao, H., Zhou, Q., Sun, P., Xu, T., Mei, C., Hao, Q., Fan, Z., & Zheng, J. (2021). An effective strategy to control thickness of Al2O3 coating layer on nickel-rich cathode materials. *Journal of Electroanalytical Chemistry (Lausanne, Switzerland)*, *880*, 114910. DOI: 10.1016/j.jelechem.2020.114910

Hua, W., Wang, S., Knapp, M., Leake, S. J., Senyshyn, A., Richter, C., Yavuz, M., Binder, J. R., Grey, C. P., Ehrenberg, H., Indris, S., & Schwarz, B. (2019). Structural insights into the formation and voltage degradation of lithium- and manganese-rich layered oxides. *Nature Communications*, *10*(1), 5365. DOI: 10.1038/s41467-019-13240-z PMID: 31772159

Huang, J., Luo, Y., Weng, M., Yu, J., Sun, L., Zeng, H., Liu, Y., Zeng, W., Min, Y., & Guo, Z. (2021). Advances and applications of phase change materials (PCMs) and PCMs-based technologies. *ES Materials & Manufacturing*, *13*, 23–39. DOI: 10.30919/esmm5f458

Jeong, M., Lee, H., Yoon, J., & Yoon, W.-S. (2019). O3-type NaNi1/3Fe1/3Mn1/3O2 layered cathode for Na-ion batteries: Structural evolution and redox mechanism upon Na (de)intercalation. *Journal of Power Sources*, *439*, 227064. DOI: 10.1016/j.jpowsour.2019.227064

Jiang, Z., Li, A., Meng, C., Chen, X., & Song, H. J. (2022). Strategies and challenges of carbon materials in the practical applications of lithium metal anode: A review. *Physical Chemistry Chemical Physics : PCCP*, *24*(43), 26356–26370. DOI: 10.1039/D2CP04032H PMID: 36286554

Kang, T., Wang, Y., Guo, F., Liu, C., Zhao, J., Yang, J., Lin, H., Qiu, Y., Shen, Y., Lu, W., & Chen, L. (2019). Self-assembled monolayer enables slurry-coating of Li anode. *ACS Central Science*, *5*(3), 468–476. DOI: 10.1021/acscentsci.8b00845 PMID: 30937374

Kato, Y., Hori, S., Saito, T., Suzuki, K., Hirayama, M., Mitsui, A., Yonemura, M., Iba, H., & Kanno, R. (2016). High-power all-solid-state batteries using sulfide superionic conductors. *Nature Energy*, *1*(4), 16030. DOI: 10.1038/nenergy.2016.30

Kim, N.-Y., Kim, I., Bornamehr, B., Presser, V., Ueda, H., Lee, H.-J., Cheong, J. Y., & Jung, J.-W. (2024). Recent advances in nanoengineering of electrode-electrolyte interfaces to realize high-performance Li-ion batteries. *Energy & Environmental Materials*, *7*(3), e12622. DOI: 10.1002/eem2.12622

Kim, T., Song, W., Son, D.-Y., Ono, L. K., & Qi, Y. (2019). Lithium-ion batteries: Outlook on present, future, and hybridized technologies. *Journal of Materials Chemistry. A, Materials for Energy and Sustainability*, *7*(7), 2942–2964. DOI: 10.1039/C8TA10513H

LaPotin, A., Zhong, Y., Zhang, L., Zhao, L., Leroy, A., Kim, H., Rao, S. R., & Wang, E. N. (2021). Dual-stage atmospheric water harvesting device for scalable solar-driven water production. *Joule*, *5*(1), 166–182. DOI: 10.1016/j.joule.2020.09.008

Li, D.-C., Muta, T., Zhang, L.-Q., Yoshio, M., & Noguchi, H. (2004). Effect of synthesis method on the electrochemical performance of LiNi1/3Mn1/3Co1/3O2. *Journal of Power Sources*, *132*(1), 150–155. DOI: 10.1016/j.jpowsour.2004.01.016

Li, H., Guo, S., & Zhou, H. (2021). In-situ/operando characterization techniques in lithium-ion batteries and beyond. *Journal of Energy Chemistry*, *59*, 191–211. DOI: 10.1016/j.jechem.2020.11.020

Li, H., Lin, Q., Wang, J., Hu, L., Chen, F., Zhang, Z., & Ma, C. (2024). A cost-effective sulfide solid electrolyte Li7P3S7.5O3.5 with low density and excellent anode compatibility. *Angewandte Chemie International Edition*, *63*(37), e202407892. DOI: 10.1002/anie.202407892 PMID: 38945831

Li, P., Zhao, Z., Fei, Y., Zhang, H., & Li, G. (2025). LiNO3-based electrolyte with fast kinetics for lithium metal batteries under practical conditions. *Advanced Energy Materials*, *15*(30), 2500882. DOI: 10.1002/aenm.202500882

Li, S., Li, L., Zhao, Y., Yang, H., Tong, H., Fan, S., Wang, Z., & Xu, W. (2024). Modulating the d-band center of single-atom catalysts for efficient Li2S2-Li2S conversion in durable lithium-sulfur batteries. *Energy Storage Materials*, *70*, 103477. DOI: 10.1016/j.ensm.2024.103477

. Li, S., Wu, F., Chen, T., Kang, K., Guo, R., Liu, C., Niu, Y., Gao, A., Zhao, R., & Wang, X. J. (2025). Progress and challenges for energy-dense and cost-effective anode-free lithium metal batteries. Energy Materials Advances, 6, Article 0168.

Li, W., Sun, X., & Yu, Y. (2017). Si-, Ge-, Sn-based anode materials for lithium-ion batteries: From structure design to electrochemical performance. *Small Methods*, *1*(3), 1600037. DOI: 10.1002/smtd.201600037

Lin, B., Wu, X., He, X., Ye, J., Song, Y., Jiang, P., Bai, L., Liao, X., Li, Y., Huo, Y., Zheng, Q., Zhang, H., & Lin, D. (2025). High-entropy doping in NASICON cathodes: Activating the V4+/V5+ redox couple and inducing a reversible single solid-solution phase reaction for advanced sodium ion batteries. *Journal of Colloid and Interface Science*, *690*, 137299. DOI: 10.1016/j.jcis.2025.137299 PMID: 40088810

Lin, Y., Wang, T., Gao, C., Zhang, X., Yu, W., Wang, M., Yang, C., & Zhang, J. (2025). Direct and low-temperature regeneration of degraded $LiFePO_4$ cathodes at ambient conditions using green and sustainable deep eutectic solvent. *Advanced Science (Weinheim, Baden-Württemberg, Germany)*, *12*(28), 2504683. DOI: 10.1002/advs.202504683 PMID: 40391653

Liu, J., Baeyens, J., Deng, Y., Wang, X., & Zhang, H. (2020). High temperature Mn2O3/Mn3O4 and Co3O4/CoO systems for thermo-chemical energy storage. *Journal of Environmental Management*, *267*, 110582. DOI: 10.1016/j.jenvman.2020.110582 PMID: 32364130

Liu, J., Zhong, S., Zhang, F., Cao, Y., Luo, D., Liao, X., Hu, D., Wang, Y., Lu, Z., & Cheng, H. J. (2025). Stabilizing 4.6 V LiCoO2 by lithium compensative additives. *ACS Applied Materials & Interfaces*, *17*(9), 13872–13880. DOI: 10.1021/acsami.4c20095 PMID: 39970084

Liu, M., Saman, W., & Bruno, F. (2012). Review on storage materials and thermal performance enhancement techniques for high temperature phase change thermal storage systems. *Renewable & Sustainable Energy Reviews*, *16*(4), 2118–2132. DOI: 10.1016/j.rser.2012.01.020

Liu, N., Yu, J., Li, Y., Liu, Y., Jabeen, S., Cheng, Y., Zhou, L., Danilov, D. L., Notten, P. H. L., & Li, H. (2025). Multifunctional carbon dots for electrochemical capacitors reviewed: Their opportunities and challenges. *Advanced Science (Weinheim, Baden-Württemberg, Germany)*, *12*(41), e08000. DOI: 10.1002/advs.202508000 PMID: 41001802

Maximov, M., Nazarov, D., Rumyantsev, A., Koshtyal, Y., Ezhov, I., Mitrofanov, I., Kim, A., Medvedev, O., & Popovich, A. (2020). Atomic layer deposition of lithium–nickel–silicon oxide cathode material for thin-film lithium-ion batteries. *Energies*, *13*(9), 2345. DOI: 10.3390/en13092345

McCulloch, M. T., Winter, A., Sherman, C. E., & Trotter, J. A. (2024). 300 years of sclerosponge thermometry shows global warming has exceeded 1.5 °C. *Nature Climate Change*, *14*(2), 171–177. DOI: 10.1038/s41558-023-01919-7

Mukhopadhyay, A., & Sheldon, B. W. (2014). Deformation and stress in electrode materials for Li-ion batteries. *Progress in Materials Science*, *63*, 58–116. DOI: 10.1016/j.pmatsci.2014.02.001

Myrthong, B., Ansari, S., & Choudhary, R. B. (2025). Bimetallic oxide-conducting polymer composites for robust supercapacitor performance and applications: A review. *Journal of Energy Storage*, *117*, 116182. DOI: 10.1016/j.est.2025.116182

Nandi, S., & Pumera, M. (2024). Transition metal dichalcogenide-based materials for rechargeable aluminum-ion batteries: A mini-review. *ChemSusChem*, *17*(9), e202301434. DOI: 10.1002/cssc.202301434 PMID: 38212248

Nguyen, V. P., Kim, J.-H., & Lee, S.-M. (2023). Strategies of boosting synergistic adsorption-catalysis effect for high-performance lithium–sulfur batteries. *Materials Today. Energy*, *38*, 101451. DOI: 10.1016/j.mtener.2023.101451

Niu, J., Wang, M., Cao, T., Cheng, X., Wu, R., Liu, H., Zhang, Y., & Liu, X. (2021). Li metal coated with Li3PO4 film via atomic layer deposition as battery anode. *Ionics*, *27*(6), 2445–2454. DOI: 10.1007/s11581-021-04030-z

Ogura, H., Yamamoto, T., & Kage, H. (2003). Efficiencies of CaO/H2O/Ca(OH)2 chemical heat pump for heat storing and heating/cooling. *Energy*, *28*(14), 1479–1493. DOI: 10.1016/S0360-5442(03)00119-1

. Okhay, O., & Tkach, A. (2022). Polyaniline-graphene electrodes prepared by electropolymerization for high-performance capacitive electrodes: A brief review. Batteries, 8(10), Article 191.

Patra, A. K. N., Jose, J. R., Sahoo, S., Chakraborty, B., & Rout, C. S. (2021). Understanding the charge storage mechanism of supercapacitors: In situ/operando spectroscopic approaches and theoretical investigations. *Journal of Materials Chemistry. A, Materials for Energy and Sustainability*, *9*(46), 25852–25891. DOI: 10.1039/D1TA07401F

Pehnt, M. (2006). Dynamic life cycle assessment (LCA) of renewable energy technologies. *Renewable Energy*, *31*(1), 55–71. DOI: 10.1016/j.renene.2005.03.002

Shen, W., He, W., Zhong, M., Zhang, J., & Guo, S. (2025). Heat generation power of graphite anode during electrochemical lithiation. *Journal of Power Sources*, *639*, 236702. DOI: 10.1016/j.jpowsour.2025.236702

Susarla, N., & Ahmed, S. (2019). Estimating cost and energy demand in producing lithium hexafluorophosphate for Li-ion battery electrolyte. *Industrial & Engineering Chemistry Research*, *58*(9), 3754–3766. DOI: 10.1021/acs.iecr.8b03752

Tang, Y., Zhang, Z., Jin, Y., Gao, S., Huang, D., Xu, S., & Zou, R. J. (2022). Kinetic and thermodynamic insights into advanced energy storage mechanisms of battery-type bimetallic metal–organic frameworks. *Chemistry of Materials*, *34*(23), 10338–10346. DOI: 10.1021/acs.chemmater.2c01798

Tokranov, A., Sheldon, B. W., Li, C., Minne, S., & Xiao, X. (2014). In situ atomic force microscopy study of initial solid electrolyte interphase formation on silicon electrodes for Li-ion batteries. *ACS Applied Materials & Interfaces*, *6*(9), 6672–6686. DOI: 10.1021/am500363t PMID: 24670933

Tran, Y. H. T., An, K., Vu, D. T. T., & Song, S.-W. (2025). High-voltage electrolyte and interface design for mid-nickel high-energy Li-ion batteries. *ACS Energy Letters*, *10*(1), 356–370. DOI: 10.1021/acsenergylett.4c02860

Wang, B., Guo, W., & Fu, Y. (2022). High-performance garnet solid-state battery enabled by improved interfaces. *Journal of Power Sources*, *542*, 231798. DOI: 10.1016/j.jpowsour.2022.231798

Wang, H., Mao, Y., Xu, P., Ding, Y., Yang, H., Li, J.-F., Gu, Y., Han, J., Zhang, L., & Mao, B.-W. (2025). Scalable copper current collectors with precisely engineered lithiophilic alloy "skins" for durable lithium-metal batteries. *Energy & Environmental Science*, *18*(5), 2622–2633. DOI: 10.1039/D4EE05862C

Wang, J., & Sun, X. (2015). Olivine LiFePO4: The remaining challenges for future energy storage. *Energy & Environmental Science*, *8*(4), 1110–1138. DOI: 10.1039/C4EE04016C

Wang, R., Mitchell, J. B., Gao, Q., Tsai, W.-Y., Boyd, S., Pharr, M., Balke, N., & Augustyn, V. (2018). Operando atomic force microscopy reveals mechanics of structural water driven battery-to-pseudocapacitor transition. *ACS Nano*, *12*(6), 6032–6039. DOI: 10.1021/acsnano.8b02273 PMID: 29767999

Wang, S., Chen, F., Zhuang, G., Wei, K., Chen, T., Zhang, X., Chen, C., & Du, P. (2023). Synthesis of an all-carbon conjugated polymeric segment of carbon nanotubes and its application for lithium-ion batteries. *Nano Research*, *16*(7), 10342–10347. DOI: 10.1007/s12274-023-5530-4

Wang, X., Li, X., Fan, H., & Ma, L. (2022). Solid electrolyte interface in Zn-based battery systems. *Nano-Micro Letters*, *14*(1), 205. DOI: 10.1007/s40820-022-00939-w PMID: 36261666

Wang, Z., Tang, Y., Zhang, L., Li, M., Shan, Z., & Huang, J. (2020). In situ TEM observations of discharging/charging of solid-state lithium-sulfur batteries at high temperatures. *Small*, *16*(28), 2001899. DOI: 10.1002/smll.202001899 PMID: 32519445

Wei, C., Tan, L., Zhang, Y., Wang, Z., Feng, J., & Qian, Y. (2022). Towards better Mg metal anodes in rechargeable Mg batteries: Challenges, strategies, and perspectives. *Energy Storage Materials*, *52*, 299–319. DOI: 10.1016/j.ensm.2022.08.014

Wei Seh, Z., Li, W., Cha, J. J., Zheng, G., Yang, Y., McDowell, M. T., Hsu, P.-C., & Cui, Y. (2013). Sulphur–TiO2 yolk–shell nanoarchitecture with internal void space for long-cycle lithium–sulphur batteries. *Nature Communications*, *4*(1), 1331. DOI: 10.1038/ncomms2327 PMID: 23299881

Wu, B., Lochala, J., Taverne, T., & Xiao, J. (2017). The interplay between solid electrolyte interface (SEI) and dendritic lithium growth. *Nano Energy*, *40*, 34–41. DOI: 10.1016/j.nanoen.2017.08.005

Wu, C., Yang, Y., Zhang, Y., Xu, H., He, X., Wu, X., & Chou, S. (2024). Hard carbon for sodium-ion batteries: Progress, strategies and future perspective. *Chemical Science (Cambridge)*, *15*(17), 6244–6268. DOI: 10.1039/D4SC00734D PMID: 38699270

Wu, D., Housel, L. M., King, S. T., Mansley, Z. R., Sadique, N., Zhu, Y., Ma, L., Ehrlich, S. N., Zhong, H., Takeuchi, E. S., Marschilok, A. C., Bock, D. C., Wang, L., & Takeuchi, K. J. (2022). Simultaneous elucidation of solid and solution manganese environments via multiphase operando extended X-ray absorption fine structure spectroscopy in aqueous Zn/MnO2 batteries. *Journal of the American Chemical Society*, *144*(51), 23405–23420. DOI: 10.1021/jacs.2c09477 PMID: 36513373

Wu, J., Li, J., & Yao, X. (2025). Exploring the potential of halide electrolytes for next-generation all-solid-state lithium batteries. *Advanced Functional Materials*, *35*(10), 2416671. DOI: 10.1002/adfm.202416671

Wu, X. W., Yamamura, T., Ohta, S., Zhang, Q. X., Lv, F. C., Liu, C. M., Shirasaki, K., Satoh, I., Shikama, T., Lu, D., & Liu, S. Q. (2011). Acceleration of the redox kinetics of VO2+/VO2+ and V3+/V2+ couples on carbon paper. *Journal of Applied Electrochemistry*, *41*(10), 1183–1190. DOI: 10.1007/s10800-011-0343-7

Xu, L., Liu, W., Hu, Y., & Luo, L. (2020). Stress-resilient electrode materials for lithium-ion batteries: Strategies and mechanisms. *Chemical Communications (Cambridge)*, *56*(87), 13301–13312. DOI: 10.1039/D0CC05359G PMID: 33034589

Xue, Z., He, D., & Xie, X. (2015). Poly(ethylene oxide)-based electrolytes for lithium-ion batteries. *Journal of Materials Chemistry. A, Materials for Energy and Sustainability*, *3*(38), 19218–19253. DOI: 10.1039/C5TA03471J

Yang, L., Lu, X., Wang, S., Wang, J., Guan, X., Guan, X., & Wang, G. (2020). Designed synthesis of nickel–cobalt-based electrode materials for high-performance solid-state hybrid supercapacitors. *Nanoscale*, *12*(3), 1921–1938. DOI: 10.1039/C9NR08156A PMID: 31907496

Yang, Z., Yan, R., Han, J., Wu, T., Wu, Q., Wei, G., Fu, Y., & Wen, M. (2025). Oxygen bridges of CoTe2/Co-O-NC enhancing adsorption-catalysis of polysulfide for stable lithium–sulfur batteries. *Advanced Functional Materials*, *35*(13), 2417834. DOI: 10.1002/adfm.202417834

Yin, J., Wu, Z., Fang, K., Zhu, Y., Zhang, K., Zhang, H., Chen, Y., Li, L., Fan, L., Dong, K., Zheng, L., Wang, Q., Huang, H., Zhang, J., Qiao, Y., & Sun, S.-G. (2025). Reassessing anionic redox in conventional layered oxide cathodes for Li-ion batteries: Ionic and covalent mechanisms. *Chemical Science (Cambridge)*, *16*(19), 8268–8281. DOI: 10.1039/D5SC00429B PMID: 40213374

Yin, X.-T., You, E.-M., Zhou, R.-Y., Zhu, L.-H., Wang, W.-W., Li, K.-X., Wu, D.-Y., Gu, Y., Li, J.-F., Mao, B.-W., & Yan, J.-W. (2024). Unraveling the energy storage mechanism in graphene-based nonaqueous electrochemical capacitors by gap-enhanced Raman spectroscopy. *Nature Communications*, *15*(1), 5624. DOI: 10.1038/s41467-024-49973-9 PMID: 38965231

Ying, H., & Han, W.-Q. (2017). Metallic Sn-based anode materials: Application in high-performance lithium-ion and sodium-ion batteries. *Advanced Science (Weinheim, Baden-Württemberg, Germany)*, *4*(11), 1700298. DOI: 10.1002/advs.201700298 PMID: 29201624

Yu, X., & Manthiram, A. (2018). Electrode–electrolyte interfaces in lithium-based batteries. *Energy & Environmental Science*, *11*(3), 527–543. DOI: 10.1039/C7EE02555F

Zhang, L., Yang, S., Fu, W., Cui, Y., Wang, J., Zhao, D., Yang, C., Wang, X., & Cao, B. (2022). Plasma-induced ε-MnO2 based aqueous zinc-ion batteries and their dissolution-deposition mechanism. *Journal of Materials Science and Technology*, *127*, 206–213. DOI: 10.1016/j.jmst.2022.03.028

Zhang, Y., Li, J., Gao, L., & Wang, M. (2020). Nitrate based nanocomposite thermal storage materials: Understanding the enhancement of thermophysical properties in thermal energy storage. *Solar Energy Materials and Solar Cells*, *216*, 110727. DOI: 10.1016/j.solmat.2020.110727

Zhang, Z., Smith, K., Jervis, R., Shearing, P. R., Miller, T. S., & Brett, D. J. L. (2020). Operando electrochemical atomic force microscopy of solid–electrolyte interphase formation on graphite anodes: The evolution of SEI morphology and mechanical properties. *ACS Applied Materials & Interfaces*, *12*(31), 35132–35141. DOI: 10.1021/acsami.0c11190 PMID: 32657567

Zheng, Q., Song, Y., Huang, W., Yang, J., Li, T., & Xu, Y. J. (2023). Challenges and strategies towards the interface between lithium anode and Li10GeP2S12 electrolyte in all-solid-state lithium metal batteries. *Energy Storage Materials*, *63*, 103038. DOI: 10.1016/j.ensm.2023.103038

Zhu, X., Dong, H., Liu, Y., Feng, Y.-H., Tang, Y., Yu, L., Xu, S.-W., Wei, G.-X., Sun, S., Liu, M., Xiao, B., Xu, R., Xiao, Y., Chou, S., & Wang, P.-F. (2024). Mitigating mechanical stress by the hierarchical crystalline domain for high-energy P2/O3 biphasic cathode materials. *ACS Nano*, *18*(46), 32003–32015. DOI: 10.1021/acsnano.4c09918 PMID: 39523605

Chapter 5
Multiscale Modeling and Machine Learning-Driven Characterization of Degradation Mechanism in Li-Ion Battery Materials

Luis A. Selis
Universidad Nacional De Ingenieria, Peru

Diego E. Galvez-Aranda
http://orcid.org/0000-0002-5427-0220
Texas A&M University, USA

ABSTRACT

Degradation in lithium-ion batteries (LIBs) arises from the coupling of phenomena that span multiple length scales from atomic-scale interatomic interactions to micrometer-scale particle morphologies and millimeter-scale electrode structures and time scales from femtosecond electronic rearrangements to year-long calendar aging. No single modeling or experimental technique is able to capture the entire spectrum. For example, bond breaking, defect migration and interfacial reactions occur at the electronic/atomic scale, while particle cracking, pore clogging, and electrolyte depletion emerge at the mesoscale, and capacity fade or impedance growth manifest at the cell/device scale. This inherently multiscale and multimodal nature of degradation demands an integrated modeling framework combining quantum mechanical, atomistic, mesoscopic, and continuum approaches.

DOI: 10.4018/979-8-3373-4402-7.ch005

1. INTRODUCTION

This chapter introduces the theoretical and computational tools used to investigate electrode degradation and interfacial characterization, with a focus on bridging scales. Presenting a short description of several computational approaches, we can start naming Density Functional Theory (DFT) which is a quantum-mechanical method to study electronic structure, chemical bonding, band structures, and material reactivity based on electron density.(Q. He, Yu, Li, & Zhao, 2019; Leung, 2020) Ab initio Molecular Dynamics (AIMD) combines DFT and molecular dynamics to simulate atomic motion under quantum-calculated forces, enabling the study of charge transfer, bond breaking, chemical reactions, and ionic migration occurring at materials interfaces.(Galvez-Aranda & Seminario, 2022) However, AIMD limitation are given by small systems and short timescales (ps–ns).(Young & Smeu, 2024) To overcome these constraints, Classical Molecular Dynamics (CMD) uses empirical or machine learning force fields to simulate larger systems over longer timescales.(Sun et al., 2020; N. Yao, Chen, Fu, & Zhang, 2022) CMD are mostly used to analyze ion transport, diffusivity, and mechanical properties.(X. He et al., 2025) CMD accuracy depends heavily on the quality of the chosen force field.

Moving from atomistic simulations, higher-scale methods enable the investigation of slower processes and larger geometries relevant to battery performance. Coarse-Grained Molecular Dynamics (CGMD) extends accessible scales by simplifying groups of atoms into clusters, reducing computational complexity while capturing mesoscopic behaviors such as polymer electrolyte dynamics and particle aggregation.(Shen, Fan, & Hall, 2021; X. Yao et al., 2025) Kinetic Monte Carlo (KMC) is useful to model events such as SEI growth, diffusion, and nucleation, representing system evolution as transitions between discrete states governed by reaction rates and extending timescales to seconds or even hours.(Gavilán-Arriazu, Mercer, Barraco, Hoster, & Leiva, 2021; J. Li, Hu, & Li, 2024) Lattice Boltzmann Method (LBM) focuses on simulating fluid and species transport within porous electrodes and polymer membranes.(Jiang, Qu, Zhou, & Tao, 2017; Shodiev et al., 2021) LBM uses distributions on a discrete lattice to balance numerical efficiency with mesoscopic realism. Finally, continuum models describe mass, charge, heat transport, mechanical deformation, and electrochemical reactions at the electrode or cell scale using partial differential equations.(A. Wang et al., 2022; Yerumoh, Cardenas-Sierra, Fernandez, & Franco)

From experimental point of view, recent advances in computer vision (CV) and machine learning (ML) are transforming the way researchers analyze data obtained from characterization techniques enabling faster, more accurate, and automated insights. In the field of microscopy, methods such as scanning electron microscopy (SEM),(Mohammed & Abdullah, 2018) transmission electron microscopy

(TEM),(Williams & Carter, 1996) 4D scanning transmission electron microscopy (4D-STEM),(Ophus, Ercius, Sarahan, Czarnik, & Ciston, 2014) and cryo-transmission electron microscopy (cryo-TEM)(Y. Liu et al., 2021) produce large high-dimensional datasets that are challenging to process using traditional approaches. ML-based image recognition, segmentation, and denoising techniques allow researchers to identify in a precise and efficient way, microstructural features, defects, and interfaces.(Finegan et al., 2022; A. G. Li, West, & Preindl, 2022) For complex experiments like 4D-STEM, which generate multi-dimensional diffraction datasets, advanced neural networks are being used to accelerate phase retrieval and enable high-resolution properties mapping at nanoscale.(Kunze et al., 2025; R. Yuan, Zhang, He, & Zuo, 2021) Similarly, in cryo-TEM, beam-sensitive materials and low signal-to-noise ratios are the main challenges, ML algorithms enhance image reconstruction and classification, opening new possibilities for exploring sensitive nanostructures and biomaterials.(Last, Voortman, & Sharp, 2023; Wu et al., 2021)

Beyond microscopy, ML and CV are also reshaping data processing for tomography and spectroscopy, which are used to analyze structure, chemistry, and performance across scales. In X-ray and ptychographic tomography, DL techniques enable rapid 3D reconstruction, artifact correction, and super-resolution imaging, allowing researchers to visualize internal structures with high fidelity and to reduce acquisition time and radiation damage.(Kunze et al., 2025) Similarly, for spectroscopic methods such as X-ray diffraction (XRD), X-ray photoelectron spectroscopy (XPS), Raman spectroscopy, and electrochemical impedance spectroscopy (EIS), ML provides tools to perform feature extraction, and automated phase identification. These approaches help to accelerate the interpretation of complex spectra, revealing chemical and structural variations that are often overlooked using conventional analysis. Collectively, the integration of CV and ML into microscopy, tomography, and spectroscopy workflows is revolutionizing materials characterization, making data analysis more efficient, accurate, scalable and providing insights about structure-property relationships.

The goal of this chapter is to show how multiscale computational and advanced characterization techniques can be used to unravel the interplay of degradation mechanisms present in Li-ion battery electrodes. By coupling multiscale computational modeling methodologies, from atomistic to continuum-scale models, we provide a robust framework to analyze electrochemical reactions, mechanical stresses, ion transport, and interfacial instabilities at the electrode-electrolyte interface. The insights produced from computational modeling are complemented with characterization techniques including microscopy, tomography, and spectrometry, all providing direct experimental evidence of the structural, chemical, and morphological changes occurring at the electrode–electrolyte interface. Collectively, these methodologies help characterize the origin and evolution of degradation phenomena by bridging scales between the atomistic to the device level.

2. DEGRADATION MECHANISMS

Li-ion batteries (LIBs) experience different kinds of degradation mechanisms during use. Each of these degradation mechanisms could contribute to active lithium consumption, increasing cell resistance, and long-term safety concerns (Figure 1). One of the most notable mechanisms is the continuous formation and thickening of the solid electrolyte interphase (SEI) on the lithium anode. As the SEI thickens, both active lithium and electrolyte components (cation, anion and solvent) continue to be consumed, disrupting ion transport properties. For lithium-metal anodes, the lithium plating on the anode surface could lead to the formation of Li dendrites, which can perforate the separator and create a short-circuit leading to structural and thermal collapse. At the particle level in electrode materials, mechanical cracking of the electrodes (especially for high-capacity anodes and cathodes, such as silicon or nickel-rich cathodes), creates fresh anode surfaces to react with the electrolyte leading to increased side reactions and further SEI growth. Additionally, gas arising from decomposed electrolytes can increase internal cell pressure and physical swelling, while electrolyte pore clogging in the separator and electrodes can reduce ion mobility and further electrochemical performance losses. The relative contribution to these battery failure mechanisms is often synergistic, as these processes can evolve over many spatial and temporal scales collectively and separately together as the long-term electrochemical and mechanical stability of lithium-ion batteries. Understanding each mechanism is essential for developing strategies to mitigate degradation and enhance battery lifetime and safety.

SEI formation: The solid electrolyte interphase forms at the anode surface due to electrolyte reduction. Its nucleation and progressive thickening consume active lithium and electrolyte, alter interfacial resistances, and can evolve heterogeneously. Modeling requires atomistic insight into chemical reactions and mesoscopic treatment for thickness and morphological evolution over many cycles.(Galvez-Aranda & Seminario, 2022; Nie & Lucht, 2014; Xu et al., 2023)

Lithium plating: Under low potentials or fast charging, metallic lithium can deposit on the anode surface. This leads to dendritic growth, short-circuit risks, and uneven current distribution. Capturing plating accurately involves coupling atomic-scale nucleation dynamics with continuum-level electric field and ion transport models.(Lin, Khosravinia, Hu, Li, & Lu, 2021; Q. Liu et al., 2016)

Particle cracking: Repeated lithiation and de-lithiation induce volumetric expansion and contraction, generating mechanical stress within active particles.

Cracks reduce electronic connectivity, expose fresh surfaces, and accelerate electrolyte decomposition. Simulation demands a combination of molecular-level stress analysis and continuum-scale fracture mechanics.(Galvez-Aranda & Seminario, 2018; Zhang, Zhao, & Guo, 2019)

Lithium Dendrite: During repeated lithium plating/stripping, especially under high current densities, metallic lithium can nucleate and form dendritic structures. These dendrites can pierce the separator, causing internal short circuits, thermal runaway, and catastrophic failure. Modeling dendrite growth requires multiscale coupling: quantum-level studies capture Li nucleation energetics, atomistic/mesoscale simulations track surface diffusion and morphology evolution, and continuum electrochemical–mechanical models describe growth under realistic cell operating conditions.(Deng, Yang, & Zhang, 2022; Kim, Choi, Cho, Kil, & Lee, 2013; Luis A. Selis & Seminario, 2018; Selis & Seminario, 2019)

Figure 1. (a) Overview of different degradation phenomena occurring in Lithium-ion batteries, (b) Pathways and consequences of cell degradation in Lithium-ion batteries. Reproduced from Mama et al., 2020, Energy Conversion and Management. (Mama, Solai, Capurso, Danlos, & Khelladi, 2025)

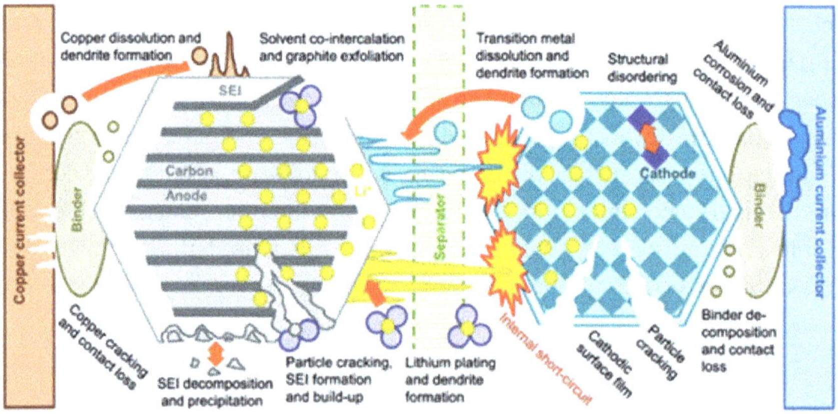

(a)

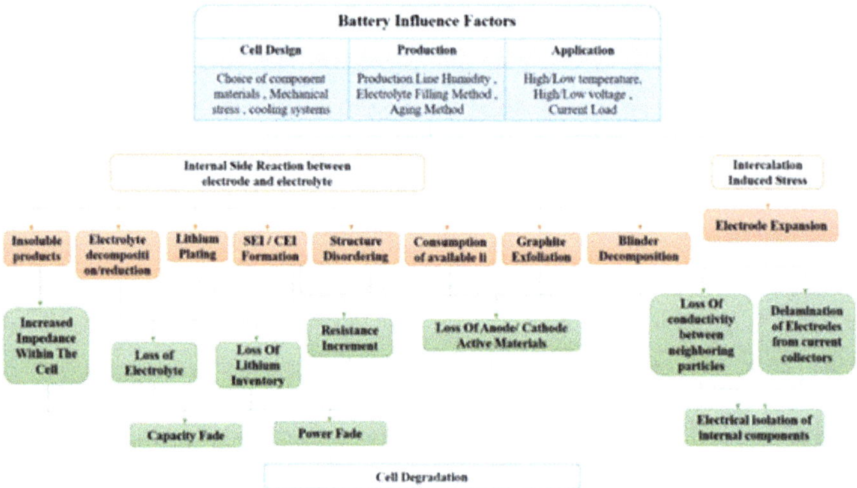

(b)

Electrolyte decomposition: Electrolyte solvents and salts degrade during cycling, producing gases, reactive intermediates, and byproducts that accelerate side reactions. This degradation modifies ionic conductivity and interfacial stability. Modeling requires atomistic insight to capture chemical reaction pathways and mesoscale treatment to account for gas accumulation and transport phenomena over multiple cycles. (Liao et al., 2024; Rinkel, Hall, Temprano, & Grey, 2020)

Insoluble product formation: Non-soluble compounds, such as Li_2CO_3, LiF, and transition-metal oxides, can precipitate within electrodes and separators, clogging pores and isolating active surfaces. These products reduce ionic and electronic conductivity and accelerate capacity fading. Accurate simulation involves combining atomistic reaction modeling with mesoscale transport and morphology evolution to predict their impact on long-term cell performance.(H. Yuan, Read, & Wang, 2019)

Structural disordering: Repeated charge and discharge, can disrupt the crystal lattice of electrode materials, causing loss of order and changes in ion transport properties. This structural disordering affects voltage profiles, mechanical integrity, and overall electrochemical stability. Understanding this mechanism requires coordinating DFT with mesoscale structural evolution models to capture both atomic-level disorder and particle-scale consequences.(Jarvis, Deng, Allard, Manthiram, & Ferreira, 2011; Manthiram & Choi, 2006; Sethuraman, Hardwick, Srinivasan, & Kostecki, 2010)

Consumption of available Li: Cyclable lithium is continuously consumed by side reactions such as SEI growth, electrolyte decomposition, and transition-metal dissolution. This loss reduces the amount of active lithium available for reversible cycling, decreasing capacity and increasing cell impedance. Modeling this phenomenon demands integration of atomistic reaction kinetics with continuum-level lithium transport across the electrode and electrolyte to predict long-term degradation trends.

3. MULTISCALE LANDSCAPE OF BATTERY DEGRADATION

The multiscale landscape of battery degradation involves a complex interplay of phenomena spanning from electronic scale to the cell level, influenced by both intrinsic material properties and extrinsic manufacturing and operational factors (Figure 2). At the electronic scale, the degradation process begins with changes in the electronic structure, density of states, and interfacial stability, impacting charge transfer and redox reactions.(Leung, 2013) At the atomic level, dendrite growth and cycling-induced structural stresses are amplified by local degradation patterns, which are defined by ion diffusion pathways, defect formation, and SEI evolution. Conductivity and ion transport are altered at nano and mesoscale by electrode particle aggregation, porosity evolution, and morphological changes, which connect microscopic phenomena to mesoscopic structural changes.

Optimization of manufacturing parameters is necessary to improve electrode reliability because they are crucial in determining mechanical stress distribution, current uniformity, and the onset of degradation on the electrode scale. These parameters include calendering pressure, coating thickness, and material homogeneity.(Hidalgo et al., 2023; Karaki, Thomitzek, Obermann, Herrmann, & Schröder, 2023) At cell and module levels, during charge/discharge cycling, it leads to capacity fade, thermal phenomena, and loss of performance, necessitating advanced monitoring methods to reduce safety risks. Finally, at the system level, battery packs must manage cross-cell variability, thermal management, and lifetime estimation under operating conditions.(Andrea, 2010; Warner, 2024) Integrating AI-powered technology and human support staff makes it possible to focus on what matters and that is developing real-time monitoring, predictive modelling, and optimization of the materials and components of the battery.(Denisart, Zapata-Dominguez, et al., 2024; Galvez-Aranda, Saravanan, & Franco, 2025; Troncoso et al., 2025)

Figure 2. Different scales and their principal studied mechanisms in battery degradation research.

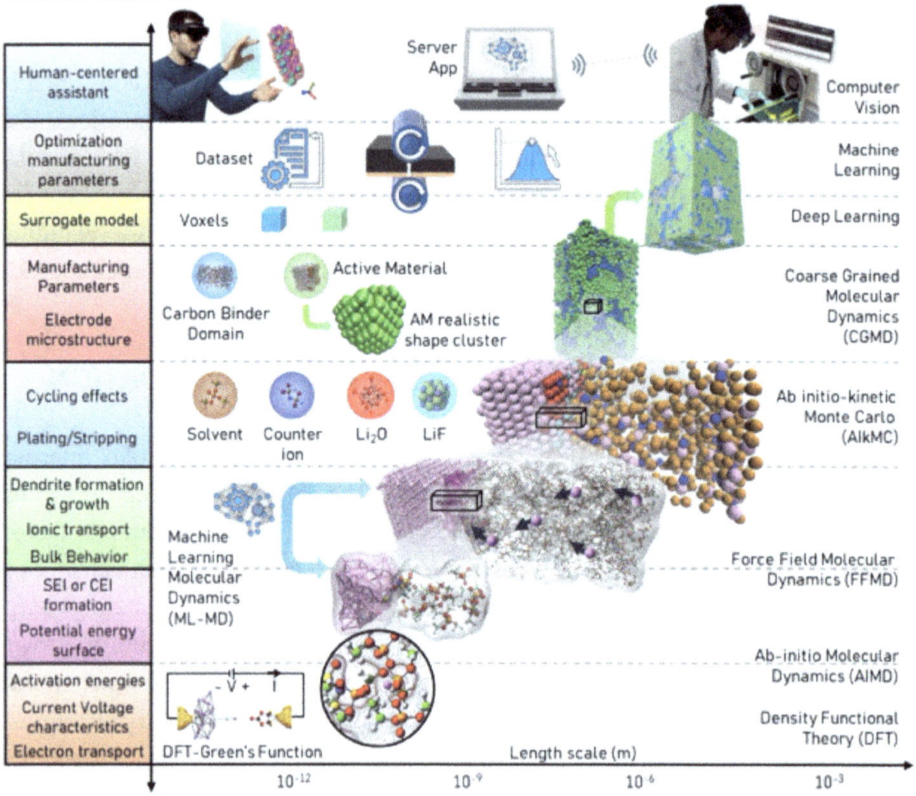

3.1. Information flow across scales

Bottom-up parameter passing

Properties calculated at smaller length and time scales—such as interfacial energies, defect formation energies, diffusion coefficients, reaction rates, elastic moduli, and thermal conductivities—serve as inputs for simulations at larger scales. For example, atomic-level calculations from DFT or AIMD can provide accurate energetics and barriers for lithium migration, which inform classical MD or mesoscale models of ion transport. These mesoscale results, in turn, parameterize continuum models of electrode porosity, tortuosity, and effective ionic conductivity. This hierarchical transfer ensures that macroscopic simulations retain fidelity to molecular and atomic realities, allowing predictions of electrode performance, degradation, and lifetime that are grounded in fundamental physics and chemistry.

Top-down constraints

Conversely, system-level requirements and operational limits impose restrictions on lower-scale models. Device-level considerations—such as allowable voltage windows, current densities, temperature limits, e-waste generated,(Aggarwal, Rana, & Akhai, 2022; Cardenas-Sierra et al., 2025; Khang & Akhai, 2025) and mechanical tolerance guide the selection of materials, interfacial chemistries, and electrode architectures. These constraints are fed down to atomistic or mesoscale simulations to ensure that the resulting models produce feasible and relevant predictions. For instance, knowledge of maximum safe charging rates can limit the exploration of lithium plating scenarios in MD simulations, while thermal constraints can guide the selection of electrolyte compositions that maintain stability under high power operation. By integrating top-down constraints, multiscale modeling ensures consistency across scales and bridges the gap between fundamental understanding and practical device design.

3.2. Verification, validation, and uncertainty quantification (VVUQ) fundamentals

Verification ensures that a computational model is implemented correctly and that the numerical algorithms used are solving the intended governing equations with the expected accuracy. This involves testing against known analytical solutions, performing grid or timestep convergence studies, and benchmarking against well-established codes. In the context of Li-ion battery modeling, verification might include checking that transport equations for ion diffusion converge properly or that quantum-mechanical solvers reproduce known energy levels and forces. The goal is to eliminate coding or numerical errors so that discrepancies between simulation and experiment can be attributed to physical approximations rather than computational artifacts.

Validation evaluates how well the predictions of a model reproduce experimental observations under relevant conditions. This step provides confidence that the model captures the essential physics and chemistry of the system being studied. For example, a validated continuum model of electrode degradation should reproduce experimentally measured capacity fade, while atomistic simulations may be validated against spectroscopic data, such as diffusion barriers from NMR or reaction pathways inferred from in situ XPS. Validation is an ongoing process, as models often need to be refined and re-validated when applied to new chemistries, electrode architectures, or operating regime.

Uncertainty quantification systematically assesses the impact of uncertainties in model parameters, boundary conditions, and physical assumptions on simulation outcomes. This includes both *aleatory uncertainty* (intrinsic variability, such as fluctuations in particle size distribution) and *epistemic uncertainty* (lack of knowledge, such as incomplete understanding of interfacial kinetics). Sensitivity analyses and probabilistic methods, such as Bayesian inference or Monte Carlo sampling, are commonly employed. In Li-ion battery degradation modeling, UQ can identify which parameters (e.g., solid electrolyte interphase growth rate, lithium plating kinetics, or diffusion coefficients) strongly influence cycle life predictions, thereby guiding both experimental characterization and model development.

Together, VVUQ provides a structured framework to ensure that multiscale and multimodal simulations of Li-ion degradation are both reliable and predictive, enabling meaningful translation of computational insights into design and engineering decisions.(Leighton & Akasapu, 2024)

4. QUANTUM-SCALE METHODS

Density Functional Theory (DFT)

Formation energies and defect chemistry quantify oxygen vacancy formation in layered oxides, site defects in NMC, and vacancy/interstitial populations that influence transition metal (TM) migration and dissolution. These quantities provide parameters for CMD and KMC simulations of TM dissolution and vacancy diffusion. Redox potentials compute Li insertion/extraction voltages and solvent/additive reduction/oxidation potentials, helping to rank SEI precursors and cathode stability windows. These results guide AIMD reaction networks and continuum side-reaction terms. Diffusion barriers (NEB) determine Li and vacancy migration barriers in bulk and at grain boundaries, providing event rates for KMC simulations of SEI growth or Li plating nucleation, and effective diffusivities for continuum models. Surface energetics, including facet-dependent surface energies and reconstructions, define Wulff shapes and reactive facets for oxygen release, electrolyte oxidation, and TM dissolution, serving as boundary conditions for AIMD and phase-field models. Examples such as cathode oxygen release and TM dissolution (e.g., NMC). DFT defect energetics and surface energies quantify the driving force for oxygen vacancy creation and TM migration, seeding CMD dissolution pathways and continuum fade terms.(Fallahzadeh & Farhadian, 2015) Regarding SEI, DFT redox potentials and initial reaction energies select SEI-forming routes,(D. Li, Shen, Zheng, & Xu, 2025) passed to AIMD and then KMC growth, informing continuum resistance growth.

Ab Initio Molecular Dynamics (AIMD)

AIMD provides a detailed picture of interfacial chemistry at finite temperature, capturing atomic motions, bond breaking/forming events, and solvation structure evolution near electrode surfaces. By tracking elementary reaction steps and intermediate lifetimes, AIMD generates quantitative data on reaction pathways, radical formation, and solvation environments, which can be directly used to construct rate catalogs for Kinetic Monte Carlo simulations. This enables the prediction of mesoscale processes such as SEI formation, growth, and composition evolution over timescales inaccessible to direct AIMD simulations. For example, AIMD identifies solvent and salt decomposition routes at the anode, generating reactive intermediates whose kinetics feed KMC to simulate SEI thickness and heterogeneity; similarly, at high-voltage cathodes, AIMD captures electrolyte oxidation channels and gas-evolving fragments, providing initial stoichiometries and reaction rates for Lattice Boltzmann modeling of bubble transport and local overpotential effects. By linking atomistic chemistry to mesoscale transport and continuum-level degradation, AIMD serves as a critical bridge in the multiscale workflow, translating first-principles insights into parameters that inform long-time, system-scale predictions of battery aging and safety risks.

Strengths and limits

Ab initio methods like DFT and AIMD provide atomistic and electronic-level accuracy, capturing bond breaking, defect formation, and interfacial reactions. Their main limitation is computational cost, which restricts system size to hundreds of atoms and timescales to picoseconds, making direct simulation of mesoscale phenomena, such as SEI growth, Li plating, or particle cracking to be impractical. Converging k-point meshes, functionals, and supercell sizes, and propagating error bars, ensures reliable extraction of diffusion barriers, reaction energies, and surface/interfacial energies for use in higher-scale models.

Parameter handoff enables multiscale coupling: barriers and prefactors feed KMC for rare-event kinetics, surface/interfacial energies inform phase-field fracture and particle shape evolution, elastic constants and moduli support finite-element chemo-mechanics, and transport tensors are upscaled to continuum models (P2D/

porous-electrode) for effective ionic and electronic transport. This approach systematically translates atomic-level fidelity into predictive mesoscopic and continuum simulations over operationally relevant scales.

5. ATOMISTIC & MESOSCOPIC DYNAMICS

Classical Molecular Dynamics (CMD)

Classical MD serves as a powerful tool to bridge atomic-scale interactions with experimentally measurable transport properties.(Bedrov et al., 2019) In the context of electrolyte transport, it enables the quantification of ionic diffusion coefficients, solvation structures, and conductivity trends as a function of concentration, temperature, and external fields. For ion conduction in solids, CMD simulations reveal the role of crystallographic defects, dopants, and dynamic lattice distortions in facilitating or hindering ion migration pathways. The method also provides unique insights into nano cracking precursors by resolving atomistic signatures of stress accumulation, dislocation nucleation, and early fracture events that can later evolve into mechanical degradation at larger scales. Equally critical is the analysis of interfacial structure, where CMD uncovers the layering of solvents, the orientation of ions at electrode surfaces, and the dynamic rearrangement of the electric double layer under charging/discharging conditions.

Coarse-Grained Molecular Dynamics (CGMD)

CGMD extends the reach of molecular simulations by reducing degrees of freedom while preserving essential chemical physics. This makes it possible to capture the role of polymer binders in ensuring cohesion, elasticity, and durability within electrode composites. At the electrode manufacturing stage, CGMD provides quantitative insight into slurry rheology, describing how particle–binder–solvent interactions govern viscosity, homogeneity, and processability. Once electrodes are formed, mesoscale morphology evolution becomes a central focus, with CGMD allowing the exploration of domain growth, aggregation, and mechanical relaxation phenomena that directly impact electrochemical performance and cycle stability.

Figure 3. (a) Computational simulation of anode particles. (b) Initial anode particle and (c) cracking of the anode particle due to lithiation. Li dendrite formation at the (d) early, (e) middle and (f) short-circuit stage.(g) Reproduced from Gerasimov et al., 2023, The Journal of Physical Chemistry C.(Gerasimov et al., 2023)

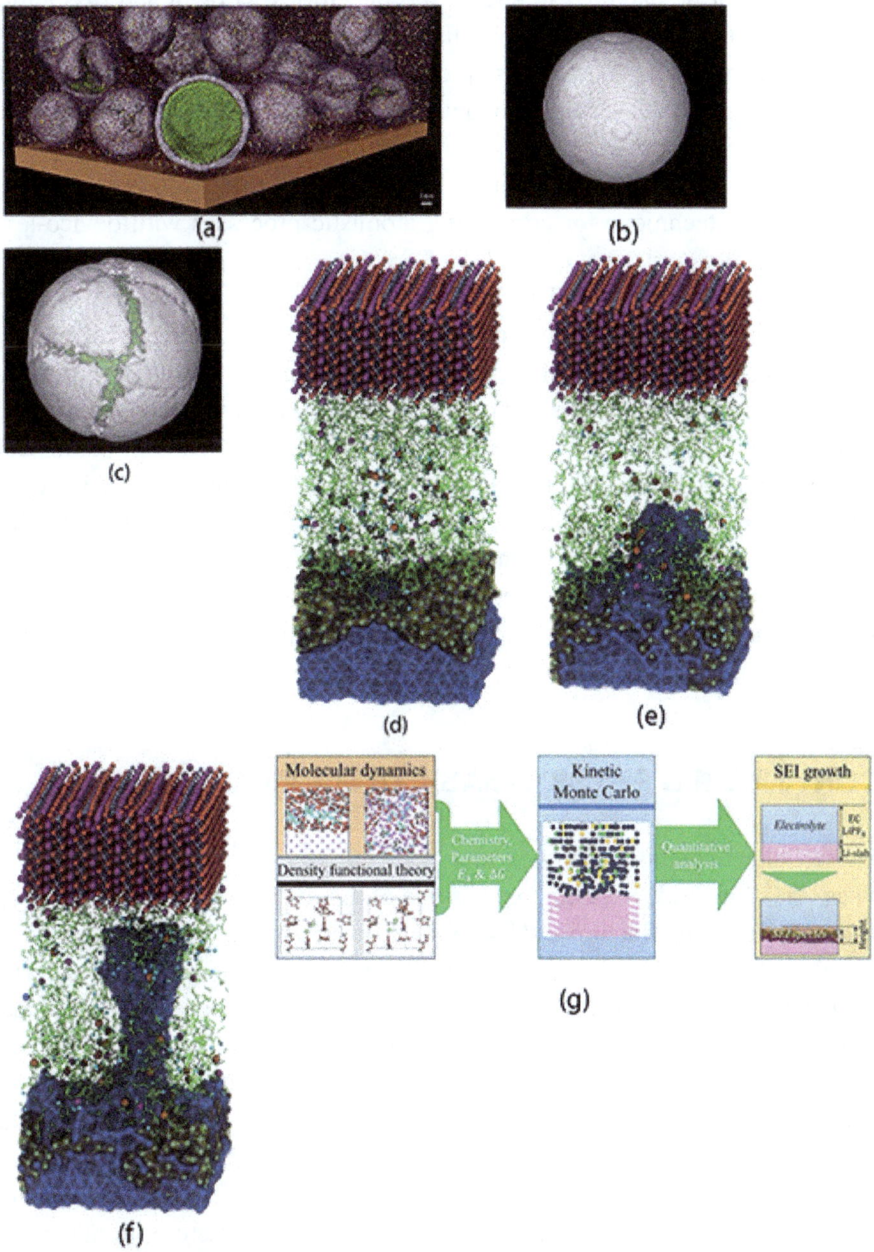

Kinetic Monte Carlo (KMC)

KMC offers a complementary approach to MD by targeting rare-event dynamics that unfold over much longer timescales. For diffusion processes, KMC tracks stochastic ion migration across complex energy landscapes, revealing emergent trends that cannot be captured in short MD trajectories. Growth and dissolution phenomena, such as the dynamic restructuring of active material surfaces, can also be studied with precision, capturing step-edge propagation and defect healing. Perhaps most notably, KMC excels in modeling nucleation processes, for instance Li plating at electrode interfaces or the progressive thickening of the SEI layer. These events, though individually rare, govern battery degradation and safety, make KMC an indispensable technique for connecting atomistic processes with device-level reliability.(Gavilán-Arriazu et al., 2021)

Lattice Boltzmann Method (LBM)

LBM addresses the mesoscale fluid dynamics of battery systems by focusing on pore-scale transport within complex electrode and separator architectures. It enables the computation of tortuosity, a key parameter linking microstructure to effective ionic and electronic conductivity, thereby providing critical input to continuum-scale models. Beyond ion transport, LBM captures multiphase flow phenomena, including liquid–gas coexistence and bubble nucleation, which are essential for understanding gas evolution in porous networks under extreme conditions such as overcharging or high-rate cycling. The method thus connects the morphological features of electrodes with their macroscopic transport properties, offering predictive capability for optimizing material design and operational protocols.

6. CONTINUUM & PHYSICS-BASED CELL MODELS

Phase-field & chemo-mechanics

Phase-field methods have emerged as one of the most powerful continuum approaches for modeling fracture and failure in electrochemical systems.(Miehe, Dal, Schänzel, & Raina, 2016) In the context of lithium-ion and solid-state batteries, these methods allow a continuous representation of cracks without explicitly tracking their interfaces, which is particularly advantageous for simulating complex crack geometries and their evolution over time. When coupled with chemo-mechanical models, phase-field simulations can explicitly capture the stress generation that arises from intercalation and deintercalation processes. Ion insertion typically leads to volumetric expansion, while extraction results in contraction, and these repeated cycles impose large stresses that may eventually exceed the fracture strength of the active material. This coupling provides a mechanistic description of how microcracks nucleate, propagate, coalesce, or even arrest under certain conditions. Recent developments also incorporate crack healing mechanisms, either through stress relaxation, chemical reconstructions, or self-healing additives in the electrolyte. Altogether, phase-field and chemo-mechanical models are critical for connecting the nanoscale physics of ion transport with the mesoscale mechanical integrity of electrodes, providing guidelines for material design strategies that mitigate stress-driven degradation.

The Pseudo-Two-Dimensional (P2D)

The porous-electrode framework, often referred to as the P2D or Doyle-Fuller-Newman model, represents one of the foundational continuum models in battery science. At its core, this framework describes the coupled phenomena of ion transport in the electrolyte, electron conduction in the solid matrix, and electrochemical kinetics at the active surface. It provides a rigorous description of how concentration gradients, potential drops, and current distribution evolve within a porous electrode under operational conditions. This balance between electrochemistry and transport has made the P2D model the basis for many batteries management system (BMS) algorithms and design optimization studies. Beyond its original formulation, the model has been extended to account for thermal effects, explicitly including Joule heating, entropic heat generation, and side reactions that contribute to temperature rise. Thermal coupling is crucial to predict hot spots and performance heterogeneity under fast-charging or high-power scenarios. Another line of extension incorporates aging mechanisms, such as the growth of the solid electrolyte interphase (SEI), lithium plating, loss of active material, or electrolyte decomposition. These processes are implemented through dynamic parameter updates or additional transport-reaction equations, enabling the model to capture long-term degradation. In doing so, the porous-electrode model evolves from a purely electrochemical description into a multi-physics predictive tool for lifetime and reliability studies.

Thermal models: heat generation, thermal runaway precursors; coupling to mechanics

Thermal models are indispensable for understanding both the safe operation and the failure pathways of batteries.(Latz & Zausch, 2015) They quantify the sources of heat within a cell, which include reversible entropic contributions, irreversible Joule heating, and reaction heat from parasitic side processes. As batteries are driven to higher power densities, the local heat generation can create steep temperature gradients, leading to accelerated degradation or even triggering runaway events. Thermal runaway precursors, such as the onset of exothermic side reactions, the breakdown of SEI layers, or gas evolution, can be identified by coupling heat transport models with electrochemistry and reaction kinetics. Detailed thermal simulations also inform the design of cooling strategies and cell packaging for thermal management at both single-cell and pack levels. A critical frontier in this area is the coupling of thermal and mechanical models, which allows for the prediction of thermal expansion, stress accumulation, and the onset of thermally driven fracture or delamination. Such coupling highlights the feedback loops between heat generation, mechanical degradation, and electrochemical performance. In addition, incorporating multi-scale effects, from particle-level hotspots to pack-level thermal propagation, enables these models to serve as a bridge between material-level physics and system-level safety design.

7. COUPLING ACROSS SCALES

At the atomistic and the molecular level, ab initio density functional theory (DFT) and classical molecular dynamics (MD) yield microscopic descriptors including ion diffusion coefficients, solvation structures, interfacial energies, elastic moduli, and local electronic conductivities. These unsmoothed values are not directly accessible to be inserted in continuum-scale models of electrodes or devices. So called homogenization and up-scale techniques are then applied for microscopic data in order to produce effective medium properties. Diffusivity MD values, for instance, may be averaged and corrected in look for tortuosity to specify the porous electrode theory (P2D) effective ionic transport coefficient. Computed DFT elastic constants on grain or interface in turn can be employed in finite-element (FE) based mechanical simulations to investigate the electrode cracking and stress evolution. (Miehe et al., 2016) Such effective parameters span scales that would otherwise be un-coupled enabling macro-scale performance predictions to be principle-based, rather than based on empirical fitting.(Gao & Xu, 2020)

Passing reaction rates and barriers

Most of such critical processes for functioning of batteries such as Li-ion intercalation, charge transfer between interfaces, surface decomposition, nucleation of new phases, and growth of lithium dendrite are events occurring on time scales long compared to atomic oscillations. To simulate these processes, multiscale coupling employs ab initio molecular dynamics (AIMD) and DFT to address activation barriers, reaction pathways, and transition-state energetics. These values are then provided as inputs to kinetic Monte Carlo (KMC) simulations, which can describe the stochastic dynamics of rare events on orders of magnitude larger scales in time. The outcomes of KMC can subsequently be used as input for mesoscale phase field (PF) models or continuum transport equations, which account for the collective morphological and electrochemical evolution in practical electrode geometries. There is no physical mechanism that is identified in the path from atomistic physics to kinetic models that couple the mesoscales and keep large scale predictions close to the energetics.

Reducing order and surrogate modeling

Although more detailed atomistic, mesoscale and continuum models offer high accuracy, they are not suitable for large-scale design exploration or applications with real-time optimization. Reduced-order modeling seeks to approximate these high-dimensional systems with low-dimensional models that can still capture the most crucial dynamics at a fraction of the computational cost. Examples are the development of P2D that approximate full electrochemical-thermal coupling, or the surrogate constitute relation to interfacial kinetics as learned from DFT or MD databases. Statistical regression or machine learning surrogate models can be applied in parallel to mimic the prediction of costly simulations or experiments and perform fast parameter sweeps of electrode designs, electrolyte compositions, or operating protocols. These models allow for sensitivity analysis, uncertainty quantification and constrained optimization and are valuable in providing guidance for both laboratory experiments and large techno-economic analyses. More specifically, these reduced models are suitable for embedding in control systems needing to make fast predictions such as for battery management systems or online diagnostics.(Arrinda, Vertiz, Sanchéz, Makibar, & Macicior, 2022)

8. MACHINE LEARNING FOR INTERATOMIC POTENTIALS & BRIDGING

The past decade has witnessed an unprecedented expansion of machine learning (ML) approaches for materials modeling, particularly in the development of interatomic potentials that combine the accuracy of quantum mechanics with the efficiency of classical molecular dynamics (CMD). The central goal is to bridge the gap between ab initio molecular dynamics (AIMD), which provides chemically accurate insights at the electronic scale but is limited to tens of picoseconds and hundreds of atoms and CMD, which enables nanosecond to microsecond timescales and systems with thousands or even millions of atoms, albeit with reduced fidelity. ML-driven interatomic potentials provide a flexible, data-driven framework to capture the essential physics of atomic interactions, enabling accurate and scalable simulations of complex degradation processes in Li-ion battery materials. Several families of ML potentials have been established, each balancing accuracy, transferability, and computational efficiency.

Figure 4. Different Machine Learning Interatomic Potential. Reproduced from Wang et al., 2024, Iscience.(G. Wang et al., 2024)

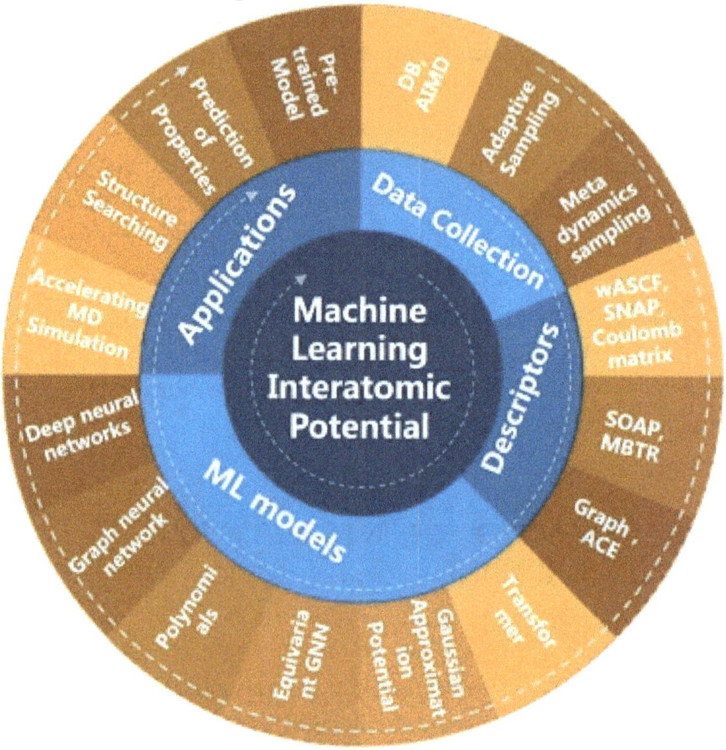

Gaussian Approximation Potentials (GAP)

GAPs are machine learning models that approximate the potential energy surface of atomic systems using ab initio data. They employ Gaussian Process Regression to incorporate prior knowledge, estimate uncertainty, and fit the model reliably. Atoms are represented through invariant descriptors, enabling predictions of energies, forces, and local properties. GAP has been applied to various materials and can be integrated with molecular dynamics simulations to improve the model in real time. (Klawohn et al., 2023)

Spectral Neighbor Analysis Potentials (SNAP)

SNAP employs spectrum components as rotationally invariant descriptors. These potentials have demonstrated strong performance in reproducing energetics, structural properties, and ionic transport, including lithium-ion diffusion in complex oxides, while providing a flexible framework for efficient fitting, validation, and active learning workflows.(Rohskopf et al., 2023)

Figure 5. FitSNAP typical workflow involving scraping configurations of atoms which serve as training data. Reproduced from Rohskopf et al., 2023, The Journal of Open Source Software.(Rohskopf et al., 2023)

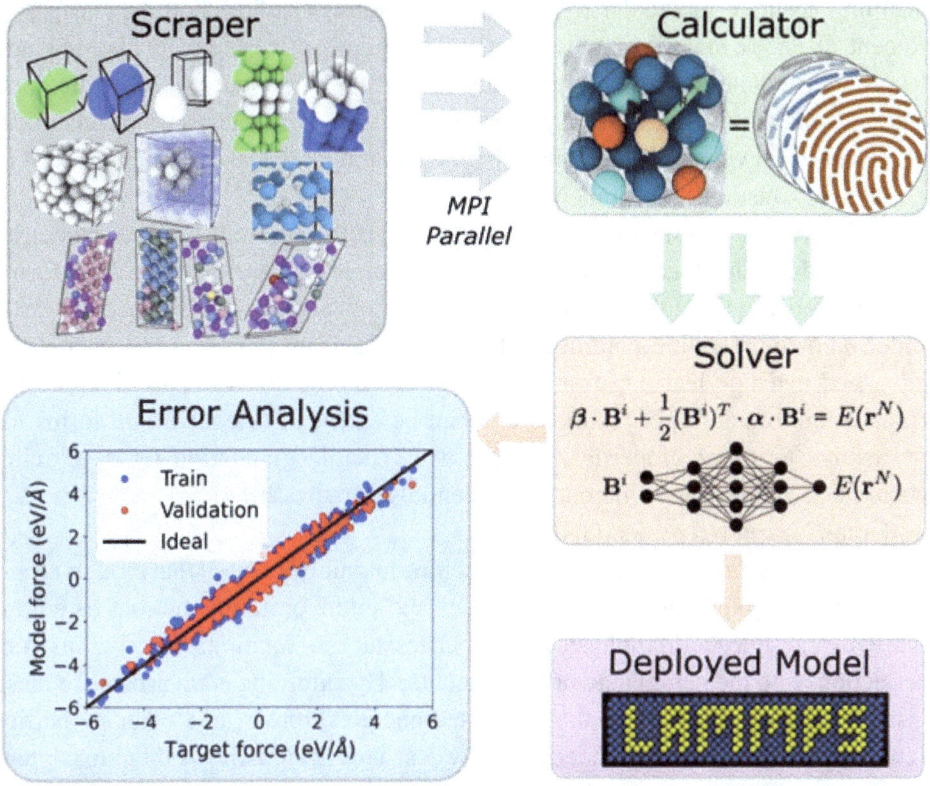

Workflow for creating a machine-learned potential using SNAP begins with collecting atomic configurations to serve as training data, managed through the Scraper class. Next, the Calculator class determines the atomic environment descriptors, which are then used by the Solver class to perform the machine learning regression. The final steps involve conducting error analysis and, if necessary, exporting the trained model for use in a molecular dynamics program.(Rohskopf et al., 2023)

Moment Tensor Potentials (MTP)

MTP employs polynomial-like tensor expansions that decay beyond a cutoff, enabling accurate modeling of multi-component systems. They can capture arbitrary local interactions and are effective for predicting structures, lattice dynamics, diffusion, including lithium diffusion in cathode materials, and thermal properties while maintaining a balance between accuracy and computational efficiency.(Novikov, Gubaev, Podryabinkin, & Shapeev, 2021)

Neural Network Potentials (NNPs)

NNPs are a class of machine learning interatomic potentials that use neural networks to model the potential energy surface of atomic systems based on ab initio data. They are trained to predict energies and forces from atomic coordinates, enabling accurate molecular dynamics simulations of solids, liquids, and gases. Recent advances, including equivariant architectures, active and delta learning, and large diverse datasets, have significantly improved their accuracy, reliability, and generalizability, making NNPs a powerful tool for studying materials and chemical processes at previously inaccessible spatial and temporal scales.(Duignan, 2024)

Atomic Cluster Expansion (ACE)

They provide a systematic and complete basis to represent local atomic environments through a hierarchical expansion of many-body correlations. ACE forms a mathematically complete set that guarantees convergence with increasing basis order, allowing flexible parametrizations ranging from simple linear models to advanced machine learning frameworks. ACE enforces fundamental invariances (translation, rotation, permutation) and can be extended to equivariant forms for vector and tensorial properties, making it a versatile foundation for developing accurate and transferable interatomic potentials.(Drautz, 2019)

Graph Neural Networks (GNNs)

They represent a new generation of machine learning potential that models atomic systems as graphs, with atoms as nodes and interatomic interactions as edges, while explicitly encoding physical symmetries such as rotations, translations, and reflections into the neural network architecture. By enforcing equivariance to these transformations, they ensure that predicted energies, forces, and other properties remain consistent with the underlying physics. This design allows them to capture complex many-body interactions, long-range effects, and multi-component systems with state-of-the-art accuracy, all while requiring relatively small training datasets. These methods have been successfully applied across molecules, condensed-phase materials, and catalytic systems, providing a highly accurate and data-efficient alternative to earlier ML potentials.(Batzner et al., 2022)

Collectively, these methods enable simulation of electrode/electrolyte interphases, defect-mediated transport, and phase transformations with accuracy far beyond traditional empirical force fields.

8.1. Active learning loops

Active learning is a central strategy for the data-driven development of machine learning interatomic potentials, providing an autonomous, iterative procedure to improve model parametrizations efficiently. The workflow consists of several steps: (1) generation of new atomic configurations, (2) identification of configurations with high uncertainty using suitable indicators, and (3) selection of the most representative configurations from the previous step. These selected configurations are then reevaluated using reference calculations (e.g., DFT or AIMD) and added to the training set, after which the model is retrained and applied until further uncertainty is detected.(Lysogorskiy, Bochkarev, Mrovec, & Drautz, 2023)

There exist various strategies for generating atomic configurations and identifying unreliable predictions. A common approach is to train ML potentials using preexisting datasets, molecular dynamics trajectories, or specifically generated configurations targeted at certain properties of interest, such as elastic constants, vibrational modes, or defect formation energies. While effective for some applications, this method is inherently limited in its coverage of configuration space and may miss regions critical for accurate predictions. A more efficient and automated approach leverages uncertainty indicators to explore configurational space adaptively. Configurations with extrapolation indicators above a given threshold are automatically selected for retraining, allowing the model to continuously improve in both accuracy and transferability.

Active learning strategies have been successfully implemented for several ML potentials, including ACE and NNPs, and can significantly reduce the computational cost associated with generating large high-fidelity datasets. By focusing on the most informative configurations, AL not only accelerates the development of robust ML potential but also ensures that simulations remain reliable when exploring new chemical environments, reaction pathways, or phase transitions.

8.2. Δ-learning, multi-fidelity training, long-range physics and charge equilibration

Δ-learning involves training ML models to predict corrections relative to a baseline potential, allowing rapid improvement of accuracy without retraining from scratch. Multi-fidelity training extends this concept by combining datasets of varying accuracy, from classical force fields to high-level ab initio calculations, balancing computational cost and precision. Incorporation of long-range physics, such as electrostatics, polarization, and charge equilibration, enables ML potentials to capture interactions beyond local atomic neighborhoods, essential for ionic systems, polar molecules, and metals.(Ko, Finkler, Goedecker, & Behler, 2021) These strategies collectively allow ML-IPs to maintain high fidelity across diverse chemical environments while keeping training datasets tractable.

8.3. Transferability, uncertainty estimation, and failure detection in deployment

Ensuring that ML interatomic potentials are transferable to unseen systems and conditions is critical for their practical deployment. Transferability requires models to generalize beyond the training domain while maintaining predictive accuracy. Uncertainty estimation techniques, including ensemble methods, Bayesian neural networks, and dropout-based approaches, provide a measure of confidence for each prediction. Automated failure detection mechanisms can flag configurations where the model may extrapolate poorly, allowing on-the-fly retraining or adaptive simulation control.(Vandermause et al., 2020) These capabilities are essential for large-scale simulations, high-throughput screening, and real-world applications where unseen chemical environments are common.

9. ML FOR CHARACTERIZATION & INVERSE PROBLEMS

ML for characterization and inverse problems is a prime mover in accelerating battery research by virtue of a capacity to offer automatic, high-resolution data in a range of experimental modalities. Advanced computer vision is used to analyze microscopy data from SEM, TEM, 4D-STEM, cryo-TEM, tomography and spectroscopy methods (XRD, XPS, Raman, EIS), facilitating comprehensive structural and chemical characterization.(Cheng et al., 2022; Zorn et al., 2022)

Through self-supervised and contrastive feature learning, ML models can learn dense representations from sparse labeled data to enable segmentation, particle tracking, and crack mapping to identify degradation pathways with high accuracy.(Fu et al., 2022; Petrich et al., 2017) Through the fusion of multimodal data fusion with physics-informed neural networks (PINNs) and Bayesian inference, researchers can fuse disparate experimental and simulation data with physical constraints for enhanced interpretability and robustness.(Hofmann et al., 2023; F. Wang, Zhai, Zhao, Di, & Chen, 2024)

Additionally, ML-informed automated parameter inference bridges the gap between continuum-scale models and experimental measurements by extracting essential kinetic, transport, and degradation parameters directly from operando data, significantly enhancing predictive modeling and real-time optimization of battery performance. This synergy between ML, multimodal characterization, and physics-based modeling creates a rich platform for the solution of inverse problems and acceleration of materials discovery and battery diagnostics.

10. IMMERSIVE ANALYTICS: VIRTUAL & MIXED REALITY FOR ELECTRODE CHARACTERIZATION

Immersive VR/MR Workspaces and Digital Twins transform the way researchers' study, analyze, and engage with advanced battery data and simulations. Mixed Reality (MR) and Virtual Reality (VR) workspaces allow seamless navigation of 3D and 4D tomograms, enabling scientists to view microstructural evolution with unprecedented clarity while overlaying mesh structures and simulation results onto experimental data for deeper insight.(Denisart, Troncoso, Loup-Escande, & Franco, 2024; Denisart, Zapata-Dominguez, et al., 2024; Franco et al., 2020)

These environments facilitate human-in-the-loop labeling since subject matter experts collaborate to mark features such as cracks, voids, or dendritic growth, and improve ML model accuracy and accelerate curation of data. With common immersive models, multidisciplinary teams can conduct collaborative review sessions and design reviews and converse with one another in real-time decision-making and hypothesis testing.

Moreover, application of digital twin-virtual replicas of battery systems that dynamically replicates real-time experimental dashboards—is presenting researchers with a stunning tool to witness, predict, and improve performance in real-time. By integrating simulations, experimental data, and visualization within fully immersive, interactive environments, VR/MR-enabled platforms enable deeper insight, accelerate innovation, and enhance collaborative problem-solving for next-generation battery research.(W. Li et al., 2020; Tang et al., 2022)

11. SOFTWARE, DATA, AND REPRODUCIBILITY

Open datasets are becoming essential in the push for reproducibility in battery materials research. Resources such as the Materials Project(Jain et al., 2013) and Novel Materials Discovery (NOMAD),(Draxl & Scheffler, 2018) offer curated computational and experimental datasets covering electronic structures, ionic conductivities as well as interfacial properties. On the experimental front, operando synchrotron radiation measurements and high-resolution electron microscopy provide extensive datasets on structural and chemical evolution during realistic cycling. Benchmark tasks derived from these datasets, such as predicting ionic conductivity, SEI composition or interfacial stability, provide an objective test for machine learning potentials and electronic-structure workflows. Data standards to support the interoperability, comparison and integration of these DFT datasets across platforms are expected to play a key role - including CIF/JSON formats, OPTIMADE APIs and new community-based ontologies. Standardized representations of atomic structures, their properties and their metadata help to build and maintain a more transparent and reproducible community.

Workflow engines are offered by software ecosystems like AiiDA(Huber et al., 2020; Pizzi, Cepellotti, Sabatini, Marzari, & Kozinsky, 2016) and FireWorks(Jain et al., 2015) that automate (multiscale) simulation execution, job submission and dependency control, as well as allows for proving memory management. The Atomic Simulation Environment (ASE) is a Python library designed to build workflows and interface with a wide range of simulation codes, although ASE is not a standard-based tracking system, it provides basic management functions and can integrate with systems that offer complete workflow tracking. Provenance tracking: to record input (files, pseudopotentials/potential files, force-representation versions, software/library versions, details about compilers and driver programs, etc.), random seeds, and execution environment metadata, which maximizes the chances that the results can eventually be confirmed and verified years after they are first published. Data infrastructures design is directed by the FAIR principles (Findable, Accessible, Interoperable, Reusable) and enhances cross-comparison of results across laboratories. With respect to HPC and accelerator scale, reproducibility faces additional challenges like differing CPU/GPU architectures, divergent compiler and library stacks as well as variations in the container or runtime environments. Standard solutions to these issues are containerized environments with immutable dependencies, deterministic build and test procedures, exhaustive regression tests, and continuous-integration pipelines. Together, all these practices improve reproducibility and make computational workflows more portable and reliable across different facilities.

12. CONCLUSIONS & ACTIONABLE GUIDANCE

This chapter shows that reliable, useful insight into Li-ion battery degradation comes from a few simple habits like start with one clear question and the observable you will use to judge it, choose the smallest chain appropriate scale tools that can answer that question, run quick numerical sanity checks and always compare model outputs to experimental data, protect ML components from blind extrapolation with active-learning or uncertainty checks, and save enough documentation so others can reproduce and audit your work.

12.1 Best-Practice Checklist

- Question and metric: state one clear research or engineering question (for example, “capacity after 1,000 cycles at 1C”) and the measurement you will use to validate it (voltage curves, EIS, tomographic porosity, etc.). Let that choice determine scales and methods.

- Minimal method chain: pick only the scales you need (electronic, atomic, mesoscale, cell) and a short, well-validated chain such as CMD → KMC → phase-field → P2D. Fewer validated links are better than many weakly connected models.
- Quick numerical checks: run a couple of convergence tests (k-points, energy cutoff, timestep) and one or two regression runs to ensure numerical stability. Record settings so others can repeat them.
- Anchor to experiment: compare model outputs, not just parameters, with at least one independent experimental observable; use multiple orthogonal measurements if possible and report ranges or confidence intervals rather than single numbers.
- Clear parameter handoffs: when moving results between scales, include units, reference state, temperature, and any factor; provide these in a small machine-readable file (CSV or JSON) to avoid ambiguity.
- ML with safeguards: train on chemically diverse data, hold out test sets, adopt active learning to label extrapolative cases, and publish uncertainty indicators and failure modes.
- Build surrogates after validation: only create reduced-order models or surrogates for design, sensitivity scans, or BMS integration once the full multiscale chain has been validated; always publish surrogate error and domain of validity.
- Preserve documentation: archive code versions, potential or pseudopotentials, random seeds, container images, and datasets, and include a short "how-to-run" so others can reproduce and verify your results.

12.2 Pitfalls

- Claiming system-level lifetime, safety, or performance from a single small-scale calculation.
- Applying ML models outside their training domain without an uncertainty flag or active-learning fallback.
- Passing parameters between models without consistent reference states, units, or temperature corrections.
- Skipping basic numerical checks or independent experimental anchors; these simple steps catch the most common mistakes.
- Treating surrogates as truth; use them only within validated domains and with documented error.

- Failing to preserve documentation or to package workflows for portability, which makes reproduction and audit difficult or even impossible.

12.3 ML Tool Applied to Multiscale Techniques

- DFT: reaction energetics, defect energies, redox windows and NEB diffusion barriers. These are the chemical inputs that higher-scale models use.
- AIMD: finite-temperature interfacial chemistry and solvation dynamics, useful for seeding reaction catalogs and short-time kinetics.
- CMD: large-cell atomistic dynamics for diffusivities, solvation structure and early mechanical precursors, when force fields or ML-IPs are validated.
- CGMD: binder and slurry behavior, mesoscale morphology and process-relevant rheology during electrode manufacture.
- KMC: long-time stochastic kinetics such as SEI growth, nucleation and plating, driven by barrier catalogs from DFT/AIMD/MD.
- LBM: pore-scale transport, tortuosity estimates and multiphase gas/liquid behavior in porous electrodes.
- Phase-field and chemo-mechanics: crack nucleation and propagation, intercalation-driven stress and morphology evolution at particle-to-electrode scales.
- P2D / DFN (porous electrode): coupled electrochemical and transport forecasts for voltage, capacity fade and lifetime when aging models are embedded.
- Thermal models: hotspot detection, thermal-runaway precursors and coupled heat–mechanics–electrochemistry effects.
- ML interatomic potentials (GAP, SNAP, MTP, NNP, ACE, equivariant GNNs): extend DFT fidelity to larger systems, enable long-timescale CMD-like runs, and reduce costs when trained with active learning and accompanied by uncertainty metrics.
- ML for characterization (computer vision, self-supervised learning, physics-informed methods): automate segmentation, fuse multimodal data and invert operando measurements into model parameters for faster, physics-consistent interpretation.

REFERENCES

Aggarwal, P., Rana, M., & Akhai, S. (2022). Briefings on e-waste hazard until COVID era in India. *Materials Today: Proceedings*, *71*, 389–393. DOI: 10.1016/j.matpr.2022.09.507

Andrea, D. (2010). *Battery management systems for large lithium-ion battery packs*. Artech house.

Arrinda, M., Vertiz, G., Sanchéz, D., Makibar, A., & Macicior, H. (2022). Surrogate Model of the Optimum Global Battery Pack Thermal Management System Control. *Energies*, *15*(5), 1695. DOI: 10.3390/en15051695

Batzner, S., Musaelian, A., Sun, L., Geiger, M., Mailoa, J. P., Kornbluth, M., Molinari, N., Smidt, T. E., & Kozinsky, B. (2022). E(3)-equivariant graph neural networks for data-efficient and accurate interatomic potentials. *Nature Communications*, *13*(1), 2453. DOI: 10.1038/s41467-022-29939-5 PMID: 35508450

Bedrov, D., Piquemal, J.-P., Borodin, O., MacKerell, A. D.Jr, Roux, B., & Schröder, C. (2019). Molecular Dynamics Simulations of Ionic Liquids and Electrolytes Using Polarizable Force Fields. *Chemical Reviews*, *119*(13), 7940–7995. DOI: 10.1021/acs.chemrev.8b00763 PMID: 31141351

Cardenas-Sierra, I., Vijay, U., Aguesse, F., Antuñano, N., Ayerbe, E., Gold, L., & Stier, S. (2025). Integration of lithium-ion battery recycling into manufacturing through digitalization: A perspective. *Journal of Power Sources*, *631*, 236158. DOI: 10.1016/j.jpowsour.2024.236158

Cheng, D., Sha, W., Xu, Z., Huang, L., Du, Y., Tang, S., Guo, Y., Cao, Y.-C., & Cheng, S. (2022). Computer vision analysis on material characterization images. *Advanced Intelligent Systems*, *4*(3), 2100158. DOI: 10.1002/aisy.202100158

Deng, J., Yang, X., & Zhang, G. (2022). Simulation study on internal short circuit of lithium ion battery caused by lithium dendrite. *Materials Today. Communications*, *31*, 103570. DOI: 10.1016/j.mtcomm.2022.103570

Denisart, L., Troncoso, J. F., Loup-Escande, E., & Franco, A. A. (2024). Breaking down the Barriers between the Digital and the Real: Mixed Reality applied to Battery Manufacturing R&D and Training. *Batteries & Supercaps*, *7*(7), e202400042. DOI: 10.1002/batt.202400042

Denisart, L., Zapata-Dominguez, D., David, X., Leclere, A., Lelong, R., Liu, C., Xu, J., Loup-Escande, E., & Franco, A. A. (2024). Combining virtual reality with mixed reality for efficient training in battery manufacturing. *Batteries & Supercaps*, *7*(1), e202300268. DOI: 10.1002/batt.202300268

Drautz, R. (2019). Atomic cluster expansion for accurate and transferable interatomic potentials. *Physical Review. B*, *99*(1), 014104. DOI: 10.1103/PhysRevB.99.014104

Draxl, C., & Scheffler, M. (2018). NOMAD: The FAIR concept for big data-driven materials science. *MRS Bulletin*, *43*(9), 676–682. DOI: 10.1557/mrs.2018.208

Duignan, T. T. (2024). The Potential of Neural Network Potentials. *ACS Physical Chemistry Au*, *4*(3), 232–241. DOI: 10.1021/acsphyschemau.4c00004 PMID: 38800721

Fallahzadeh, R., & Farhadian, N. (2015). Molecular dynamics simulation of lithium ion diffusion in LiCoO2 cathode material. *Solid State Ionics*, *280*, 10–17. DOI: 10.1016/j.ssi.2015.07.001

Finegan, D. P., Squires, I., Dahari, A., Kench, S., Jungjohann, K. L., & Cooper, S. J. (2022). Machine-learning-driven advanced characterization of battery electrodes. *ACS Energy Letters*, *7*(12), 4368–4378. DOI: 10.1021/acsenergylett.2c01996

Franco, A. A., Chotard, J. N., Loup-Escande, E., Yin, Y., Zhao, R., Rucci, A., & Ciger, J. (2020). Entering the augmented era: Immersive and interactive virtual reality for battery education and research. *Batteries & Supercaps*, *3*(11), 1147–1164. DOI: 10.1002/batt.202000120

Fu, T., Monaco, F., Li, J., Zhang, K., Yuan, Q., Cloetens, P., Pianetta, P., & Liu, Y. (2022). Deep-Learning-Enabled Crack Detection and Analysis in Commercial Lithium-Ion Battery Cathodes. *Advanced Functional Materials*, *32*(39), 2203070. DOI: 10.1002/adfm.202203070

Galvez-Aranda, D. E., Saravanan, S., & Franco, A. A. (2025). Human-Centered Holographic Assistant using Computer Vision to Interact with Battery Manufacturing Machinery in a Predictive Manner. *Batteries & Supercaps*, *8*(12), 2500292. DOI: 10.1002/batt.202500292

Galvez-Aranda, D. E., & Seminario, J. M. (2018). Simulations of a LiF solid electrolyte interphase cracking on silicon anodes using molecular dynamics. *Journal of the Electrochemical Society*, *165*(3), A717–A730. DOI: 10.1149/2.0991803jes

Galvez-Aranda, D. E., & Seminario, J. M. (2022). Li-metal anode in a conventional Li-ion battery electrolyte: Solid electrolyte interphase formation using ab initio molecular dynamics. *Journal of the Electrochemical Society*, *169*(3), 030502. DOI: 10.1149/1945-7111/ac55c8

Gao, X., & Xu, J. (2020). Multiscale Modeling of Electro-Chemo-Mechanical Degradation in Si/C Core–Shell Anode for the Lithium-Ion Battery of High Energy Density. *Journal of Electrochemical Energy Conversion and Storage*, *18*(2), 020903. Advance online publication. DOI: 10.1115/1.4048704

Gavilán-Arriazu, E. M., Mercer, M., Barraco, D., Hoster, H. E., & Leiva, E. P. M. (2021). Kinetic Monte Carlo simulations applied to Li-ion and post Li-ion batteries: A key link in the multi-scale chain. *Progress in Energy*, *3*(4), 042001. DOI: 10.1088/2516-1083/ac1a65

Gerasimov, M., Soto, F. A., Wagner, J., Baakes, F., Guo, N., Ospina-Acevedo, F., Röder, F., Balbuena, P. B., & Krewer, U. (2023). Species Distribution During Solid Electrolyte Interphase Formation on Lithium Using MD/DFT-Parameterized Kinetic Monte Carlo Simulations. *The Journal of Physical Chemistry. C, Nanomaterials and Interfaces*, *127*(10), 4872–4886. DOI: 10.1021/acs.jpcc.2c05898

He, Q., Yu, B., Li, Z., & Zhao, Y. (2019). Density functional theory for battery materials. *Energy & Environmental Materials*, *2*(4), 264–279. DOI: 10.1002/eem2.12056

He, X., Zhang, Y., Li, H., Zhou, M., Wang, W., Wang, R., Jiang, K., & Wang, K. (2025). Improving the reliability of classical molecular dynamics simulations in battery electrolyte design. *Journal of Energy Chemistry*, *101*, 34–41. DOI: 10.1016/j.jechem.2024.09.038

Hidalgo, M., Apachitei, G., Dogaru, D., Faraji-Niri, M., Lain, M., Copley, M., & Marco, J. (2023). Design of experiments for optimizing the calendering process in Li-ion battery manufacturing. *Journal of Power Sources*, *573*, 233091. DOI: 10.1016/j.jpowsour.2023.233091

Hofmann, T., Hamar, J., Rogge, M., Zoerr, C., Erhard, S., & Schmidt, J. P. (2023). Physics-informed neural networks for state of health estimation in lithium-ion batteries. *Journal of the Electrochemical Society*, *170*(9), 090524. DOI: 10.1149/1945-7111/acf0ef

Huber, S. P., Zoupanos, S., Uhrin, M., Talirz, L., Kahle, L., Häuselmann, R., Gresch, D., Müller, T., Yakutovich, A. V., Andersen, C. W., Ramirez, F. F., Adorf, C. S., Gargiulo, F., Kumbhar, S., Passaro, E., Johnston, C., Merkys, A., Cepellotti, A., Mounet, N., & Pizzi, G. (2020). AiiDA 1.0, a scalable computational infrastructure for automated reproducible workflows and data provenance. *Scientific Data*, *7*(1), 300. DOI: 10.1038/s41597-020-00638-4 PMID: 32901044

Jain, A., Ong, S. P., Chen, W., Medasani, B., Qu, X., Kocher, M., Brafman, M., Petretto, G., Rignanese, G.-M., Hautier, G., Gunter, D., & Persson, K. A. (2015). FireWorks: A dynamic workflow system designed for high-throughput applications. *Concurrency and Computation*, *27*(17), 5037–5059. DOI: 10.1002/cpe.3505

Jain, A., Ong, S. P., Hautier, G., Chen, W., Richards, W. D., Dacek, S., Cholia, S., Gunter, D., Skinner, D., Ceder, G., & Persson, K. A. (2013). Commentary: The Materials Project: A materials genome approach to accelerating materials innovation. *APL Materials*, *1*(1), 011002. Advance online publication. DOI: 10.1063/1.4812323

Jarvis, K. A., Deng, Z., Allard, L. F., Manthiram, A., & Ferreira, P. J. (2011). Atomic structure of a lithium-rich layered oxide material for lithium-ion batteries: Evidence of a solid solution. *Chemistry of Materials*, *23*(16), 3614–3621. DOI: 10.1021/cm200831c

Jiang, Z., Qu, Z., Zhou, L., & Tao, W. (2017). A microscopic investigation of ion and electron transport in lithium-ion battery porous electrodes using the lattice Boltzmann method. *Applied Energy*, *194*, 530–539. DOI: 10.1016/j.apenergy.2016.10.125

Karaki, H., Thomitzek, M., Obermann, T., Herrmann, C., & Schröder, D. (2023). Optimizing the Microstructure and Processing Parameters for Lithium-Ion Battery Cathodes: A Use Case Scenario with a Digital Manufacturing Platform. *Energy Technology (Weinheim)*, *11*(5), 2201032. DOI: 10.1002/ente.202201032

Khang, A., & Akhai, S. (2025). *E-Waste and Lithium-ion Battery Recycling Insights for Sustainable Transportation Driving Green Transportation System Through Artificial Intelligence and Automation: Approaches, Technologies and Applications*. Springer.

Kim, S.-H., Choi, K.-H., Cho, S.-J., Kil, E.-H., & Lee, S.-Y. (2013). Mechanically compliant and lithium dendrite growth-suppressing composite polymer electrolytes for flexible lithium-ion batteries. *Journal of Materials Chemistry. A, Materials for Energy and Sustainability*, *1*(16), 4949–4955. DOI: 10.1039/c3ta10612h

Klawohn, S., Darby, J. P., Kermode, J. R., Csányi, G., Caro, M. A., & Bartók, A. P. (2023). Gaussian approximation potentials: Theory, software implementation and application examples. *The Journal of Chemical Physics*, *159*(17), 174108. Advance online publication. DOI: 10.1063/5.0160898 PMID: 37929869

Ko, T. W., Finkler, J. A., Goedecker, S., & Behler, J. (2021). A fourth-generation high-dimensional neural network potential with accurate electrostatics including non-local charge transfer. *Nature Communications*, *12*(1), 398. DOI: 10.1038/s41467-020-20427-2 PMID: 33452239

Kunze, S., Nam, C., Kim, H., Chung, J., Hong, E., Song, J., Choi, H., & Lim, J. (2025). X-ray imaging methods for multiscale characterization of batteries. *Bulletin of the Korean Chemical Society*, *46*(4), 360–380. DOI: 10.1002/bkcs.70009

Last, M. G., Voortman, L. M., & Sharp, T. H. (2023). Measuring cryo-TEM sample thickness using reflected light microscopy and machine learning. *Journal of Structural Biology*, *215*(2), 107965. DOI: 10.1016/j.jsb.2023.107965 PMID: 37100102

Latz, A., & Zausch, J. (2015). Multiscale modeling of lithium ion batteries: Thermal aspects. *Beilstein Journal of Nanotechnology*, *6*, 987–1007. DOI: 10.3762/bjnano.6.102 PMID: 25977870

Leighton, M., & Akasapu, U. (2024). An Application of Uncertainty Quantification to Efficiency Measurements and Validating Requirements through Correlating Simulation and Physical Testing Results. *Sensors (Basel)*, *24*(15), 4867. DOI: 10.3390/s24154867 PMID: 39123913

Leung, K. (2013). Electronic structure modeling of electrochemical reactions at electrode/electrolyte interfaces in lithium ion batteries. *The Journal of Physical Chemistry. C, Nanomaterials and Interfaces*, *117*(4), 1539–1547. DOI: 10.1021/jp308929a

Leung, K. (2020). DFT modelling of explicit solid–solid interfaces in batteries: Methods and challenges. *Physical Chemistry Chemical Physics : PCCP*, *22*(19), 10412–10425. DOI: 10.1039/C9CP06485K PMID: 32073055

Li, A. G., West, A. C., & Preindl, M. (2022). Towards unified machine learning characterization of lithium-ion battery degradation across multiple levels: A critical review. *Applied Energy*, *316*, 119030. DOI: 10.1016/j.apenergy.2022.119030

Li, D., Shen, C., Zheng, Y., & Xu, J. (2025). Electrochemo-Mechanical Degradation and Failure of Active Particles in High Energy Density Batteries: A Review. *Small*, *21*(8), 2407740. DOI: 10.1002/smll.202407740 PMID: 39776184

Li, J., Hu, X., & Li, T. (2024). Simulation of Solid Electrolyte Interphase Growth for Lithium Batteries Based on Kinetic Monte Carlo. *Energy Material Advances, 5*, 0137.

Li, W., Rentemeister, M., Badeda, J., Jöst, D., Schulte, D., & Sauer, D. U. (2020). Digital twin for battery systems: Cloud battery management system with online state-of-charge and state-of-health estimation. *Journal of Energy Storage*, *30*, 101557. DOI: 10.1016/j.est.2020.101557

Liao, Y., Zhang, H., Peng, Y., Hu, Y., Liang, J., Gong, Z., Wei, Y., & Yang, Y. (2024). Electrolyte degradation during aging process of lithium-ion batteries: Mechanisms, characterization, and quantitative analysis. *Advanced Energy Materials*, *14*(18), 2304295. DOI: 10.1002/aenm.202304295

Lin, X., Khosravinia, K., Hu, X., Li, J., & Lu, W. (2021). Lithium plating mechanism, detection, and mitigation in lithium-ion batteries. *Progress in Energy and Combustion Science*, *87*, 100953. DOI: 10.1016/j.pecs.2021.100953

Liu, Q., Du, C., Shen, B., Zuo, P., Cheng, X., Ma, Y., Yin, G., & Gao, Y. (2016). Understanding undesirable anode lithium plating issues in lithium-ion batteries. *RSC Advances*, *6*(91), 88683–88700. DOI: 10.1039/C6RA19482F

Liu, Y., Ju, Z., Zhang, B., Wang, Y., Nai, J., Liu, T., & Tao, X. (2021). Visualizing the sensitive lithium with atomic precision: Cryogenic electron microscopy for batteries. *Accounts of Chemical Research*, *54*(9), 2088–2099. DOI: 10.1021/acs.accounts.1c00120 PMID: 33856759

Lysogorskiy, Y., Bochkarev, A., Mrovec, M., & Drautz, R. (2023). Active learning strategies for atomic cluster expansion models. *Physical Review Materials*, *7*(4), 043801. DOI: 10.1103/PhysRevMaterials.7.043801

Mama, M., Solai, E., Capurso, T., Danlos, A., & Khelladi, S. (2025). Comprehensive review of multi-scale Lithium-ion batteries modeling: From electro-chemical dynamics up to heat transfer in battery thermal management system. *Energy Conversion and Management*, *325*, 119223. DOI: 10.1016/j.enconman.2024.119223

Manthiram, A., & Choi, J. (2006). Chemical and structural instabilities of lithium ion battery cathodes. *Journal of Power Sources*, *159*(1), 249–253. DOI: 10.1016/j.jpowsour.2006.04.028

Miehe, C., Dal, H., Schänzel, L.-M., & Raina, A. (2016). A phase-field model for chemo-mechanical induced fracture in lithium-ion battery electrode particles. *International Journal for Numerical Methods in Engineering*, *106*(9), 683–711. DOI: 10.1002/nme.5133

Mohammed, A., & Abdullah, A. (2018). *Scanning electron microscopy (SEM): A review.* Paper presented at the Proceedings of the 2018 international conference on hydraulics and pneumatics—HERVEX, Băile Govora, Romania.

Nie, M., & Lucht, B. L. (2014). Role of lithium salt on solid electrolyte interface (SEI) formation and structure in lithium ion batteries. *Journal of the Electrochemical Society*, *161*(6), A1001–A1006. DOI: 10.1149/2.054406jes

Novikov, I. S., Gubaev, K., Podryabinkin, E. V., & Shapeev, A. V. (2021). The MLIP package: Moment tensor potentials with MPI and active learning. *Machine Learning: Science and Technology*, *2*(2), 025002. DOI: 10.1088/2632-2153/abc9fe

Ophus, C., Ercius, P., Sarahan, M., Czarnik, C., & Ciston, J. (2014). Recording and using 4D-STEM datasets in materials science. *Microscopy and Microanalysis*, *20*(S3), 62–63. DOI: 10.1017/S1431927614002037

Petrich, L., Westhoff, D., Feinauer, J., Finegan, D. P., Daemi, S. R., Shearing, P. R., & Schmidt, V. (2017). Crack detection in lithium-ion cells using machine learning. *Computational Materials Science*, *136*, 297–305. DOI: 10.1016/j.commatsci.2017.05.012

Pizzi, G., Cepellotti, A., Sabatini, R., Marzari, N., & Kozinsky, B. (2016). AiiDA: Automated interactive infrastructure and database for computational science. *Computational Materials Science*, *111*, 218–230. DOI: 10.1016/j.commatsci.2015.09.013

Rinkel, B. L., Hall, D. S., Temprano, I., & Grey, C. P. (2020). Electrolyte oxidation pathways in lithium-ion batteries. *Journal of the American Chemical Society*, *142*(35), 15058–15074. DOI: 10.1021/jacs.0c06363 PMID: 32697590

Rohskopf, A., Sievers, C., Lubbers, N., Cusentino, M., Goff, J., Janssen, J., & Sargsyan, K. (2023). FitSNAP: Atomistic machine learning with LAMMPS. *Journal of Open Source Software*, *8*(84), 5118. DOI: 10.21105/joss.05118

Selis, L. A., & Seminario, J. M. (2018). Dendrite formation in silicon anodes of lithium-ion batteries. *RSC Advances*, *8*(10), 5255–5267. DOI: 10.1039/C7RA12690E PMID: 35542415

Selis, L. A., & Seminario, J. M. (2019). Dendrite formation in Li-metal anodes: An atomistic molecular dynamics study. *RSC Advances*, *9*(48), 27835–27848. DOI: 10.1039/C9RA05067A PMID: 35530483

Sethuraman, V. A., Hardwick, L. J., Srinivasan, V., & Kostecki, R. (2010). Surface structural disordering in graphite upon lithium intercalation/deintercalation. *Journal of Power Sources*, *195*(11), 3655–3660. DOI: 10.1016/j.jpowsour.2009.12.034

Shen, K.-H., Fan, M., & Hall, L. M. (2021). Molecular dynamics simulations of ion-containing polymers using generic coarse-grained models. *Macromolecules, 54*(5), 2031–2052. DOI: 10.1021/acs.macromol.0c02557

Shodiev, A., Primo, E., Arcelus, O., Chouchane, M., Osenberg, M., Hilger, A., Manke, I., Li, J., & Franco, A. A. (2021). Insight on electrolyte infiltration of lithium ion battery electrodes by means of a new three-dimensional-resolved lattice Boltzmann model. *Energy Storage Materials, 38*, 80–92. DOI: 10.1016/j.ensm.2021.02.029

Sun, Y., Yang, T., Ji, H., Zhou, J., Wang, Z., Qian, T., & Yan, C. (2020). Boosting the optimization of lithium metal batteries by molecular dynamics simulations: A perspective. *Advanced Energy Materials, 10*(41), 2002373. DOI: 10.1002/aenm.202002373

Tang, H., Wu, Y., Cai, Y., Wang, F., Lin, Z., & Pei, Y. (2022). Design of power lithium battery management system based on digital twin. *Journal of Energy Storage, 47*, 103679. DOI: 10.1016/j.est.2021.103679

Troncoso, J. F., Zanotto, F. M., Galvez-Aranda, D. E., Zapata Dominguez, D., Denisart, L., & Franco, A. A. (2025). The ARTISTIC battery manufacturing digitalization initiative: From fundamental research to industrialization. *Batteries & Supercaps, 8*(1), e202400385. DOI: 10.1002/batt.202400385

Vandermause, J., Torrisi, S. B., Batzner, S., Xie, Y., Sun, L., Kolpak, A. M., & Kozinsky, B. (2020). On-the-fly active learning of interpretable Bayesian force fields for atomistic rare events. *npj Computational Materials, 6*(1), 20. DOI: 10.1038/s41524-020-0283-z

Wang, A., O'Kane, S., Planella, F. B., Le Houx, J., O'Regan, K., Zyskin, M., & Howey, D. A. (2022). Review of parameterisation and a novel database (LiionDB) for continuum Li-ion battery models. *Progress in Energy, 4*(3), 032004. DOI: 10.1088/2516-1083/ac692c

Wang, F., Zhai, Z., Zhao, Z., Di, Y., & Chen, X. (2024). Physics-informed neural network for lithium-ion battery degradation stable modeling and prognosis. *Nature Communications, 15*(1), 4332. DOI: 10.1038/s41467-024-48779-z PMID: 38773131

Wang, G., Wang, C., Zhang, X., Li, Z., Zhou, J., & Sun, Z. (2024). Machine learning interatomic potential: Bridge the gap between small-scale models and realistic device-scale simulations. *iScience, 27*(5), 109673. DOI: 10.1016/j.isci.2024.109673 PMID: 38646181

Warner, J. T. (2024). *The handbook of lithium-ion battery pack design: Chemistry, components, types, and terminology*. Elsevier.

Williams, D. B., & Carter, C. B. (1996). *The transmission electron microscope Transmission electron microscopy: a textbook for materials science*. Springer. DOI: 10.1007/978-1-4757-2519-3

Wu, J.-G., Yan, Y., Zhang, D.-X., Liu, B.-W., Zheng, Q.-B., Xie, X.-L., Liu, S.-Q., Ge, S.-X., Hou, Z.-G., & Xia, N.-S. (2021). Machine learning for structure determination in single-particle cryo-electron microscopy: A systematic review. *IEEE Transactions on Neural Networks and Learning Systems, 33*(2), 452–472. DOI: 10.1109/TNNLS.2021.3131325 PMID: 34932487

Xu, Y., Jia, H., Gao, P., Galvez-Aranda, D. E., Beltran, S. P., Cao, X., & Li, S. (2023). Direct in situ measurements of electrical properties of solid–electrolyte interphase on lithium metal anodes. *Nature Energy, 8*(12), 1345–1354. DOI: 10.1038/s41560-023-01361-1 PMID: 38249622

Yao, N., Chen, X., Fu, Z.-H., & Zhang, Q. (2022). Applying classical, ab initio, and machine-learning molecular dynamics simulations to the liquid electrolyte for rechargeable batteries. *Chemical Reviews, 122*(12), 10970–11021. DOI: 10.1021/acs.chemrev.1c00904 PMID: 35576674

Yao, X., Hafiz, H., Chen, L., Cai, W., Mao, Z., & Hu, S. (2025). Coarse-grained molecular dynamics modelling of lithium-ion battery electrode drying: A sensitivity analysis. *Journal of Manufacturing Processes, 144*, 54–59. DOI: 10.1016/j.jmapro.2025.03.117

Yerumoh, E., Cardenas-Sierra, I., Fernandez, F., & Franco, A. A. 3D Resolved Computational Modeling to Simulate the Electrolyte Wetting of a Lithium-Ion Battery Cell with 18650 Format. *Batteries & Supercaps, n/a*(n/a), 2500434. doi:DOI: 10.1002/batt.202500434

Young, J., & Smeu, M. (2024). *Applications of Ab Initio Molecular Dynamics for Modeling Batteries Computational Design of Battery Materials*. Springer.

Yuan, H., Read, J. A., & Wang, Y. (2019). Capacity loss of non-aqueous Li-Air battery due to insoluble product formation: Approximate solution and experimental validation. *Materials Today. Energy, 14*, 100360. DOI: 10.1016/j.mtener.2019.100360

Yuan, R., Zhang, J., He, L., & Zuo, J.-M. (2021). Training artificial neural networks for precision orientation and strain mapping using 4D electron diffraction datasets. *Ultramicroscopy, 231*, 113256. DOI: 10.1016/j.ultramic.2021.113256 PMID: 33773843

Zhang, Y., Zhao, C., & Guo, Z. (2019). Simulation of crack behavior of secondary particles in Li-ion battery electrodes during lithiation/de-lithiation cycles. *International Journal of Mechanical Sciences*, *155*, 178–186. DOI: 10.1016/j.ijmecsci.2019.02.042

Zorn, M., Ionescu, C., Klohs, D., Zähl, K., Kisseler, N., Daldrup, A., & Flamme, S. (2022). An approach for automated disassembly of lithium-ion battery packs and high-quality recycling using computer vision, labeling, and material characterization. *Recycling*, *7*(4), 48. DOI: 10.3390/recycling7040048

Chapter 6
Artificial Intelligence and Machine Learning Application for Materials Design Methodology

Minghe Qu
http://orcid.org/0009-0009-8139-9423
Tongren University, China

Linkun Liao
http://orcid.org/0009-0002-4731-4675
Zhejiang Red Flag Machinery Corporation, China

Shenshen Li
Jining University, China

Qiaoli Li
Hubei Aerospace River Chemical Co., Ltd., China

Xiangyang Lin
Nanjing University of Science and Technology, China

ABSTRACT

Artificial intelligence (AI) and machine learning (ML) have demonstrated powerful transformative capabilities across numerous fields. Especially in new material design, AI applications not only enhance efficiency but also drive innovation. Meanwhile, ML—with its ability to automate complex tasks—has emerged as a new pathway to accelerate material discovery. This chapter first introduces the fundamental concepts of AI and materials design, detailing the underlying principles and algorithms of AI

DOI: 10.4018/979-8-3373-4402-7.ch006

applications while demonstrating their practical effectiveness through case studies. Secondly, this chapter describes the general process and the application of ML in materials science, covering applications such as material prediction, classification, and auxiliary microstructural characterization, which enhance computational simulation efficiency, and reviews the multi-objective optimization and inverse the design for materials. Finally, this chapter analyzes the challenges and prospects of AI an

INTRODUCTION

In today's era of rapid development of science and technology, the design and research and development of materials, as the material basis for the progress of human society, have always been a key factor to promote the development of various fields. The traditional material design method mainly relies on the trial-and-error method, and explores new materials through a large number of experiments and experience accumulation, which is not only time-consuming and labor-intensive, but also inefficient, making it hard to meet the demands of modern society for the rapid development and diversification of new materials (Fang et al., 2008). With the rapid development of science and computing technologies, especially the continuous advancement of artificial intelligence (AI) and machine learning (ML) technology, it has brought new opportunities and changes to material design (Bhuvaneswari et al., 2021). AI and ML can process and analyze massive material data to automatically identify the complex relationships between material composition, structure, and properties, enabling efficient material design and innovative development. This article will explore the application of artificial intelligence and machine learning in materials design in detail, including the basic concepts, application principles, algorithms, practical examples, as well as challenges and future prospects.

CURRENT RESEARCH TRENDS IN ARTIFICIAL INTELLIGENCE AND MACHINE LEARNING

In recent years, there have been remarkable breakthroughs in the fields of artificial intelligence (AI) and machine learning (ML), especially in deep learning and generative AI (Bhuvaneswari et al., 2021). These technological advancements are not only driving innovation in the field of technology, but also profoundly impacting human daily life and work patterns.

The Development of Artificial Intelligence and Machine Learning

AI, as one of the hottest topics in the field of science and technology in recent years, is not actually a new technology. As early as 1956, the concept of artificial intelligence was proposed at a summer symposium held at Dartmouth College in the United States. After more than half a century of development, it failed to achieve widespread practical application. In 2012, deep learning, as an important branch of machine learning, achieved remarkable progress, and Professor Jeffrey Hinton and his students from the University of Toronto proposed a deep learning neural network model based on machine learning, which made a major breakthrough in the image recognition challenge (Grzybowski, Pawlikowska–Łagód, & Lambert, 2024). This breakthrough ignited a new wave of development boom in the field of AI, and promoted rapid progress in many fields such as speech recognition, image recognition, and natural language processing. AI technology continues to be iteratively upgraded, especially around 2022, generative AI has ushered in a full-scale explosion. ChatGPT, launched by OpenAI in the United States, has achieved stronger content generation (AIGC) capabilities through larger-scale models. In 2024, OpenAI launched Sora, a large text-to-video model, to promote generative AI from unimodal to multimodal, which can understand multiple forms of information (such as images and text) at the same time. At the same time, China's AI models have also formed a vibrant and diverse development ecosystem. 2025 will be a critical year for AI to move towards wider and deeper applications, and AI agents are expected to become the highlight of AI development and application this year. AI agents have significant advantages such as autonomy, perception ability, decision-making ability, and action ability, and can automatically call the corresponding intelligent model functions and take corresponding actions by understanding the assigned tasks and perceiving the corresponding execution environment. Through AI agents, various AI technologies will be applied more conveniently, more widely and more effectively, which will lead to fundamental changes in many fields. From discrimination to generation, from single-modal to multi-modal, from dedicated to general-purpose, it is an important symbol of the new stage of intelligent development (Jordan & Mitchell, 2015). AI is penetrating all aspects of life through natural language-based interactions. From smart assistants such as mobile phone apps and smart speakers, to embodied intelligence such as intelligent connected cars and intelligent robots, to digital twins that can provide personalized services, and more AI-native products and services, they are accelerating towards us.

Machine Learning Drives Artificial Intelligence

ML, as the core technology of artificial intelligence, has shown explosive growth around the world in recent years (Jain, 2024). Machine learning refers to the training of models on data, enabling computers to learn patterns from data and make predictive judgments or decisions. As the mainstream technology route of AI development, this route allows machines to obtain intelligence through learning instead of relying on preset rules, which reflects the essence of learning for the formation and development of intelligence. It can be said that biological intelligence and machine intelligence are inseparable from learning, and all successes and breakthroughs come from learning. After all, "being born to know" is a trait of a very small number of geniuses, and "learning to know" is applicable to most people. Humans have also taken the route of learning when developing artificial intelligence. Machine learning is primarily implemented through neural networks. Neural networks, as specific models of machine learning, process data by simulating the structure and function of the human brain. The human brain is made up of neurons and synapses, and neurons transmit signals through synapses to achieve functions such as information processing, learning, and memory. Artificial neural networks simulate this structure, using an electronic computing system composed of integrated circuit chips, with artificial neurons and their connections (i.e., model parameters) instead of biological neurons and synapses, and learning and predicting by inputting data and adjusting weights, laying the material foundation for AI. In short, machine learning is the core driver of artificial intelligence, which helps humans make predictions, judgments, and generate the content they need by extracting valuable information from massive amounts of data, thereby improving task processing efficiency and freeing humans from heavy, repetitive, low-skill, and dangerous labor (Jordan & Mitchell, 2015). The 2024 Nobel Prize in Physics was awarded to John Hopfield and Jeffrey Hinton for their foundational discoveries and inventions of machine learning through artificial neural networks. Looking forward to the future, machine learning will surely bring greater surprises to human society through continuous development.

BASIC CONCEPTS, APPLICATION PRINCIPLES AND ALGORITHMS OF AI

Artificial Intelligence

Artificial intelligence (AI) is a new technical science that studies and develops theories, methods, technologies and application systems for simulating, extending and expanding human intelligence. AI is designed to enable computer systems to think, learn, reason, and make decisions like humans so as to accomplish a variety of complex tasks. AI covers several fields such as machine learning, natural language processing, computer vision, expert systems, etc. Among them, machine learning is one of the core branches of artificial intelligence, which mainly studies how to make computers automatically learn rules from data to make predictions and decisions about unknown data.

Material Design

Material design refers to the method of designing the composition and structure of new materials based on the existing understanding of the relationship between material composition, structure, process and performance, and carrying out "structural design" and "performance prediction" on the computer, so as to "design materials according to the specified performance". Its goal is to predict the properties and behavior of materials before experiments through theoretical calculations and simulations, guide experimental design, reduce the number of experiments, and enhance the efficiency and success rate of material research and development. Materials design involves the cross-integration of multiple disciplines such as materials science, physics, chemistry, and computer science, and is a complex system engineering.

Principles of Application

The application principle of artificial intelligence in material design is as follows: first, collect and process massive material data to establish mathematical models that correlate material composition, structure, processes, and performance; second, train and optimize these models by using AI algorithms; finally, realize material property prediction and structure design through the trained models. Specifically, it is necessary to obtain a large amount of material data, including the composition, preparation processes, microstructures, mechanical properties, physical properties, and chemical properties of the materials. These data can come from experimental data, literature, databases, etc. Then, these data are preprocessed, including data cleaning, normalization, dimensionality reduction and other steps, to remove noise and redundant information, and improve the quality and usability of the data. Next, select appropriate artificial intelligence algorithms, such as machine learning algorithms, deep learning algorithms, etc., to establish a material performance prediction model or material structure design model. Through the training of the model, the model can learn the internal rules and patterns in the material data. Finally, the trained model is used to predict the material properties and design the material structure, which provides guidance for the research and development of new materials.

Figure 1. Artificial intelligence in material design.

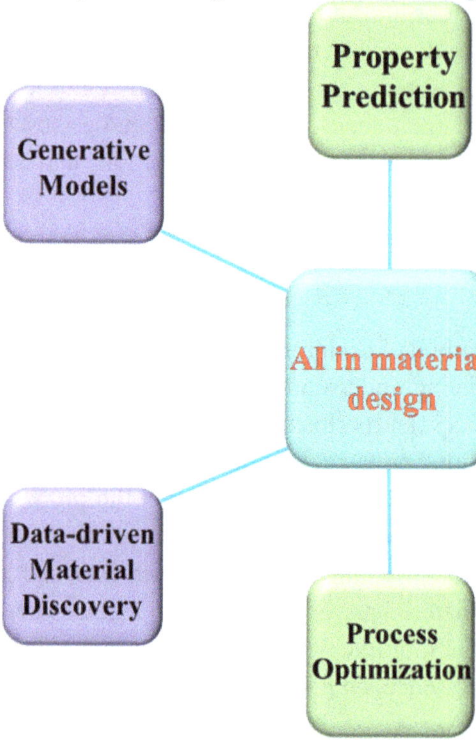

Commonly Used Algorithms

Machine learning algorithms

(1) Linear regression: It is a simple and effective regression algorithm used to model the linear relationship between independent and dependent variables. In material design, it can be used to predict the linear relationship between material properties and factors such as composition and process.

(2) Decision tree: It is a classification and regression algorithm based on tree structure, which establishes decision rules through recursive partitioning of data. Decision trees can visually display the classification and regression process of data, and have certain applications in material classification and performance prediction.

(3) Random forest: It is an ensemble learning algorithm that improves the predictive performance and generalization ability of a model by building multiple decision trees and voting or averaging them. Random forests perform well in processing high-dimensional data and noise data, and have been widely used in material design.

Deep learning algorithms

Deep learning, as an important branch of machine learning, automatically learns hierarchical feature representations from data through multi-layer neural networks, demonstrating strong application potential in fields such as materials science (Sarker, 2021). The following are several commonly used deep learning algorithms in material design:

(1) Artificial Neural Network (ANN): An algorithm that mimics the structure and function of neurons in the human brain, consisting of a large number of neurons connected to each other. It has strong nonlinear fitting ability, can handle complex material data, and has great potential in material performance prediction and structural design (Lu, Wang, Ma, Cheng, & Zhou, 2025).

(2) Convolutional Neural Networks (CNNs):These are deep learning algorithms specially used to process image data, extracting image features through convolution operations and pooling operations. In materials science, CNNs can be used to process images of the microstructure of materials, such as scanning electron microscopy (SEM) images, transmission electron microscopy (TEM) images, etc., to achieve the analysis and identification of the microstructure of materials.

(3) Recurrent Neural Network (RNNs): These are deep learning algorithms used to process sequence data, which memorize historical information in sequences through the cyclic connection of hidden layers. In material design, RNNs can be used to process time-series data, such as changes in process parameters during material preparation, to dynamically predict material properties.

(4) Graph Neural Networks (GNNs): These are a class of deep learning algorithms specifically designed to handle graph-structured data. Unlike Convolutional Neural Networks (CNNs) that process Euclidean data (such as images) and Recurrent Neural Networks (RNNs) that handle sequential data, GNNs can effectively learn the representations of nodes in a graph through information propagation and neighbor aggregation mechanisms. In material design, the molecular structure, crystal structure, and interatomic connectivity of materials are essentially graph-structured data, and GNNs can directly learn on these structures to achieve accurate prediction of material properties and characteristics (Waikhom & Patgiri, 2023). Lu et al. (2025) systematically reviewed the development of GNNs and their applications in the industrial field, providing an important reference for the application of GNNs in materials science.

Figure 2. Commonly used algorithms in materials science.

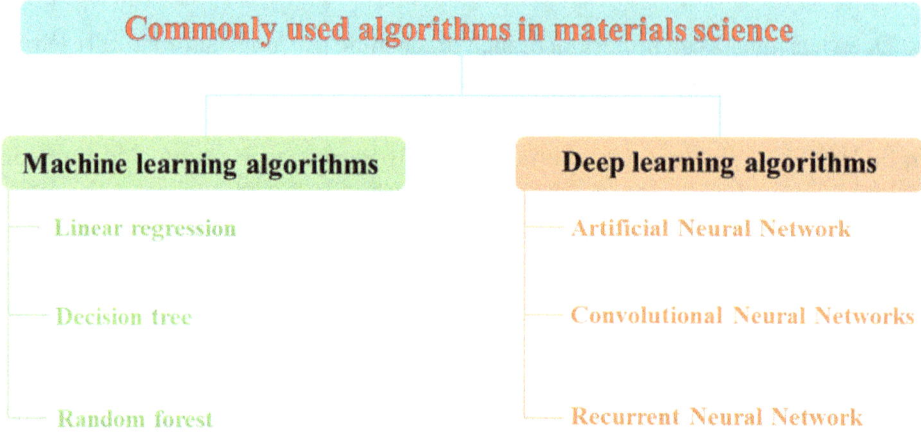

PRACTICAL EXAMPLES OF AI AND ML IN MATERIALS DESIGN

Design of New Battery Materials

With the rapid development of electric vehicles and renewable energy, there is a growing demand for high-performance battery materials (Juan, Dai, Yang, & Zhang, 2021). Traditional R&D of battery materials mainly relies on trial and error, with long R&D cycles and high costs. Artificial intelligence has played an important role in the design of new battery materials. For example, by collecting a large amount of battery material data, including the composition, crystal structures, and electrochemical properties of the materials, the machine learning algorithm is used to establish a battery material performance prediction model to predict performance indicators such as capacity, cycle life and safety of battery materials with different compositions and structures (Liu, Guo, Zou, Li, & Shi, 2020). Then, based on the predicted results, a new battery material with excellent performance is designed. Studies have shown that the use of artificial intelligence technology can shorten the R&D cycle of battery materials by more than half and reduce R&D costs. In recent years, artificial intelligence has made remarkable progress in the field of battery material design, encompassing various energy storage systems such as lithium-ion batteries (Mao et al., 2021; Xu et al., 2024) and sodium-ion batteries (Meng et al., 2024). It is particularly noteworthy that Yan et al. (2025) successfully accelerated the discovery of high-voltage electrolytes for lithium-metal batteries using machine learning methods, providing new insights for the development of next-generation high-energy-density batteries.

High-Performance Alloy Design

Alloy is an important engineering material, which is widely used in aerospace, automotive, energy and other fields. The design of high-performance alloys needs to consider a variety of performance indicators, such as strength, toughness, high temperature resistance, corrosion resistance, etc. Traditional alloy design methods struggle to meet multiple performance requirements simultaneously, but artificial intelligence can process a large amount of alloy data to establish complex relational models between alloy composition, processes and performance, so as to achieve multi-objective optimization design of alloy properties. For example, machine learning algorithms are used to optimize the composition and heat treatment process of aluminum alloys, and aluminum alloy materials with high strength, high toughness and good corrosion resistance are designed for structural components in the aerospace field (Liu et al., 2022). Furthermore, in the field of high-entropy alloys, Sulley et al. (2024) employed active learning methods to achieve efficient discovery of high-entropy alloys, demonstrating the immense potential of machine learning in the design of complex alloy systems.

Nanomaterial Design

Nanomaterials have unique physical, chemical and mechanical properties, and have broad application prospects in electronics, energy, medical and other fields (Han et al., 2020). The design of nanomaterials requires precise control of nanoparticle size, shape, surface structure and other parameters, which is difficult to achieve by traditional methods. Artificial intelligence can analyze the structure and performance data of nanomaterials, establish the mapping relationships between the structures and properties of nanomaterials, and guide the design and preparation of nanomaterials. For example, deep learning algorithms are used to optimize the structure of nanocatalysts to improve the activity and selectivity of catalysts, which can be applied to the catalytic process of chemical reactions.

THE GENERAL PROCESS OF MACHINE LEARNING IN MATERIALS DESIGN

The application of machine learning in materials design typically follows a standardized process involving data collection, data preprocessing, model building, model training, and model evaluation (Liu et al., 2017).

Data Collection

Data collection is the first step in the application of machine learning in materials science, which requires the collection of a large amount of material-related data, including the composition, preparation process, microstructure, and properties of materials. Material data, characterized by high dimensionality, multimodality, and nonlinearity, inherently falls within the typical scope of big data (Zhang et al., 2018). Data sources can include experimental results, literature, and databases. During the data collection process, attention should be paid to the accuracy, completeness and representativeness of the data to ensure that the data can reflect the true properties of the material.

Data Preprocessing

Data preprocessing is the process of cleaning, transforming, and normalizing the collected data to enhance data quality and usability. Effective data preprocessing methods are particularly crucial when dealing with high-dimensional, multi-source, and heterogeneous material big data (Zhang et al., 2018). Data cleaning is mainly to remove noise, outliers, and missing values from data. Data transformation is the transformation of data into a form suitable for processing by machine learning algorithms, such as converting categorical data into numerical data; normalization involves scaling data to a specific range to avoid dimensional differences between different features from affecting model training.

Model Building

Model building is the selection of appropriate machine learning algorithms to build material performance prediction or structural design models based on preprocessed data. When choosing an algorithm, you need to consider the characteristics of the data, the type of problem (e.g., classification, regression, clustering), and the performance requirements of the model. For example, for high-dimensional nonlinear data, deep learning algorithms may perform better than traditional machine learning algorithms.

Model Training

Model training involves using training data to train the constructed model and adjust its parameters, enabling the model to learn the internal rules and patterns in the data. In the training process, appropriate training strategies, such as cross-validation and regularization, should be adopted to avoid model overfitting and underfitting and improve the generalization ability of the model.

Model Evaluation

Model evaluation involves using test data to assess the trained model and measure its performance indicators, such as accuracy, recall, and mean square error(MSE). Through model evaluation, the effectiveness and reliability of the model can be judged, which provides a basis for the optimization and improvement of the model.

Figure 3. The general process of machine learning in materials science (Liu et al., 2017).

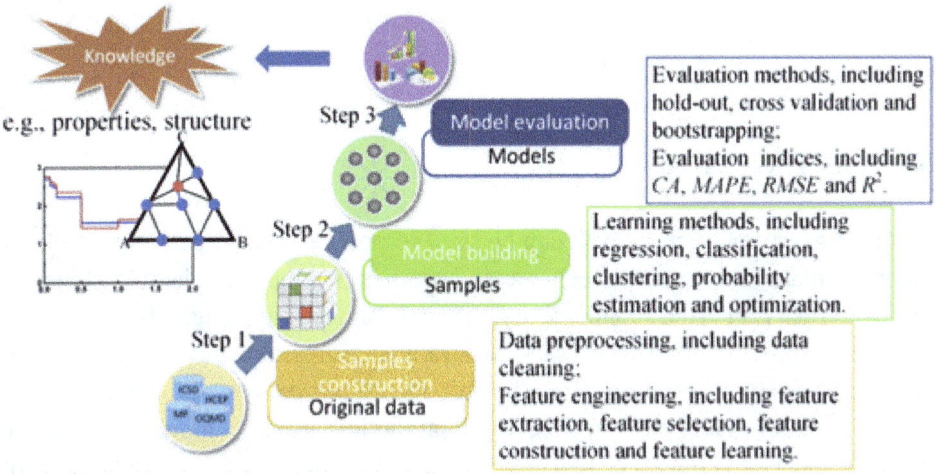

THE APPLICATION OF ML IN MATERIALS SCIENCE

The application of ML in materials science is becoming increasingly widespread, encompassing various aspects ranging from material property prediction to microstructural characterization (Mobarak et al., 2023). The specific content of these applications will be elaborated upon below

Figure 4. The application of machine learning in materials science (Liu et al., 2017).

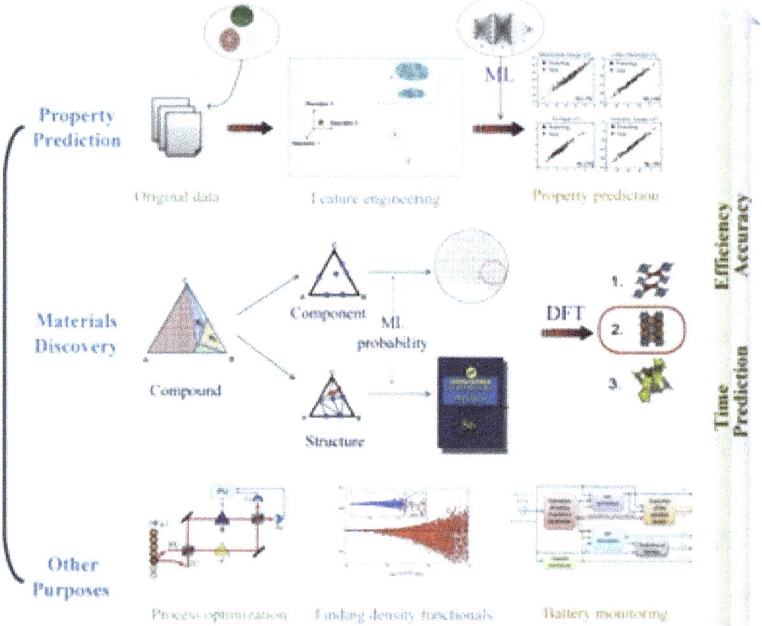

Material Prediction

(1) Property prediction: ML can be used to predict various performance indicators of materials, such as mechanical properties (strength, hardness, toughness, etc.), physical properties (conductivity, thermal conductivity, magnetism, etc.), and chemical properties (corrosion resistance, oxidation resistance, etc.) (Lee, Yoo, & Kang, 2020). By establishing a relationship model between material composition, structure, process, and performance, rapid prediction of material properties can be achieved, providing reference for the design and development of new materials. For example, using machine learning algorithms to predict the strength of ceramic materials, optimizing the composition and preparation processes of ceramic materials based on the predicted results.

(2) Structural prediction: ML can also be used to predict the microstructure and crystal structure of materials (Han et al., 2020). For example, by analyzing diffraction data of materials and using machine learning algorithms to predict the crystal structure of materials, new methods can be provided for structural analysis of materials. In addition, machine learning can also predict the microstructural evolution of materials under different preparation processes, guiding the preparation process of materials.

Classification and Identification

(1) Material Classification: Machine learning possesses the capability to categorize materials into distinct groups based on their composition, structure, properties, and other defining features. For instance, categorizing metal materials, ceramic materials, and polymer materials facilitates the management and research of these materials. Furthermore, machine learning can differentiate among various varieties of the same material, such as different grades of steel and ceramic types.

(2) Material identification: Machine learning can be used to identify the composition and structure of materials. For example, by analyzing the spectral data of materials (such as X-ray diffraction spectra, infrared spectra, Raman spectra, etc.), machine learning algorithms can be used to identify the composition and structure of materials, enabling rapid materials identification. This method has important application value in material testing and quality control.

Auxiliary Microscale Characterization

The performance of materials largely depends on their microstructure, so characterizing the microstructure of materials is an important aspect of materials science research. Machine learning can assist in microscale characterization, improving the efficiency and accuracy of characterization. For example, using machine learning algorithms to analyze scanning electron microscopy (SEM) and transmission electron microscopy (TEM) images, automatically identifying and measuring the microstructural characteristics of materials, such as grain size, phase distribution, defect density, etc. In addition, machine learning can also be used to reconstruct the three-dimensional microstructure of materials, providing more accurate models for simulating and predicting their properties. Ragone et al. (2023) systematically reviewed the application of deep learning in the modeling of material science microscopy imaging, demonstrating the great potential of machine learning in automatically analyzing SEM and TEM images and reconstructing the three-dimensional microstructure of materials.

Phase Transition Research and Phase Diagram Construction

Phase transition is an important physical phenomenon in materials science, such as solid-liquid phase transition, solid phase transition, etc. The phase transition process involves changes in material composition, structure, and properties, which have a significant impact on the preparation and application of materials. Machine learning can be used to study the process of phase transition, analyze the thermodynamic and kinetic laws of phase transition, predict the conditions for phase transition and the resulting product structure. In addition, machine learning can also assist in constructing phase diagrams of materials. By analyzing a large amount of phase transition data, mathematical models of phase diagrams can be established to predict the phase composition and equilibrium state of materials under different compositions and temperatures.

Process Optimization and Service Behavior Evaluation

(1) Process optimization: In the material design and preparation process, the preparation process parameters have a significant impact on the properties of the material. Machine learning can be used to optimize material design and preparation processes, by establishing a relationship model between process parameters and material properties, finding the optimal combination of process parameters, and improving the efficiency and quality of material preparation (Xu et al., 2024). For example, in the process of metal heat treatment, machine learning algorithms are used to optimize parameters such as heat treatment temperature, holding time, and cooling rate to achieve the best mechanical properties of the metal material.

(2) Service behavior evaluation: The service behavior of materials during use, such as fatigue, corrosion, wear, etc., directly affects the service life and reliability of materials. Machine learning can be used to evaluate the service behavior of materials by analyzing monitoring data during use, establishing a service behavior prediction model, predicting the remaining life and failure probability of materials, and providing a basis for material maintenance and replacement (Mao et al., 2021). For example, using machine learning algorithms to analyze the fatigue data of aircraft engine blades, predict the remaining life of the blades, and ensure the safe operation of aircraft engines.

MULTI OBJECTIVE OPTIMIZATION AND REVERSE DESIGN OF MATERIALS

Multi-Objective Optimization

In material design, it is often necessary to consider multiple performance indicators simultaneously, such as strength, toughness, high temperature resistance, cost, etc. These performance indicators may have conflicting relationships and require multi-objective optimization. Machine learning can be combined with optimization algorithms to achieve multi-objective optimization design of materials (X. Liu et al., 2022). For example, using optimization algorithms such as genetic algorithms, combined with machine learning models, to search for optimal solutions in the design space of material composition and structure, and find material design schemes that simultaneously meet multiple performance indicators. Multi objective optimization can improve the comprehensiveness and practicality of material design, meeting the diverse needs of materials in different fields. When the two work together, machine learning models can quickly evaluate the performance of different design schemes, providing decision-making basis for optimizing algorithms; optimization algorithms constantly explore new design solutions and expand the boundaries of material properties. In practical applications, researchers train machine learning models to predict the performance of materials under different composition ratios and processing techniques by constructing databases with thousands of sets of experimental data. Genetic algorithms are guided by these predicted results and iteratively search in a vast design space. After hundreds or even thousands of evolutionary iterations, they ultimately find material design solutions that meet multiple performance indicators simultaneously. This multi-objective optimization strategy significantly enhances the comprehensiveness and practicality of material design. It can not only meet the stringent requirements for high-performance materials in high-end manufacturing fields such as aerospace, automotive manufacturing, and energy equipment, but also customize new materials that meet specific functional requirements for emerging fields such as electronic information and biomedicine, promoting the advancement of materials science and engineering towards intelligence and precision.

Reverse Design

Reverse design refers to the design method of inferring the composition and structure of materials based on the required material properties (Pilania, 2021). Traditional material design predicts material properties based on material composition and structure, while reverse design seeks suitable material composition and structure based on performance requirements. Machine learning has important application value in reverse design. By establishing a reverse mapping model between material properties, composition, and structure, reverse derivation from performance to composition and structure can be achieved. For example, in the aerospace industry, there is a demand for ultra-high strength alloys. Based on machine learning algorithms and combined with performance indicators such as target yield strength and fracture toughness, research teams reverse engineer alloy compositions containing specific proportions of Al, Zn, and Mg elements. At the same time, they optimize multi-stage age-hardening heat treatment process parameters to improve the comprehensive performance of materials by more than 30% compared to traditional designs.

CHALLENGES AND FUTURE PROSPECTS OF AI AND ML IN MATERIAL DESIGN

Despite the immense potential of artificial intelligence and machine learning in material design, their widespread application still faces numerous challenges, while also presenting new opportunities for development (Morgan & Jacobs, 2020). The following will elaborate on these challenges and future prospects.

Challenges Faced

Despite the rapid development of machine learning, it still faces challenges such as data quality, Limitations of Algorithms, demand for computing resources model, and Interdisciplinary talent shortage (Liu et al., 2017).

(1) Data quality issue: Data is the foundation of artificial intelligence and machine learning applications, and the quality of data directly affects the performance and reliability of models. In the field of materials science, data collection faces many challenges, such as incomplete and inaccurate experimental data, inconsistent literature data, and poor standardization of databases (Jain, 2024). In addition, material data has characteristics such as high dimensionality, multimodality, and nonlinearity, which place higher requirements on data preprocessing and feature extraction.

(2) Limitations of Algorithms: Currently, although artificial intelligence and machine learning algorithms have achieved certain results in material design, there are still some limitations (Jain, 2024). For example, traditional machine learning algorithms have limited performance in dealing with high-dimensional nonlinear data, while deep learning algorithms, although having strong nonlinear fitting capabilities, require a large amount of training data and have poor interpretability of the model. In addition, different material problems have different characteristics and require the selection of appropriate algorithms and models, which poses challenges to researchers' ability to select algorithms and construct models.

(3) The demand for computing resources: Artificial intelligence and machine learning algorithms, especially deep learning algorithms, require substantial computing resources for model training and optimization. In material design, processing massive material data and complex model calculations requires high-performance computer hardware and cloud computing platforms, which may lead to insufficient computing resources for some research institutions and enterprises.

(4) Interdisciplinary talent shortage: The application of artificial intelligence and machine learning in material design requires interdisciplinary knowledge and skills, involving multiple disciplines such as materials science, computer science, mathematics, etc. At present, there is a relative shortage of interdisciplinary professionals, and there are few researchers and engineers who understand both materials science

and are familiar with artificial intelligence and machine learning technologies, which to some extent restricts the development of this field.

Future Prospects

(1) Combining data-driven and physical models: In the future, the application of artificial intelligence and machine learning in material design will pay more attention to the combination of data-driven approaches and physical models (Lu et al., 2025). Traditional physical models are based on the fundamental theories and principles of materials science and have strong interpretability, but may have limitations when addressing complex material issues. The data-driven approach can extract patterns from large amounts of data, but lacks physical meaning. Combining the two can fully leverage their respective advantages and establish more accurate and reliable material design models.

(2) Deep integration with experimental technology: With the continuous development of experimental technology, such as high-throughput experimental technology, in-situ observation technology, etc., a large amount of real-time experimental data will be generated. Artificial intelligence and machine learning can be deeply integrated with these experimental techniques to achieve automated control of the experimental process and real-time analysis of data, accelerating the development process of materials. For example, using machine learning algorithms to analyze high-throughput experimental data in real time, guiding subsequent experimental steps, and enabling autonomous material R&D.

(3) Multi scale simulation and design: The performance of materials is determined by the structure and behavior of multiple scales (atomic scale, nanoscale, micrometer scale, macroscopic scale). In the future, artificial intelligence and machine learning will play an important role in multi-scale simulation and design. By establishing multi-scale models, cross scale correlations and designs from the atomic scale to the macroscopic scale can be achieved, improving the accuracy and effectiveness of material design. For example, Tong et al. (2020) successfully achieved accelerated material design and discovery by combining machine learning potential with structural prediction methods, demonstrating the great potential of the integration of data-driven and physical models.

(4) Green and environmentally friendly material design: With the global emphasis on environmental protection and sustainable development, the design of green and environmentally friendly materials will become a future development trend. Artificial intelligence and machine learning can be used to screen and design green and eco-friendly materials with features such as low energy consumption, low pollution, and recyclability, such as biodegradable polymer materials and eco-friendly coating materials, to provide support for achieving sustainable development. Big data

technology has demonstrated significant potential in promoting green applications (Wu et al., 2016), and the further development of artificial intelligence and machine learning will provide stronger technical support for green material design.

(5) Intelligent material design platform: In the future, it is expected to establish an intelligent material design platform that integrates technologies such as artificial intelligence, machine learning, material databases, and computational simulation to achieve full process automation and intelligence in material design (Lu et al., 2025). Researchers and engineers can quickly input performance requirements through this platform, which automatically generates material design schemes, conducts performance prediction and performs optimization, greatly improving the efficiency and innovation capability of material research and development.

CONCLUSION

AI and ML, as two of the most promising technologies today, have demonstrated powerful transformative power in the field of new material design. They not only enhance the efficiency of material design, but also drive the innovative R&D of new materials. By introducing the basic concepts, application principles, and algorithms of artificial intelligence and material design, as well as practical application cases, we can observe the widespread application of AI and ML in material prediction, classification and identification, auxiliary microscale characterization, phase transition research and phase diagram construction, process optimization, and service behavior evaluation. Meanwhile, multi-objective optimization and reverse design of materials have also provided new ideas and methods for material design. However, artificial intelligence and machine learning also face challenges in new material design, such as data quality, algorithm limitations, computing resource requirements, and interdisciplinary talent shortages. In the future, with the continuous advancement of technology and the deepening of interdisciplinary research, artificial intelligence and machine learning will achieve a deeper integration with materials science, ushering in broader development prospects in the combination of data-driven approaches and physical models, deep integration with experimental technology, multi-scale simulation and design, green and eco-friendly material design, and intelligent material design platforms. I believe that in the near future, AI and ML will bring more breakthroughs and innovations to new material design, promote the rapid development of materials science and engineering, and make greater contributions to the progress of human society.

REFERENCES

Bhuvaneswari, V., Priyadharshini, M., Deepa, C., Balaji, D., Rajeshkumar, L., & Ramesh, M. (2021). Deep learning for material synthesis and manufacturing systems: A review. *Materials Today: Proceedings*, *46*, 3263–3269. DOI: 10.1016/j.matpr.2020.11.351

Fang, S. F., Wang, M. P., Qi, W. H., & Zheng, F. (2008). Hybrid genetic algorithms and support vector regression in forecasting atmospheric corrosion of metallic materials. *Computational Materials Science*, *44*(2), 647–655. DOI: 10.1016/j.commatsci.2008.05.010

Grzybowski, A., Pawlikowska–Łagód, K., & Lambert, W. C. (2024). A history of artificial intelligence. *Clinics in Dermatology*, *42*(3), 221–229. DOI: 10.1016/j.clindermatol.2023.12.016 PMID: 38185196

Han, Y., Tang, B., Wang, L., Bao, H., Lu, Y., Guan, C., Zhang, L., Le, M., Liu, Z., & Wu, M. (2020). Machine-learning-driven synthesis of carbon dots with enhanced quantum yields. *ACS Nano*, *14*(11), 14761–14768. DOI: 10.1021/acsnano.0c01899 PMID: 32960048

Jain, A. (2024). Machine learning in materials research: Developments over the last decade and challenges for the future. *Current Opinion in Solid State and Materials Science*, *33*, 101189. DOI: 10.1016/j.cossms.2024.101189

Jordan, M. I., & Mitchell, T. M. (2015). Machine learning: Trends, perspectives, and prospects. *Science*, *349*(6245), 255–260. DOI: 10.1126/science.aaa8415 PMID: 26185243

Juan, Y., Dai, Y., Yang, Y., & Zhang, J. (2021). Accelerating materials discovery using machine learning. *Journal of Materials Science and Technology*, *79*, 178–190. DOI: 10.1016/j.jmst.2020.12.010

Lee, B., Yoo, J., & Kang, K. (2020). Predicting the chemical reactivity of organic materials using a machine-learning approach††Electronic supplementary information (ESI) available. See DOI: . *Chemical Science, 11*(30), 7813-7822. doi:https://doi.org/DOI: 10.1039/D0SC01328E

Liu, X., Xu, P., Zhao, J., Lu, W., Li, M., & Wang, G. (2022). Material machine learning for alloys: Applications, challenges and perspectives. *Journal of Alloys and Compounds*, *921*, 165984. DOI: 10.1016/j.jallcom.2022.165984

Liu, Y., Guo, B., Zou, X., Li, Y., & Shi, S. (2020). Machine learning assisted materials design and discovery for rechargeable batteries. *Energy Storage Materials*, *31*, 434–450. DOI: 10.1016/j.ensm.2020.06.033

Liu, Y., Zhao, T., Ju, W., & Shi, S. (2017). Materials discovery and design using machine learning. *Journal of Materiomics*, *3*(3), 159–177. DOI: 10.1016/j.jmat.2017.08.002

Lu, H., Wang, L., Ma, X., Cheng, J., & Zhou, M. (2025). A survey of graph neural networks and their industrial applications. *Neurocomputing*, *614*, 128761. DOI: 10.1016/j.neucom.2024.128761

Mao, J., Miao, J., Lu, Y., & Tong, Z. (2021). Machine learning of materials design and state prediction for lithium ion batteries. *Chinese Journal of Chemical Engineering*, *37*, 1–11. DOI: 10.1016/j.cjche.2021.04.009

Meng, K., Bai, K., & Sun, S. (2024). Artificial intelligence driven design of cathode materials for sodium-ion batteries using graph deep learning method. *Journal of Energy Storage*, *101*, 113809. DOI: 10.1016/j.est.2024.113809

Mobarak, M. H., Mimona, M. A., Islam, M. A., Hossain, N., Zohura, F. T., Imtiaz, I., & Rimon, M. I. H. (2023). Scope of machine learning in materials research—A review. *Applied Surface Science Advances*, *18*, 100523. DOI: 10.1016/j.apsadv.2023.100523

Morgan, D., & Jacobs, R. (2020). Opportunities and Challenges for Machine Learning in Materials Science. *Annual Review of Materials Research*, *50*(1), 71–103. DOI: 10.1146/annurev-matsci-070218-010015

Pilania, G. (2021). Machine learning in materials science: From explainable predictions to autonomous design. *Computational Materials Science*, *193*, 110360. DOI: 10.1016/j.commatsci.2021.110360

Ragone, M., Shahabazian-Yassar, R., Mashayek, F., & Yurkiv, V. (2023). Deep learning modeling in microscopy imaging: A review of materials science applications. *Progress in Materials Science*, *138*, 101165. DOI: 10.1016/j.pmatsci.2023.101165

Sarker, I. H. (2021). Deep learning: A comprehensive overview on techniques, taxonomy, applications and research directions. *SN Computer Science*, *2*(6), 420. DOI: 10.1007/s42979-021-00815-1 PMID: 34426802

Sulley, G. A., Raush, J., Montemore, M. M., & Hamm, J. (2024). Accelerating high-entropy alloy discovery: Efficient exploration via active learning. *Scripta Materialia*, *249*, 116180. DOI: 10.1016/j.scriptamat.2024.116180

Tong, Q., Gao, P., Liu, H., Xie, Y., Lv, J., Wang, Y., & Zhao, J. (2020). Combining Machine Learning Potential and Structure Prediction for Accelerated Materials Design and Discovery. *The Journal of Physical Chemistry Letters*, *11*(20), 8710–8720. DOI: 10.1021/acs.jpclett.0c02357 PMID: 32955889

Waikhom, L., & Patgiri, R. (2023). A survey of graph neural networks in various learning paradigms: Methods, applications, and challenges. *Artificial Intelligence Review*, *56*(7), 6295–6364. DOI: 10.1007/s10462-022-10321-2

Wu, J., Guo, S., Li, J., & Zeng, D. (2016). Big data meet green challenges: Big data toward green applications. *IEEE Systems Journal*, *10*(3), 888–900. DOI: 10.1109/JSYST.2016.2550530

Xu, G., Jiang, M., Li, J., Xuan, X., Li, J., Lu, T., & Pan, L. (2024). Machine learning-accelerated discovery and design of electrode materials and electrolytes for lithium ion batteries. *Energy Storage Materials*, *72*, 103710. DOI: 10.1016/j.ensm.2024.103710

Yan, Y., Hai, F., Wang, B., Cao, W., Li, M., Wang, C., Li, N., & Zhao, D. (2025). Machine learning accelerates high-voltage electrolyte discovery for lithium metal batteries. *Energy Storage Materials*, *79*, 104312. DOI: 10.1016/j.ensm.2025.104312

Zhang, Q., Yang, L. T., Chen, Z., & Li, P. (2018). A survey on deep learning for big data. *Information Fusion*, *42*, 146–157. DOI: 10.1016/j.inffus.2017.10.006

Chapter 7
Progress of Industrial Waste Resource Utilization in China

Jia Song
http://orcid.org/0000-0003-2369-401X
Tongren Univesity, China

Qiankun Zhang
Huanghe Science and Technology University, China

Chenglong Lu
Hubei Polytechnic University, China

Yinfeng Zhang
Hubei Polytechnic University, China

Song Liu
http://orcid.org/0009-0000-7410-2550
Tongren University, China

ABSTRACT

The scale of global industrial waste generation has reached alarming levels. As a byproduct of human industrial activities, industrial waste has long been a critical challenge in balancing economic growth with environmental sustainability. China's transition from a "waste disposal" mindset to a "resource utilization" paradigm has become not only an environmental imperative but also a strategic economic opportunity. Beyond environmental and economic benefits, this transformation aligns with the global trend in the circular economy. This chapter provides a systematic introduction to the classification and characteristics of industrial wastes, the his-

DOI: 10.4018/979-8-3373-4402-7.ch007

torical development of resource utilization in China, typical case studies, policy framework in China, challenges and future prospects. The chapter offers detailed descriptions of various processes, operational parameters, practical outcomes and economic benefits. This chapter is intended for researchers, designers and production specialists engaged in environmental protection and resource recycling.

1. UNIT INTRODUCTION

Industrial waste refers to solid and semi-solid waste materials discarded at certain times and locations during production and construction, which cannot be utilized under specific temporal and spatial condition and pollutes the environment. Here, "semi-solid" denotes viscous mixtures intermediate between solid and liquid states, consisting of solid particles and liquid, such as sludge from wet dust collection, wastewater, and residual sludge from biological treatment processes (Castro et al., 2019), which are also classified as solid waste. Solid waste represents one of the primary forms of pollutants. Preventing environmental pollution from solid waste is a crucial task in China's environmental protection efforts (Li et al., 2025). As a byproduct of human industrial activities, industrial waste has long been a critical challenge in balancing economic growth with environmental sustainability. Globally, the scale of industrial waste generation has reached alarming levels. According to the United Nations Environment Programme (UNEP), over 30 billion tons of industrial solid waste are produced annually-an amount equivalent to filling 3 million standard shipping containers each day. This crisis is exacerbated by rapid urbanization, population growth, and the expansion of heavy industries such as steel, chemicals, and electronics. As the world's largest industrial nation, China accounts for nearly 30% of global industrial waste output, with annual industrial solid waste generation exceeding 4 billion tons. The issue of industrial waste pollution is becoming increasingly prominent and has even become a factor constraining local economic development in China.

The environmental and economic implications of unmanaged industrial waste in China are multifaceted. This has exerted tremendous pressure on the environment, making solid waste pollution an increasingly prominent issue and a constraint on local economic development in some regions. Landfills, once the primary disposal method, now face saturation. The major industrial provinces like Hebei and Shandong report that 60% of their existing landfill sites will reach their capacity by 2025. Meanwhile, improper disposal, such as the dumping of alkaline steel slag into rivers or the open-air storage of heavy metal-laden chemical residues, has led to soil contamination, groundwater pollution, and biodiversity loss. For instance, a 2021 study in Environmental Science & Technology highlighted that the concentrations of cadmium and lead in farmland soils near electrical wastes (e-waste) recycling hubs in Guangdong were 10-15 times higher than the national safety limits, posing severe health risks to local communities (Manikandan et al., 2023).

Nevertheless, it is important to note that waste is often relative in nature. Temporally, a material may be considered waste under existing technological and economic conditions. However, as technology advances and resources become increasingly scarce, materials once deemed waste may transform into valuable resources for the present and future. Spatially or procedurally, a substance may lack utility in one specific process or context but retain potential value in others. Under appropriate conditions, waste from one process can serve as raw material for another. For example, fly ash from thermal power plants can be used to produce construction materials and plastic products.

The Chinese government and all sectors of society have fully recognized the severity of this problem and are taking comprehensive measures. China's transition from a "waste disposal" mindset to a "resource utilization" paradigm has become an environmental imperative. The "Double Carbon" goals (carbon peak by 2030, carbon neutrality by 2060) and the 14th Five-Year Plan for Circular Economy (2021-2025) explicitly frame the resource utilization of industrial waste as a cornerstone of sustainable development. By converting waste into raw materials (for example, transforming steel slag into cement and recovering lithium from spent batteries as shown in Fig. 1), China aims to reduce its dependence on original resources (e.g., limestone for cement, cobalt for batteries) (Jayamuthunagai et al., 2025). A 2022 report by the National Development and Reform Commission (NDRC) estimates that full utilization of China's industrial waste reduced annual CO_2 emissions by 800 million tons, equivalent to the total emissions of Turkey.

Figure 1. Composition of organic materials, metals and ceramics in e-waste

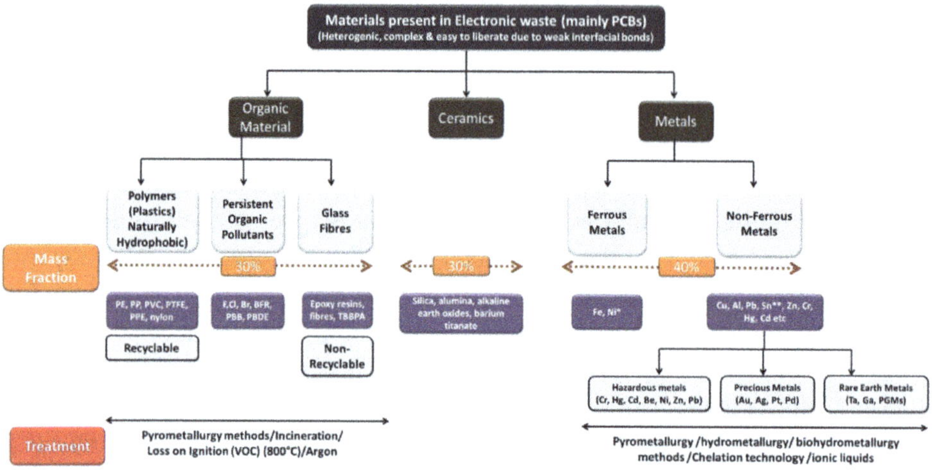

Beyond environmental and economic benefits, this transformation aligns with the global trend in the circular economy. The Circular Economy Action Plan of European Union and Sound Material-Cycle Society Initiative of Japan have set precedents for waste resource utilization models, while the unique market scale and industrial structure in China offer a laboratory for large-scale innovation. For chemistry professionals, understanding this transformation is critical: it bridges fundamental concepts (such as pyrometallurgy and hydrolysis) with practical applications (such as waste salt purification and lithium recovery), while also demanding proficiency in policy analysis and interdisciplinary collaboration.

2. CLASSIFICATION, CHARACTERISTICS AND TREATMENT OF INDUSTRIAL WASTES IN CHINA

2.1 Classification of Solid Wastes

Industrial waste in China, shaped by its vast industrial landscape and extremely large production scale, presents remarkable diversity in both composition and risk profiles. To effectively address its utilization, a systematic classification based on industry origin and chemical properties proves particularly crucial. This framework not only organizes the wide variety of waste types but also illuminates their unique challenges and potential for resource recovery, laying a critical foundation for subsequent discussions on technological solutions.

Solid waste primarily originates from two main sources: waste newly generated during production processes, known as production waste, and solid waste generated during the circulation or after the use/consumption of products entering the market, known as domestic waste. Production waste includes both industrial and agricultural waste. Industrial waste mainly comes from three major sectors: metallurgy, coal mining, and thermal power generation, followed by industries such as chemical engineering, petroleum, and nuclear energy. Agricultural waste mainly stems from farming, forestry, and livestock breeding activities. Domestic waste primarily refers to municipal solid waste (MSW).

Given its wide sources, diverse types, and complex composition, solid waste can be classified from different perspectives. It is commonly categorized based on its source. According to this classification, solid waste can be broadly divided into five major categories: industrial waste, agricultural waste, municipal solid waste, construction waste and hazardous waste. Among these, industrial solid waste can be further subdivided into various types.

(1) Industrial Solid Waste

Industrial solid waste refers to solid materials discarded during production activities in industrial and mining enterprises. It mainly includes mining and mineral processing residues, fuel residues, metallurgical slag, chemical and petrochemical wastes, radioactive wastes and various non-conforming or discarded products from different sectors.

(2) Agricultural Waste

This mainly includes waste generated during the harvesting and processing of crops, such as rice straw, stalks, chaff, as well as manure from poultry and livestock.

(3) Municipal Solid Waste (MSW)

This primarily refers to household waste from urban residents, commercial waste, and waste generated during municipal maintenance and management. Examples include such as food waste, various scraps and residues, discarded utensils, feces, and sewage sludge.

(4) Construction Waste

Construction waste exhibits distinct regional clustering patterns in both production and distribution. In terms of composition, construction waste primarily consists of engineering spoil, accounting for 67.2% of total production at 1.62×10^4 tons, followed by demolition waste (1.30×10^4 tons or 5.4%). Other categories including engineering slurry, construction debris, and renovation waste collectively account for 27.4%, highlighting the dominant role of earth excavation and building demolition activities in urban infrastructure development and urban renewal projects. Geographically, construction waste generation is highly concentrated in economically active regions with rapid urbanization. Provinces such as Guangdong, Zhejiang, Fujian, Sichuan, and Anhui collectively account for 52.9% of national production, underscoring the significant impact of large-scale urban construction on total waste generation. This distribution pattern correlates closely with regional fixed asset investment scales, real estate development intensity, and urban renewal progress, providing critical data support for strategic planning of construction waste resource utilization facilities and policy formulation.

(5) Hazardous Wastes

In terms of waste categories, the primary types include HW11 distillation residues (both stillage and distillate), HW34 waste acid, HW18 incineration disposal residues, HW48 non-ferrous metal mining and smelting wastes, and HWO8 waste mineral oil with mineral oil content. These five waste categories collectively account for 50.2% of total waste generation. Notably, HW34 waste acid and HW48 non-ferrous metal mining and smelting wastes exhibit significant regional concentration (Gao et al., 2023). The top 10 cities producing HW34 waste acid (e.g., Panzhihua City, Binzhou City) account for 58.0% of national total emissions, while the top 10 cities generating HW48 non-ferrous metal waste (e.g., Chifeng City, Qujing City) contribute 55.5%. This highly concentrated distribution pattern underscores the dominant role of heavily polluting industries such as chemical and metallurgical sectors in hazardous waste production. Geographically, hazardous waste generation is predominantly clustered in industrial hubs, with Shandong Province, Jiangsu Province, Inner Mongolia Autonomous Region, Zhejiang Province, and Guangdong Province collectively accounting for 40.2% of national total emissions. This spatial distribution aligns closely with manufacturing clusters along eastern coastal regions and resource-intensive industries in central and western China. The structural characteristics of hazardous waste and their correlation with regional industrial layouts provide critical insights for implementing targeted classification management and recycling strategies.

2.2 Classification of Industrial Wastes in China

Industrial solid waste refers to solid waste products generated during industrial production processes and manufacturing operations, including residues, dust particles, debris, and sludge. Industrial solid waste is primarily categorized into six sectors by industry. The schematic diagram of classification is shown in Fig. 2 (Li et al., 2025). The situation in China demands serious attention and requires effective resolution. In China, only about 40% of the annual industrial waste generated is utilized; the majority of the remainder is still disposed of through simple stockpiling or indiscriminate dumping. As a global industrial powerhouse, China faces a particularly severe crisis: the nation accounts for nearly 30% of global industrial waste output, with annual production exceeding 4 billion tons of industrial solid waste. Within the massive volume, metallurgical, chemical, mining, and electronic wastes emerge as the primary categories, each characterized by distinct characteristics and chemical behaviors.

Figure 2. Classification of industrial wastes in China

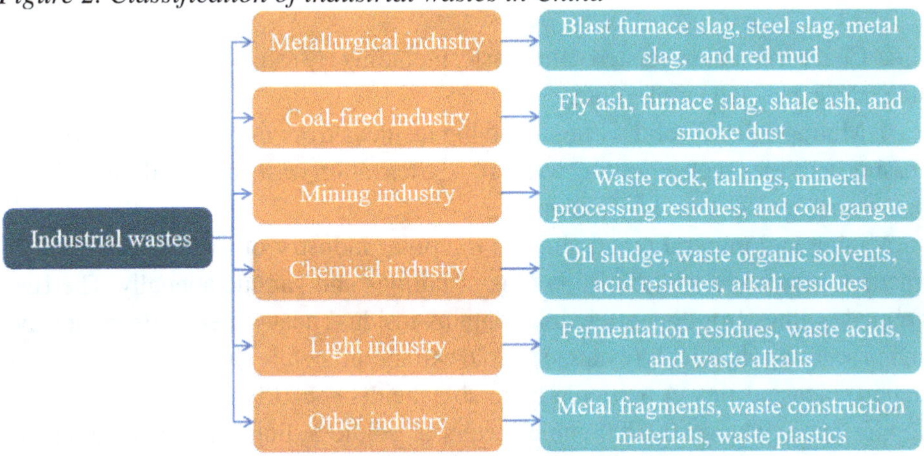

① Metallurgical industrial solid waste, mainly consisting of various residues discharged from ferroalloy smelting industries such as blast furnace slag, steel slag, metal slag, as well as dust particles and red mud.

Metallurgical industries, including steel and non-ferrous metal production, contribute approximately 25% of total industrial solid waste in China. Steel slag, a byproduct of steelmaking processes (Fig. 3) like blast furnaces and basic oxygen furnaces, leads this category with an annual production over 1.2 billion tons, and 70% of which is blast furnace slag. Copper slag, generated from copper smelting, contains residual copper, iron, and silica, while aluminum slag, a byproduct of aluminum smelting, comprises salt slag (rich in NaCl and KCl) and black slag (alumina and aluminum carbide). Although these metallurgical wastes are viewed as byproducts, their mineralogical composition harbors significant potential for resource recovery.

Figure 3. Schematic illustration of steel-making processes and associated reactions

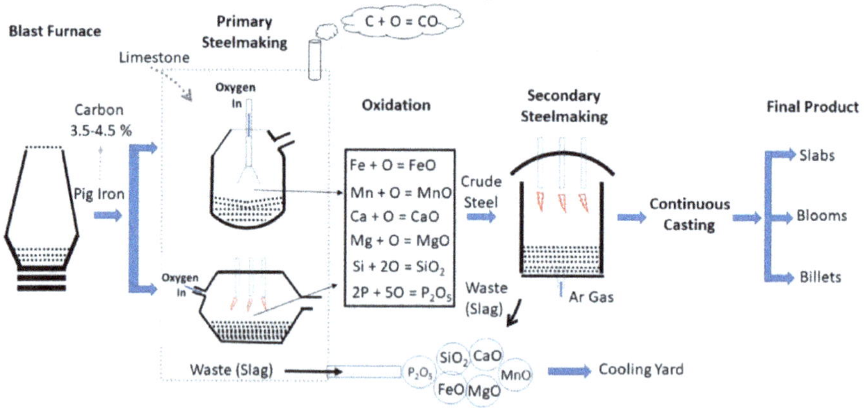

② Coal combustion-related solid waste, including fly ash, coal-fired ash, furnace slag, shale ash, and smoke dust.

③ Mining-related solid waste, involving residues from mining and mineral processing activities such as mining waste rock, excavation waste rock, tailings, mineral processing residues, and coal gangue.

Mining and mineral processing, as the pillar industry of China's resource extraction, generate over 1.5 billion tons of tailings and gangue annually. The fine-grained tailings from beneficiation contains residual heavy metals (lead, arsenic) and processing chemicals (cyanide in gold extraction), while gangue or waste rock is rich in silica and clay. A particularly notable incident occurred in 2021 at a Yunnan copper mine, where 500,000 cubic meters of acidic tailings slurry contaminated the Nanpan River, underscores the geological hazards of these wastes, particularly their potential to generate acid mine drainage through sulfide mineral oxidation (Mahmoud et al., 2017.

④ Chemical industrial solid waste, including oil sludge, tar shale residues, calcium carbide residues, waste organic solvents, acid residues, alkali residues, chromium residues, phosphorus residues, mercury residues, discarded catalysts, and pharmaceutical waste.

The chemical industry, spanning petrochemicals, basic chemicals, and pharmaceuticals, adds another 20% to China's industrial waste burden. Its wastes are marked by high diversity and hazard. For instance, soda ash residue emerges from sodium carbonate production, carrying calcium chloride brine and unreacted minerals. PVC waste includes production residues contaminated with vinyl chloride monomer (VCM) and post-consumer materials, while phenolic resin waste, a byproduct of resin manufacturing, contains high levels of phenol and formaldehyde. These wastes often pose direct environmental threats due to their solubility, toxicity, and reactivity.

Figure 4. Common e-waste: (a) classification, (b) proportion

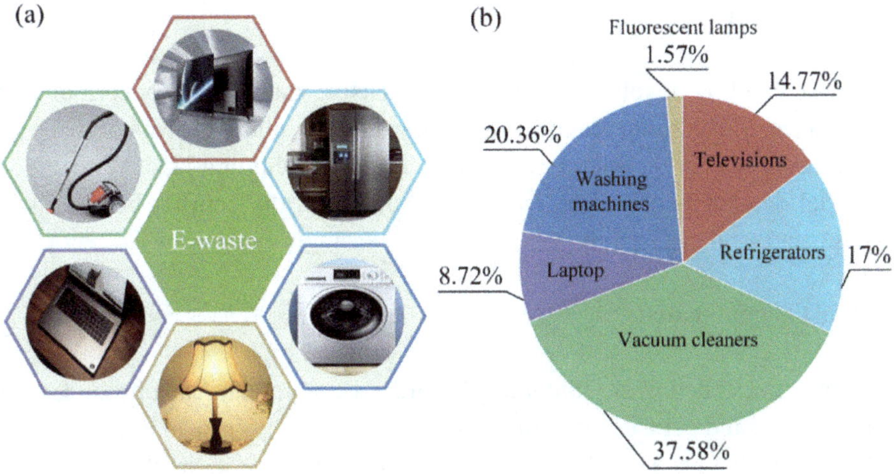

The rapid development of electronics manufacturing and consumption has further swelled the electronic and e-waste in industrial scrap of China (Fig. 4). Printed circuit boards (PCBs), derived from discarded devices, contain copper (up to 20% by weight), gold (0.04-0.2 ppm) (Syed, 2012), and rare earth elements, making them economically valuable. However, these materials also contain toxic components like brominated flame retardants (e.g., tetrabromobisphenol A) and heavy metals (Pd from solder, Cd from coatings), which leach into the environment at dangerous levels. As another major source of e-waste, spent lithium-ion batteries (LIBs) offer high recovery potential for lithium, cobalt, nickel, and manganese but require careful handling to avoid toxic leakage (Muravyov et al., 2025).

⑤ Light industrial solid waste, such as fermentation residues, waste acids, and waste alkalis.

⑥ Other industrial solid waste products that are non-compliant or discarded, including metal fragments, waste glass, waste ceramics, papermaking residues, waste construction materials, waste plastics, and waste chemical fibers.

2.3 Pollution of Industrial Wastes in China

Direct disposal of solid waste without proper treatment not only poses significant hazards to the environment but also results in the wasteful loss of potentially recyclable resources. The detrimental effects of solid waste on the human environment are primarily manifested in the following five aspects:

(1) Land Occupation

The stockpiling of solid waste requires extensive land area, a concern of global significance. Furthermore, landfills and dumps in the suburban areas of many Chinese cities have encroached upon vast stretches of agricultural land. With ongoing industrial development and increasing consumption, the conflict between waste disposal and land availability is becoming increasingly acute.

(2) Soil Contamination

Firstly, stockpiled waste containing hazardous components can introduce pathogens, viruses, and parasites into the soil. Consuming crops grown in contaminated soil or water that has percolated through it poses direct or indirect health risks to humans. Secondly, industrial solid waste can disrupt the ecological balance within soil. Soil hosts a diverse community of microorganisms forming a crucial subsystem of the ecosystem, essential for natural biochemical cycles like carbon and nitrogen fixation. Industrial solid waste, particularly hazardous types, undergoes weathering, leaching by rain and snowmelt, and erosion by surface runoff. This process can generate high temperatures, release toxins, or trigger other reactions that kill soil microbes, disrupt soil ecology, and impair the growth of vegetation and crops. For instance, in the suburbs of a city in southwestern China, the long-term application of municipal waste as fertilizer has led to soil mercury concentrations eight times above background levels, with copper and lead increasing by 87% and 55% respectively, causing significant harm to crop production.

(3) Water Pollution

Solid waste can be transported by rainfall or surface runoff into rivers and lakes, polluting surface water. Simultaneously, leachate from waste percolates through the soil, contaminating groundwater aquifers. In Germany, leachate from waste dumps polluted groundwater in the Rhine region, forcing some water treatment plants to shut down and others to reduce output. In China, the Hanjiawazi landfill in Harbin involved the direct burial of untreated waste, resulting in severe groundwater contamination (Dembele et al., 2022). Levels of manganese, iron, phenol, mercury, total bacterial count, and total coliform bacteria far exceeded safety standards—manganese by more than three times, mercury by over twenty times, total bacteria by 4.3 times, and coliform bacteria by 11 times.

(4) Air Pollution

Certain organic solid wastes decompose under suitable temperature and moisture conditions due to microbial activity, releasing harmful gases. Finely powdered waste materials can become airborne with strong winds, dispersing into the atmosphere. For example, a landfill near Washington, D.C. in US, reportedly smoldered continuously for over two decades. Furthermore, incineration of solid waste without accompanying gas purification systems can also contribute significantly to air pollution.

(5) Impacts on Public Hygiene and Urban Aesthetics

When rates of comprehensive utilization and safe treatment of solid waste are low, large quantities of industrial residue and municipal waste often accumulate in neglected urban areas. This not only detracts from the city's appearance but also severely compromises public hygiene, posing a latent threat to human health.

Global industrial waste generation has surged to an alarming level, with the United Nations Environment Programme estimating over 30 billion tons produced annually. The global situation regarding solid waste discharge is concerning. In industrialized nations, urbanization and rising consumption levels have led to a rapid increase in municipal solid waste generation. Annual growth rates range from 3% to 10% in developed countries and 2% to 3% in developing countries. Of the world's annual industrial solid waste output, the United States accounts for approximately 1/4 and Japan for about 1/7.

The industrial sources of these industrial solid wastes significantly influence their pollution types. Beyond their industrial sources, the chemical properties of these wastes dictate their behavior, risks, and suitability for recovery. For example, steel slag is composed of calcium silicates (C_2S, C_3S), calcium aluminates (C_3A), and free lime (CaO). While its cementitious performance makes it a promising substitute for clinker in cement production, unreacted free lime and magnesia pose expansion risks, causing up to 30% volume increase over time and structural damage in concrete. Additionally, as shown by TCLP and GB 5085.3-2007 tests, trace heavy metals (chromium, vanadium) from ore impurities can leach beyond regulatory limits if improperly stabilized.

Chemical waste salts, such as sodium sulfate and sodium chloride from soda ash and PVC production, exhibit high salinity (total dissolved solids > 30%) and solubility. Direct land-filling leads to salt migration, rendering adjacent soils infertile. This issue was highlighted by a 2022 Environmental Pollution study, where soil electrical conductivity near a soda ash plant in Shandong reached 12 dS/m (normal: <2 dS/m), rendering 80% of farmland unusable. These salts also adsorb heavy metals (lead, cadmium) from upstream processes. As seen in Jiangsu's PVC plants, waste salts contained lead at 150 mg/kg lead and cadmium at 80 mg/kg, exceeding hazardous waste identification limits.

E-waste, a composite of valuable and toxic elements, exemplifies the complexity of industrial waste (Chauhan et al., 2018). For instance, PCBs balance high metal recovery potential with acute environmental risks, while copper and gold drive economic incentives (Ming et al., 2025), brominated flame retardants and heavy metals leach at levels surpassing global regulations (e.g., TBBPA concentrations in leachate reaching 5 mg/L, exceeding the EU's reach limit of 0.1 mg/L).

Though often dismissed as "low-grade," mining tailings present dual challenges. Iron ore tailings, composed of quartz (50-70%), hematite (10-20%), and clay minerals, are geotechnically unstable and prone to acid mine drainage when sulfide minerals (pyrite) react with water and oxygen, producing sulfuric acid and dissolved heavy metals. This risk was tragically demonstrated in Yunnan's 2021 tailings dam failure, which released 500,000 cubic meters of acidic slurry into local waterways.

Classifying industrial waste by industry origin and chemical nature thus serves dual purposes: it provides guidance for development of tailored utilization technologies (e.g., cement substitution for steel slag, hydrometallurgy for e-waste) and establishes a basis for targeted regulatory strategies (e.g., stricter leaching standards for chemical wastes, acid drainage control for tailings). For chemistry professionals, this classification is the first step toward designing efficient, safe, and economically viable resource recovery processes, which bridges fundamental chemical principles with the urgent demands of China's circular economy.

2.4 Industrial Waste Treatment in China

Figure 5. The output of various types of solid waste in China in 2023

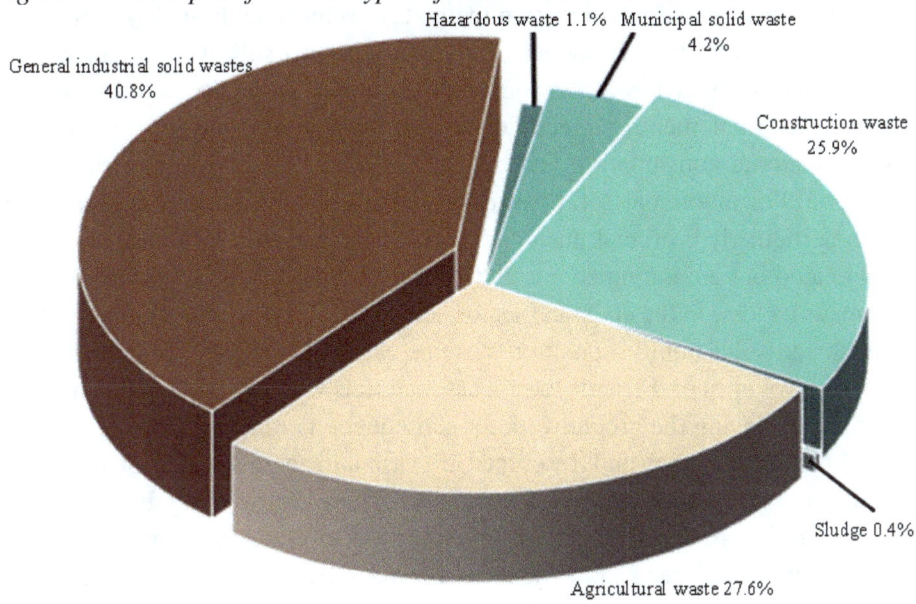

Solid waste management is not only crucial for efficient resource recycling but also a key component in achieving the "dual carbon" goals and building a "zero-waste society". It plays a vital role in driving comprehensive green transformation of economic and social development. According to 2023 data presented in Fig. 5, China's total solid waste generation reached 9.32 billion tons, with general industrial solid waste accounting for the largest share followed by agricultural solid waste. Although significant progress has been made in resource utilization and environmentally safe disposal of various solid wastes, overall performance still requires substantial improvement. Notably, there exist notable disparities in resource utilization rates and disposal efficiency across different waste categories, with underlying constraints exhibiting diverse characteristics depending on waste types.

Historically, solid waste management primarily relied on end-of-pipe disposal. During the initial phase of China's 15th Five-Year Plan, a fundamental paradigm shift occurred in solid waste governance, emphasizing coordinated efforts spanning the entire lifecycle from source generation to final disposal to establish a closed-loop, synergistic management model. Aligning with international best practices and national strategic priorities for high-quality waste recycling and risk mitigation, this transformation signifies that solid waste management has evolved beyond simple emission control. It now prioritizes source reduction through comprehensive analysis of waste's resource-environmental attributes and disposal characteristics. By addressing current generation trends and technological limitations in utilization/ disposal processes, customized strategies are implemented through waste-specific approaches and diversified utilization methods to achieve high-quality recycling.

(1) Source Reduction Strategies. Effective industrial solid waste management requires addressing issues at the source through precise quality-based reduction. This involves implementing green design and clean production practices to control raw material consumption and by-product ratios, thereby minimizing waste generation. For municipal solid waste, optimizing the "three-stream separation" system-particularly is crucial. In construction waste disposal, technologies such as prefabricated building components and BIM modeling can be leveraged to minimize material loss. In the chemical sector, technologies such as centralized storage facilities and solid-liquid separation systems can reduce waste production while decreasing volumes of difficult-to-process materials with high moisture content or heavy metals, laying the groundwork for subsequent treatment processes.

(2) Enhancing Energy and Resource Utilization. In energy utilization processes, organic solid waste such as waste oil and solvents is treated using grate-fluidized bed incineration systems integrated with ultra-low emission technologies to optimize thermal energy recovery efficiency. Biomass waste is converted into syngas and biochar through pyrolysis-gasification technology, which complements waste heat from industrial equipment like cement kilns, forming an integrated "furnace-kiln-boiler-pipeline network" multi-energy system within industrial parks. Energy utilization not only enhances waste value but also reduces disposal costs while providing clean energy support. For resource utilization, material recycling serves as the primary approach. Bulk inorganic solid wastes including fly ash, steel slag, and desulfurization gypsum can be transformed into high-value construction materials like microcrystalline glass, fused mineral wool, and mineralized bricks through mineral phase modification and alkali activation technologies (Chandel et al., 2023). Construction waste processed via intelligent sorting and mineralization carbon fixation technology enables production of recycled aggregates and concrete, achieving 50% resource recovery rates. Fly ash stabilizes heavy metals and chlorides through vitrification or alkali curing processes for safe reuse. Hazardous waste containing rare metals is recovered through acid leaching-extraction-electroplating processes, creating closed-loop resource cycles. Agricultural solid waste is returned to fields via pyrolysis biochar, while livestock manure is processed into organic fertilizers and substrates to replenish soil nutrients through circular utilization. Through coordinated implementation of these technologies, solid waste recycling rates continue to rise, realizing the goal of "waste-to-resource" transformation.

(3) Prioritize enhancing secondary pollutant control and regional coordination capabilities. During solid waste treatment, the secondary pollutants—including

incineration ash, leachate, and hazardous gases emitted during combustion—can be effectively mitigated. These pollutants pose significant threats to environmental health. Recent advances in artificial intelligence and big data technologies have enabled intelligent and dynamic optimization of secondary pollutant control in solid waste management systems. Intelligent control systems leveraging machine learning and data analytics can monitor real-time pollutant generation and emissions during waste processing, continuously refining treatment processes based on data feedback. With ongoing technological breakthroughs, solid waste management is poised to achieve genuine "zero pollution" goals, providing robust technical support for building green, low-carbon societies and "zero-waste communities." Additionally, regional capacity coordination and demonstration park development play crucial roles in solid waste governance. Establishing cross-regional collaborative governance mechanisms facilitates the integration of mature coastal treatment technologies with resource sharing from high-production areas in central and western regions, promoting knowledge and expertise exchange. Such coordinated approaches enhance overall resource utilization efficiency, mitigate regional governance bottlenecks, and elevate comprehensive solid waste management standards.

Through coordinated advancement in three key areas (source reduction, energy resource recovery, and enhanced secondary pollutant control), solid waste governance will achieve full-chain closed-loop management. This transition marks a shift from traditional end-of-pipe treatment models to a new era characterized by end-to-end reduction, circular utilization, and green disposal. Such an integrated system not only enhances resource recovery efficiency but also promotes integrated governance encompassing waste reduction, resource utilization, low-carbon practices, and risk control, providing technical support for achieving carbon peaking and carbon neutrality goals.

3. HISTORICAL DEVELOPMENT OF RESOURCE UTILIZATION IN CHINA

The evolution of industrial waste resource utilization in China unfolds as a dynamic narrative interwoven with environmental imperatives, policy innovation, and technological progress. Over the past four decades, the nation has traversed a remarkable trajectory from a "waste disposal-first" mindset to a "circular economy-driven" paradigm, reflecting both its industrial growth and global leadership in scaling sustainable solutions.

In the 1980s and 1990s, rapid industrialization of steel, chemical, and mining sectors generated explosive waste volumes, yet limited environmental awareness and regulatory gaps left waste largely unmanaged. The primary method of disposal was land-filling: by the 1990s, about 30 million tons of steel slag produced annually by steel mills were directly dumped near the plants, occupying land and risking contamination. Chemical waste salts containing heavy metals and salts were often discharged directly into rivers or left uncovered, causing localized pollution. Moreover, policy efforts were nascent. China's first Solid Waste Pollution Prevention and Control Law(1995) focused on "control" rather than "utilization," while technical capabilities were rudimentary, as crushing and screening were standard practices, yielding a dismal 20% utilization rate by 2000, according to State Environmental Protection Administration data.

The 2010s marked a turning point, driven by mounting environmental crises (smog, soil contamination) and resource scarcity. Policy innovation took center stage. The 2010 12th Five-Year Plan elevated "resource conservation" as a national priority, catalyzing landmark regulations: the 2016 revised Solid Waste Law introduced extended producer responsibility (EPR) for electronics and batteries; the 2015-2020 Resource Utilization Development Planset a 70% utilization target by 2020 and 100 circular economy demonstration bases; and the 2018 "Zero Waste Cities" pilot program (launched in 11 cities like Shenzhen) promoted integrated industrial park management. Technologically, industry-academia collaboration thrived: CAS partnered with Baosteel to develop "slag grinding + mineral activation," enabling 20% slag substitution in cement. Tsinghua University advanced membrane filtration for waste salt purification, cutting heavy metal leaching by 90%. By 2020, utilization rates surged to 62%, with steel slag at 55% and e-waste recovery at 35%.

Since 2020, China has shifted to systemic circularity, aligning with its "Double Carbon" goals (carbon peak 2030, neutrality 2060). Current policies emphasizes cross-sector integration, digitalization, and green chemistry. The 2021-2025 14th Five-Year Plan for Circular Economy called for "full industrial chain" approaches, co-locating waste generators and users (e.g., steel mills paired with cement plants). The amended 2020 Solid Waste Law mandated utilization reports for large enterprises, while the 2022 Green Manufacturing System Standard tied waste utilization to corporate sustainability evaluations. Technological breakthroughs accelerated: the aluminum dross calcination-leaching (via reaction $2Al + 3H_2SO_4 \rightarrow Al_2SO_{4\,3} + 3H_2$) now recovers over 1 million tons of alumina annually; Zhejiang's 2023 project purified PVC waste salts to 98% sodium chloride, meeting food-grade standards; and AI-powered sorting robots in Shanghai achieved 95% accuracy in e-waste identification, with hydrometallurgy recovering 95% of lithium from spent batteries.

Case studies from different provinces also exemplify this progress. Shandong's Zibo National Circular Economy Demonstration Park in Shandong Province co-processes steel slag, coal gangue, and fly ash into building materials, cutting landfill use by 40% and annual CO_2 emissions by 2 million tons. With the support of subsidies and blockchain traceability (Lim et al., 2025), the lithium battery recycling cluster in Jiangsu recovers 90% of critical metals (Li, Co, Ni) from EV batteries (Baniasadi et al., 2019), supplying 30% of China's cobalt demand.

The journey of China offers critical insights: policy drives innovation, scale enables cost-effective solutions, and circularity demands collaboration. As the nation advances toward 2030, the focus of China shifts to deep decarbonization (e.g., CO_2 mineralization from waste calcination) and global technology export (e.g., sharing waste salt purification with Southeast Asia). This evolution positions China not just as a waste manager, but as a pioneer in redefining industrial sustainability.

4. TYPICAL CASE STUDIES

The transformation of China's industrial waste into resources hinges on cutting-edge technologies that marry chemical principles, engineering innovation, and environmental science. Tailored to the unique properties of metallurgical, chemical, and electronic wastes, these advancements are turning once problematic byproducts into valuable inputs, marking a leap from basic disposal to high-value recovery. In essence, China's industrial waste resource utilization technologies have evolved from basic treatment to high-value recovery, propelled by chemical innovation, policy support, and cross-sector collaboration. These breakthroughs not only address environmental challenges but also cement China's leadership in the global circular economy technology.

(1) As metallurgical wastes, the utilization of steel slag and aluminum slag has achieved remarkable progress as 65% and 90%.

Steel slag is primarily composed of calcium silicates (C_2S, C_3S), calcium aluminates (C_3A), and free lime (CaO), and is now a key candidate for cement substitution. Challenges like expansion from unreacted CaO/MgO and heavy metal leaching. Similarly, the variability of steel slag (depending on furnace type, raw materials, and cooling conditions) increases activation difficulty. Consider Baosteel's Steel Slag Comprehensive Utilization Project, a landmark initiative by China's largest steel producer. Baosteel once encountered the significant landfill issue (30 million tons of steel slag annually, enough to fill 12,000 Olympic-standard swimming pools) and the risk of soil/water pollution caused by unreacted free lime (CaO) expansion. By 2015, the company shifted its focus to innovative research, partnering with the Chinese Academy of Sciences (CAS) to develop a "slag grinding + mineral activation" solution. The process involves grinding slag into particles <45 μm to increase surface area, then activating it with gypsum and fly ash (coal combustion byproducts) to trigger hydration. This converts calcium silicates (C_2S, C_3S) in the slag into calcium-silicate-hydrate (C-S-H) gel.Identical to clinker in cement, this binder functions as a proprietary additive that neutralizes free lime, eliminating expansion risks (Sun et al., 2024). The "slag grinding + mineral activation" technology of Baosteel has been adopted in over 30 cement plants, replacing 20-30% of clinker in cement, slashing annual CO_2 emissions by 5 million tons (Wang et al., 2025). In road construction, non-reactive slag fractions reduce pavement thickness by 20% and extend the service life by 5-8 years, with Shanghai alone consuming 1.2 million tons annually. By 2023, 85% of Baosteel's slag was utilized, reducing annual landfill use by 25 million tons and cutting CO_2 emissions by 5 million tons (equivalent to planting 270 million trees). Economically, the project supplies 20% of Shanghai's cement plants' clinker demand, reducing reliance on imported limestone and saving ¥3 billion ($420 million) annually in raw material costs. Technically, the "Baosteel Model" has been replicated in 15 steel plants nationwide, elevating China's steel slag utilization rate from 25% (2010) to 55% (2023).

GEM Co., Ltd.'s "low-temperature plasma leaching" recovers 98% of lithium from low-grade e-waste black mass, cutting energy use by 40%. Tsinghua University's AI-driven process optimization boosts steel slag utilization by 10-15% via real-time composition analysis.

Aluminum slag rich in alumina (Al_2O_3) bound in salt slag (NaCl/KCl) and black slag (Al_4C_3, AlN) undergoes calcination-leaching to recover alumina efficiently (Zhu et al., 2022). The slag is heated to 800-900 °C to volatilize NaCl/KCl, leaving an alumina-rich residue. Treating this with dilute sulfuric acid (60-80 °C) to reactively dissolve alumina via the reaction of $Al_2O_3 + 3H_2SO_4 \rightarrow Al_2SO_3 + 3H_2O$, then purification and precipitation to yield $Al(SO_4)_3$, which is calcined back to alumina (Li et al., 2024). This process recovers 90% of alumina, with China's annual recycled output exceeding 1.5 million tons (He et al., 2021).

(2) In chemical industry, chemical wastes with high salinity and toxicity demands precision technologies to neutralize hazards, and the utilization of wastes can reach 90%.

Soda ash residue (NaCl/$CaCl_2$ brine) and PVC waste salts (Na_2SO_4, NaCl) undergo heavy metal removal through nanofiltration (NF) and ion exchange (IE) (Ghasemi & Azizi, 2018). NF membranes (1-10 nm pores) reject Pb^{2+}, Cd^{2+}, and divalent ions while allowing Na^+ and Cl^- passage, achieving 98% salt purity in Zhejiang's 2023 project, reducing heavy metal content from 150 mg/kg to <1 mg/kg. IE resins with sulfonic acid groups selectively capture Pb^{2+} with 99.9% efficiency, enabling purified salts for detergent production. Phenolic resin waste (30-50% phenol/formaldehyde) is converted to fuel oil via thermal decomposition at 400-600 °C with ZnO catalysts, yielding 70-80% fuel (42 MJ/kg calorific value) and reducing landfill volume by 90%. Tianjin plant processes 50,000 tons annually, producing 35,000 tons of fuel for industrial boilers. Current hydrometallurgical methods recover ~90% of copper but struggle with rare earth elements, losing 30-40% due to their low concentrations and complex bonding with organic matrices.

E-wastes like PCBs and LIBs require precision to recover critical metals (Fig. 6) while minimizing toxicity. PCBs (20% Cu, 0.04-0.2 ppm Au, toxic BFRs) (Rodrigues et al., 2015) are dissolved with acid leaching (2 M $H_2SO_4 + 0.5\% H_2O_2$) to recover 95% Cu, or bioleaching with acidithiobacillus ferrooxidans for 80-90% Cu recovery at low energy cost (Garg et al., 2019; Yaashikaa et al., 2022). Spent LIBs (15-20% Li, 10-15% Co and 5-10% Ni) undergo direct regeneration (Zhang et al., 2024). The process involves heating the black mass (crushed cathodes) to 300-500°C to decompose binders, followed by Li_2CO_3 leaching to recover Li, cutting costs by 30% (Shen & Zhang, 2025; Shokri et al., 2025). Solvent extraction (SX) with P507 extractant selectively binds with Li^+ at pH 2-3, stripping with HCl to produce high-purity LiCl (>95% recovery), supporting China's dominance in LIB recycling (70% global cobalt supply for EVs).

Figure 6. General scheme for different processes for LIB recycling.

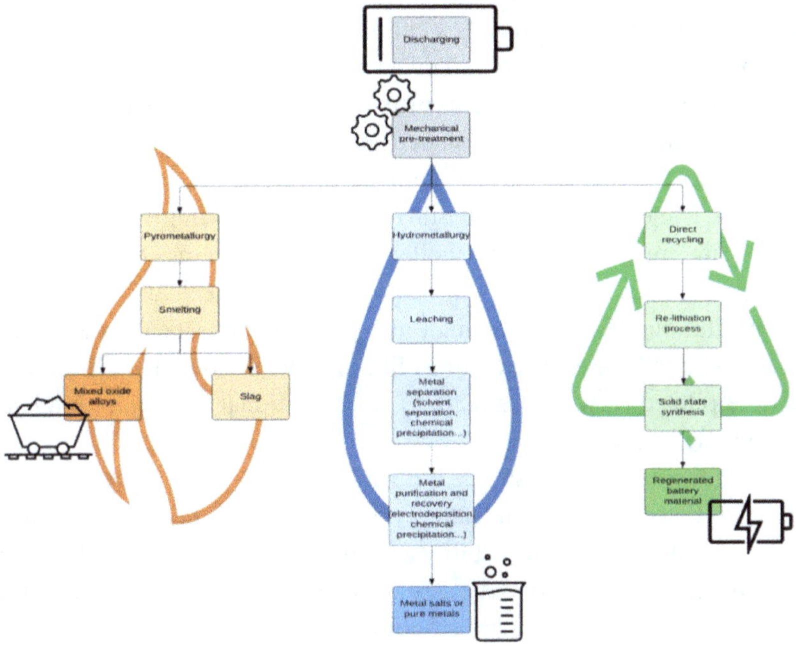

A standout example is Jiangsu's Spent LIB Recycling Cluster, born from the urgent need to address China's growing battery waste crisis. With over 5 million EVs sold annually, the total amount of waste LIB in China is expected to reach 12 million tons by 2030. As a home to 30% of China's EV battery manufacturers, Jiangsu pioneered a cluster-based model integrating hydrometallurgy, direct regeneration, and digital traceability.

The cluster's approach begins with disassembling spent LIBs and leaching cathodes (rich in lithium, cobalt, nickel) using a mixed acid solution ($H_2SO_4 + H_2O_2$). Extractants like P507 (2-ethylhexyl phosphonic acid) selectively bind these metals, followed by purification through solvent extraction (Shi et al., 2025). For direct regeneration technology, black mass (crushed cathodes) undergoes heat treatment (300-500°C) to decompose polymer binders (PVDF), then by leaching with lithium carbonate (Li_2CO_3) to recover lithium as lithium carbonate (Li_2CO_3) (Shokri et al., 2025). This process skips energy-intensive refining steps, cutting costs by 30% (Li D et al., 2025). A blockchain platform, co-developed with Alibaba, tracks each battery from production to recycling, ensuring compliance with China's Extended Producer Responsibility (EPR) laws and guaranteeing 95% material recovery rates (Lim et al., 2025). By 2023, the cluster processed 200,000 tons of spent LIBs annually, recovering 90% of lithium, cobalt, and nickel-supplying 30% of China's cobalt demand and 20% of its lithium needs (Cheng et al., 2024). Environmentally, recycling 1 ton of LIBs saves 7 tons of water, 1.5 tons of coal, and reduces CO_2 emissions by 4 tons compared to mining virgin materials. Economically, the cluster generates ¥5 billion ($700 million) in annual revenue and creates 10,000 jobs.

These technological advancements thrive on synergies between policy, industry, and academia and underscore three shared success factors. First, policy-technology synergy: Baosteel's progress was accelerated by the 2020 Solid Waste Law's utilization targets, while Jiangsu's cluster thrived under the 14th Five-Year Plan's "full industrial chain" mandate. Second, collaborative ecosystems: Both parties established cooperative networks involving enterprises (Baosteel, EV battery makers), research institutions (CAS, Tsinghua), and governments (local EPR regulators). Third, scalability advantages: Solutions like slag activation and LIB direct regeneration were piloted (Igwilo et al., 2025) before national scaling, minimizing risk and maximizing impact.

China's industrial waste resource utilization journey comes alive through real-world projects that exemplify innovation, policy alignment, and technical ingenuity. All these cases illustrate China's ability to transform industrial waste from an environmental liability into an economic asset, one that aligns with global sustainability goals while setting new benchmarks for circular economy innovation.

5. POLICY FRAMEWORK: FROM REGULATION TO INCENTIVIZATION

The advancement of industrial waste resource utilization in China is propelled by a dynamic interplay of policy frameworks, standardized regulations, and innovative industrial practices, three interconnected pillars that have propelled the nation from waste disposal to circular economy implementation. This synergy is rooted in regulatory evolution, technical standardization, and cross-sector collaboration, effectively addressing domestic environmental challenges while setting global benchmarks for sustainable industrial development.

China's policy approach to industrial waste has undergone a profound transformation, shifting from reactive regulation to proactive incentivization in alignment with its "Double Carbon" goals (carbon peak by 2030, neutrality by 2060). The core of this shift is the 2020 revised Law of the People's Republic of China on the Prevention and Control of Environmental Pollution by Solid Waste, which redefined waste management as a resource utilization challenge rather than a pollution control issue. As a key provision of the law, Extended Producer Responsibility (EPR), mandates electronics, battery, and packaging manufacturers to assume financial and managerial responsibilities for the recycling and reuse of their products after disposal. For example, battery firms must now ensure 90% recovery of critical metals (lithium, cobalt, nickel) from discarded LIBs by 2025 (Li J et al., 2025). The law also enforces utilization targets, requiring provincial governments to establish annual industrial solid waste utilization rates with a national goal of achieving 70% by 2025 (up from 62% in 2020), which greatly encourages the local innovation. Zhejiang Province introduced subsidies for enterprises exceeding 80% utilization, driving a 25% surge in recycling rates within two years. Additionally, the law intensifies stricter penalties for illegal disposal, raising fines from 5-20 times the value of waste to 1-10 times, drastically reducing unregulated dumping. As seen in Guangdong, the 2022 crackdown cut illegal e-waste disposal by 40%. Complementing national laws, sector-specific policies like the 14th Five-Year Plan for Circular Economy (2021-2025) actively promote "full industrial chain integration," encouraging co-location of waste generators and users (e.g., steel mills paired with cement plants to use slag as raw material). The expanded "Zero Waste Cities" Pilot Program, currently covering 113 cities, further incentivizes industrial parks to submit utilization plans, rewarding those achieving 90% utilization with tax breaks. The revised "Circular Economy Promotion Law" in 2024 introduces "green product certification" for recycled materials, mandating government projects prioritize them. By mid-2025, recycled steel and aluminum comprised 22% of state infrastructure, up from 12% in 2020. A national carbon trading pilot now rewards waste projects with credits (e.g., Baosteel earns ¥150 /$21 per ton of CO_2 saved), improving ROI (the return on investment) by 25%.

Standards are critical for ensuring technical feasibility and environmental safety. Since 2015, China has developed over 50 national and industry standards spanning waste characterization, processing techniques, and emission controls. For example, the General Rules for Storage of Solid Waste (GB/T 18599-2020) specifies moisture control, leakage prevention, and heavy metal monitoring for storage facilities, reducing improper storage incidents by 60%. The Technical Guidelines for Industrial Waste Resource Utilization (GB/T 39198-2020) provides evaluation metrics like "resource recovery ratio," guiding enterprises (such as a Jiangsu chemical plant) to adjust waste salt purification lines to achieve a 95% recovery threshold, doubling byproduct value. The Emission Limits for Air Pollutants from Waste Incineration (GB/T 40048-2021) tightened dioxin and heavy metal limits, promoting the adoption of cleaner technologies like plasma gasification for hazardous waste. These standards also align with global norms, and China's Recycling of Waste Lithium-Ion Batteries (GB/T 36587-2018) influenced the EU's 2023 Battery Regulation, enhancing its role in global battery supply chains.

Policy and standards have catalyzed innovative industrial practices that blend technology, management, and collaboration. The "Yangtze River Economic Belt" initiative links eastern tech hubs (Shanghai's cement clusters) with western resource bases (Sichuan's lithium mines), where eastern firms provide tech and funding, while western partners supply waste and labor. In Guizhou, a Baosteel-local miner joint venture converted 5 million tons of slag into construction materials, creating 8,000 jobs and cutting clinker imports by 30%. Similar models are spreading across the Yellow River Basin.

Taking the "Enterprise + Park" integration model serving as a example. Zibo National Circular Economy Demonstration Zone in Shandong Province co-locates steel mills, coal mines, and building material factories, with slag used in cement, coal gangue in bricks, and fly ash in road aggregates. By 2023, the park cut landfill use by 40%, reduced annual CO_2 emissions by 2 million tons, and generated ¥1.2 billion in membrane purification to recover 100,000 tons of caustic soda annually. This model has been replicated in 15 petrochemical parks, boosting national alkali recovery from 35% to 65%. As technology advances, digital tools are revolutionizing management. AI-powered sorting robots in Shanghai's e-waste hubs (developed by Tsinghua University) achieve 95% accuracy in identifying PCBs (Magoda et al., 2024) and LIBs (Sethurajan & Gaydardzhiev, 2021), cutting labor costs by 70%. Blockchain, via Alibaba's "Green Chain" platform, tracks LIB origins, processing, and end-use. Statistics show 80% of electric vehicle manufacturers now require traceable blockchain materials to ensure the Environmental, Social, and Governance standards and enhance consumer trust.

These synergies have yielded measurable outcomes. Between 2015 and 2023, China's industrial solid waste utilization rate rose from 50% to 68%, reducing annual landfill demand by 1.2 billion tons. CO_2 emissions from waste processing dropped by 300 million tons, which is equivalent to removing 130 million cars. Economically, the waste utilization industry grew to ¥2.3 trillion ($310 billion) in 2023, creating 3 million jobs. Despite these encouraging achievements, challenges persist. Addressing challenges such as uneven policy enforcement across regions, high upfront costs of advanced technologies, and the need for stricter standards for emerging wastes (e.g., perovskite solar panels), requires deeper digital integration (Wen et al., 2025), expanded PPP models, and global knowledge sharing, areas where China continues to lead.

China's approach to industrial waste resource utilization is a blueprint for sustainable industrial development. By aligning policy, standards, and practice, China not only tackles its environmental challenges but also contributes global solutions. This fully demonstrating that the combination of chemical technology and visionary governance can transform waste into wealth.

6. CHALLENGES AND FUTURE PERSPECTIVES

China's remarkable progress in industrial waste resource utilization, not only demonstrates a testament to policy innovation and technological prowess, but also unfolds against a complex landscape of persistent challenges and emerging opportunities. As China accelerates its transition to a circular economy, a deeper understanding of these development bottlenecks and their breakthrough paths reveals both the resilience of China's model and the global lessons it offers for sustainable industrial development.

(1) Despite breakthroughs in steel slag utilization and LIB recycling (Lee & Pandey, 2012), scaling these solutions remains fraught with technical, economic, and systemic barriers. The technical challenges in complex waste streams and costs for smaller plants are particularly prominent.

(2) Market incentives often misalign with environmental goals. Though China's Extended Producer Responsibility (EPR) laws mandate producers to fund recycling, recycled materials frequently lag virgin resources in price and quality. Recycled cathode materials for LIBs sell for 15-20% less than newly blends, but impurities in waste streams create quality inconsistencies that deter buyers. Only 35% of EV manufacturers prioritize recycled materials till 2023. This "chicken-and-egg" dilemma, where low demand stifles investment, while underdeveloped capacity keeps prices high, is now dragging down the industry's growth.

(3) Regional disparities further challenge national targets. Eastern provinces like Jiangsu and Guangdong, with dense industrial clusters and advanced research institutions, lead in utilization, their steel slag rates exceed 65%, and LIB recycling efficiency reaches 95%. In contrast, western provinces (e.g., Gansu, Qinghai) lag significantly due to insufficient technical clusters and skilled labor (less than 10% of technicians hold advanced degrees) (Kaksonen et al., 2020).

Against these challenges, China is pioneering solutions that transform obstacles into catalysts. Technological leaps in green chemistry and digitalization are redefining resource recovery (Kovačević et al., 2024). Additionally, policy evolution is coordinating a large number of incentives. Besides, regional collaboration is narrowing development gaps.

China's practices holds global relevance. As the UN projects a 70% waste increase by 2050, China's experience offers actionable insights. Looking ahead, China aims to hit 70% utilization by 2030. This requires doubling R&D investment (¥200 billion/$28 billion by 2027) for breakthroughs like bioleaching (Moazzam et al., 2021; Singh et al., 2025) and CO_2 mineralization, expanding international cooperation (via the Belt and Road Circular Economy Network), and fostering public participation (e.g., "Green Electronics" labels to drive demand).

Essentially, China's transformation of industrial waste is not only an environmental milestone, but also a testament to innovation, policy foresight, and collective action. As the world grapples with its own waste crises, China's experience demonstrates that "waste" will become a relic of the past, bringing hope for the future of a circular economy.

7. PROFESSIONAL ENGLISH SKILLS DEVELOPMENT

Mastering professional English is far more than just a linguistic exercise; it is a critical tool for engaging with global research., fostering cross-disciplinary collaboration, and advancing China's industrial waste resource utilization agenda. This skill set bridges technical expertise with clear, impactful communication, enabling professionals to contribute to both local solutions and global sustainability efforts. In industrial waste resource utilization, key terms span waste classification, processing technologies, and circular economy concepts. Proficiency requires more than memorization; it demands contextual understanding.

Academic reading and critical analysis are equally vital. Technical literature on industrial waste often features complex data, methodologies, and conclusions. Developing this skill involves deconstructing research papers into abstracts, methodologies, results, and discussions. For instance, when reading a study on lithium recovery from spent batteries, one must identify the objective, methodology, results, and implication. Practicing data analysis on China's waste utilization rate charts (2010-2023) to identify trends, outliers, and policy-driven acceleration can significantly enhance the ability to extract actionable insights.

Technical writing serves as a bridge between research and practice. Clear and concise writing is essential for drafting research papers, policy briefs, and industry reports. For example, an abstract on "Steel Slag in Cement Production" might structure background, methods, results, and conclusions as follows: "Steel slag, a smelting byproduct, faces storage challenges. This study evaluates replacing 20% of clinker with activated slag. Using grinding and gypsum activation, we achieved 42 MPa compressive strength, matching conventional cement, while reducing annual CO_2 emissions by 5 million tons." Additionally, technical descriptions should translate policy brief findings into actionable recommendations, such as proposing training programs and mobile recycling units to address rural e-waste risks.

Translation and cross-cultural communication are increasingly important as China's industrial waste expertise globalizes. First, it is necessary to accurately translate policy texts. The technical manuals demand familiarity with domain-specific terminology in target languages, whether French or German. Second, cross-cultural collaboration should emphasize data-driven arguments as well as highlight policy impacts. When collaborating with European researchers on China's battery recycling model, the communication approach also needs to be adjusted accordingly.

Ultimately, proficient professional English is a catalyst for innovation. It connects China's industrial waste expertise to global knowledge networks, enabling professionals to lead international research, shape sustainability initiatives, and drive circular economy practices. By mastering vocabulary, critical reading, technical writing, and cross-cultural communication, learners transform technical breakthroughs into tangible global impact, ensuring China's leadership in industrial waste resource utilization resonates far beyond its borders.

REFERENCES

Baniasadi, M., Vakilchap, F., Bahaloo-Horeh, N., Mousavi, S. M., & Farnaud, S. (2019). Advances in bioleaching as a sustainable method for metal recovery from e-waste: A review. *Journal of Industrial and Engineering Chemistry*, *76*, 75–90. DOI: 10.1016/j.jiec.2019.03.047

Castro, C., Urbieta, M. S., Plaza Cazón, J., & Donati, E. R. (2019). Metal biorecovery and bioremediation: Whether or not thermophilic are better than mesophilic microorganisms. *Bioresource Technology*, *279*, 317–326. DOI: 10.1016/j.biortech.2019.02.028 PMID: 30755320

Chandel, S. S., Singh, P. K., Katiyar, P. K., & Randhawa, N. S. (2023). A review on environmental concerns and technological innovations for the valorization of steel industry slag. *Mining, Metallurgy & Exploration*, *40*(6), 2059–2086. DOI: 10.1007/s42461-023-00886-z

Chauhan, G., Jadhao, P. R., Pant, K. K., & Nigam, K. D. P. (2018). Novel technologies and conventional processes for recovery of metals from waste electrical and electronic equipment: Challenges & opportunities - A review. *Journal of Environmental Chemical Engineering*, *6*(1), 1288–1304. DOI: 10.1016/j.jece.2018.01.032

Cheng, Q., Wang, Z., Wang, Y., Li, J.-T., & Fu, H. (2024). Recent advances in preferentially selective Li recovery from spent lithium-ion batteries: A review. *Journal of Environmental Chemical Engineering*, *12*(3), 112903. DOI: 10.1016/j.jece.2024.112903

Dembele, S., Akcil, A., & Panda, S. (2022). Technological trends, emerging applications and metallurgical strategies in antimony recovery from stibnite. *Minerals Engineering*, *175*, 107304. DOI: 10.1016/j.mineng.2021.107304

Gao, W., Zhou, W., Lu, X., Liu, X., Su, H., Li, C., & Wang, H. (2023). Comprehensive utilization of steel slag: A review. *Powder Technology*, *422*, 118449. DOI: 10.1016/j.powtec.2023.118449

Garg, H., Nagar, N., Ellamparuthy, G., Angadi, S. I., & Gahan, C. S. (2019). Bench scale microbial catalysed leaching of mobile phone PCBs with an increasing pulp density. *Heliyon*, *5*(12), e02883. DOI: 10.1016/j.heliyon.2019.e02883 PMID: 31872109

Ghasemi, S. M., & Azizi, A. (2018). Alkaline leaching of lead and zinc by sodium hydroxide: Kinetics modeling. *Journal of Materials Research and Technology*, *7*(2), 118–125. DOI: 10.1016/j.jmrt.2017.03.005

He, L., Shi, L., Huang, Q., Hayat, W., Shang, Z., Ma, T., Wang, M., Yao, W., Huang, H., & Chen, R. (2021). Extraction of alumina from aluminum dross by a non-hazardous alkaline sintering process: Dissolution kinetics of alumina and silica from calcined materials. *The Science of the Total Environment*, *777*, 146123. DOI: 10.1016/j.scitotenv.2021.146123 PMID: 33684763

Igwilo, C. N., Chime, T. O., Ude, C. N., Ikechukwu, O. M., Anezichukwu, A. F., Ikechukwu, N. A., Omotioma, M., Chukwudi, E. B., Chisom, M. K., Ifeanyichukwu, O. T., & Scholastica, C. O. (2025). Examining the efficiency of microbe-assisted metal extraction: A review of bio-hydrometallurgical leaching techniques. *Hybrid Advances*, *9*, 100407. DOI: 10.1016/j.hybadv.2025.100407

Jayamuthunagai, J., Flora, R. M. N., Senthilkumar, K., & Bharathiraja, B. (2025). Review on sustainable strategies for lithium recovery from spent lithium-ion batteries. *Future Batteries*, *7*, 100105. DOI: 10.1016/j.fub.2025.100105

Kaksonen, A. H., Deng, X., Bohu, T., Zea, L., Khaleque, H. N., Gumulya, Y., Boxall, N. J., Morris, C., & Cheng, K. Y. (2020). Prospective directions for biohydrometallurgy. *Hydrometallurgy*, *195*, 105376. DOI: 10.1016/j.hydromet.2020.105376

Kovačević, A., Tolazzi, M., Sanadar, M., & Melchior, A. (2024). Hydrometallurgical recovery of metals from spent lithium-ion batteries with ionic liquids and deep eutectic solvents. *Journal of Environmental Chemical Engineering*, *12*(4), 113248. DOI: 10.1016/j.jece.2024.113248

Lee, J. C., & Pandey, B. D. (2012). Bio-processing of solid wastes and secondary resources for metal extraction - A review. *Waste Management*, *32*(1), 3–18. DOI: 10.1016/j.wasman.2011.08.010 PMID: 21925857

Li, C., Liu, H., Xu, Y., Li, W., Long, J., Meng, L., Huang Q. (2025) Curent sitation and development trend of solid waste geeration, ulization and disposal in China. *Environmental Sanitation Engineering*, 33(6): 28-36, 50.

Li, C., Liu, W., Jiao, F., Dong, L., Liu, S., & Qin, W. (2024). Intensifying recovery of metallic aluminum in secondary aluminum dross by alkali roasting-water leaching. *Transactions of Nonferrous Metals Society of China*, *34*(12), 4063–4074. DOI: 10.1016/S1003-6326(24)66658-6

Li, D., Zhi, X., Zhang, L., Xie, F., & Wu, M. (2025). Direct regeneration of spent lithium-ion batteries: Advancing from powder to cell. *Journal of Energy Chemistry*, *111*, 767–789. DOI: 10.1016/j.jechem.2025.08.013

Li, J., Nazari, S., Ma, X., Wei, N., & He, Y. (2025). A critical review on the advances in thermal reduction technology for recycling spent lithium-ion batteries. *Next Materials*, *8*, 100641. DOI: 10.1016/j.nxmate.2025.100641

Lim, J., Jang, Y., Lee, J., Lee, C., Jbari, O., Kwon, K., & Chung, E. (2025). Hydrometallurgical process of spent lithium-ion battery recycling Part. 2 Recovery of valuable metals from the cathode active material leachates: Review and cost analysis. *Hydrometallurgy*, *236*, 106516. DOI: 10.1016/j.hydromet.2025.106516

Magoda, K., Nomngongo, P.N., & Mekuto, L. (2024) Two-step bio-dissolution of metals from printed circuit boards using acidophilic iron - and sulfur-oxidizing mesophiles. *Recycling*, 9(1).

Mahmoud, A., Cézac, P., Hoadley, A. F. A., Contamine, F., & D'Hugues, P. (2017). A review of sulfide minerals microbially assisted leaching in stirred tank reactors. *International Biodeterioration & Biodegradation*, *119*, 118–146. DOI: 10.1016/j.ibiod.2016.09.015

Manikandan, S., Inbakandan, D., Valli Nachiyar, C., & Namasivayam, S. K. R. (2023). Towards sustainable metal recovery from e-waste: A mini review. *Sustainable Chemistry for the Environment*, *2*, 100001. DOI: 10.1016/j.scenv.2023.100001

Ming, W., Hu, S., Zhang, X., Ma, J., He, W., Xu, Y., Wang, H., Xu, Y., Xing, W., Chen, Z., & Li, X. (2025). Review: Biorecovery of precious metals from waste printed circuit boards. *Minerals Engineering*, *228*, 109306. DOI: 10.1016/j.mineng.2025.109306

Moazzam, P., Boroumand, Y., Rabiei, P., Baghbaderani, S. S., Mokarian, P., Mohagheghian, F., Mohammed, L. J., & Razmjou, A. (2021). Lithium bioleaching: An emerging approach for the recovery of Li from spent lithium ion batteries. *Chemosphere*, *277*, 130196. DOI: 10.1016/j.chemosphere.2021.130196 PMID: 33784558

Muravyov, M., Panyushkina, A., & Fomchenko, N. (2025). Two-step ferric and biological leaching as an environmentally friendly method for improving nickel sulfide concentrate processing. *Process Safety and Environmental Protection*, *195*, 105850. DOI: 10.1016/j.psep.2025.106850

Rodrigues, M. L. M., Leão, V. A., Gomes, O., Lambert, F., Bastin, D., & Gaydardzhiev, S. (2015). Copper extraction from coarsely ground printed circuit boards using moderate thermophilic bacteria in a rotating-drum reactor. *Waste Management*, *41*, 148–158. DOI: 10.1016/j.wasman.2015.04.001 PMID: 25899037

Sethurajan, M., & Gaydardzhiev, S. (2021). Bioprocessing of spent lithium ion batteries for critical metals recovery - A review, Resources. *Resources, Conservation and Recycling, 165*, 105225. DOI: 10.1016/j.resconrec.2020.105225

Shen, Y., & Zhang, S. (2025). Recent advances in sustainable reductive roasting for spent lithium-ion batteries recycling. *Energy Storage Materials, 79*, 104358. DOI: 10.1016/j.ensm.2025.104358

Shi, H., Luo, Y., Deng, Y., Dai, J., Zhang, J., & Ou, L. (2025). Recovery of valuable metals from spent lithium-ion batteries based on a green and efficient leaching system of dimethyl-β-propionic acid thiophene (DMPT). *Hydrometallurgy, 237*, 106551. DOI: 10.1016/j.hydromet.2025.106551

Shokri, Z., Mousavi, S. M., & Omidkhah, M. R. (2025). Advancing sustainable lithium recovery from spent laptop lithium-ion battery cathodes through high pulp density bioleaching followed by Li_2CO_3 precipitation: Process intensification and environmental insights. *Journal of Environmental Chemical Engineering, 13*(3), 117014. DOI: 10.1016/j.jece.2025.117014

Singh, S., Santhiya, D., & Sharma, J. G. (2025). An overview of bioprocess based recovery and recycling techniques for battery waste management: Transitioning from environmental hazard to sustainability. *Minerals Engineering, 233*, 109598. DOI: 10.1016/j.mineng.2025.109598

Sun, J., Hou, S., Guo, Y., Zhou, J., Cui, Y., & Bao, J. (2024). Preparation of steel slag-based cementitious materials compounded with GBFS and phosphogypsum: Mix design, hydration, and microstructure characteristic. *Journal of Thermal Analysis and Calorimetry, 149*(23), 13785–13798.

Wang, M., Xu, J., Li, T., Liu, H., & Qu, L. (2025). Comprehensive Performance Evaluation of Steel Slag-Slag-Desulfurization Gypsum Ternary Solid Waste Cementitious Material Based on Principal Component Analysis. *Buildings (Basel, Switzerland), 15*(4), 645. DOI: 10.3390/buildings15040645

Wen, T., Bu, Y., & Zhang, H. (2025). Research advances and futureprospects of artificial intelligence technology in the integratedmanagement of solid waste. *Journal of Ecology and Rural Environment, 41*(9), 1122–1133.

Yaashikaa, P. R., Priyanka, B., Kumar, P. S., Karishma, S., Jeevanantham, S., & Indraganti, S. (2022). A review on recent advancements in recovery of valuable and toxic metals from e-waste using bioleaching approach. *Chemosphere, 287*, 132230. DOI: 10.1016/j.chemosphere.2021.132230 PMID: 34826922

Zhang, S., Gu, K., Lu, B., Han, J., & Zhou, J. (2024). Hydrometallurgical Processes on Recycling of Spent Lithium-lon Battery Cathode: Advances and Applications in Sustainable Technologies. *Wuli Huaxue Xuebao*, *40*(10), 2309028. DOI: 10.3866/PKU.WHXB202309028

Zhu, X., Yang, J., Yang, Y., Huang, Q., & Liu, T. (2022). Pyrometallurgical process and multipollutant co-conversion for secondary aluminum dross: A review. *Journal of Materials Research and Technology*, *21*, 1196–1211. DOI: 10.1016/j.jmrt.2022.09.089

Chapter 8
Globally Sustainable Manganese Metal Production and Use

Zinan Xie
http://orcid.org/0000-0001-9234-3980
Tongren University, China

Jiancheng Shu
Southwest University of Scaience and Technology, China

Huang Zhou
North Sichuan Medical College, China

Yong Yang
South Manganese Group Limited, China

ABSTRACT

This chapter provides a comprehensive analysis of globally sustainable manganese production and use, addressing key challenges and strategies across the entire value chain. Manganese, critical for steel, batteries, and alloys, faces issues of resource depletion, environmental degradation, and supply chain vulnerabilities due to geographically concentrated reserves. Sustainable solutions emphasize responsible mining (precision extraction, water management, renewable energy, community engagement), low-carbon smelting (hydrogen reduction, plasma technology, biomass reductants), energy-saving and consumption-reduction electrodeposition process of manganese, sustainability of green manganese materials in end-use applications and circular economy practices. Policies and industry collaboration are highlighted as essential drivers for promoting transparency, ethical sourcing, and green innovation.

DOI: 10.4018/979-8-3373-4402-7.ch008

The integration of life-cycle assessment, eco-design, and advanced recycling technologies is urged to achieve a resilient, low-carbon manganese industry supporting global sustainability goals.

1. GLOBAL MANGANESE RESOURCES AND SUSTAINABLE SUPPLY CHAIN

Manganese (Mn) is a vital raw material extensively utilized in various industrial applications, such as steel production, battery manufacturing, and the development of specialty alloys (Wang et al., 2024; Yang et al., 2025). Its importance in strengthening steel and acting as a key component in battery cathodes renders it indispensable for infrastructure development and the clean-energy transition (Wu et al., 2024; Wu et al., 2025). However, the increasing global demand for manganese has raised concerns regarding resource depletion, environmental degradation, and supply chain resilience (He et al., 2021a; He et al., 2021b). Achieving a sustainable manganese supply chain necessitates a comprehensive approach that involves responsible resource extraction, ethical sourcing, low-carbon technologies, and circular economy practices, as shown in Figure 1.

Figure 1. Structure of manganese industry chain

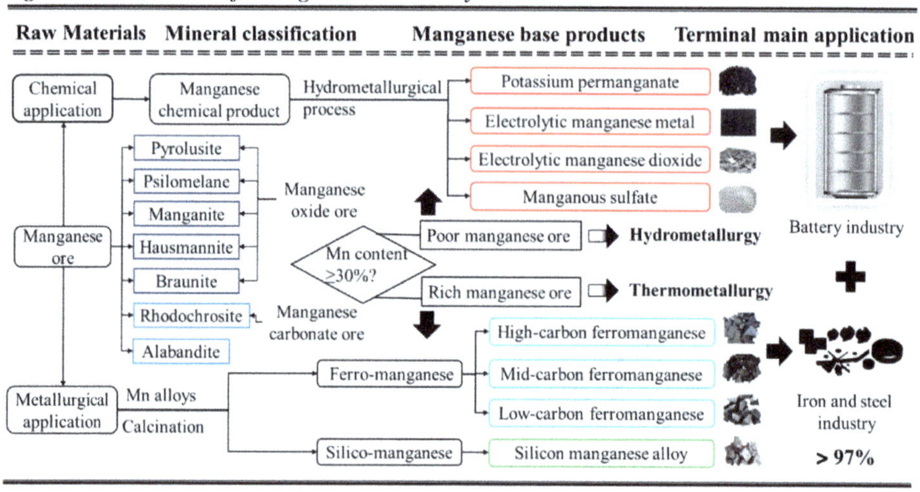

Manganese resources are distributed unevenly across the globe, with a significant concentration in a few key regions. South Africa, which is home to the vast Kalahari Basin, possesses the world's largest manganese reserves (Tao et al., 2024). It is followed by other major producers such as Australia, Gabon, and Brazil. These countries account for the majority of the global manganese supply, as shown in Figure 2. China has established the world's most complete and largest manganese processing and consumption industrial chain. Despite its limited domestic ore production, China is the world's largest producer and consumer of refined manganese. Its industrial chain has developed in depth, extending from electrolytic manganese metal and manganese alloys to high-value-added products such as positive electrode materials for new energy batteries. Domestic clusters have been formed in multiple regions, including Tongren in Guizhou, Chongzuo in Guangxi, and Changyang in Hubei. It is moving towards an upgraded full industrial chain model of 'mining-deep processing-new materials-comprehensive utilization.

Sustainable mining practices play a critical role in minimizing the environmental and social impacts associated with manganese extraction. Conventional mining operations, particularly open-pit mining, often result in habitat destruction, soil degradation, and water pollution. Additionally, the energy-intensive nature of manganese extraction and refining processes contributes significantly to carbon emission. To mitigate these impacts, mining companies can adopt precision mining technologies, remote sensing systems, and automated equipment to reduce material wastage and optimize resource utilization (Farjana et al., 2019). Furthermore, implementing advanced water management systems, employing tailings management technologies, and promoting land reclamation and reforestation projects can minimize ecological damage. Beyond environmental considerations, ensuring fair labor practices, engaging with local communities, and investing in social development initiatives are essential components of responsible mining. Compliance with international standards such as the Initiative for Responsible Mining Assurance (IRMA) and the Extractive Industries Transparency Initiative (EITI) further reinforces transparency and ethical governance in the manganese supply chain.

Building a sustainable manganese supply chain also requires the integration of circular economy principles. Instead of relying solely on primary manganese ore extraction, emphasis must be placed on the recovery of manganese from secondary sources, including steel scrap, spent batteries, and industrial waste. Advanced recycling technologies, such as hydrometallurgical leaching and pyrometallurgical smelting, enable the extraction of manganese from end-of-life products. Establishing robust collection and sorting systems for used batteries and scrap materials is crucial to enhancing the supply of secondary manganese. Additionally, implementing closed-loop recycling systems within manufacturing processes can further reduce manganese material losses and reliance on virgin manganese ore.

Promoting low-carbon technologies within the manganese value chain is another essential aspect of sustainable development. The conventional smelting of manganese ore is a carbon-intensive process, primarily using coal-based energy sources. Transitioning to hydrogen-based manganese ore reduction technologies (Wang et al.,

2024b), plasma smelting (Islam et al., 2024; Statritama et al., 2024), and electric arc furnaces (EAFs) (Yang et al., 2025; Marzoughi et al., 2025) powered by renewable energy can significantly reduce greenhouse gas emissions. Moreover, carbon capture, utilization and storage (CCUS) (Shu et al., 2025; Sun et al., 2025) technology can effectively capture carbon dioxide emitted during manganese ore leaching, thereby mitigating its environmental impact. Furthermore, developing technologies such as manganese ore crushing, enhanced leaching, intelligent tank operation, passivation equipment, large-scale treatment of electrolytic manganese solid waste, and low-carbon preparation of high-value-added manganese-based products can further drive the high-quality development of the manganese industry chain.

Ensuring transparency and traceability within the manganese supply chain is equally crucial for promoting responsible sourcing and preventing the proliferation of conflict manganese minerals. Digital solutions, such as blockchain technology and supply chain traceability platforms, offer end-to-end visibility into the origins, transportation, and processing of manganese materials. Certification programs such as the OECD Due Diligence Guidance for Responsible Supply Chains and the Responsible Minerals Initiative (RMI) further guarantee that companies comply with ethical sourcing standards. Implementing these traceability mechanisms not only strengthens accountability but also cultivates consumer confidence in sustainable manganese products.

Looking ahead, the sustainable development of the manganese supply chain demands coordinated efforts from governments, industries, and research institutions. Policymakers can play a pivotal role by strictly enforcing environmental regulations, incentivizing green technologies in the manganese industry, and supporting R&D in material recycling and process optimization. Simultaneously, fostering international cooperation between manganese-producing and consuming nations is essential to promote knowledge sharing, fair trade, and supply chain resilience. To achieve this, a holistic strategy must be adopted that integrates responsible mining practices, low-carbon technologies, and circular economy principles. By enhancing transparency, promoting ethical procurement, and investing in clean production and recycling solutions, we can significantly reduce environmental footprints and ensure long-term resource security. Collaborative efforts across the entire manganese industry chain will not only accelerate the global transition to a low-carbon economy but also enable sustainable management of manganese resources for future generations.

Figure 2. Proportion of global manganese reserves distribution

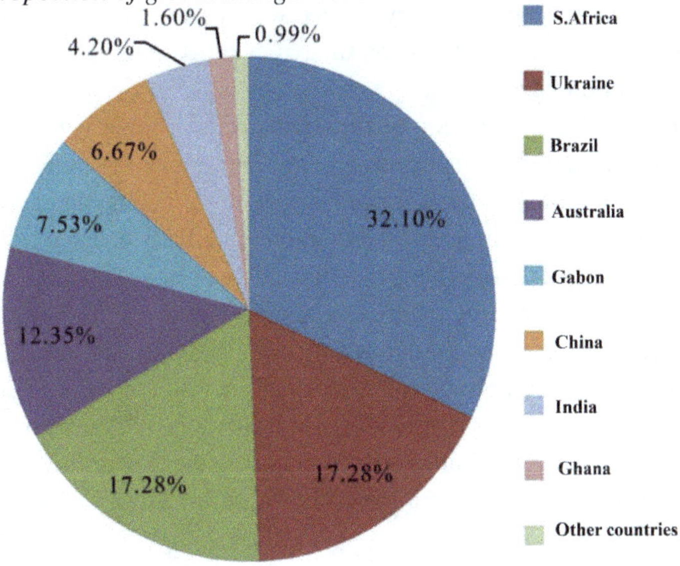

2. SUSTAINABLE MANGANESE MINING AND GREEN MINING

As a key component in steel production, battery manufacturing, and various industrial applications, manganese is essential for infrastructure development and the clean energy transition. However, traditional mining practices often lead to significant environmental degradation, including deforestation, soil erosion, water pollution, and carbon emission (Wu et al., 2024; Wang et al., 2024). To address these challenges, sustainable manganese mining is crucial for reducing the environmental and social impacts of resource extraction while ensuring the long-term availability of this critical material. By integrating innovative solutions, reducing resource consumption, and promoting ethical operations, the manganese industry can significantly minimize its environmental footprint while supporting local communities.

Conventional manganese mining, particularly via open-pit and underground methods, has resulted in severe ecological consequences. Large-scale excavation disrupts natural ecosystems, causing habitat destruction and biodiversity loss. Additionally, acid mine drainage (AMD) from exposed ores contaminates water sources, introducing toxic metals into the surrounding ecosystems (Nahum et al., 2024). Mining operations also consume vast quantities of water and energy, as shown in Figure 3, contributing to water scarcity and greenhouse gas emissions (Wu et al., 2024). Particulate emissions and dust pollution pose further health risks to workers and nearby populations. These severe challenges highlight the urgency of transforming the manganese industry chain towards green and responsible models. Only through sustainable practices can the balance between resource demand and ecological protection be achieved.

The adoption of green mining technologies offers viable solutions to reduce the environmental impact of manganese extraction. Precision mining, enabled by remote sensing technologies such as drones, satellite imaging, and LiDAR, allows for targeted extraction, minimizing land disruption and waste generation. Additionally, eco-friendly extraction methods like bioleaching (Han et al., 2024; Mohanty et al., 2018) and hydrometallurgy (Moro et al., 2023; He et al., 2025) can efficiently recover manganese from low-grade ores or mining waste, significantly reducing energy consumption and pollution emissions. Renewable energy integration also plays a significant role in sustainable mining operations. Transitioning to solar, wind, or hydropower for mining activities and using electric or hydrogen-powered equipment can substantially cut carbon emission and reliance on fossil fuels (Soulamidis et al., 2024).

In terms of water resource management, closed-loop recycling systems and zero-emission treatment technologies drastically reduce water usage and prevent water pollution (Shu et al., 2017; Li et al., 2024; Olii et al., 2024; Han & Wu, 2019). Additionally, innovative waste management strategies—such as using heap leaching and slurry backfilling instead of traditional tailings dams—not only enhance safety but also convert tailings and by-products from electrolytic manganese into construction materials, promoting resource recycling and building a circular mining economy (Wang et al., 2024b; Wang et al., 2025; Luo et al., 2024).

Sustainable mining also encompasses broad social, economic dimensions, and regulatory frameworks. Manganese mining enterprises must fully implement environmental, social, and governance (ESG) principles. By establishing transparent information disclosure mechanisms, deepening stakeholder collaboration, and following the process of free, prior, and informed consent, they can protect community rights and obtain social operating permits. Investing in community development projects such as education, healthcare, and infrastructure can promote long-term regional prosperity. Simultaneously, adhering to international labor standards, providing safety training, and promoting workplace equality are key to achieving fair labor practices.

To promote and enforce sustainability, regulatory frameworks and industry standards play a crucial role. The Initiative for Responsible Mining Assurance (IRMA) and the Extractive Industries Transparency Initiative (EITI) provide independent third-party certification and promote transparency in mining operations. Additionally, the OECD Due Diligence Guidance for Responsible Supply Chains offers a comprehensive framework for companies to identify, prevent, and mitigate the environmental and social impacts of their activities. Governments are also introducing stricter regulations on emissions, waste management, and land restoration to ensure responsible resource extraction. Compliance with these standards not only enhances corporate accountability but also strengthens consumer confidence in sustainably sourced manganese.

Looking ahead, accelerating the adoption of sustainable manganese mining requires deep collaboration among governments, industries, and research institutions. Policymakers can incentivize green technology investments and low-carbon innovations through fiscal support, R&D funding, and carbon pricing mechanisms. Enterprises should actively adopt traceability tools like blockchain to enhance supply chain transparency, while developing circular economy models that recycle manganese resources from manganese-based solid waste, spent batteries, and industrial residues—reducing reliance on primary ores. By integrating precision mining, clean energy, closed-loop water management, and ethical operations, the manganese industry can significantly reduce its environmental footprint while strengthening resource security and social inclusion. Global cooperation and continuous innovation will ultimately position manganese mining as a key pillar in the transition to a low-carbon circular economy, safeguarding natural resources and ecological balance for future generations.

Figure 3. Sources of electrolytic manganese wastewater

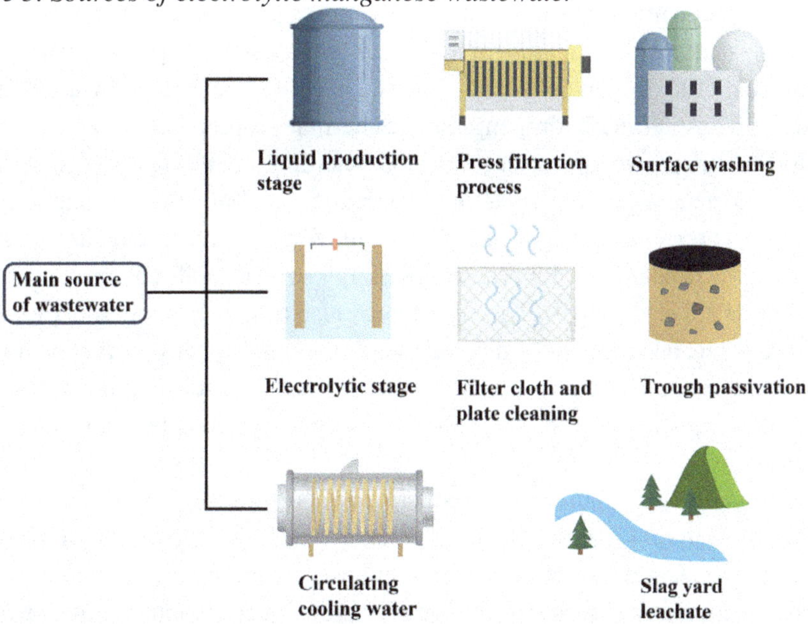

3. LOW CARBON MANGANESE SMELTING TECHNOLOGY

Decarbonization in low-carbon manganese smelting processes is a critical link to reducing the carbon footprint of the metallurgical industry and achieving global climate goals. The production of manganese metal products from pyrolusite ore mainly relies on high-temperature kilns, using coke or coal as reducing agents, leading to significant greenhouse gas emissions; similarly, the smelting of manganese carbonate ore to produce manganese-based products also results in large amounts of greenhouse gas emissions. To address this challenge, the manganese industry must transition to low-carbon technologies, which is of decisive significance for reducing direct emissions, improving energy efficiency, and ensuring the long-term sustainability of the manganese supply chain.

Currently, multiple low-carbon technological pathways have demonstrated transformative potential. One of the most promising alternatives to conventional smelting is hydrogen-based reduction (Marzoughi et al., 2025). This process involves using green hydrogen (H_2) as a reductant replaces carbon-based reductants to reduce manganese ore (MnO_2) into metallic manganese, producing water vapor (H_2O) instead of carbon dioxide (CO_2) (Wang et al., 2024c; Ernst et al., 2024; Ma et al., 2024). Hydrogen reduction not only eliminates direct carbon emission but also offers higher reaction efficiency compared to carbon-based reduction. Furthermore, when hydrogen is produced using renewable energy sources, such as solar or wind power, the overall environmental impact is significantly reduced. However, challenges remain in the large-scale implementation of hydrogen-based smelting, primarily because of the high cost of hydrogen production and the need for specialized furnace designs capable of withstanding hydrogen's unique thermal properties.

Another innovative approach is plasma smelting technology, which utilizes a plasma arc to generate extremely high temperatures, enabling the efficient reduction of manganese ore (Zulhan et al., 2024). Plasma furnaces operate using electricity, making them compatible with renewable energy sources and reducing reliance on fossil fuels (Pani et al., 2019). This technology also offers advantages in processing low-grade manganese ores and industrial waste materials, thereby promoting resource circularity. Despite its benefits, plasma smelting requires substantial energy input, making the availability of low-carbon electricity a prerequisite for its sustainable application.

Additionally, using biomass as reductants is considered a near-carbon-neutral transitional solution (Zhou et al., 2022; Zhang et al., 2012). Biomass-derived materials, such as charcoal or bio-coke, can replace conventional coke in the smelting process. As biomass absorbs CO_2 during its growth, its combustion is considered carbon-neutral. Utilizing agricultural and forestry residues as biomass reductants not only reduces net emissions but also adds value to waste materials. However, the variability in biomass quality and its competition with other applications, such as bioenergy production, remain challenges to its widespread use.

For existing manganese smelting facilities, implementing carbon capture, utilization, and storage (CCUS) offers a transitional pathway toward decarbonization. CCUS technology captures CO_2 emissions from smelting operations, preventing them from being released into the atmosphere. The captured carbon can then be stored underground or utilized in various industrial applications, such as concrete production or synthetic fuel generation. While CCUS can achieve high levels of emission reduction, it requires substantial infrastructure investment and may face challenges related to the long-term storage of captured carbon.

Moreover, the integration of renewable energy sources into manganese smelting operations is a critical aspect of achieving low-carbon production. Electric arc furnaces (EAFs) and plasma furnaces powered by renewable electricity, such as solar, wind, or hydropower, can further reduce indirect carbon emission. The integration of renewable energy not only enhances the sustainability of the smelting process but also alleviates the volatility of fossil fuel prices, providing economic resilience to the manganese industry.

However, large-scale application of these technologies still faces systemic challenges. Hydrogen-based smelting encounters difficulties in high-cost green hydrogen and specialized equipment retrofitting; plasma technology requires extremely high energy consumption and heavy reliance on affordable, stable green electricity supply; biomass reductants are constrained by the stability and sustainability of raw material supply. Although CCUS technology demonstrates strong applicability, it faces bottlenecks including high energy consumption for carbon capture, challenging site selection for storage, and massive investment requirements. Therefore, overcoming these obstacles cannot rely on a single technology but requires systematic integration and collaborative innovation through "renewable energy + novel manganese metallurgical processes" to seek solutions.

Accelerating the transformation of low-carbon manganese smelting technology urgently requires the combined efforts of governments, industry, and research institutions. Policymakers can create a favorable market and innovation environment through means such as carbon pricing, green subsidies, and R&D funding. Enterprises need to actively explore pilot projects to drive technological iteration and cost reduction, and use blockchain and other methods to enhance traceability and transparency of green products. In summary, by firmly deploying low-carbon manganese smelting technology, building a green energy system, and strengthening policy guidance, the manganese industry will not only significantly reduce its environmental footprint but also provide critical support for the global industrial low-carbon future.

4. MANGANESE ELECTRODEPOSITION PROCESS AND ENERGY-SAVING AND CONSUMPTION- REDUCTION REGULATION

As an important metallurgical and chemical raw material, electrolytic manganese metal (EMM) has made outstanding and highly significant contributions to China's national economy. In 2022, China produced 1.5 million tons of electrolytic manganese metal, accounting for 98% of the global total output (Wu et al., 2024; Tao et al., 2024). EMM production primarily employs the electrolysis process of manganese sulfate solution, with the electrolysis process being the core. The direct - current energy consumption per ton of manganese ranges from 5,800 to 6,200 kW·h, and the direct-current energy consumption for electrolysis accounts for nearly 90% of the total energy consumption in EMM production. Owing to issues such as low current efficiency, high cell voltage, and environmental pollution, manganese electrolytic deposition encounters severe challenges in both technology and commercialization. On the other hand, the metal manganese electrolytic deposition process exhibits characteristics such as openness, fluidity, non-equilibrium, and non-linearity Moreover, the variable valence characteristics of the manganese element endow the electrode reaction process with rich non-linear kinetic mechanisms, which are likely to induce non-linear behaviors such as electrochemical oscillation and metal fractal, resulting in reduced current efficiency and increased electrolysis energy consumption (Fan et al., 2013; Fan et al., 2015; Bai et al., 2017), as shown in Figure 4.

The primary raw materials for EMM production are manganese carbonate ore, pyrolusite, and high-manganese slag from blast furnace smelting. After processes such as powder making, leaching, and impurity removal, a qualified manganese sulfate solution is obtained (Duan et al., 2010; Lu et al., 2014). This qualified solution is electrolyzed in an electrolytic cell, and metal manganese is deposited on the cathode. After subsequent treatments such as passivation, cleaning, drying, and peeling, high-purity metal manganese products are produced. The electrolytic deposition process is the core of EMM production, accounting for nearly 90% of the total process energy consumption. Improving current efficiency is the most concerning issue in production practice. The level of current efficiency is directly related to the quality of manganese output, environmental protection, and economic benefits. The main factors affecting the current efficiency of manganese electrolytic deposition include electrolyte composition (such as Mn^{2+} concentration, $(NH_4)_2SO_4$ concentration, electrolyte pH value, etc.), operating conditions (including electrolyte temperature, current density, electrolysis time, etc.), and electrode materials. After decades of exploration and development, the electrolyte composition and operating conditions used for industrial manganese electrolytic deposition are relatively fixed at present. Meanwhile, research on related areas such as electrolyte additives, electrode materials, and electrolysis methods has remained active and continues to yield new innovations.

In the process of manganese electrolytic deposition, compounds of Group VIA elements (S, Se, and Te) are commonly used as they have similar beneficial effects on manganese electrolytic deposition (Zmuidznaviciene et al., 2023; Song et al., 2024). The role of selenium compounds is derived from their reduced or decomposed products. The main functions are as follows: (a) increasing the hydrogen evolution overpotential; (b) promoting the crystallization of stable α-Mn; (c) improving the leveling effect of the electrolyte; (d) enhancing current efficiency. (Jiao et al., 2016; Reyes-Morales et al., 2021). Researchers typically focus on the influence of electrolytic additives on the morphology and crystal type of cathodically deposited manganese (Weng et al., 2024), while paying less attention to their impact on anode reactions and the control of anode by-products during electrolysis. Since the cathode and anode reactions in the electrolysis process are within the same cyclic system, studying the influence of electrolytic additives on anode reactions may help improve the overall current efficiency (Xie et al., 2019).

Electrodes are another important component of the electrolysis process. Compared to cathode materials, industrial applications place more stringent requirements on anode materials. Anode materials suitable for industrial production must meet three basic performance requirements: good electrical conductivity, efficient electro- catalytic activity, and stability for continuous operation. Currently, the three main types of anode materials widely used in the two major industrial fields of electrochemistry and electrometallurgy are lead-based, titanium-based, and composite anodes. Lead-based anode materials have been widely used in industrial electrolytic production for over a century because of their good electrical conductivity, oxygen evolution activity, strong corrosion resistance in acidic systems, low cost, and the characteristics of easy processing and forming (Mohammadi & Alfantazi, 2015; Mohammadi & Alfantazi, 2016;). However, lead-based anodes also encounter issues such as high electrolytic cell voltage, lead pollution, and by-product anode mud during use. To reduce energy consumption and pollutant emissions, modifying lead-based anodes (Tang et al., 2023) or developing new alternative anodes has become a current research focus for anode materials used in electrolysis (Ye et al., 2021; Zhong et al., 2023; Lyu et al., 2024).

In traditional electrolytic deposition processes, direct current (DC) power supplies are commonly used to drive non-spontaneous chemical reactions. To simplify operations, constant current or constant potential electrolysis modes have dominated research in this field. In practical manganese electrolytic deposition processes, industry predominantly employs constant current electrolysis. Typically, a thick diffusion layer forms at the cathode-solution interface, reducing the concentration of metal ions at the cathode surface and causing concentration polarization, which slows down the rate of metal deposition. On the other hand, during continuous DC power supply, high cathode current density leads to dendritic growth—branch-like protrusions—at the cathode edges. This increases the actual surface area of the cathode, thereby reducing the cathode current density during electrolytic deposition and consequently decreasing the current efficiency. Pulse mode provides additional control parameters. By regulating potential or current through pulsed electrolysis, the interface characteristics between the electrode and the electrolyte can be uniquely adjusted, resulting in electrochemical behaviors that are distinctly different from those of steady - state methods. This enables the regulation of the selectivity and overall efficiency of the electrolytic reaction (Zhou et al., 2020). Studies have discovered that introducing pulse and chaotic circuit electrolysis into the manganese electrolytic deposition process and using an externally controlled power supply to regulate internal electrochemical oscillations and metal fractal nonlinear behaviors provides a new method to further adjust the chemical properties and surface morphology of anodically deposited MnO_2, as well as inhibit the growth of manganese dendrites at the cathode edges (Xie et al., 2022; Yang et al., 2023; Yang et al., 2025), as shown in Figure 4. This approach helps reduce electrolytic energy consumption and enhance current efficiency.

Figure 4. Nonlinear kinetic behavior of electrode reactions and process energy-saving and consumption-reduction regulation in manganese electrolytic deposition

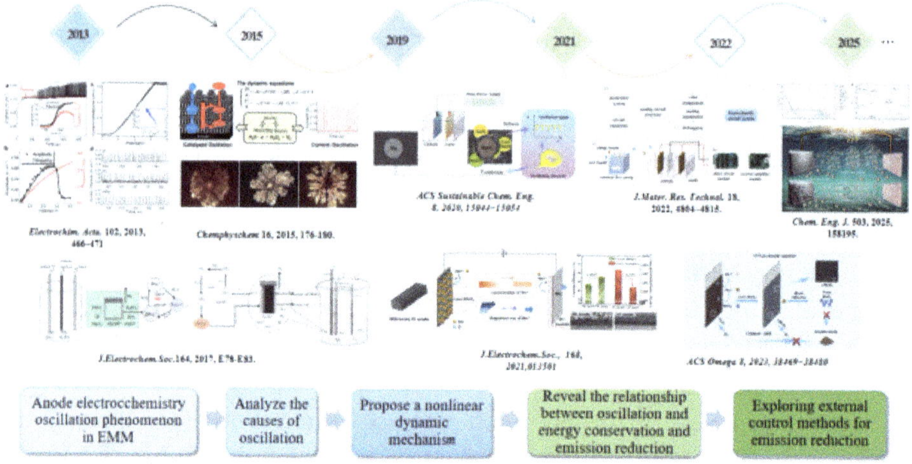

5. SUSTAINABILITY OF GREEN MANGANESE MATERIALS IN END-USE APPLICATIONS

Manganese materials play a vital role in various end-use sectors, including steel manufacturing, advanced batteries, catalysts, and high-performance alloys. They serve as an indispensable component of modern industrial systems. With the global acceleration towards carbon neutrality and green transformation, the sustainable utilization of manganese materials has received increasing attention. Despite their performance advantages, conventional manganese-based products often entail high energy consumption, significant carbon emission, and inefficient resource use during their production, application, and disposal stages. Therefore, promoting the green and sustainable deployment of manganese materials in terminal applications is critical for building a low-carbon and circular economy.

In the steel industry, manganese is widely employed as a deoxidizer and alloying element to enhance mechanical strength and wear resistance (Pani et al., 2016; Yang et al., 2025). However, manganese ferroalloys used in blast furnace-based steelmaking are typically produced via carbon-intensive reduction methods, which leads to considerable greenhouse gas emissions and solid waste generation. The adoption of electric arc furnace (EAF) technology powered by renewable energy sources offers a more sustainable alternative. In addition, microalloying design can reduce the manganese input without compromising performance, while recycling manganese from steel scrap and valorizing manganese-containing slag can further improve resource efficiency and contribute to a closed-loop metallurgical system.

Driven by the transition to renewable energy, manganese applications in battery technologies have grown rapidly, particularly in lithium-ion batteries (e.g., NMC, LMO) and emerging sodium-ion systems (Su et al., 2024; Bing et al., 2025; Wang et al., 2023). Manganese is considered a highly promising material due to its abundance, low cost, and relatively low toxicity, making it an attractive substitute for critical metals such as cobalt and nickel. Current research focuses on developing manganese-rich layered oxides and sodium manganese oxides as cathode materials to reduce reliance on scarce elements (Ding et al., 2023). Concurrently, the establishment of efficient recycling systems, the development of low-energy synthesis routes (e.g., hydrothermal or bio-assisted methods), and the use of non-toxic precursors and solvents are essential for enabling a sustainable battery supply chain.

In catalysis and chemical industries, manganese-based materials have demonstrated excellent redox activity and environmental compatibility in applications such as selective catalytic reduction (SCR), organic synthesis, and water treatment (Li et al., 2025; Yashnik et al., 2025; Peng et al., 2024). The development of low-toxicity, recyclable manganese catalysts—such as magnetically recoverable MnO_x systems and supported catalysts—alongside ambient condition and aqueous-phase reactions, is advancing the green chemistry agenda. Moreover, utilizing industrial by-products, such as electrolytic manganese residue, electrolytic manganese anode slime, as precursor materials for catalyst synthesis offers a promising strategy for waste valorization and material circularity.

Manganese is also critical in the development of advanced alloys for lightweight and corrosion-resistant applications in sectors such as automotive and aerospace (Sorokina et al., 2023). Future trends emphasize the design of high-performance, low-manganese alloys that reduce manganese consumption while maintaining desired mechanical properties. In parallel, recycling manganese from end-of-life alloy components (e.g., scrapped vehicles or aircraft) helps minimize raw material extraction and associated carbon emission.

Furthermore, the fractal behavior of electrolytic manganese cathodes—specifically the dendritic, non-periodic self-organized growth structures that emerge during cathodic electrodeposition of metallic manganese—represents a critical nonlinear phenomenon in the electrolytic manganese industry (Gebara et al., 2020). Research and applications focus on modulating fractal growth to reduce energy consumption, enhance current efficiency, and optimize electrode material performance, which are key technological areas driving the industry's transition toward green and efficient development. Studies have demonstrated that replacing traditional DC or constant-current power supply modes with pulsed electric fields effectively regulates fractal structures. For instance, optimizing pulse frequency (e.g., 100-200 Hz) and duty cycle can suppress excessive dendritic growth, resulting in denser deposition layers that reduce electrolytic power consumption and improve current efficiency (Zhou et al., 2020). Conversely, understanding and controlling fractal behavior has driven the design of functionalized electrode materials. Precise regulation of electrodeposition processes (e.g., through pulsed technology) enables the transformation of metallic manganese from forming smooth deposition layers to growing layers with specific fractal structures (e.g., dendrites or arrays). These fractal-structured metals, characterized by high specific surface area and conductivity, can be grown on fibrous or mesh substrates to develop flexible electrode materials. Such materials, due to their large specific surface area and high chemical/mechanical sensitivity, exhibit application potential in specialized fields such as energy conversion (e.g., batteries, capacitors) and healthcare (e.g., biosensors)(Zhou et al., 2023; Zhou et al., 2024), as shown in Figure 5.

In conclusion, the green transformation of manganese materials in end-use applications is not only a strategic response to environmental challenges but also a crucial pathway for enhancing industrial competitiveness. Through collaborative efforts in green process innovation, resource recovery, and life-cycle governance, the efficient, safe, low-carbon, and sustainable use of manganese can be achieved, making a significant contribution to the global transition towards clean manufacturing and energy systems.

Figure 5. Fractal growth control of manganese via pulsed electric field and its application in multifunctional monitoring system

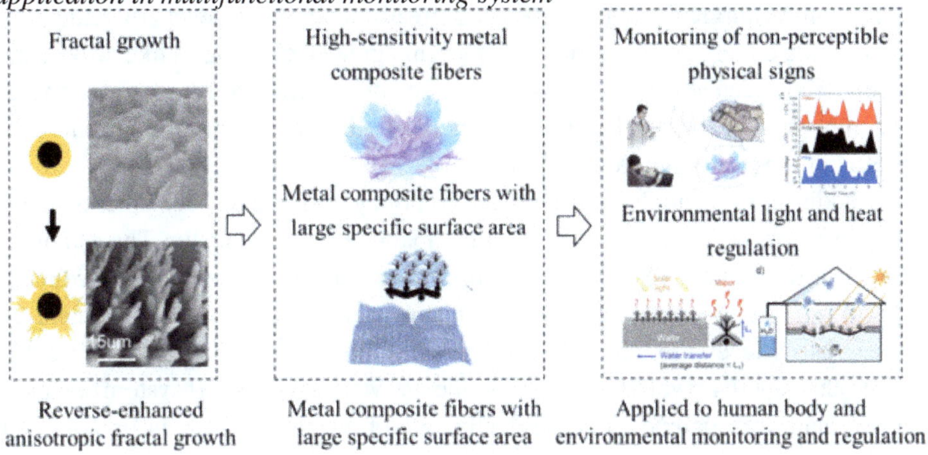

6. CIRCULAR ECONOMY AND MANGANESE RESOURCE RECOVERY

Establishing a circular economy system for manganese resources is a core strategy to achieve sustainable management and ensure supply chain security. With the growing demand for manganese in steel, batteries, and specialty chemicals, the traditional linear model of extraction, use, and disposal has exacerbated resource depletion, environmental pressures, and supply chain risks. To address this, a transition to a closed-loop system is imperative. By prioritizing comprehensive strategies that encompass manganese resource recovery and recycling, we can enhance utilization efficiency while minimizing environmental impacts throughout the entire life cycle.

This transformation hinges on technological innovation and large-scale implementation of two critical recycling pathways. First, efficient manganese recovery from metallurgical by-products. The substantial amounts of slag, dust, and tailings generated during ferromanganese and silicomanganese alloy production represent valuable secondary resources. By employing advanced hydrometallurgical techniques such as acid leaching, bioleaching, and solvent extraction, combined with physical separation and thermal pretreatment processes, manganese can be selectively extracted with high selectivity, providing high-quality recycled materials for downstream applications. Second, recycling in the battery industry is crucial. With the widespread adoption of electric vehicles and energy storage systems, a significant amount of spent lithium-ion and sodium-ion batteries containing manganese-rich cathode materials will be produced. While mainstream pyrometallurgical smelting and hydrometallurgical leaching processes already enable co-recovery of manganese with lithium, cobalt, and other metals, emerging green technologies like direct cathode material regeneration and low-eutectic solvent leaching further reduce energy consumption and chemical usage, enhancing both economic and environmental benefits of the recycling process.

In addition to technological innovation, policy and market instruments play a pivotal role in promoting circularity in the manganese value chain. The establishment of a manganese resource recycling system ultimately requires coordinated support from policy tools, market mechanisms, and full life cycle management. At the regulatory level, the Extended Producer Responsibility (EPR) system incentivizes products with recyclable designs, while the Material Passport system enhances supply chain transparency. Economically, carbon pricing and green procurement standards guide capital investment toward manganese recycling infrastructure and low-carbon technologies. Furthermore, circular economy principles must be embedded across the entire life cycle of manganese—from mine to market. This includes promoting eco-design of manganese-based products, reducing processing losses, and building industrial symbiosis networks that transform byproducts into raw materials for subsequent stages. Through life cycle assessment (LCA) to quantify recycling benefits and global coordination of standards and cooperation, a resilient manganese recycling system can be established, providing a solid foundation for sustainable global industrial development.

In summary, integrating circular economy principles into manganese resource management is essential for achieving material security, environmental sustainability, and economic resilience. By combining technological advancements in resource recovery with supportive regulatory frameworks and market incentives, a regenerative manganese system can be realized, contributing meaningfully to the global transition toward sustainable and circular industrial ecosystems.

7. FUTURE PROSPECTS AND INNOVATIONS

In the context of the global transition towards a low-carbon economy and sustainable development, the manganese industry faces both significant opportunities and challenges. As a critical material in steel manufacturing, energy storage systems, and high-performance alloys, the demand for manganese is expected to grow steadily. However, to mitigate issues related to resource depletion, environmental degradation, and carbon emission, the industry must actively promote technological innovation and industrial transformation. The future development of the manganese sector will focus on the following aspects:

Precision and Intelligent Innovation in Manganese Ore Exploration and Mining: Future manganese exploration and mining will achieve deep integration of Artificial Intelligence (AI) and big data, advancing toward precision and sustainability. In exploration, geological data-based ore-seeking models and intelligent prediction systems will become core components. Through "manganese ore big data prediction analysis" and smart exploration technologies, these systems will accurately locate hidden ore bodies and guide drilling operations. The mining phase will extensively adopt modular, low-disruption green equipment, such as drone-delivered portable drilling rigs, to obtain rock cores with minimal ecological impact. Building on this foundation, specialized AI models trained with industry knowledge will serve as the "intelligent mine brain." Leveraging digital twin technology, they will simulate mine operations, dynamically optimize mining plans, predict equipment failures, and coordinate drones and smart drilling rigs, achieving a fundamental shift from "experience-driven" to "data and algorithm-driven" operations.

Precision and Data-Driven Optimization in Manganese Crushing and Concentration: The core objectives in crushing and concentration are to maximize manganese resource utilization and reduce waste residue at the source. Future concentrators will become highly intelligent "precision resource sorting centers," fully data-driven in operation. Advanced sensor networks will monitor manganese ore particle size and grade in real-time, feeding data streams into a central AI optimization platform. Through machine learning models, this platform can dynamically adjust crusher parameters and process conditions for magnetic separation and flotation, enabling adaptive production. The entire process functions like a precision "digestive system," aiming to "maximize resource utilization" from the initial stage, laying the foundation for efficient, low-carbon operations across the industrial chain.

Intelligent Optimization of Manganese Ore Leaching Electrolysis and Smart Solid Waste Treatment: In the future, electrolysis workshops will deploy comprehensive IoT sensors to collect real-time data points including cell temperature, pH levels, and current density across thousands of parameters. AI models analyzing these live data alongside historical process databases can achieve precise temperature control and dynamic equilibrium of solution chemistry, significantly reducing energy consumption. For the massive volume of electrolytic manganese solid waste, research has demonstrated that machine learning (such as BP neural networks optimized by particle swarm optimization) can build predictive models to accurately forecast heavy metal leaching behaviors under different calcination temperatures and material ratios. This enables intelligent recommendations for optimal harmless treatment and resource recovery parameters, minimizing environmental risks. Simultaneously, a broader circular economy model will be implemented through big data platforms that match multi-industry solid waste characteristics, promoting collaborative utilization of electrolytic manganese waste with red mud, calcium carbide slag, and waste incineration fly ash. These materials will be transformed into high-value-added products like permeable bricks and roadbed materials, achieving "waste-to-waste" solutions and full-component resource recovery.

Innovation in Manganese Product Preparation and Smart Industrial Synergy: By establishing a "Material Genome Engineering" database integrating synthesis parameters of various manganese oxide precursors, crystal structures, and performance data of final products (e.g., lithium manganate, lithium iron phosphate), machine learning algorithms can accelerate new material development cycles and reverse-engineer optimal preparation processes meeting specific electrochemical performance requirements. On the production side, the AI scheduling system analyzes customer orders, equipment status, raw material inventory, and energy consumption data to create optimized production plans, dramatically cutting delivery times.

Looking ahead, the entire manganese industry chain will evolve into a data-driven intelligent collaborative network. This integrated industrial cloud platform will consolidate mineral exploration models, real-time data from ore processing and smelting, customer demand for product manufacturing, and material matching information for solid waste treatment. The platform's specialized industry large-scale models will enable global optimization and simulation, achieving end-to-end coordination across resource exploration, production scheduling, energy management, solid waste recycling, and product innovation. Ultimately, this will drive the manganese industry's comprehensive transformation toward a sustainable development model characterized by green, low-carbon, high-efficiency, and high-value operations.

8. TOWARDS A SMART, GREEN, CIRCULAR AND RESILIENT MANGANESE INDUSTRY

The manganese industry stands at a historical turning point, transitioning comprehensively from a traditional resource-dependent model to a modern green and intelligent one. Faced with surging global energy transition demands, highly concentrated resource geographical distribution, and severe carbon reduction pressures, its core future goal is to build a sustainable manganese industrial system that is resource-closed-loop, low-carbon and clean, intelligent and efficient, and resilient.

At the mining end, the future will rely on "artificial intelligence & big data" to achieve precise prediction of manganese ore geological exploration and dynamic optimization of mining plans. Combined with intelligent equipment and renewable energy, this will minimize ecological disturbance to the maximum extent. In the high-carbon emission smelting part, the key lies in energy substitution centered on green electricity and green hydrogen, vigorously developing disruptive technologies such as hydrogen-based direct reduction and plasma smelting, and coupling carbon capture, utilization, and storage to achieve carbon reduction. Meanwhile, through Internet of Things and digital twin technologies, the entire process of crushing, leaching, and electrolysis will be optimized to achieve intelligent production with reduced consumption and increased efficiency.

The future manganese industry will elevate the development of "urban mine" resources to a strategic height equivalent to primary ore mining. This means establishing an efficient recycling system covering solid waste from electrolytic manganese, scrap steel, spent batteries, manganese-containing alloys, and various industrial slag. Through advanced technologies such as green hydrometallurgy and direct regeneration, all manganese resources will be incorporated into the closed-loop of "resources - products - regenerated resources". Traceability technologies such as blockchain will ensure transparency and credibility in this recycling process, ultimately significantly reducing absolute dependence on primary manganese ore resources and environmental footprint.

Ultimately, the combined force of technological innovation, policy mechanisms, and global cooperation drives this transformation. It requires enterprises to make forward- looking investments to overcome key technological bottlenecks and innovate business models. Meanwhile, it also depends on governments to shape a fair and effective market environment through policy tools such as carbon pricing, green procurement, and extended producer responsibility. More importantly, facing global challenges, comprehensive cooperation is needed in manganese resource governance, technical standards, and circular economy models. When data intelligence permeates the entire industrial chain and circular networks connect every link, the manganese industry can transcend its role as a mere raw - material supplier and evolve into a key sustainable material cornerstone supporting global green infrastructure and clean - energy systems. While ensuring its own resource security, it will make an irreplaceable contribution to global carbon - neutrality goals and the green evolution of industrial civilization.

REFERENCES

Bai, H., Qing, S. L., Yang, D. P., Zhang, Y. X., Fan, X., & Tao, C. Y. (2017). Periodic potential oscillation during oxygen evolution catalyzed by manganese oxide at constant current. *Journal of the Electrochemical Society*, *164*(4), E78–E83. DOI: 10.1149/2.1241704jes

Bing, D., Qinglin, C., Yuji, W., Yijia, S., Rukang, Y., & Qiyuan, L. (2025). Selectively and deeply extracting manganese from modified manganese-rich slag to prepare lithium manganese oxide battery materials. *Separation and Purification Technology*, *377*, 134374. Advance online publication. DOI: 10.1016/j.seppur.2025.134374

Ding, C., Chen, Z., Cao, C., Liu, Y., & Gao, Y. (2023). Advances in Mn-based electrode materials for aqueous sodium-ion batteries. *Nano-Micro Letters*, *15*(1), 192. Advance online publication. DOI: 10.1007/s40820-023-01162-x PMID: 37555908

Ernst, M. S., Tangstad, M., & Du Preez, S. P. (2024). Pre-reduction of Nchwaning manganese ore in CO/CO_2, H_2/H_2O, and H_2 atmospheres. *Minerals Engineering*, *216*, 108854. Advance online publication. DOI: 10.1016/j.mineng.2024.108854

Fan, X., Hou, J., Sun, D. G., Xi, S. Y., Liu, Z. H., Du, J., Luo, J. L., & Tao, C. Y. (2013). Mn-oxides catalyzed periodic current oscillation on the anode. *Electrochimica Acta*, *102*, 466–471. DOI: 10.1016/j.electacta.2013.03.175

Fan, X., Yang, D. P., Ding, L. F., Du, J., & Tao, C. Y. (2015). Periodic current oscillation catalyzed by δ-MnO_2 nanosheets. *Chemphyschem : A European Journal of Chemical Physics and Physical Chemistry*, *16*(1), 176–180. DOI: 10.1002/cphc.201402623 PMID: 25382485

Farjana, S. H., Huda, N., Mahmud, M. A. P., & Lang, C. (2019). A global life cycle assessment of manganese mining processes based on EcoInvent database. *The Science of the Total Environment*, *688*, 1102–1111. DOI: 10.1016/j.scitotenv.2019.06.184 PMID: 31726541

Gebara, P., Gebara, M., & Owczarek, A. (2020). Fractal structures in electrodeposition process. *Acta Physica Polonica A*, *138*(2), 287–290. DOI: 10.12693/APhysPolA.138.287

Han, F., & Wu, L. (2019). Resource utilization of electrolytic manganese residues. In Industrial solid waste recycling in Western China (pp. 127–164). DOI: 10.1007/978-981-13-8086-0_3

Han, X. W., Zhang, W. B., Li, J. J., Liu, X. Y., Yuan, X. Y., Ma, X. J., & Long, J. P. (2024). Salinity gradient energy harvesting via Na-Mn(PO_4)-based electrochemical system. *ACS Sustainable Chemistry & Engineering*, *12*(21), 8010–8021. DOI: 10.1021/acssuschemeng.3c08508

He, D., Shu, J., Wang, R., Chen, M., Wang, R., Gao, Y., Liu, R., Liu, Z., Xu, Z., Tan, D., Gu, H., & Wang, N. (2021). A critical review on approaches for electrolytic manganese residue treatment and disposal technology: Reduction, pretreatment, and reuse. *Journal of Hazardous Materials*, *418*, 126235. Advance online publication. DOI: 10.1016/j.jhazmat.2021.126235 PMID: 34126381

He, S., Jiang, D., Hong, M., & Liu, Z. (2021). Hazard-free treatment and resource utilisation of electrolytic manganese residue: A review. *Journal of Cleaner Production*, *306*, 127224. Advance online publication. DOI: 10.1016/j.jclepro.2021.127224

He, S., Wang, T., Liu, Z., Jiang, T., & Liu, Z. (2025). Mechanochemistry enhanced low-temperature molten salt synergistic treatment of low-grade pyrolusite and waste salts from the electrolytic manganese metal industry. *Chemical Engineering Journal*, *506*, 159338. Advance online publication. DOI: 10.1016/j.cej.2025.159338

Islam, I., Tandersen, D., Taimullah, A., Sata, Y. A., Hendrawan, Y., Hidayat, T., & Zulhan, Z. (2024). Hydrogen plasma smelting reduction for fast production of green ferronickel. *International Journal of Hydrogen Energy*, *94*, 300–309. DOI: 10.1016/j.ijhydene.2024.11.022

Jiao, P. P., Xu, F. Y., Li, J. H., Duan, N., Chen, G. Y., & Jiang, L. H. (2016). The inhibition effect of SeO_2 on hydrogen evolution reaction in $MnSO_4$-$(NH_4)_2SO_4$ solution. *International Journal of Hydrogen Energy*, *41*(2), 784–791. DOI: 10.1016/j.ijhydene.2015.11.034

Li, X. T., Li, X. Y., Yang, J., Cao, Z. F., Li, C. X., Xue, J. R., Ma, X., & Wang, S. (2024). Highly efficient Mn^{2+}, Mg^{2+}, and NH_4^+ recovery from electrolytic manganese residue via leaching, solvent extraction, coprecipitation, and atmospheric oxidation. *Journal of Hazardous Materials*, *472*, 134430. Advance online publication. DOI: 10.1016/j.jhazmat.2024.134430 PMID: 38718502

Li, Y. Y., Jia, Z. G., Xu, L. X., Sheng, G. H., & Wan, C. (2025). Synergistic removal of tetracycline hydrochloride by nano-MnOx/NCF composite fiber catalysts via heterogeneous peroxymonosulfate activation. *Inorganic Chemistry Communications*, *172*, 113682. Advance online publication. DOI: 10.1016/j.inoche.2024.113682

Lu, J. M., Dreisinger, D., & Glück, T. (2014). Manganese electrodeposition: A literature review. *Hydrometallurgy*, *141*, 105–116. DOI: 10.1016/j.hydromet.2013.11.002

Luo, W. J., Li, B., Yang, G., Xu, M. X., Pang, C. H., Kow, K. W., & Wu, T. (2024). Utilisation of electrolytic manganese residue as a sulphate activator in producing concrete blocks with high-volume fly ash. *Journal of Cleaner Production*, *434*, 139813. Advance online publication. DOI: 10.1016/j.jclepro.2023.139813

Lyu, K., Zhang, W. J., Chen, Y. Q., & Wang, C. Y. (2024). Energy-efficient and advanced electrowinning of metallic manganese within a novel H^+ exchange membrane cell using a Ti/IrO_2-RuO_2-SiO_2 anode. *Separation and Purification Technology*, *347*, 127489. Advance online publication. DOI: 10.1016/j.seppur.2024.127489

Ma, K., Su, Z. J., Zhang, Y. B., & Jiang, T. (2024). Effect mechanism of manganese on hydrogen-based reduction behavior of magnetite pellet. *Journal of Materials Research and Technology*, *30*, 8304–8315. DOI: 10.1016/j.jmrt.2024.05.205

Marzoughi, O., & Tafaghodi, L. (2025). Hydrogen as a sustainable reducing agent for recycling electric arc furnace dust: A kinetic review. *Journal of Sustainable Metallurgy*, *11*(1), 232–247. DOI: 10.1007/s40831-024-00975-3

Mohammadi, M., & Alfantazi, A. (2015). The performance of Pb-MnO_2 and Pb-Ag anodes in Mn(II)-containing sulphuric acid electrolyte solutions. *Hydrometallurgy*, *153*, 134–144. DOI: 10.1016/j.hydromet.2015.02.009

Mohammadi, M., & Alfantazi, A. (2016). Evaluation of manganese dioxide deposition on lead-based electrowinning anodes. *Hydrometallurgy*, *159*, 28–39. DOI: 10.1016/j.hydromet.2015.10.023

Mohanty, S., Ghosh, S., Bal, B., & Das, A. P. (2018). A review of biotechnology processes applied for manganese recovery from wastes. *Reviews in Environmental Science and Biotechnology*, *17*(4), 791–811. DOI: 10.1007/s11157-018-9482-1

Moro, K., Haubrich, F., Martin, M., Grimmer, M., Hoth, N., & Schippers, A. (2023). Reductive leaching behaviour of manganese and cobalt phases in laterite and manganese ores. *Hydrometallurgy*, *220*, 106101. Advance online publication. DOI: 10.1016/j.hydromet.2023.106101

Nahum, A. B. R., Pereira, W. V. D., Martins, G. C., Dias, Y. N., Ribeiro, P. G., Salomão, G. N., Gastauer, M., Caldeira, C. F., Fernandes, A. R., & de Souza, E. S.. (2024). Properties and environmental quality of the overburden and tailings of manganese mining in the eastern Amazon. *Environmental Research*, *262*, 119965. Advance online publication. DOI: 10.1016/j.envres.2024.119965 PMID: 39265760

Olii, M. R., Hambali, R., Djafar, L., Ichsan, I., & Saliko, M. (2024). Innovative porous mortar filters: Wastewater purification for clean water. *Water Science and Technology*, *89*(6), 1526–1538. DOI: 10.2166/wst.2024.073 PMID: 38557716

Pani, S., Dash, N., Mohapatra, B. K., & Singh, S. K. (2019). Siliceous manganese ore from eastern India: A potential resource for ferrosilicon-manganese production. *High-Temperature Materials and Processes*, *38*(2019), 425–435. DOI: 10.1515/htmp-2018-0081

Pani, S., Singh, S. K., & Mohapatra, B. K. (2016). Bog manganese ore: A resource for high manganese steel making. *Journal of the Minerals Metals & Materials Society*, *68*(6), 1525–1534. DOI: 10.1007/s11837-016-1935-9

Peng, D., Wang, Q., Zang, S. H., & Mo, L. Y. (2024). Catalytic complete oxidation of ethyl acetate on MnO/MgAlO catalysts. *Kinetics and Catalysis*, *65*(4), 386–395. DOI: 10.1134/S0023158423601134

Reyes-Morales, Q. L., Reyes-Cruz, V. E., Trujillo-Estrada, A., Cobos-Murcia, J. A., Urbano-Reyes, G., & Pérez-Labra, M. (2021). Selenium and sulphur reactions involved in manganese reduction from sulphate solutions. *RSC Advances*, *11*(41), 25542–25550. DOI: 10.1039/D1RA03679C PMID: 35478895

Shu, J. C., Liu, R. L., Liu, Z. H., Qiu, J., Chen, H. L., & Tao, C. Y. (2017). Simultaneous removal of ammonia nitrogen and manganese from wastewater using nitrite by electrochemical method. *Environmental Technology*, *38*(3), 370–376. DOI: 10.1080/09593330.2016.1194482 PMID: 27249226

Shu, Y., Yang, Z. Y., Pan, W. K., Xiang, D. X., Wu, J. G., Gao, L. H., Jiao, X., He, H., Zhang, Z. X., & Chen, G. (2025). Achieving carbon neutrality through conversion of CO_2 to cyclic carbonates using bifunctional benzimidazolium-based pyridine hypercrosslinked polymers. *Applied Surface Science*, *686*, 162164. Advance online publication. DOI: 10.1016/j.apsusc.2024.162164

Song, X. S., Shen, Q. C., Liu, B., Li, J., & Zhu, D. W. (2024). Research on the use of low-selenium additives in the electrodeposition of manganese metal and co-production of electrolytic manganese. *Electrochemistry (Tokyo, Japan)*, *92*(7), 077004. Advance online publication. DOI: 10.5796/electrochemistry.24-00045

Sorokina, S. A., Vorob'ev, R. A., Gorshunov, M. G., & Chernigin, M. A. (2023). Use of chromium-manganese corrosion-resistant steel in district heating network applications. *Metal Science and Heat Treatment*, *65*(3–4), 159–166. DOI: 10.1007/s11041-023-00908-z

Su, Z. H., Guo, Z. H., Xie, H. Y., Qu, M. Z., Wang, H., & Peng, G. C. (2024). Multifunctional surface modification to enhance the electrochemical performance of lithium-rich manganese-based cathode materials. *Electrochimica Acta*, *500*, 144727. Advance online publication. DOI: 10.1016/j.electacta.2024.144727

Sun, Z., Sun, G. Y., Salazar-Tio, R., Fager, A., & Crouse, B. (2025). A comprehensive review of CO_2 geological storage projects with regulatory requirements of computational modeling. *Fuel, 386*, 134209. Advance online publication. DOI: 10.1016/j.fuel.2024.134209

Tang, B., Yang, F., Chen, C., Shi, C., Deng, W., Li, J., & Jiang, L. (2023). A novel lead-based pseudo-dimensional stable anode toward efficient and clean extraction of metallic manganese. *Journal of Cleaner Production, 386*, 135806. Advance online publication. DOI: 10.1016/j.jclepro.2022.135806

Tao, M., Qiu, X. P., & Lu, D. M. (2024). Life cycle assessment of electrolytic manganese metal production. *The Science of the Total Environment, 949*, 174862. Advance online publication. DOI: 10.1016/j.scitotenv.2024.174862 PMID: 39038680

Wang, F., Long, G. C., & Zhou, J. L. (2024). Enhanced green remediation and refinement disposal of electrolytic manganese residue using air-jet milling and horizontal-shaking leaching. *Journal of Hazardous Materials, 465*, 133419. Advance online publication. DOI: 10.1016/j.jhazmat.2023.133419 PMID: 38183942

Wang, K. T., Yi, M., Sun, Y. X., Yang, Y., Lu, C. J., Fujita, T., & Cui, X. M. (2025). Fabrication of electrolytic manganese residue/granulated blast furnace slag-based geopolymer microspheres and enhanced synergic adsorption of Cr(III)-containing wastewater. *Separation and Purification Technology, 355*, 129555. Advance online publication. DOI: 10.1016/j.seppur.2024.129555

Wang, L., Zhou, W. W., Lin, X. F., & Xie, Y. L. (2024). TiO_2-doping boosting resource utilization of high-manganese electrolytic manganese residue for superior performance glass-ceramics. *Ceramics International, 50*(23), 49889–49905. DOI: 10.1016/j.ceramint.2024.09.333

Wang, M., Meng, Y., Xu, Y., Chen, N., Chuai, M., Yuan, Y., Sun, J., Liu, Z., Zheng, X., Zhang, Z., Li, D., & Chen, W. (2023). Aqueous all-manganese batteries. *Energy & Environmental Science, 16*(11), 5284–5293. DOI: 10.1039/D3EE01679J

Wang, R. C., Zhang, S. L., Guo, Y. J., Liu, J. Y., Zhang, F. L., & Dang, C. E. (2024). Techno-economic-environmental analysis based on life cycle carbon accounting and pollutants audit for cleaner production of electrolytic manganese. *Process Safety and Environmental Protection, 191*, 1010–1023. DOI: 10.1016/j.psep.2024.09.035

Wang, R. F., Yuan, S., Li, Y. J., Gao, P., & Li, R. (2024). Hydrogen-based mineral phase transformation mechanism investigation of pyrolusite ore. *International Journal of Minerals Metallurgy and Materials, 31*(11), 2445–2457. DOI: 10.1007/s12613-023-2819-y

Weng, W., Lan, Y., Zhang, R., Chi, X., Tan, W., Tang, D., Chen, H., & Zhong, S. (2024). Gluconic acid as the supplementary additive realizing high-efficiency production of metal manganese with decreased dosage of toxic selenium dioxide. *Electrochimica Acta*, *499*, 144708. Advance online publication. DOI: 10.1016/j.electacta.2024.144708

Wu, D. Y., Zhang, Z. H., & Wang, Z. H. (2024). Impact of manganese mining on potentially toxic elements pollution and bioaccumulation in the Xiaojiang River. *Environmental Geochemistry and Health*, *46*(10). Advance online publication. DOI: 10.1007/s10653-024-02171-4 PMID: 39167155

Wu, T., Ma, B., An, Y., Chen, Y., & Wang, C. (2024). Improvement of manganese electrolytic process and secondary resources recovery of manganese: A review. *Process Safety and Environmental Protection*, *186*, 895–909. DOI: 10.1016/j.psep.2024.03.097

Wu, W., Zhao, B., Liang, Q., Lei, M., Zhang, B., & Zhang, F. (2025). Effect of direct alloying of manganese ore-cored wires on the composition and quality of molten steel. *Ironmaking & Steelmaking*, *52*(7), 739–751. DOI: 10.1177/03019233241307487

Xie, Z., Liu, Z., Tao, C., Li, C., & Chang, J. (2022). Production of electrolytic manganese metal using a new hyperchaotic circuit system. *Journal of Materials Research and Technology*, *18*, 4804–4815. DOI: 10.1016/j.jmrt.2022.04.141

Xie, Z. N., Liu, Z. H., Tao, C. Y., Li, C. B., & Chang, J. (2022). Production of electrolytic manganese metal using a new hyperchaotic circuit system. *Journal of Materials Research and Technology*, *18*, 4804–4815. DOI: 10.1016/j.jmrt.2022.04.141

Xie, Z. N., Liu, Z. H., Zhang, X. J., Yang, L. R., Chang, J., & Tao, C. Y. (2019). Electrochemical oscillation on anode regulated by sodium oleate in electrolytic metal manganese. *Journal of Electroanalytical Chemistry (Lausanne, Switzerland)*, *845*, 13–21. DOI: 10.1016/j.jelechem.2019.05.012

Yang, H., Liu, X., Guo, H., He, Y., Shi, J., Liu, H., Hao, J., & Liu, S. (2025). Effect of laser surface remelting on the microstructure and properties of high manganese steel coating. *Surface and Coatings Technology*, *496*, 131667. Advance online publication. DOI: 10.1016/j.surfcoat.2024.131667

Yang, J., Li, C., Zhang, Q., Wu, Z., Liu, P., Liu, Z., Tao, C., Zheng, G., Yang, Y., & Wei, H. (2025). Improving the current efficiency of electrodeposited manganese metal using variable current electrolysis and its mechanism. *Chemical Engineering Journal*, *503*, 158195. Advance online publication. DOI: 10.1016/j.cej.2024.158195

Yang, J., Li, C. B., Zhang, Q., Huang, K. Y., Liu, P. Q., Liu, Z. H., Li, Z. N., Meng, T., Zhang, X., Wu, J. Y., Tao, C., & Zheng, G. (2023). A new approach of electrolytic metal manganese with lower energy consumption and fewer spherical dendrites based on a hyperchaotic circuit. *International Journal of Bifurcation and Chaos in Applied Sciences and Engineering*, *33*(14), 2350173. Advance online publication. DOI: 10.1142/S0218127423501730

Yang, J., Wu, Z., Li, C., Zhang, Q., Liu, P., Zheng, G., Liu, Z., Li, J., Tao, C., Yang, Y., & Wei, H. (2025). Enhancing manganese electrodeposition efficiency at high current densities utilizing a novel variable gradient chaotic current technique. *Journal of Cleaner Production*, *503*, 145373. Advance online publication. DOI: 10.1016/j.jclepro.2025.145373

Yang, L. Z., Feng, Z., Hu, H., Wei, G. S., Xue, B. T., Guo, Y. F., & Jiang, T. (2025). A review on production and application of direct reduced iron in gas-based shaft furnace-electric arc furnace route. *Journal of Iron and Steel Research International*, *32*(3), 485–518. DOI: 10.1007/s42243-024-01377-1

Yashnik, S. A., Surovtsova, T. A., & Salnikov, A. V. (2025). Low-loaded Pt–Pd diesel oxidation catalyst: MnO loading effects. *Topics in Catalysis*, *68*(3–4), 357–378. DOI: 10.1007/s11244-024-02023-6

Ye, W. Q., Xu, F. Y., Jiang, L. H., Duan, N., Li, J. H., Zhang, F. L., Zhang, G., & Chen, L. J. (2021). A novel functional lead-based anode for efficient lead dissolution inhibition and slime generation reduction in zinc electrowinning. *Journal of Cleaner Production*, *284*, 124767. Advance online publication. DOI: 10.1016/j.jclepro.2020.124767

Zhao, Y. N., Zhu, G. C., & Cheng, Z. (2010). Thermal analysis and kinetic modeling of manganese oxide ore reduction using biomass straw as reductant. *Hydrometallurgy*, *105*(1–2), 96–102. DOI: 10.1016/j.hydromet.2010.08.004

Zhong, H. D., Zhang, Q., Liu, Z. H., Du, J., & Tao, C. Y. (2023). Ti/TiO_2 anodes for efficient electrodeposition of manganese metal and anode slime generation reduction. *ACS Omega*, *8*(41), 38469–38480. DOI: 10.1021/acsomega.3c05273 PMID: 37867691

Zhou, H., Cao, H. B., Que, Y. T., Chen, Y. F., Fu, J. J., Wang, H. Y., Zheng, Y. Y., He, X., Yang, Y. X., & Liu, J. (2024). Hierarchical Cu dendrites@CuO multifunctional nanowire mesh for solar-thermal clean water production. *Desalination*, *572*, 117114. Advance online publication. DOI: 10.1016/j.desal.2023.117114

Zhou, H., Xue, J., Che, Q. L., Kong, L. F., Zhang, N. N., Tao, C. Y., & Fan, X. (2023). Rollable and ventilated net-based solar thermal water evaporator for casting on water surface. *Small Structures*, *4*(7), 2300008. Advance online publication. DOI: 10.1002/sstr.202300008

Zhou, H., Zhang, N. N., Bai, H., Ming, H., Zhang, Q., Du, J., Fan, X., & Tao, C. Y. (2020). A pulse modulatable self-oscillation kinetics for water oxidation at large current on manganese catalyst. *Electrochimica Acta*, *337*, 135798. Advance online publication. DOI: 10.1016/j.electacta.2020.135798

Zhou, Y. Y., Yin, G. D., Zeng, X., Zhao, J. F., & Yao, G. D. (2022). Potential application of carbohydrate biomass in hydrometallurgy: One-pot reduction of metal oxides/salts under mild hydrothermal conditions. *RSC Advances*, *12*(32), 20747–20754. DOI: 10.1039/D2RA01493A PMID: 35919188

Zmuidzinaviciene, N., Griskonis, E., & Sulcius, A. (2023). The corrosion of Mn coatings electrodeposited from a sulphate bath with Te(VI) additive and influence of phosphate post-treatment on corrosion resistance. *Coatings*, *13*(9), 1617. Advance online publication. DOI: 10.3390/coatings13091617

Chapter 9
Design Principles and Prospects of Advanced Materials for Water Pollution Control

Jiming Huang
http://orcid.org/0009-0007-7779-3677
Tongren University, China

Chaowu Wang
Guizhou Kingray New Materials Co., Ltd., China

Ping Xue
Hubei University of Science and Technology, China

ABSTRACT

Water pollution now involves complex mixtures (heavy metals, POPs, pharmaceuticals, microplastics) that conventional treatment trains often fail to remove when pollutants are dilute, recalcitrant, or transformation-prone. This chapter reviews material-enabled strategies built on (i) semiconductor photocatalysts, connecting band structure, charge dynamics, and ROS generation to performance gains from doping/defects, plasmonic coupling, and Type-II, Z-scheme, and S-scheme heterojunctions; and (ii) MOFs/COFs as modular porous platforms for high-affinity adsorption, ion exchange, and catalysis toward metals and organic micropollutants. It also links synthesis routes to morphology, stability, scalability, and recoverability (e.g., magnetic composites), integrates case studies on dyes/pharmaceuticals, metal removal, and microplastic mitigation, and critiques translation barriers

DOI: 10.4018/979-8-3373-4402-7.ch009

(cost, long-term durability, fouling, ecotoxicity, regeneration). Future directions emphasize responsive materials, AI-assisted discovery, hybrid process integration, and circular-economy resource recovery.

1. INTRODUCTION

The availability of clean and safe water is a fundamental prerequisite for human health, ecosystem stability, and socioeconomic development. However, the world is facing an escalating water crisis, a multifaceted challenge characterized by both quantitative scarcity and qualitative degradation. The relentless pace of global industrialization, the intensification of agricultural practices, and rapid urbanization have led to the unprecedented discharge of a vast and complex array of pollutants into aquatic environments. This contamination of freshwater resources poses a severe threat to the delicate balance of ecosystems and presents significant, long-term risks to public health. Consequently, the development of efficient, sustainable, and economically viable water purification technologies has become one of the most critical scientific and engineering challenges of the 21st century.

The spectrum of aquatic pollutants is remarkably diverse, encompassing inorganic and organic compounds of varying toxicity and persistence. Key contaminant groups include heavy metal ions such as hexavalent chromium (Cr(VI)), lead (Pb(II)), and arsenic (As(V)), which are non-biodegradable and can accumulate in living organisms, causing severe health issues. Another major category is persistent organic pollutants (POPs), which includes synthetic dyes from the textile industry, pesticides from agricultural runoff, and various industrial chemicals. These compounds are often recalcitrant to natural degradation processes and can exhibit carcinogenic, mutagenic, or endocrine-disrupting properties. Furthermore, the rise of emerging contaminants, such as pharmaceuticals, personal care products, and antibiotics, presents a new frontier of challenges. The presence of these substances, even at trace concentrations, can lead to the development of antibiotic-resistant microorganisms and disrupt aquatic ecosystems.

Traditional water and wastewater treatment methodologies, which often rely on physical processes like coagulation, flocculation, sedimentation, and biological processes such as the activated sludge method, have been the cornerstone of sanitation for decades. While effective for removing suspended solids and bulk organic matter, these conventional techniques often fall short in eliminating dissolved, non-biodegradable, and low-concentration micropollutants. Moreover, these processes can be energy-intensive and may generate secondary pollution in the form of large volumes of contaminated sludge, which requires further costly and complex disposal procedures. This technological gap has catalyzed a global research effort to develop more robust and sustainable water purification strategies, with a strong focus on advanced materials and innovative chemical processes.

Among the most promising next-generation solutions are Advanced Oxidation Processes (AOPs), which are characterized by the in-situ generation of highly reactive and non-selective chemical species, primarily hydroxyl radicals (•OH) (Singh et al., 2018). More recently, sulfate radical ($\bullet SO_4^-$)-based AOPs have also gained significant attention due to their high oxidative potential and selectivity (Wang & Zhu, 2024; Wang et al., 2019). These radicals can mineralize a wide spectrum of organic pollutants into benign end-products like carbon dioxide and water. Heterogeneous photocatalysis, a subset of AOPs, has emerged as a particularly attractive green technology because it can utilize solar energy to drive these degradation reactions. This process hinges on the development of efficient semiconductor photocatalysts, materials that can absorb photons to generate electron-hole pairs, which in turn initiate the redox reactions responsible for pollutant destruction.

For decades, research in semiconductor photocatalysis was dominated by inorganic metal oxides, with titanium dioxide (TiO_2) being the most extensively studied material due to its chemical stability, low cost, and high photoactivity. However, the practical application of traditional photocatalysts like TiO_2 is hampered by several intrinsic limitations, including a wide bandgap that restricts its activation to the ultraviolet (UV) portion of the solar spectrum and a high rate of recombination of photogenerated electron-hole pairs, which lowers quantum efficiency. To overcome these challenges, extensive research has focused on engineering new photocatalytic materials with enhanced visible-light absorption and improved charge separation efficiency. This has led to the exploration of novel classes of materials, such as polymeric graphitic carbon nitride (g-C_3N_4), which possesses a suitable bandgap for visible light absorption, low cost, and high stability (Reddy et al., 2019). Similarly, conjugated polymers like polyaniline (PANI) have shown promise due to their tunable electrochemistry and visible-light photoresponse.

A powerful strategy to enhance photocatalytic performance involves the fabrication of heterostructures, where two or more semiconductor materials are coupled to promote charge separation and transfer. Advanced heterojunction designs, such as the S-scheme, have been developed to maximize the redox potential of the charge carriers, thereby boosting the generation of reactive oxygen species (ROS). Furthermore, integrating photocatalysts with other functional materials, such as graphene and its derivatives, can provide a high-surface-area support, facilitate charge transport, and enhance pollutant adsorption, leading to synergistic improvements in overall efficiency (Singh et al., 2018). The incorporation of magnetic components into these composites also offers a practical advantage by enabling easy separation and recycling of the catalyst from the treated water using an external magnetic field.

In parallel with degradative approaches like photocatalysis, adsorption remains a cornerstone technology for water purification due to its simplicity, high efficiency, and versatility in removing a wide range of pollutants. The effectiveness of an adsorbent is largely determined by its surface area, pore structure, and surface chemistry. While conventional adsorbents like activated carbon have been widely used, the quest for materials with higher capacity, selectivity, and regenerability has driven the exploration of novel porous materials. In this context, Metal-Organic Frameworks (MOFs) have emerged as a revolutionary class of crystalline porous materials. Constructed from metal ions or clusters linked by organic ligands, MOFs exhibit unprecedentedly high specific surface areas, tunable pore sizes, and easily functionalizable frameworks.

These unique properties make MOFs exceptionally promising candidates for the adsorptive removal of various water contaminants, including heavy metal ions, organic dyes, and pharmaceuticals. The ability to tailor the pore environment and introduce specific functional groups allows for the design of MOFs with high selectivity towards target pollutants. Beyond their role as passive adsorbents, MOFs are also highly versatile platforms for catalysis (Li, 2019). Many MOFs exhibit semiconductor-like behavior and can function directly as photocatalysts for pollutant degradation or for other important reactions like CO_2 reduction and water splitting. The well-defined structure of MOFs facilitates the establishment of clear structure-activity relationships, which is often challenging for conventional amorphous catalysts.

The synergy between different classes of advanced materials has led to the development of sophisticated composites and hybrid systems. MOFs can be integrated with other semiconductors to form highly efficient heterojunction photocatalysts (Wang & Zhu, 2024; Dong et al., 2024). These composites leverage the light-harvesting and charge-generating properties of the semiconductor with the high surface area and catalytic sites of the MOF (Li, 2019). For instance, MOFs can act as co-catalysts to enhance charge separation or as hosts to confine and stabilize photoactive species. The challenge of using powdered MOFs, which can be difficult to recover, is being addressed by immobilizing them on or within macroscopic supports like aerogels or by using advanced manufacturing techniques like 3D printing to create structured, easily separable adsorbents and catalysts.

Beyond the well-established fields of semiconductor photocatalysts and MOFs, the materials science toolbox for water purification continues to expand. Two-dimensional (2D) materials like MXenes have shown great potential for membrane-based separation and adsorption due to their unique lamellar structures and hydrophilic surfaces. Porous boron nitride nanosheets have demonstrated remarkable capacity for absorbing oils and organic solvents. Hydrogels, with their unique three-dimensional polymeric networks and high water content, are being engineered as platforms for solar-driven water evaporation and purification, offering a low-energy pathway to freshwater production.

Despite the tremendous progress in designing advanced materials for water remediation, significant challenges remain on the path to widespread practical implementation. Issues such as the scalability and cost of material synthesis, long-term stability under harsh operational conditions, potential ecotoxicity of nanomaterials, and the complexity of treating real-world wastewater containing multiple competing pollutants must be addressed. The future of this field will depend on a multidisciplinary approach that integrates rational materials design with principles of green chemistry, sustainable engineering, and the circular economy. This includes developing materials from abundant, non-toxic precursors, designing processes that minimize energy consumption and waste generation, and creating systems that allow for the recovery and reuse of both the material and valuable resources from the wastewater.

This review aims to provide a comprehensive and critical analysis of the state-of-the-art in advanced materials for the remediation of water pollution. We will begin by surveying the key classes of materials that have shown significant promise, with a particular focus on semiconductor photocatalysts and porous materials like Metal-Organic Frameworks. We will then delve into the diverse synthesis and modification strategies employed to enhance their performance, stability, and applicability, highlighting recent progress in creating complex heterostructures and composites. The fundamental mechanisms governing pollutant removal, including photocatalytic degradation, adsorption, and other catalytic processes such as Fenton-like reactions, will be elucidated. Finally, we will critically assess the current challenges and outline future perspectives, emphasizing the critical need to bridge the gap between laboratory-scale innovations and real-world applications to forge a sustainable path toward global water security. Building on the challenges and opportunities outlined above, the next section surveys the principal advanced material platforms for water purification and frames their functional roles within modern treatment trains.

2. KEY ADVANCED MATERIALS FOR WATER PURIFICATION

Advanced materials have become central to modern water pollution control because they can be engineered to deliver specific interfacial functions—selective adsorption, catalytic oxidation or reduction, size/charge-based separation, and electrically driven regeneration—at levels that are difficult to achieve with conventional media. To provide a coherent framework for material selection, this section organizes the discussion around a small number of material platforms and the practical questions that determine whether a platform is deployable: (i) what the material does (capture, destroy, or both), (ii) how it performs in realistic water matrices, (iii) whether it can be shaped, recovered, and regenerated, and (iv) what residual streams are created and how they are managed.

Two families dominate the advanced-materials landscape discussed here: semiconductor photocatalysts used in light-driven advanced oxidation processes (AOPs) and porous frameworks (MOFs and COFs) designed for selective adsorption and, increasingly, catalytic transformation. Semiconductor photocatalysis is often framed as a green approach because it can harness light energy to generate reactive oxygen species (ROS) that degrade organics (Garcia-Segura & Brillas, 2017; Surana et al., 2024). Porous frameworks, by contrast, are designed at the level of building blocks so that pore architecture, binding-site chemistry, and hydrolytic stability can be tuned for targeted removal of metals, oxyanions, and organic micropollutants (Li et al., 2019; Dutta et al., 2024). A third, cross-cutting group—carbonaceous materials, nanocomposites, and bio-based media—should not be treated as an afterthought. These materials frequently act as practical enablers: they stabilize catalysts, provide conductive pathways for electron transfer, create hierarchical porosity, or enable low-cost shaping into pellets, membranes, fibers, and hydrogels. In practice, many of the most field-promising systems are hybrids that combine the strengths of multiple material classes while mitigating their individual weaknesses.

2.1 Photocatalytic Semiconductor Platforms

Semiconductor photocatalysis is a light-driven AOP in which a semiconductor absorbs photons with energy at or above its bandgap, generating electron–hole pairs. When these carriers reach the surface before recombining, they can initiate redox reactions that produce ROS such as hydroxyl radicals (•OH) and superoxide ($•O_2^-$), enabling broad-spectrum degradation of organic contaminants. Because these oxidants are highly reactive and often non-selective, the value of photocatalysis is greatest for persistent organics that are poorly removed by biological treatment or that require destructive, rather than separative, approaches. However, real-world deployment depends on more than intrinsic activity: photon delivery, catalyst recovery, and long-term stability under realistic water chemistry frequently dominate system feasibility.

Benchmark oxides remain foundational because they are chemically robust and relatively inexpensive. Titanium dioxide (TiO_2) is the canonical photocatalyst, valued for stability and affordability, but its wide bandgap (~3.2 eV) restricts absorption primarily to UV light and its quantum efficiency is limited by rapid electron–hole recombination. Strategies such as non-metal doping (e.g., N, C, S) and defect engineering can extend absorption into the visible range and modify carrier dynamics (Devi & Kavitha, 2013). Metal deposition and plasmonic coupling can further enhance visible response, but these approaches introduce cost and potential leaching concerns. A particularly productive line of work is heterojunction engineering: coupling TiO_2 with complementary semiconductors can create built-in electric fields or directional band alignment that promotes charge separation (Qi et al., 2024).

ZnO provides a useful comparison point to TiO_2 because it has a similar bandgap but higher electron mobility, which can improve charge transport. In practice, ZnOs susceptibility to photocorrosion under certain conditions and its UV-limited absorption motivate the same solution set used for TiO_2: doping, surface modification, and heterojunction construction. Importantly, claims of high performance often rely on powder suspensions; for engineering translation, the more relevant demonstrations are immobilized configurations (coatings, composites, or supported nanoparticles) and performance tests in realistic matrices where carbonates, chlorides, and natural organic matter compete for radicals and suppress apparent kinetics.

Visible-light-active oxides provide another pathway to practical solar utilization. WO_3 is notable for a narrower bandgap (~2.4–2.8 eV), which enables visible absorption and makes it attractive for solar-driven applications (Dutta et al., 2021). WO_3 also exhibits strong stability under acidic conditions, which is important for industrial effluents. Yet pristine WO_3 has a key limitation: its conduction-band potential is often insufficiently negative to reduce O_2 efficiently to superoxide, constraining ROS pathways and limiting mineralization in some cases. The field has responded by building heterojunctions that pair WO_3 with partners that provide stronger reduction power (e.g., g-C_3N_4), thereby preserving visible absorption while restoring redox capability and suppressing recombination. This example illustrates a broader design rule: the most useful photocatalyst designs treat band-edge positions, carrier lifetime, and interfacial transfer as coupled variables rather than optimizing a single parameter in isolation.

Earth-abundant, narrow-bandgap semiconductors such as hematite (α-Fe_2O_3, ~2.1 eV bandgap) are appealing for cost and visible absorption, but their poor carrier mobility and short diffusion lengths make charge collection difficult. Nanostructuring, conductive scaffolds, and interfacial engineering (including photoelectrochemical architectures) can alleviate these limitations. In water treatment, hematites strongest role may be in structured electrodes or coupled photoelectrochemical systems where external bias or built-in fields guide carriers to reactive interfaces and improve robustness over long operation.

Bismuth-based semiconductors (e.g., $BiVO_4$, Bi_2WO_6, and bismuth oxyhalides) illustrate how crystal structure can be exploited to improve charge separation. Layered motifs can facilitate spatial separation of electrons and holes, improving photocatalytic response under visible light (Meng & Zhang, 2016; Liu et al., 2020). Composite design is again central: ternary systems and junctions can broaden absorption and promote directional carrier flow, while surface engineering can increase adsorption of target pollutants and thus improve apparent kinetics. For practical relevance, these systems should be evaluated not only in model-dye tests but also against trace micropollutants in complex water matrices, with explicit reporting of transformation products and mineralization indicators.

Metal-free photocatalysts, especially graphitic carbon nitride (g-C_3N_4), have gained prominence because they avoid metal leaching risks while offering visible-light absorption (bandgap ~2.7 eV) and chemical stability (Reddy et al., 2019). The main bottlenecks of pristine g-C_3N_4 are limited surface area, incomplete visible utilization, and high recombination. Practical enhancement strategies fall into three merged categories rather than many subtopics: (i) morphology control (nanosheets, porous structures) that increases accessible surface and shortens diffusion lengths; (ii) electronic tuning via doping, defects, and donor–acceptor architectures that improve absorption and internal charge separation; and (iii) composite formation with oxides, carbons, or co-catalysts that accelerate surface reactions and provide conductive transport. Exfoliated or ultrathin g-C_3N_4 nanosheets can increase active-site availability and improve performance by reducing bulk recombination (Huang et al., 2018), while halogen modification can tune photoredox behavior (Lan et al., 2016).

From an engineering viewpoint, however, the key question is not "which semiconductor-based design remains most robust in laboratory batch tests," but "which semiconductor-based design can be implemented in a reactor with stable, reproducible performance." Photocatalysts must be immobilized or otherwise recovered; they must tolerate abrasion and cleaning; and they must operate under realistic irradiation and hydraulic conditions. Strategies include coating catalysts on glass fibers or ceramics, integrating catalysts into membranes or monoliths, and designing photoelectrochemical electrodes. When compared on a system basis, designs that simplify recovery and reduce operator burden often outperform higher-activity powders that require complex downstream separation.

2.2 Porous Frameworks and Engineered Sorbents

Porous materials complement photocatalysts by offering strong, selective capture of contaminants, especially metals and oxyanions, where destructive pathways can be slow, energy-intensive, or impractical. Metal–organic frameworks (MOFs) and covalent organic frameworks (COFs) are distinguished from conventional adsorbents because their performance can be designed from first principles. MOFs consist of metal nodes connected by organic linkers, producing crystalline networks with high surface area and tunable pore size. COFs are fully covalent networks built from light elements, offering an all-organic route to ordered porosity and, in many cases, improved chemical stability (Li et al., 2019; Dutta et al., 2024). In the original material, frameworks were presented with multiple micro-headings; here the discussion is consolidated around the three design variables that most directly predict field performance: pore architecture, binding-site chemistry, and hydrolytic stability.

Pore architecture governs accessibility and mass transport. Micropores can deliver high affinity and size selectivity, but they may suffer from diffusion limitations in flow-through operation, particularly for larger organics. Introducing hierarchical porosity (micro–meso combinations) can improve kinetics by providing fast transport channels while retaining high internal surface area. In a practical sense, the best framework is not necessarily the one with the highest BET surface area, but the one that maintains accessible porosity in real water under repeated cycles and that can be shaped into pellets, coatings, or membranes without severe pore blockage. This shift—from surface area as a headline metric to accessibility and durability as decision metrics—is essential for translational relevance.

Binding-site chemistry determines selectivity and regenerability. For cationic metals, inner-sphere complexation dominates when framework ligands provide donor atoms (N, O, S) capable of coordination. "Soft" metals often bind strongly to sulfur-containing sites, while "hard" metals prefer oxygen donor motifs. For oxyanions (e.g., chromate, arsenate, phosphate), selectivity must be engineered against competition from background sulfate and bicarbonate. A rational design strategy is to combine positive internal surfaces with accessible metal centers capable of ligand exchange and with hydrogen-bond donor arrays that stabilize tetrahedral anion geometries. Crucially, selectivity claims require competitive, multi-ion testing; single-analyte isotherms are not sufficient to predict performance in groundwater or industrial waters where multiple ions compete for the same sites.

Hydrolytic stability is the governing constraint for many frameworks. Ultralarge surface area is irrelevant if the framework hydrolyzes, collapses, or leaches metal ions under realistic pH and ionic strength. Stability can be improved by selecting strong nodes and robust linkages, and by post-synthetic modifications that increase hydrophobicity or reduce defect-driven hydrolysis. Composite formation provides an additional stabilization route: embedding frameworks in carbon, polymer, or ceramic scaffolds can protect them from mechanical attrition while maintaining pore access.

Frameworks are not only adsorbents; they can also be catalytic platforms and catalyst precursors. Metal nodes can activate oxidants such as peroxymonosulfate or persulfate, or they can support photoactive components in hybrid systems. Controlled conversion of MOFs into derivative materials (e.g., metal/metal-oxide nanoparticles embedded in porous carbon) can yield robust, conductive catalysts that retain high surface area and improve electron transfer. Such derivatives are especially relevant for hybrid AOPs such as photo-Fenton, electro-Fenton, and persulfate activation, where conductivity and redox cycling influence kinetics.

COFs deserve special attention because linkage chemistry can be chosen to balance crystallinity, pore order, and water stability. Hydrolytically robust linkages enable operation in aqueous environments across a broader pH range, while donor–acceptor architectures can introduce internal charge transfer that extends light absorption—relevant both for adsorption and for COF-based photocatalysis. For adsorption applications, COFs can present dense arrays of functional motifs along pore walls, enabling high affinity and potentially cooperative binding. Yet these advantages only translate into performance if synthesis and processing yield accessible porosity and sufficient structural integrity.

Regeneration and life-cycle implications close the gap between materials design and deployable technology. Framework regeneration can be achieved by pH switching, chelating washes, ionic-strength adjustments, solvent rinsing for hydrophobic organics, or thermal/oxidative steps for strongly bound species. Each regeneration route creates tradeoffs: mild regenerants preserve structure but may leave residual loading; aggressive regenerants improve capacity recovery but can accelerate degradation or create secondary waste streams. Practical evaluation should therefore include multi-cycle tests, leaching analysis, and performance in representative real waters.

2.3 Carbonaceous, Composite, and Bio-based Materials

Carbon-based materials occupy a special role because they bridge adsorption, catalysis, and process engineering. Activated carbon remains a workhorse for broad-spectrum organic adsorption, while graphene-related materials and carbon dots are frequently introduced to enhance conductivity, provide high-surface-area scaffolds, and facilitate electron transfer. Strategies such as coupling graphene with semiconductors can improve charge separation, increase adsorption of target molecules near catalytic sites, and enhance disinfection performance (Singh et al., 2018). Carbon dots offer tunable surface chemistry and photophysical properties, enabling sensitization of photocatalysts or direct photocatalytic degradation pathways under visible light (Kaur et al., 2024).

A major practical value of carbonaceous materials is their ability to activate oxidants in AOPs. Carbon-based catalysts can activate peroxymonosulfate to generate reactive radicals and non-radical oxidizing species, offering a route to oxidative degradation that complements semiconductor photocatalysis (Hao et al., 2023). Because many oxidant-activation processes can proceed without direct photon input, they may be easier to implement in enclosed reactors or in low-light settings. However, reagent consumption and by-product formation become important decision factors; thus, designs should be benchmarked using reagent-normalized metrics and screening for transformation products.

Nanocomposites should be viewed less as a separate topic and more as the dominant implementation pattern for advanced materials. Composite design addresses the core translation barriers identified across platforms: recovery, stability, and performance in complex matrices. Embedding catalytic sites within a conductive aerogel can improve mass transport and facilitate electron transfer in electrochemical AOPs. A representative example is a single-atom NiCo alloy embedded in a graphene aerogel for electro-Fenton degradation of emerging contaminants (Dong et al., 2024). The broader principle is consistent: hybridizing catalytic centers with conductive, porous scaffolds can increase utilization of active sites while enabling structured forms that are compatible with reactors.

Bio-based materials extend sustainability and cost advantages and can provide unique functionalities. Cellulose-based superabsorbent hydrogels offer biodegradable, renewable platforms for pollutant capture and water management, and they can host functional additives such as nanoparticles or chelating groups (Omidian et al., 2024). Bio-derived carbons (e.g., biochar) are also notable as low-cost supports for catalyst immobilization and as adsorbents with tunable surface functionalization. For deployment, performance consistency is essential: feedstock variability can significantly alter pore structure and surface chemistry. Standardized characterization and quality-control specifications therefore become part of the technology, not an optional add-on.

2.4 Guidelines for Engineering Selection and Deployment

In practical projects, advanced materials are chosen to fill a role within a treatment train rather than to act as standalone solutions. Photocatalysts are most valuable when destructive removal is required for persistent organics; frameworks and sorbents are most valuable when selective capture is needed for metals, oxyanions, and difficult-to-oxidize species; and carbonaceous/composite/bio-based media often enable shaping, stability, and hybridization. Selection should be driven by contaminant class, water-matrix complexity, operational constraints, and residual management pathways.

Performance should be benchmarked using metrics that scale. For adsorption-driven processes, batch isotherms provide screening information but are insufficient for design; column tests and breakthrough curves under representative competing ions are essential for estimating bed volumes treated to a defined effluent criterion. For photocatalysis and oxidant activation, time-resolved kinetics should be accompanied by energy and reagent accounting (energy per treated volume, oxidant dose per mass removed) and by evidence that the process is truly destructive rather than producing partially oxidized by-products. When possible, mineralization indicators (e.g., TOC reduction) and targeted transformation-product screening should be reported.

Shaping and immobilization are first-order constraints. Powder catalysts and high-surface-area sorbents may deliver impressive intrinsic performance yet be impractical due to pressure drop, abrasion, particle release, and recovery needs.

Common deployment forms include coated substrates, pellets and granules for packed beds, fibers for low-pressure modules, and composite membranes or monoliths for integrated separation and reaction. Shaping can reduce apparent kinetics by introducing diffusion limitations; therefore, design should explicitly address mass-transfer resistances.

Durability and safety requirements become more stringent when treated water is intended for reuse or drinking. Leaching of metals, ligands, or additives must be evaluated under relevant pH, ionic strength, oxidant exposure, and cleaning protocols. For nanomaterials, particle-release testing under hydraulic shear and backwashing conditions is important to avoid secondary contamination. These requirements often shift the best material choice toward systems with robust supports and well-defined regeneration procedures, even when intrinsic activity is modest.

Having established the major material platforms and their functional roles, the following section examines synthesis, modification, and shaping strategies that ultimately determine stability, recoverability, and scalability in practical deployment.

3. SYNTHESIS, MODIFICATION, AND SHAPING STRATEGIES FOR ADVANCED WATER-TREATMENT MATERIALS

Advanced materials rarely fail in the laboratory because of a lack of intrinsic reactivity; they fail in the field because the synthesis route, post-treatment steps, and final physical form were not designed for the realities of water treatment. Real waters contain natural organic matter, competing ions, suspended solids, and disinfectants; reactors impose hydrodynamic shear, pressure drop constraints, and long run-times; and operators need media that can be regenerated, handled, and disposed of safely. For these reasons, synthesis should not be treated as a purely chemical exercise. It is the first stage of process engineering, because it controls surface chemistry, porosity, electronic structure, mechanical strength, and the probability of leaching or particle release.

This section merges closely related themes into four integrated discussions: (i) cross-platform design logic for synthesis and functionalization, (ii) how modification strategies tune selectivity and kinetics while preserving stability, (iii) shaping and immobilization methods that convert powders into deployable forms, and (iv) scale-up and reporting practices that make material claims actionable for pilots. The aim is practical: to link the chemistry of making materials to the engineering of using them.

3.1 A Strategic Framework for Functional and Stable Material Synthesis

Across material families—photocatalysts, porous sorbents, membranes, and electrochemical electrodes—synthesis routes differ, but the governing logic is surprisingly consistent. Water treatment demands a combination of accessible active sites, transport pathways for water and solutes, and a stable backbone that survives hydrolysis, oxidation, abrasion, and repeated regeneration. The synthesis route sets these attributes through choices about precursors, solvents, thermal history, and structure-directing agents.

For inorganic photocatalysts (e.g., TiO_2, doped oxides, and composite semiconductors), common bottom-up routes include sol–gel synthesis, hydrothermal or solvothermal growth, flame or spray pyrolysis, and controlled calcination. Sol–gel methods can deliver high surface area and tunable crystallinity, but residual organics and poor phase control can create unstable surface states. Hydrothermal routes offer control over morphology (nanorods, nanosheets, hierarchical structures) and can promote specific facets, yet they may introduce synthesis additives that require thorough removal. Calcination improves crystallinity and charge transport but can collapse pores or sinter nanoparticles, reducing accessible surface area. In practice, synthesis conditions should be chosen not only for high activity but also for manufacturability and reproducibility; small variations in temperature ramp or precursor purity can meaningfully change band-edge positions, defect density, and long-term stability.

Porous frameworks and functional sorbents (including MOFs, COFs, functionalized polymers, and bio-based carbons) typically rely on condensation chemistry and self-assembly. Here, the central trade-off is between high porosity and water stability. Highly porous structures may be vulnerable to hydrolysis if metal–ligand bonds or linkages are weak. Synthesis choices that improve stability—stronger coordination nodes, more robust covalent linkages, post-synthetic crosslinking, or hydrophobic pore environments—can reduce pore accessibility if not carefully managed. A practical design rule is to treat water stability tests (multi-day soaking, pH cycling, oxidant exposure) as an explicit synthesis target rather than a late-stage validation step.

Carbon-based materials provide a useful bridge between high performance and scalable production. Activated carbon and engineered biochars can be produced at scale, and their functionality is often introduced through activation (steam/CO_2, chemical agents), heteroatom incorporation (N, S, P), or surface oxidation/reduction. The synthesis lever is not only surface chemistry but also pore-size distribution. For trace organics and PFAS, micropores can drive adsorption capacity, while mesopores often govern kinetics in real waters where foulants block access. As a result, activation routes should be selected to create hierarchical porosity and to minimize ash or unstable surface groups that could leach.

For membranes and coatings, synthesis is best interpreted as interfacial engineering. Whether the active layer is formed by interfacial polymerization, sol–gel coating, layer-by-layer assembly, or grafting of functional groups, the outcome must balance selectivity with permeability and fouling resistance. Many highly selective coatings fail because they are mechanically weak or chemically unstable under cleaning conditions. Therefore, the synthesis endpoint should include compatibility with backwashing, chemical cleaning, and long-term hydraulic cycling.

3.2 Modification for Performance

Modification is often presented as an open-ended toolbox—doping, defect engineering, functional group grafting, composite formation, and surface texturing. In water treatment, the most valuable modifications are those that create a predictable shift in system-level behavior: selective binding, faster mass transfer, or more efficient reaction pathways. Modifications that improve these behaviors but introduce leaching, instability, or hard-to-control synthesis variability often reduce practical relevance.

For photocatalysts, doping and heterojunction engineering are used to extend light absorption into the visible spectrum and to reduce electron–hole recombination. Non-metal doping, transition-metal doping, and semiconductor coupling can each shift optical properties, but they can also create mid-gap states that accelerate photocorrosion or generate unwanted by-products. Practical screening should therefore pair activity data with stability evidence (multi-cycle tests, post-run crystal phase and surface analysis), and should include realistic water matrices where radical scavengers and co-contaminants can dominate performance.

For sorbents, functionalization is most useful when it creates a clear binding motif that resists competition. Examples include grafting amine, quaternary ammonium, thiol, or chelating groups; introducing fluorophilic domains for PFAS; or tailoring pore polarity to target specific neutral organics. These modifications should be evaluated using competitive adsorption because effective selectivity in real water can collapse when natural organic matter occupies binding sites. Regenerability is the second practical criterion: strongly binding groups may require aggressive eluents that create difficult residual streams, so modification should be co-optimized with a feasible regeneration pathway (salt, pH swing, solvent-minimized thermal regeneration, or electro-regeneration).

Composite formation unifies adsorption and catalysis. Capture-and-destroy designs—where a sorbent is coupled to a catalyst or redox-active component—can reduce secondary waste and simplify treatment trains. Yet composites introduce interfaces that can fail mechanically or chemically. Good composite design requires matching chemical compatibility and preserving pore accessibility across phases. A defensible approach is to quantify both performance and integrity: capacity or kinetics, plus particle release, leaching, and mechanical attrition under agitation or flow.

Electrochemical materials (capacitive deionization electrodes and electrocatalysts) follow a distinct modification logic: conductivity, wettability, and redox stability are often more important than maximizing surface area. Surface functionalization can improve ion selectivity and wettability, while conductive additives or graphitized domains reduce resistive losses. However, unstable binders and corrosion of current collectors can dominate failure. Electrochemical modification should therefore be validated using long-cycle tests with realistic voltage windows and post-run analysis to identify degradation pathways.

3.3 Shaping and immobilization: translating powders into deployable media and modules

The most decisive step after synthesis is shaping. Many advanced materials are discovered as powders, yet most treatment processes require structured forms: packed beds, cartridges, coated reactors, membranes, or electrodes. Shaping alters mass transfer, pressure drop, photon penetration, and durability, and it can change the apparent ranking of candidate materials. Consequently, shaping should be treated as part of the material design space rather than a late-stage packaging step.

For adsorption systems, deployable forms include granules, pellets, monoliths, and composite beads. Granulation typically uses binders that must preserve pore accessibility while increasing mechanical strength. The primary risks are pore blockage, binder leaching, and reduced kinetics due to diffusion limits. Best practice is to report both intrinsic capacity and shaped-form performance in flow-through tests, because diffusion and tortuosity can shift breakthrough behavior dramatically. Attrition tests and repeated backwash simulations further indicate whether a medium is likely to generate fines in operation.

For photocatalysts, immobilization on supports (glass, ceramics, fibers, stainless steel meshes) addresses catalyst recovery and reduces particle release. Yet immobilization introduces optical and hydrodynamic constraints: thick films can self-shade, while thin films can limit active mass. Support selection should therefore reflect reactor design. Flat-plate reactors favor uniform coatings; packed photoreactors benefit from coated beads or structured packing; and photocatalytic membrane reactors integrate catalyst function with separation, but require co-design of light delivery, fouling control, and chemical stability.

Membrane-based systems represent shaping at the module level. When advanced materials are used as coatings or interlayers, durability under crossflow shear and cleaning chemicals is crucial. Surface-grafted functionalities that improve hydrophilicity or charge selectivity can reduce fouling, but only if they remain attached. Layer-by-layer coatings and sol–gel layers can be effective, yet they must be tested for long-term stability in oxidants and variable pH. Reporting the evolution of permeability and rejection over extended cycles provides stronger evidence than a single initial data point.

Electrochemical systems add architectural requirements: porous flow-through electrodes, carbon felts, structured current collectors, and binder-free monoliths. Architecture determines hydraulic resistance and contact efficiency, while binder and collector stability influence lifetime. In practical terms, a high-performance electrode maintains capacity and low resistance over many cycles without producing particulate debris or dissolved contaminants. Cycling protocols and post-run characterization (surface chemistry, morphology, mechanical integrity) should therefore be standard.

3.4 Scale-up and reporting practices that make synthesis outcomes actionable

Scale-up is where many promising materials stall. The obstacles are usually concrete: expensive precursors, long solvothermal steps, hazardous solvents, or products that are difficult to wash, activate, and shape. A translationally relevant synthesis plan should document yield, major cost drivers, solvent recovery options, and the feasibility of continuous or semi-continuous production routes. In many cases, spray-drying, extrusion, and roll-to-roll coating are more scalable than batch crystallization or multi-step solvent exchange (Shannon et al., 2008).

Reproducibility is equally important. Small differences in activation temperature, washing protocol, or drying method can change surface groups and pore accessibility, leading to conflicting performance across laboratories. Reporting should therefore include key synthesis parameters (temperatures, times, pH, precursor ratios), post-treatment steps (washing solvents, activation conditions), and quality-control metrics such as surface area, pore-size distribution, and composition. When materials are modified or formed into composites, authors should quantify component loadings and provide evidence that active sites remain accessible.

Because water treatment is a systems problem, the most useful synthesis studies connect material properties to process conditions. For sorbents, this means column evidence (or at least kinetic parameters compatible with column modeling), regeneration performance, and mass balance. For photocatalysts, it means photon flux, reactor geometry, and energy-normalized metrics. For membranes, it means extended fouling and cleaning cycles. For electrochemical systems, it means stable cycling in realistic electrolytes with clear reporting of voltage windows and energy consumption (Xu et al., 2014; Fujishima & Honda, 1972).

Finally, synthesis agendas should include safe-by-design checks. Even when the pollutant is removed, the treatment can be undermined if the material releases metals, ligands, nanoparticles, or persistent by-products. Minimal screening can include leaching tests under relevant pH and oxidant exposure, abrasion tests for particle release, and targeted by-product screening for catalytic processes. These checks provide the evidence base that allows utilities and industrial decision makers to consider advanced materials for pilots (Shannon et al., 2008). These synthesis and shaping choices directly influence interfacial chemistry and mass transport, which in turn govern the fundamental pollutant-removal mechanisms discussed in the next section.

4. MECHANISMS OF POLLUTANT REMOVAL

Mechanistic understanding is essential for moving advanced materials from promising laboratory prototypes to reliable water-treatment components. Mechanisms explain why a material works, what will cause it to fail, and how it should be integrated into reactors, columns, or membranes. This section merges closely related concepts into two connected themes: (i) reaction-driven removal, with photocatalysis as the primary example, and (ii) capture-driven removal, centered on adsorption and its variants. Rather than treating these mechanisms as isolated categories, the discussion emphasizes shared design variables—interfacial chemistry, mass transfer, and the influence of real-water matrices—that control performance and practical relevance. (Xu et al., 2014; Fujishima & Honda, 1972)

4.1 Reaction-Driven Removal: Photocatalysis and Reactive-Oxygen Pathways

Photocatalytic degradation is often presented as a three-step sequence—light absorption, charge separation, and surface redox reactions—and this framing remains a useful starting point. Upon irradiation with photons of sufficient energy, electrons are excited from the valence band to the conduction band, leaving holes in the valence band. The central efficiency bottleneck is not the creation of charges but their survival: electron–hole pairs recombine rapidly unless the material structure and surface chemistry promote separation and migration to reactive sites. This is why heterojunction design, defect control, and co-catalyst placement are among the most common strategies for improving photocatalytic performance. The photocatalytic process for degrading organic pollutants (OPs) can be summarized in three key steps, as illustrated in **Figure 1** (Xu et al., 2014; Fujishima & Honda, 1972).

Figure 1. Schematic illustration of the fundamental mechanism of semiconductor photocatalysis. Upon light irradiation, electron-hole pairs are generated. These charge carriers migrate to the surface to produce reactive oxygen species (ROS) such as hydroxyl radicals (•OH) and superoxide radicals (•O_2^-), which then degrade organic pollutants into harmless products like CO_2 and H_2O.

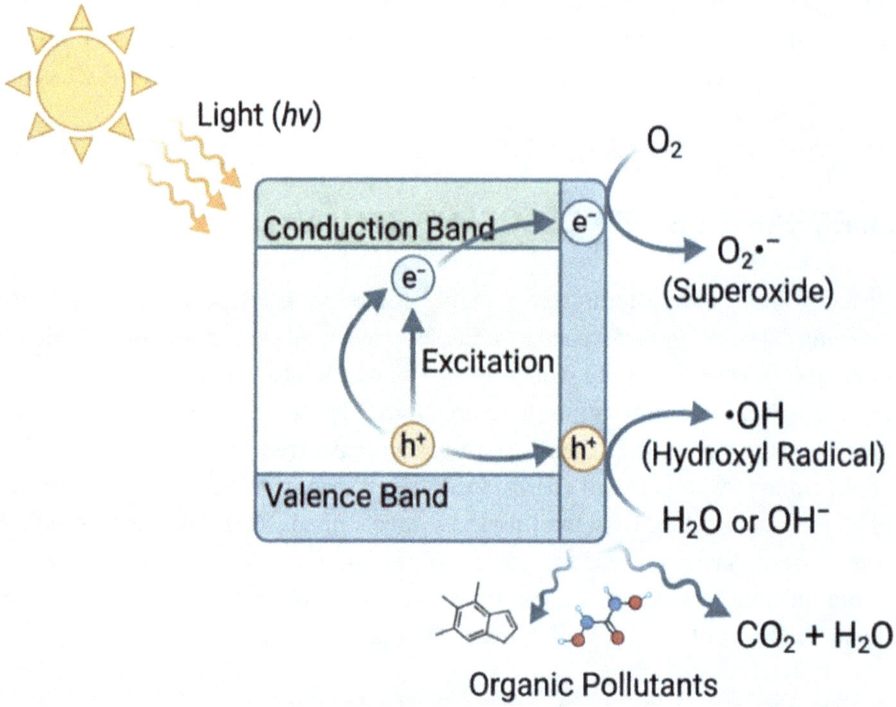

Once charges reach the surface, they drive redox chemistry that creates reactive oxygen species (ROS) and other short-lived oxidants. Holes can oxidize adsorbed organics directly or react with water/hydroxide to form hydroxyl radicals (•OH). Electrons can reduce dissolved oxygen to form superoxide radicals (•O_2^-), which can subsequently generate additional oxidizing species. These ROS attack organic pollutants and, under favorable conditions, can progress from partial transformation to mineralization into CO_2 and H_2O. From a practical standpoint, mineralization is often more relevant than parent-compound disappearance, because partial oxidation can generate intermediates that remain toxic or persistent. Therefore, mechanistic work should connect charge-carrier behavior to system-level outcomes such as

TOC reduction, identifiable intermediates, and the fate of heteroatoms (e.g., N, S, halogens) where applicable.

Modern heterojunction concepts refine the classical model by engineering charge-transfer pathways. Type-II junctions can separate charges but may reduce redox potentials, while Z-scheme and S-scheme designs aim to preserve the strongest oxidizing holes and the strongest reducing electrons. In an S-scheme heterojunction, a built-in electric field and band bending promote recombination of "low-value" charges while retaining carriers with high redox power, improving the driving force for degradation. This mechanistic intent matters operationally: higher redox potential can counteract inhibition from carbonate, chloride, and natural organic matter that scavenge radicals in real waters. However, stronger oxidizing conditions can also increase catalyst corrosion or accelerate formation of by-products, so stability and by-product screening must accompany activity claims. The development of advanced heterojunctions, such as the S-scheme, has refined this model (**Figure 2**).

Figure 2. Comparison of charge transfer mechanisms in different semiconductor heterojunctions. (a) A Type-II heterojunction separates electrons and holes into different materials, but with reduced redox potentials. (b) A direct Z-scheme and (c) an S-scheme heterojunction preserve the strongest reduction and oxidation potentials by facilitating the recombination of useless charge carriers, leading to enhanced photocatalytic efficiency.

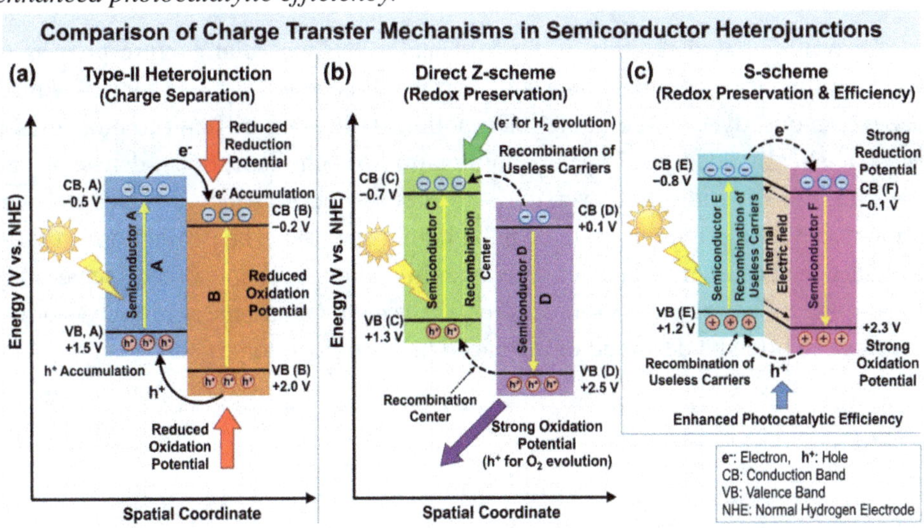

For practical relevance, photocatalysis should be discussed as an engineered process rather than only a material property. Three cross-cutting factors dominate field performance. First, photon delivery determines the usable reaction volume: the same catalyst can appear highly active in a shallow batch reactor but underperform in deeper or turbid systems where light attenuation is severe. Second, mass transfer couples with surface kinetics; when pollutants must diffuse through boundary layers or through immobilized catalyst films, intrinsic kinetics may not control the overall rate. Third, water-matrix effects can flip the dominant reaction pathway. For instance, bicarbonate and natural organic matter can quench •OH and shift oxidation toward less reactive species, lowering apparent rates even if the catalyst is unchanged. A mechanistic section that supports translation should therefore specify how catalyst design interacts with reactor optics, hydrodynamics, and the expected matrix composition.

Finally, photocatalysis often benefits from hybridization with capture processes. Adsorption can concentrate trace organics near reactive sites, while catalytic degradation can regenerate sorbents and reduce secondary waste. This "capture-and-destroy" logic is mechanistically coherent: adsorption enriches interfacial coverage, improving the probability that short-lived ROS react with the target rather than being wasted in bulk solution. The practical advantage is also significant—hybrids can reduce chemical dosing and simplify residual management—but they require careful design to avoid pore blockage and to ensure that catalytic components do not leach during operation (Xu et al., 2014; Fujishima & Honda, 1972).

4.2 Capture-driven removal

Adsorption-based removal is fundamentally an interfacial capture process in which pollutants accumulate on a solid surface. It is widely applied for heavy metals, oxyanions, dyes, and organic micropollutants that are difficult to biodegrade. Mechanistically, adsorption spans a continuum from physisorption—driven by van der Waals forces and often reversible—to chemisorption, which involves stronger ionic or covalent interactions and can be highly specific. The distinction is useful for practice because it influences regenerability: strongly chemisorbed species may require aggressive regeneration or may render the media single-use, while physisorbed species can often be desorbed or displaced under milder conditions.

Figure 3. Schematic representation of pollutant removal using porous materials like MOFs and COFs. The high surface area and tunable pores allow for efficient capture of pollutants (e.g., heavy metals, organic dyes) through mechanisms such as pore filling, surface complexation, and π-π stacking.

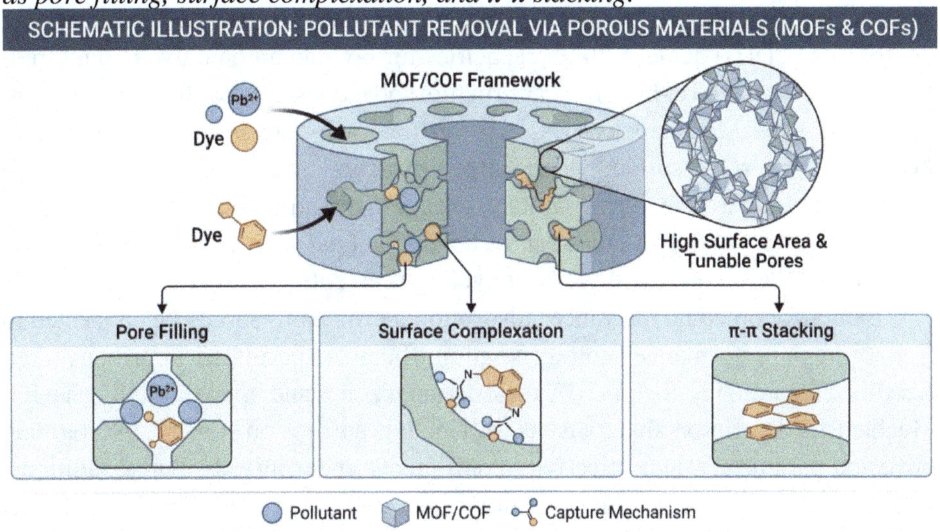

Porous materials such as activated carbon, MOFs, and COFs (**Figure 3**) are effective adsorbents primarily because they combine high surface area with tunable pore environments. Their removal performance usually reflects multiple concurrent mechanisms: (i) pore filling, where molecules are physically accommodated in micro- and mesopores; (ii) surface complexation, in which functional groups or open metal sites coordinate metal ions; (iii) ion exchange, where charged pollutants replace counter-ions; and (iv) π–π interactions, where aromatic pollutants associate with aromatic linkers or graphitic domains. Importantly, these mechanisms are not additive in a simple way: increasing functional-group density can enhance binding but also reduce pore volume; increasing microporosity can raise capacity but slow kinetics in real waters where foulants block access (Furukawa et al., 2013; Diercks & Yaghi, 2017).

A key practical theme is competition. Real waters contain natural organic matter, inorganic anions (e.g., sulfate, phosphate, bicarbonate), and mixtures of organics. These species compete for binding sites and can reduce effective capacity or shift selectivity. Mechanistic discussions should therefore include competitive adsorption and matrix effects as first-class topics, not as footnotes. For example, a sorbent designed for oxyanion removal may show excellent capacity in single-solute tests but perform poorly when phosphate competes strongly for the same sites. Similarly, aromatic pollutants may be displaced when dissolved organic matter occupies π–π interaction domains. A practitioner-oriented mechanism section should explicitly state which binding motifs dominate and which competitors are most disruptive (Ahmad et al., 2014).

The section you provided includes a useful example demonstrating how structure and reactivity couple in adsorption: hierarchical calcined Ni/Mg/Al layered double hydroxides (LDHs) achieved high capacities for both an anionic dye (Congo red) and Cr(VI), attributed to hierarchical structure and reconstruction that created abundant active sites. Mechanistically, LDHs illustrate why "adsorption" is often more dynamic than a static surface picture. Reconstruction, surface hydroxylation, and ion exchange can evolve during contact, changing the number and type of binding sites. In practice, such dynamics can be beneficial (self-activation) or problematic (structure collapse), so stability under cycling is essential.

To improve practical relevance, adsorption mechanisms should be connected to flow-through performance and regeneration. Batch isotherms are informative, but fixed-bed operation is governed by mass-transfer zones and breakthrough behavior. Mechanistic descriptors that translate well include adsorption kinetics, intraparticle diffusion parameters, and selectivity coefficients under mixed-solute conditions. Likewise, the feasibility of regeneration is not merely an operational detail; it is part of the mechanism because it reveals whether binding is primarily physical, electrostatic, or covalent. Reporting should therefore include at least a small number of regeneration cycles with mass balance: how much pollutant is desorbed, whether capacity declines, and whether the media releases components that could create secondary contamination.

Taken together, photocatalysis and adsorption represent two ends of a spectrum: reaction-driven destruction and capture-driven separation. Modern advanced materials increasingly blur this boundary by combining adsorption with catalytic or electrochemical regeneration, or by integrating functional sorbents into membranes and structured reactors. Mechanistic clarity matters most when it guides engineering decisions: which species will inhibit performance, what operational window preserves stability, and how residuals should be managed. A mechanism section that makes these links explicit provides the strongest foundation for selecting materials that will succeed beyond laboratory conditions (Xu et al., 2014; Fujishima & Honda, 1972). With the mechanistic framework in place, the subsequent section interprets how these principles translate into performance across application domains and representative case studies under realistic water matrices.

5. APPLICATIONS AND CASE STUDIES

Advanced materials are now used across a wide range of water-treatment objectives, from the polishing removal of trace organic micropollutants to the capture of toxic metal(loid)s and the control of emerging particle pollutants such as microplastics. A recurring lesson from the past decade is that successful applications rarely depend on a single material property (e.g., the highest adsorption capacity or the fastest apparent degradation rate). Instead, success depends on how a material is embedded in a treatment train, how it behaves under realistic water matrices, and whether the process can be operated and maintained at reasonable cost while managing residuals responsibly (Yang et al., 2024; Shannon et al., 2008).

To enable a coherent synthesis across studies and contaminant classes, this section consolidates closely related topics into three integrated discussions. First, it summarizes application archetypes for major contaminant classes (organics, metal(loid)s/ oxoanions, and emerging contaminants) using a shared process-engineering lens. Second, it expands the microplastics discussion into a practical removal versus degradation framework and highlights hybrid capture-and-destroy module concepts. Third, it provides a case-study template and decision criteria that help translate laboratory material performance into pilot- and field-ready evidence (Yang et al., 2024; Shannon et al., 2008).

5.1 Application archetypes across contaminant classes

In practice, advanced materials enter water treatment through a limited set of unit-process roles. The most common roles are: (i) capture media used in packed beds or cartridges (adsorbents, ion exchangers, and functionalized porous solids), (ii) reactive materials that transform pollutants (photocatalysts, electrocatalysts, and oxidant-activating composites), and (iii) selective barriers or interfaces (membranes and functional coatings). The key to synthesis across case studies is to evaluate each material in the role it is expected to play, rather than comparing materials across incompatible configurations (Xu et al., 2014; Fujishima & Honda, 1972).

Organic pollutants (dyes, phenols, pesticides, pharmaceuticals, and personal-care products) are often addressed with catalytic destruction because complete removal by adsorption alone can generate large volumes of spent media. Photocatalysis and related advanced oxidation approaches are attractive because they can mineralize or transform organics into less harmful products, ideally toward CO_2 and H_2O. However, in real waters the reaction environment is frequently dominated by light delivery, radical scavenging by natural organic matter (NOM) and bicarbonate, and catalyst recovery. This is why many practically relevant demonstrations use immobilized catalysts (coated supports, catalytic membranes, or structured packings) and report energy-normalized metrics. In electrochemical advanced oxidation, a key translational advantage is the ability to generate reactive species on demand and to integrate oxidation with regeneration of catalytic surfaces.

A representative example illustrating the power of engineered nanocomposites is the electro-Fenton degradation of emerging organics using a single-atom NiCo alloy embedded in a graphene-aerogel structure (Dong et al., 2024). The practical importance of this case is not merely high removal rates; it also demonstrates how architecture (a conductive, porous scaffold) can combine mass transfer, electron transport, and catalytic site accessibility in a single material form. When evaluating such systems, it is essential to distinguish parent-compound disappearance from detoxification and mineralization, and to verify stability across repeated cycles under realistic electrolyte compositions.

For heavy metals, adsorption remains the dominant strategy because many targets are not readily destroyed and must be captured with high selectivity. Advanced porous materials, including MOFs and COFs, offer tunable binding motifs that can sequester oxo-pollutants such as chromate and arsenate at low concentrations (Dutta et al., 2024). Here, the most actionable evidence for engineers is not a single batch capacity but fixed-bed breakthrough under competing ions, regeneration performance, and leaching resistance. Selective binding often relies on electrostatics, coordination interactions, ion exchange, and pore-size/polarity matching; these interactions can be strongly disrupted by NOM and high ionic strength, making competitive testing essential.

For emerging contaminants, the application logic often shifts from "bulk removal" to "risk-driven polishing" because target concentrations can be extremely low and regulated endpoints may be stringent. Emerging contaminants include PFAS, endocrine disruptors, antibiotic residues, disinfection by-product precursors, and particulate pollutants. In these contexts, advanced materials are frequently used as polishing steps downstream of conventional treatment, or as selective modules within decentralized systems. Hybrid trains—capture followed by targeted destruction of concentrates—are particularly attractive because they can reduce secondary waste while maintaining reliable compliance.

Across all contaminant classes, the best-performing material in the laboratory may fail in application if it cannot be shaped into a stable form or if regeneration produces problematic residual streams. Therefore, application-oriented case studies should report: (i) the treated-water objective (compliance limit, reuse target, or toxicity endpoint), (ii) the operational configuration (batch, continuous flow, membrane module, packed bed), (iii) stability and regeneration protocols, and (iv) residual management and by-product screening. These items enable meaningful synthesis across studies.

5.2 Microplastics and related emerging particles

Microplastics and nanoplastics are increasingly treated as a distinct class of contaminants because they behave differently from dissolved chemicals. They span diverse sizes and shapes (fibers, fragments, films, beads), polymer types (e.g., PE, PP, PS, PET, PVC), and additive packages. They can also act as vectors by sorbing hydrophobic organic pollutants and binding metals, and they may host biofilms, creating mobile ecological niches. Their presence in urban waters is driven by fragmentation of larger plastics, tire and textile abrasion, and direct release from consumer and industrial sources (Yang et al., 2024).

From an engineering perspective, the central question is whether a process should prioritize removal, degradation, or a controlled combination of both. Removal is typically achieved through physical separation—screening, sedimentation, coagulation–flocculation, filtration, and membrane processes. These approaches can provide high capture efficiency, but they mainly transfer the pollutant to another phase (sludge or concentrate) without eliminating polymer mass. Degradation aims to chemically transform polymers, ideally reducing persistence and hazard. Yet partial oxidation can produce smaller fragments or soluble oligomers that may be more mobile or bioavailable. Therefore, a practical framework for material-enabled microplastics control must explicitly evaluate both capture efficiency and the risk profile of transformation products (Elimelech & Phillip, 2011; Yang et al., 2024).

Advanced oxidation processes (AOPs) have been proposed as promising tools for microplastics mitigation (Kumar et al., 2024). Photocatalysis can generate reactive oxygen species (ROS) that initiate surface oxidation and chain scission, potentially breaking polymer chains into smaller molecules (Surana et al., 2024). In practice, polymer oxidation is often surface-limited: ROS must reach the particle surface, and the reaction is inhibited by NOM and radical scavengers. Consequently, reported degradation rates vary widely with irradiation intensity, catalyst–particle contact frequency, and whether oxidant promoters such as H2O2 or persulfate are used. These realities motivate integrated designs that increase repeated contact between retained particles and reactive sites.

A particularly translational direction is the capture-and-destroy concept, which merges separation and degradation within a single module. Catalytic membranes can retain microplastics while simultaneously generating ROS at or near the membrane surface, increasing the probability that retained particles experience repeated oxidative attack. Similarly, porous materials can pre-concentrate additives and hydrophobic co-contaminants associated with plastics, while photocatalysts or redox-active components target both the polymer surface and the sorbed chemicals. Such hybrid approaches align with an engineering viewpoint in which unit operations are intentionally combined to address both the physical and chemical aspects of the problem (Xu et al., 2014; Fujishima & Honda, 1972).

Analytical evaluation is a major barrier to synthesis across microplastics studies. Because particle types and measurement techniques vary, comparability requires standardized metrics. Removal studies should report size-resolved capture efficiency, mass balance across phases (influent, effluent, sludge/concentrate), and the fate of retained particles. Degradation studies should provide polymer identification (e.g., spectroscopic signatures), indicators of oxidation (such as increased carbonyl features), molecular-weight changes where feasible, shifts in particle-size distribution, and quantification of released additives or dissolved organic carbon. Importantly, disappearance of visible particles is not sufficient evidence of risk reduction; the formation of smaller fragments or soluble products may require downstream treatment and toxicity-informed evaluation (Yang et al., 2024).

Finally, realistic water matrices matter. In real waters, microplastics co-occur with NOM, salts, and co-pollutants that can coat surfaces, alter hydrophobicity, and influence coagulation and membrane performance. Oxidative treatments can generate oxygenated by-products (aldehydes, ketones, carboxylic acids) from polymer chains; halides and other ions can also influence secondary reactive species and by-product spectra. Therefore, practically relevant studies should include long-duration tests that capture how particle properties and by-products evolve under sustained operation and cleaning cycles (Elimelech & Phillip, 2011; Yang et al., 2024).

5.3 Case-study template and translation criteria for pilots

A case study becomes most valuable when it can be used as a design precedent. To make advanced-material demonstrations actionable for pilots, case studies should clearly specify the decision context and provide enough information for scale-up calculations. The following template consolidates common translation needs across adsorption, catalytic oxidation, membranes, and electrochemical approaches (Elimelech & Phillip, 2011; Shannon et al., 2008).

Case-study template (recommended reporting items): 1) Treatment objective and constraints: pollutant(s), influent range, target effluent criterion, required throughput, footprint limits, and acceptable operator burden. 2) Configuration: batch versus continuous, hydraulic residence time or bed contact time, module geometry, and whether the material is suspended, immobilized, or shaped. 3) Water matrix: pH, ionic strength, alkalinity, major ions, DOC/UV254, and co-contaminants relevant to competition or radical scavenging. 4) Performance metrics: for packed beds, bed volumes to breakthrough; for membranes, flux and rejection with fouling evolution; for catalytic processes, kinetics plus energy/reagent-normalized metrics; for electrochemical systems, energy per treated volume and stable cycling. 5) Stability and regeneration: multi-cycle tests with explicit regeneration steps, mass balance across cycles, and evidence for minimal leaching or particle release. 6) Residuals and by-products: concentrate/spent media management, by-product screening, and disposal or destruction pathway (Elimelech & Phillip, 2011).

When these items are reported, case studies across different contaminant classes become comparable. For example, photocatalytic systems for trace organics can be compared meaningfully when photon flux, reactor geometry, and energy accounting are disclosed; adsorbents for metal(loid)s can be compared when column breakthrough and regeneration are reported under competing ions; and microplastics strategies can be evaluated on both size-resolved capture efficiency and transformation-product risk (Xu et al., 2014; Fujishima & Honda, 1972).

From a practical standpoint, the strongest case studies also identify failure modes and mitigation strategies. Common failure modes include: (i) rapid loss of activity due to fouling or surface poisoning, (ii) mechanical attrition and fines generation, (iii) leaching of active components or functional groups, (iv) loss of selectivity in real-water matrices, and (v) regeneration that produces hazardous or difficult residuals. Reporting these issues is not a weakness; it accelerates translation by defining realistic design targets.

In summary, advanced materials can make the largest near-term contribution when they are treated as components of engineered systems rather than isolated powders. The most promising application pathways are likely to be hybrid: selective capture paired with targeted destruction of concentrates, catalytic membranes that integrate separation and reaction, and electrochemical systems that enable on-demand regeneration. By consolidating case studies around common reporting items and system-level metrics, the field can move toward demonstrations that are directly usable by designers, regulators, and operators (Elimelech & Phillip, 2011; Shannon et al., 2008).

The application evidence also reveals recurring bottlenecks and implementation gaps, motivating the translational challenges and future research directions summarized in the next section.

6. CHALLENGES AND FUTURE PERSPECTIVES

Advanced materials—photocatalysts, functional sorbents, porous frameworks, catalytic membranes, and electrochemical electrodes—have delivered impressive laboratory results for removing trace organics, metals, nutrients, and persistent contaminants. Yet the distance from a promising batch experiment to a robust, field-ready unit process remains substantial. The barriers are not only scientific; they are also economic, operational, and regulatory. This section consolidates the most important translation challenges and outlines future directions that are both technically credible and practically relevant. In line with the chapters systems perspective, the discussion emphasizes how materials must be designed for realistic water matrices, regeneration, residual management, and long-term reliability.

6.1 Translational Challenges

Five challenges repeatedly determine whether an advanced material progresses beyond the lab: (i) cost and scalability, (ii) durability and stability, (iii) safety and ecotoxicity, (iv) performance under complex matrices, and (v) recovery and regeneration. These themes are tightly coupled: a synthesis route that boosts performance may increase cost; a modification that improves selectivity may reduce regenerability; and a catalyst that is highly active in clean water may be inhibited in real wastewater by radical scavengers or competing solutes.

Cost and scalability. Many high-performing materials rely on expensive precursors, multi-step synthesis, and energy-intensive treatments. Even when a material is intrinsically effective, a treatment system must compete with established technologies (e.g., activated carbon, ion exchange, conventional membranes) whose supply chains are mature. Practical scale-up therefore requires synthesis pathways with high yield, minimal solvent burden, and compatibility with continuous production. A useful rule of thumb is to treat the final shaped form—not the powder—as the real product, because shaping, binders, and module fabrication often dominate cost.

Long-term stability and durability. Field operation stresses materials through months of exposure to oxidants, disinfectants, variable pH, and hydraulic shear. Photocatalysts may undergo photocorrosion or surface deactivation; sorbents and porous frameworks can experience hydrolysis, pore collapse, or leaching of metals/ligands; and functional coatings may detach under cleaning and crossflow shear. Durability is therefore a first-order performance metric. A translationally relevant study should include multi-cycle operation with realistic regeneration steps and post-run characterization that identifies failure modes (e.g., loss of crystallinity, changes in surface functional groups, or mechanical attrition).

Safety, ecotoxicity, and public acceptance. The potential environmental and health impacts of nanomaterials and reactive catalysts remain an important concern. Even low levels of particle release or metal leaching can undermine adoption in drinking-water contexts. Safety-by-design approaches—immobilization on stable supports, robust binders, and conservative choices of dopants—can reduce risk, but must be validated with leaching and abrasion tests under representative operational conditions. Beyond intrinsic toxicity, regulators and utilities increasingly expect evidence that transformation products and residuals are managed responsibly.

Performance in complex water matrices. A major reason for disappointing field performance is that most laboratory studies use deionized water or simplified synthetic matrices. Real waters contain natural organic matter (which competes for adsorption sites and scavenges radicals), high ionic strength (which affects electrostatic interactions and double-layer behavior), suspended solids (which foul surfaces), and co-contaminants (which create competitive binding and reaction pathways). Materials should therefore be benchmarked under representative matrices and realistic concentrations, including mixtures of pollutants. Reporting should explicitly quantify matrix effects—for example, capacity loss due to organic matter competition or reduced oxidation rates in the presence of bicarbonate.

Catalyst recovery and regeneration. Powder recovery remains a practical bottleneck for many nanomaterials. Magnetic separation can help, but it introduces its own issues (magnetic core stability, coating integrity, and added synthesis steps). Immobilization on supports and integration into membranes or structured reactors can avoid recovery steps, but may reduce apparent kinetics due to mass-transfer limitations. Regeneration is equally critical: a sorbent that cannot be regenerated without severe capacity loss simply shifts the problem to spent-media disposal. Regeneration protocols should be designed to minimize chemical consumption, avoid hazardous eluents, and provide a clear pathway for handling concentrates or regenerant streams.

Practical implication: These challenges suggest that "best-in-class material" claims should be reframed as "best-in-class material–process combinations." For translation, performance must be reported in a way that connects to unit-process design: bed volumes to breakthrough for fixed beds, flux and rejection under fouling/cleaning cycles for membranes, energy-normalized metrics for catalytic and electrochemical processes, and explicit mass balance and residual pathways for all systems.

6.2 Future Directions and Research Priorities for Practical Impact

Future progress is likely to be driven by integration and design for operation rather than incremental improvements in single-point performance. Five directions are especially promising: intelligent and responsive materials, data-driven material discovery, hybrid integrated treatment trains, circular economy and resource recovery, and improved utilization of solar energy. The strongest opportunities sit at interfaces—materials with embedded sensing and actuation, catalytic membranes that couple reaction and separation, and modular processes that manage concentrates while enabling regeneration.

Intelligent and responsive materials. "Smart" materials that respond to pH, redox conditions, temperature, or pollutant signatures could enable more precise treatment. In practice, responsiveness is most valuable when it reduces operator burden: automatic switching between adsorption and regeneration modes, self-reporting of exhaustion state, or selective capture triggered by water-quality changes. To be field-relevant, responsive behavior must be reversible, stable over many cycles, and compatible with realistic hydraulic operation.

AI- and data-assisted design. Machine learning and high-throughput computation can accelerate the identification of candidate compositions, structures, and functional groups. In water treatment, the key is to predict not only intrinsic affinity or catalytic activity, but also stability and manufacturability. Data-driven tools are particularly useful when they can encode practical constraints—precursor availability, synthesis temperature limits, and known degradation pathways—so that suggested materials have a plausible path to scale-up. The sections source text highlights the growing role of AI-assisted design (Wan et al., 2024).

Hybrid and integrated systems. Combining unit processes can create synergies that overcome the limitations of individual technologies. For example, coupling photocatalysis with membrane filtration can simultaneously degrade pollutants and separate catalysts, improving continuous operation while limiting particle release. Similarly, sonophotocatalysis can enhance mass transfer and radical generation, although the energy intensity must be justified for the intended application (Theerthagiri et al., 2021). More broadly, integrated designs should be evaluated as systems: does integration reduce chemical feeds, decrease downtime, or simplify residual handling? If not, added complexity may outweigh benefits.

Circular economy and resource recovery. A major conceptual shift is moving from "pollutant removal" to "resource recovery." Advanced materials can be designed to selectively capture valuable species such as phosphorus, nitrogen, or strategic metals from waste streams. For impact, recovery materials should be paired with regeneration approaches that yield usable products (e.g., concentrated phosphate streams) rather than mixed hazardous wastes. This aligns with circular economy principles and can change the economics of treatment by creating value streams (Yang, 2018).

Harnessing solar energy and low-carbon operation. Photocatalysis remains attractive because it can, in principle, use sunlight to drive oxidation or reduction. Practical progress requires improving quantum efficiency under real solar spectra, stabilizing catalysts against photocorrosion, and designing reactors that deliver photons efficiently at scale. Beyond degradation, solar-driven processes such as water splitting for hydrogen and CO_2 reduction are often discussed as sustainability pathways, but their relevance to water treatment depends on coupling to real waste streams and demonstrating robust operation (Fajrina & Tahir, 2018; Wang, 2019). Taken together, the preceding analysis sets the stage for a concise synthesis of key messages and practical implications for advancing material-enabled water treatment.

7. CONCLUSION

Advanced materials have ushered in a new era for water pollution control, offering powerful and versatile tools to address the limitations of conventional technologies. Semiconductor photocatalysts and porous materials like MOFs and COFs stand out for their exceptional capabilities in degrading persistent organic pollutants and sequestering toxic heavy metals. Through sophisticated synthesis and modification strategies—including heterojunction engineering, defect creation, and green chemistry approaches—researchers are continuously pushing the boundaries of material performance, enhancing efficiency, and expanding the utility of these materials to tackle emerging contaminants like microplastics.

However, the path to widespread practical application is fraught with challenges, including high costs, long-term stability concerns, and potential ecotoxicity. Overcoming these obstacles requires a multidisciplinary approach that integrates materials science, chemical engineering, environmental science, and economics. The future of water treatment lies in the development of sustainable, intelligent, and multifunctional material-based systems. By embracing principles of the circular economy, leveraging the power of artificial intelligence for material discovery, and designing integrated systems that not only purify water but also recover valuable resources, we can transition from a linear model of "use and discard" to a sustainable, circular management of our most precious resource. The continued innovation in advanced materials science is not just a scientific pursuit but a critical component in ensuring global water security for generations to come.

REFERENCES

Ahmad, M., Rajapaksha, A. U., Lim, J. E., Zhang, M., Bolan, N., Mohan, D., Vithanage, M., Lee, S. S., & Ok, Y. S. (2014). Biochar as a sorbent for contaminant management in soil and water: A review. *Chemosphere*, *99*, 19–33. DOI: 10.1016/j.chemosphere.2013.10.071 PMID: 24289982

Devi, L., & Kavitha, R. (2013). A review on non metal ion doped titania for the photocatalytic degradation of organic pollutants under UV/solar light: Role of photogenerated charge carrier dynamics in enhancing the activity. *Applied Catalysis B: Environment and Energy*, *140-141*, 559–587. DOI: 10.1016/j.apcatb.2013.04.035

Diercks, C S, & Yaghi, O M (2017). The atom, the molecule, and the covalent organic framework. Science, 355(6328), eaal1585. DOI: 10.1126/science.aal1585

Dong, C., Chen, Y., Yang, C., Li, P., Zhang, Y., Wang, P., Wang, Y., Fang, W., Lin, L., & Li, X. (2024). Pinning-effect single-atom NiCo alloy embedded graphene-aerogel in electro-fenton process for rapid degradation of emerging contaminants. *Applied Catalysis B: Environment and Energy*, *357*, 124286–124286. DOI: 10.1016/j.apcatb.2024.124286

Dutta, S., Fajal, S., & Ghosh, S. K. (2024). Heavy Metal-Based Toxic Oxo-Pollutants Sequestration by Advanced Functional Porous Materials for Safe Drinking Water. *Accounts of Chemical Research*, *57*(17), 2546–2560. DOI: 10.1021/acs.accounts.4c00348 PMID: 39163541

Dutta, V., Sharma, S., Raizada, P., Thakur, V., Khan, A., Saini, V., Asiri, A., & Singh, P. (2021). An overview on WO_3 based photocatalyst for environmental remediation. *Journal of Environmental Chemical Engineering*, *9*(1), 105018–105018. DOI: 10.1016/j.jece.2020.105018

Elimelech, M., & Phillip, W. A. (2011). The future of seawater desalination: Energy, technology, and the environment. *Science*, *333*(6043), 712–717. DOI: 10.1126/science.1200488 PMID: 21817042

Fajrina, N., & Tahir, M. (2018). A critical review in strategies to improve photocatalytic water splitting towards hydrogen production. *International Journal of Hydrogen Energy*, *44*(2), 540–577. DOI: 10.1016/j.ijhydene.2018.10.200

Fujishima, A., & Honda, K. (1972). Electrochemical photolysis of water at a semiconductor electrode. *Nature*, *238*(5358), 37–38. DOI: 10.1038/238037a0 PMID: 12635268

Furukawa, H., Cordova, K. E., O'Keeffe, M., & Yaghi, O. M. (2013). The chemistry and applications of metal-organic frameworks. *Science, 341*(6149), 1230444. DOI: 10.1126/science.1230444 PMID: 23990564

Garcia-Segura, S., & Brillas, E. (2017). Applied photoelectrocatalysis on the degradation of organic pollutants in wastewaters. *Journal of Photochemistry and Photobiology C, Photochemistry Reviews, 31*, 1–35. DOI: 10.1016/j.jphotochemrev.2017.01.005

Hao, L., Zhang, J., Liu, J., Min, Y., & Chen, C. (2023). Applications of Carbon-Based Materials in Activated Peroxymonosulfate for the Degradation of Organic Pollutants: A Review. *Chemical Record : An Official Publication of the Chemical Society of Japan), 23*(12), e202300203. Advance online publication. DOI: 10.1002/tcr.202300203 PMID: 37639150

Huang, D., Li, Z., Zeng, G., Zhou, C., Xue, W., Gong, X., Yan, X., Chen, S., Wang, W., & Cheng, M. (2018). Megamerger in photocatalytic field: 2D g-C_3N_4 nanosheets serve as support of 0D nanomaterials for improving photocatalytic performance. *Applied Catalysis B: Environment and Energy, 240*, 153–173. DOI: 10.1016/j.apcatb.2018.08.071

Kaur, I., Batra, V., Bogireddy, N. K. R., Baveja, J., Kumar, Y., & Agarwal, V. (2024). Chemical- and green-precursor-derived carbon dots for photocatalytic degradation of dyes. *iScience, 27*(2), 108920. DOI: 10.1016/j.isci.2024.108920 PMID: 38352227

Kumar, A., Indhur, R., Bux, F., & Kumari, S. (2024). Recent advances in mechanistic insights into microplastics mitigation strategies via emerging advanced oxidation processes: Legislation, challenges, and future direction. *The Science of the Total Environment, 957*, 177150. DOI: 10.1016/j.scitotenv.2024.177150 PMID: 39486547

Lan, Z., Zhang, G., & Wang, X. (2016). A facile synthesis of Br-modified g-C_3N_4 semiconductors for photoredox water splitting. *Applied Catalysis B: Environment and Energy, 192*, 116–125. DOI: 10.1016/j.apcatb.2016.03.062

Li, X., Wang, B., Cao, Y., Zhao, S., Wang, H., Feng, X., Zhou, J., & Ma, X. (2019). Water Contaminant Elimination Based on Metal–Organic Frameworks and Perspective on Their Industrial Applications. *ACS Sustainable Chemistry & Engineering, 7*(5), 4548–4563. DOI: 10.1021/acssuschemeng.8b05751 PMID: 33659101

Liu, X., Gu, S., Zhao, Y., Zhou, G., & Li, W. (2020). BiVO4, Bi_2WO_6 and Bi_2MoO_6 photocatalysis: A brief review. *Journal of Materials Science and Technology, 56*, 45–68. DOI: 10.1016/j.jmst.2020.04.023

Meng, X., & Zhang, Z. (2016). Bismuth-based photocatalytic semiconductors: Introduction, challenges and possible approaches. *Journal of Molecular Catalysis A Chemical*, *423*, 533–549. DOI: 10.1016/j.molcata.2016.07.030

Omidian, H., Akhzarmehr, A., & Chowdhury, S. D. (2024). Advancements in Cellulose-Based Superabsorbent Hydrogels: Sustainable Solutions across Industries. *Gels (Basel, Switzerland)*, *10*(3), 174. Advance online publication. DOI: 10.3390/gels10030174 PMID: 38534592

Qi, K., Imparato, C., Almjasheva, O., Khataee, A., & Zheng, W. (2024). TiO2-based photocatalysts from type-II to S-scheme heterojunction and their applications. *Journal of Colloid and Interface Science*, *675*, 150–191. DOI: 10.1016/j.jcis.2024.06.204 PMID: 38968635

Reddy, K. R., Reddy, C. V., Nadagouda, M. N., Shetti, N. P., Jaesool, S., & Aminabhavi, T. M. (2019). Polymeric graphitic carbon nitride (g-C_3N_4)-based semiconducting nanostructured materials: Synthesis methods, properties and photocatalytic applications. *Journal of Environmental Management*, *238*, 25–40. DOI: 10.1016/j.jenvman.2019.02.075 PMID: 30844543

Shannon, M. A., Bohn, P. W., Elimelech, M., Georgiadis, J. G., Marinas, B. J., & Mayes, A. M. (2008). Science and technology for water purification in the coming decades. *Nature*, *452*(7185), 301–310. DOI: 10.1038/nature06599 PMID: 18354474

Singh, P., Shandilya, P., Raizada, P., Sudhaik, A., Rahmani-Sani, A., & Hosseini–Bandegharaei, A. (2018). Review on various strategies for enhancing photocatalytic activity of graphene based nanocomposites for water purification. *Arabian Journal of Chemistry*, *13*(1), 3498–3520. DOI: 10.1016/j.arabjc.2018.12.001

Surana, M., Pattanayak, D. S., Yadav, V., Singh, V. K., & Pal, D. (2024). An insight decipher on photocatalytic degradation of microplastics: Mechanism, limitations, and future outlook. *Environmental Research*, *247*, 118268. DOI: 10.1016/j.envres.2024.118268 PMID: 38244970

Theerthagiri, J., Lee, S. J., Karuppasamy, K., Arulmani, S., Veeralakshmi, S., Ashokkumar, M., & Choi, M. Y. (2021). Application of advanced materials in sono-photocatalytic processes for the remediation of environmental pollutants. *Journal of Hazardous Materials*, *412*, 125245. DOI: 10.1016/j.jhazmat.2021.125245 PMID: 33545645

Wan, X, Xiao, Z, Tian, Y, Chen, M, Liu, F, Wang, D, Liu, Y, Bartolo, PJDS, Yan, C, Shi, Y, Zhao, RR, Qi, HJ, & Zhou, K (2024). Recent Advances in 4D Printing of Advanced Materials and Structures for Functional Applications. Advanced materials (Deerfield Beach, Fla.), 36(34), e2312263. DOI: 10.1002/adma.202312263

Wang, L, & Zhu, W (2024). Organic Donor-Acceptor Systems for Photocatalysis. Advanced science (Weinheim, Baden-Wurttemberg, Germany), 11(10), e2307227. DOI: 10.1002/advs.202307227

Wang, X., Faungnawakij, K., & Chareonpanich, M. (2019). Editorial: Photocatalysis – From Solar Power to Sustainable Chemical Production. *ChemCatChem, 11*(24), 5838–5841. DOI: 10.1002/cctc.201902248

Xu, H., Ouyang, S., Liu, L., Reunchan, P., Umezawa, N., & Ye, J. (2014). Recent advances in TiO_2-based photocatalysis. *Journal of Materials Chemistry. A, Materials for Energy and Sustainability*, *2*(32), 12642–12642. DOI: 10.1039/C4TA00941J

Yang, S., Zhang, F., Ding, H., He, P., & Zhou, H. (2018). Lithium Metal Extraction from Seawater. *Joule*, *2*(9), 1648–1651. DOI: 10.1016/j.joule.2018.07.006

Yang, Z., Li, Y., & Zhang, G. (2024). Degradation of microplastic in water by advanced oxidation processes. *Chemosphere*, *357*, 141939. DOI: 10.1016/j.chemosphere.2024.141939 PMID: 38621489

Chapter 10
Biomass Resource Facilities and Biomass Conversion Processing for Fuels and Chemicals

Yan Tan
http://orcid.org/0009-0003-9778-5851
Tongren University, China

Lei Guo
http://orcid.org/0000-0001-7849-9583
Tongren University, China

Ziyang Wang
Tongren University, China

Valentine Chikaodili Anadebe
http://orcid.org/0000-0002-7559-446X
Alex Ekwueme Federal University, Nigeria

ABSTRACT

Biomass-derived fuels and chemicals have attracted considerable attention for their benefits, including economical, ecofriendly, nontoxic, and renewable. This chapter will focused on the preparation and characterization of biomass-derived carbon dots, biomass fuels, and biochar. Part I review details the synthesis of carbon dots from green sources such as plants parts and waste. Renewable sources such as biofuels show a potential usage with low carbon dioxide emission, less polluting and can be created from biomass in the forms of organic waste. Part II aims to review the various sources of biofuels and the methods used in the production, and discuss

DOI: 10.4018/979-8-3373-4402-7.ch010

its advantages and sustainability. Biochar, a carbon-rich material produced from the pyrolysis of organic biomass, has gained significant attention as a potential solution for sustainable green remediation practices. Part III aims delving into the diverse methods of biochar production, highlighting their influence on physical and chemical properties.

1 INTRODUCTION

In the context of the era of sustainable development, the conversion and utilization of biomass resources have become a global concern. Biomass resources capture solar energy through photosynthesis and store it in the form of chemical energy. Its conversion and utilization not only help to alleviate the energy crisis and reduce dependence on fossil fuels, but also plays an irreplaceable and important role in environmental protection and ecological balance.

Biomass-derived carbon dots are emerging as an emerging product of biomass conversion. It is usually prepared from natural biomass such as ripe fruits, plants and waste peels. These carbon dots have unique optical and chemical properties and show great potential in numerous fields (Kuang et al., 2023; Zheng et al., 2023).

Biofuels, as an important representative of renewable energy, refer broadly to solid, liquid or gaseous fuels composed of or extracted from biomass. Among them, biodiesel and bioethanol are the most common. Ethanol can be made by fermenting sugars from grains, straw or beet sugar, while biodiesel is made from vegetable or animal oils. Since the 1970s, countries around the world have stepped up their R&D and promotion of biofuels due to fluctuations in traditional energy prices, environmental protection needs, and global climate change. Biofuels are characterized by a variety of raw materials, such as crop residues, forestry residues, and animal and poultry manure, etc. They are also available in a variety of forms, including liquid, solid, and gaseous fuels, and can be widely used in many fields, such as transportation, heating, and power generation, and can achieve near-zero carbon dioxide emissions during their entire life cycle.

Biochar is a porous carbon made from organic waste such as animal manure and plant roots and stems processed under specific conditions. It is known as "black gold" and has many excellent properties. In agriculture, biochar improves soil water storage and nutrient capacity, protects soil microorganisms, and fixes carbon in the soil for hundreds of years while reducing greenhouse gas emissions such as methane and nitrous oxide. Studies have shown that biochar could even offset 12% of global carbon emissions. In the production process, about one-third of the feedstock is converted to biochar, another one-third can be used to generate electricity, and another one-third can form crude oil, whose by-products are of significant value in terms of energy generation and environmental protection. These three types of biomass conversion products, from microscopic carbon points to macroscopic fuels and biochar, play key roles in different fields, and together build up a multifaceted pattern of biomass resource conversion and utilization, which provides strong support for the sustainable development of human society.

Figure 1. Reuse of biomass resources.

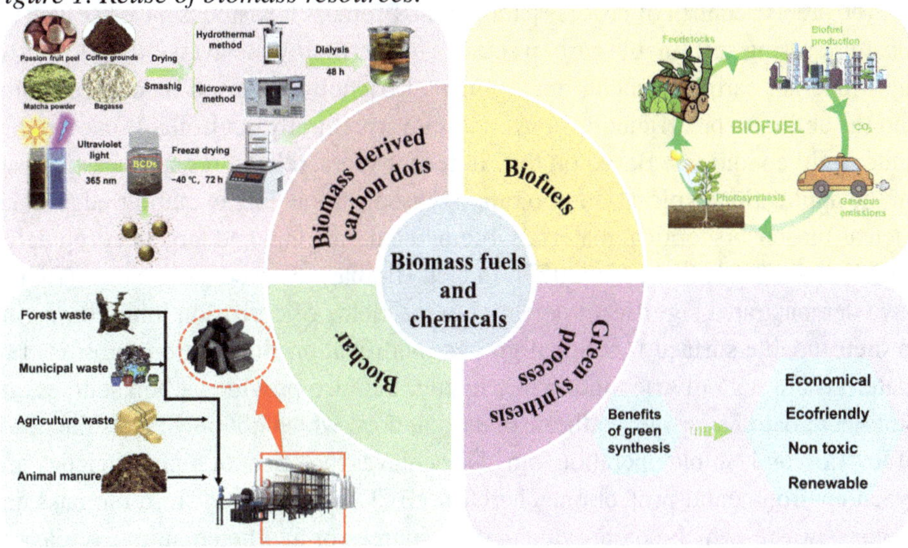

2 BIOMASS-DERIVED CARBON DOTS

Carbon dots (CDs) have attracted extensive attention in recent years in multiple fields such as biomedicine, catalysis, sensing, optoelectronic devices and environmental remediation, due to their unique physicochemical properties including excellent water solubility, abundant surface functional groups, wide availability of raw materials and environmentally friendly synthetic routes. CDs were accidentally found during the purification of single-walled carbon nanotubes. Based on their intrinsic carbon nuclear structure characteristics and the sources of synthetic precursors, the current mainstream classification system divides them into four categories: graphene quantum dots (GQDs), carbonized polymer dots (CPDs), carbon nanodots (CNDs), and carbon quantum dots (CQDs). CDs are mainly composed of three elements: carbon, oxygen, and nitrogen, and the relative atomic content of C/O/N is significantly influenced by the synthesis strategy, which depends on whether the bottom-up or top-down approach is adopted. Typical bottom-up methods include pyrolysis, hydrothermal, chemical oxidation, and microwave-assisted methods; while the mainstream top-down methods cover laser ablation and electrochemical methods. It should be noted that the type of precursor and reaction conditions jointly determine the elemental composition, surface functional group distribution, and optical properties of the final product. (He et al., 2022; Onyeachu et al., 2021). The current mainstream bottom-up synthesis strategy is highly dependent on the selection of carbon sources and the precise control of process parameters. Biomass feedstocks have become the most promising sustainable carbon sources due to their wide availability, low cost, high inherent carbon content, environmental friendliness, good biocompatibility, and the ability to be efficiently converted into structurally controllable carbon dots under mild conditions. Based on this, in recent years, related research has focused on the application exploration of biomass-based carbon dots in cutting-edge fields such as biosensors, optical materials, biomedical imaging, and targeted drug delivery. (E et al., 2021; Liu et al., 2020). In fact, Biomass-derived carbon dots (BCDs) have demonstrated significant potential in the fields of corrosion inhibitor, owing to their tunable surface functional groups modification. In addition, biomass raw materials are not only rich in organic matter, but also provide carbon sources and other heteroatoms for the synthesis of carbon dots, which not only meets the needs of low cost and simple operation, but also achieves the effect of high efficiency and green environmental protection. Therefore, BCDs synthesized from biomass and biomass waste, which can provide carbon sources or and heteroatoms, or easy to doping to achieves the high efficiency of corrosion inhibitor.

This section systematically summarizes the synthesis of BCDs and their functional applications, with particular emphasis on matcha as a sustainable and nitrogen- and carbon-rich precursor. Matcha is a widely available food-grade biomass rich in natural organic components, including polyphenols, chlorophyll, free amino acids, and plant proteins, which collectively serve as intrinsic carbon and heteroatom (such as nitrogen and oxygen) sources for BCDs formation. Therefore, a series of readily accessible, low-cost, and environmentally friendly biomass feedstocks, including matcha powder, spent coffee grounds, passion fruit peels, and sugarcane bagasse, can all be used as scalable green precursors for the synthesis of high-performance BCD.

2.1 Methods Used for the Synthesis of BCDs

CDs were synthesized by a typical hydrothermal method or microwave method (Edison et al., 2016), using biomass or waste as carbon sources and doping with nitrogen sources to achieve rich functionalized group. In this bottom-up approach, carbon sources and nitrogen sources are pyrolyzed or carbonized to form nanoparticles with nano-size range. Figure 2 shows the typical flow chart for the fabrication of

BCDs. Briefly, the desired amounts of biomass matter, desired volume of nitrogen sources matter, usually using amines or amino acids were added in deionized water and mixed via stirring and ultrasound. Then the mixed solution was transferred to reacting in autoclave of hydrothermal methods or chamber of microwave method. After the reaction, it was cooled to room temperature and solution was centrifuged. Then continuously filtered with a filter film to remove the sediment, and dialyzed with a dialysis bag. Finally, freeze dried to obtain BCDs. The BCDs aqueous solution has fluorescent under ultraviolet light (UV), so it can be preliminarily determined that the CDs were successfully prepared.

Figure 2. Schematic diagram of the biomass carbon dots preparation process.

2.2 Properties of BCDs

(1) Optical properties

The primary optical characteristic of BCDs is fluorescence emissions in the ultraviolet to visible region excited under UV irradiation, as shown in Fig.2. The distinct fluorescence property of BCDs is due to the surface defects and quantum size effects. The absorption peak in the UV region, are attributed to the π-π^* and n-π^* transition of C=C and C=O or C=N bonds, respectively. Another appealing characteristic is phosphorescence, which is produced by intersystem crossing (transition from S_1 to T_n state) and radiative transition (transition from T_1 to S_0). It has been established that C=O and C=N moieties contribute to the formation of phosphorescence (Kim et al., 2018).

(2) Corrosion inhibition properties

In addition to optical properties, BCDs also have quantum size effects, cost effectiveness, rich functional groups, low toxicity, excellent water solubility and so on. These unique attributes have made BCDs attractive for applications in green corrosion protection (Dong et al., 2025; X. Wu et al., 2022). Usually, the corrosion inhibition properties were evaluated from electrochemical impedance spectroscopy (EIS) test and potentiodynamic polarization (PDP) curve in electrochemical test. The corrosion rate and corrosion efficiency can be further tested by weight loss experiment. The corrosion inhibition mechanism of BCDs on carbon steel sample can discussed via experimental results and analysis. The adsorption of BCDs on the carbon steel arises from the abundant functional groups on the BCDs and nanoscale effects aggregate and spread over the metal surface through electrostatic and van der Waals forces, forming a protective layer that reduces the exposed area of carbon steel in the acidic environment.

Such as:

In order to meet the social demand for green chemistry, a new type of environmentally friendly carbon dots (PEI-MCDs) were successfully synthesized by hydrothermal method using MT powder from biomass resources as the carbon source and polyethyleneimine (PEI) as the nitrogen source,in order to achieve efficient corrosion inhibition performance on mild steel under acidic conditions. The experimental results of electrochemical tests (Fig. 3(a-c)), weight loss experiments (Fig. 3d) and corrosion monitoring (Fig. 3e) of PEI-MCDs on mild steel in 1 M HCl solution, respectively, are shown in Fig. 3.

The Nyquist and Bode plots for PEI-MCDs shown in Fig. 3a and 3b show that the diameter of the impedance are expands with increasing PEI-MCDs concentration, indicating that adsorption of PEI-MCDs attenuates charge transfer and gradually inhibits steel corrosion. Combined with the electrochemical impedance spectroscopy datas (Table 1), it can be seen that the value of R_p increased with the increase of PEI-MCDs concentration, indicating that the inhibitor molecules were effectively adsorbed on the surface of the samples, which reduced the interfacial electronic charge transfer (Wu et al., 2025). The corrosion inhibition efficiency of PEI-MCDs at 10 mg/L was as high as 95.4%. In general, a lower C_{dl} value means a thicker adsorption layer. It can be seen that 10 mg/L PEI-MCDs has the lowest C_{dl} value. Therefore, trace amounts of PEI-MCDs can form a thicker adsorption film, resulting in better resistance to corrosion and better protection. Where R_p denotes the polarization resistance and C_{dl} denotes the capacitance value of the electric double layer.

Weight loss experiments were used to further analyze the corrosion rate (C_R) and inhibition efficiency (η_w) of PEI-MCDs. As shown in Fig. 3d, the C_R decreased sharply with the increase of PEI-MCDs concentration, while the ηw increased gradually. When the added concentration was 10 mg/L, the C_R decreased to 0.38 mg cm^{-2} h^{-1}, and the η_w was as high as 97.2%. The corrosion rate (C_R) and corrosion inhibition efficiency (η_w) are obtained from equations 1 and 2 (Y. Wu et al., 2022):

$$C_R = \frac{m_0 - m_1}{St} \tag{1}$$

$$\eta_W = \frac{C_R^0 - C_R^i}{C_R^0} \times 100 \tag{2}$$

where m_0 and m_1 are the initial weight of the steel block and the weight after the immersion experiment, correspondingly, S and t are the sample's surface area and time of immersion, respectively. *C*0 R and *Ci* R respectively, represent the corrosion rates without and with the addition of PEI-MCDs.

Fig. 3e shows the surface morphology of the steel samples during the immersion study phase as a function of immersion time. With the time lapse from 0 h to 24 h in the blank solution, the corrosion damage on the surface of the sample increased rapidly and was accompanied by the generation of a large number of bubbles. Finally, it was completely covered with corrosion products and severely corroded. In contrast, after the addition of PEI-MCD, the corrosion damage on the steel surface slowed down; the surrounding bubbles were also gradually reduced, hardly any corrosion products were produced, the overall corrosion was slight and the surface remained relatively smooth. Therefore, it can be concluded that PEI-MCDs showed excellent inhibition of mild steel corrosion, effectively inhibiting the anodic dissolution of Fe and hydrogen evolution reaction.

Figure 3. Nyquist and Bode plots of (a, b) PEI-MCDs; (c) Equivalent circuit model; (d) Results of weight loss measurements; (e) Surface morphology variation of samples in blank and inhibited solutions as a function of immersion time.

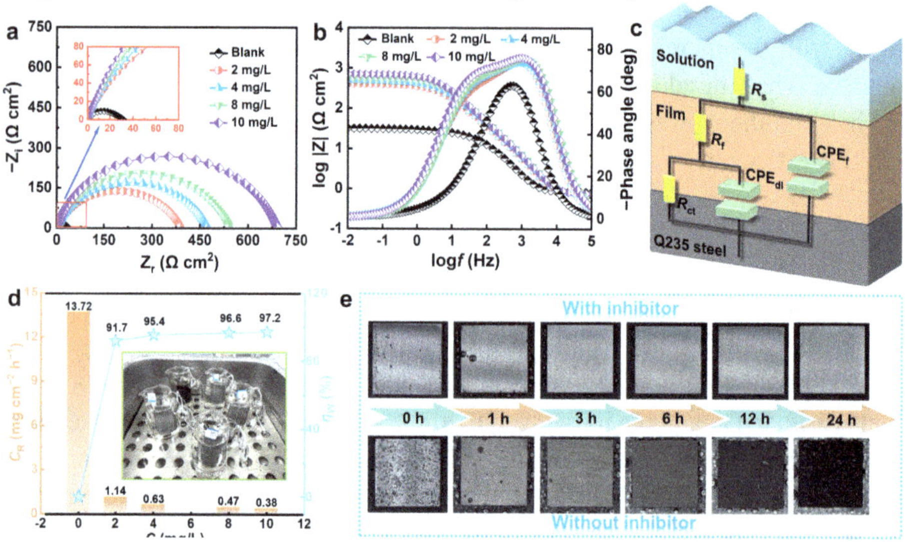

Table 1. Electrochemical parameters obtained from EIS curves of PEI-MCDs.

Inhibitors	*C* (mg/L)	R_s (Ω cm²)	CPE_f Y_0 (10^{-6} $S s^n$ cm^{-2})	CPE_f *n*	R_f (Ω cm²)	CPE_{dl} Y_0 (10^{-6} $S s^n$ cm^{-2})	CPE_{dl} *n*	C_{dl} (μF cm^{-2})	R_{ct} (Ω cm²)	R_p (Ω cm²)	χ^2 (10^{-3})	η_{EIS} (%)
	Blank	0.832	76.0	1.00	5.5	2336	0.458	653.1	26.1	31.6	1.79	/
PEI-MCDs	2	0.761	30.3	1.00	28.9	204.7	0.708	48.7	361.0	389.9	1.75	91.9
	4	0.744	27.4	1.00	24.5	174.1	0.701	37.7	451.1	475.6	1.32	93.4
	8	0.725	21.1	1.00	33.7	101.9	0.741	25.7	507.3	541.0	4.21	94.2
	10	0.741	23.3	1.00	36.7	91.63	0.753	23.0	654.8	691.5	2.39	95.4

(3) Antibacterial properties

BCDs has ultra-small size, typically <10 nm in size, facilitates bacterial adsorption, disrupts cellular structure, and induces bacterial death (Zhao et al., 2021). In addition, the surface charge properties and structural features of BCDs have a significant regulatory effect on their antibacterial activity. A large number of studies have shown that positively charged surface functional groups, especially amino and amide groups, can enhance their binding ability to negatively charged bacterial cell membranes through electrostatic attraction, and further disrupt membrane integrity, thereby mediating key antibacterial effects at the molecular level. (Travlou et al., 2018). Here, we select spent coffee grounds as precursors and uses a bottom-up hydrothermal method to prepared CDs. The structure and composition of the prepared CDs were characterized, and it was found that they have good fluorescence performance and stability. The average size of the CDs is 2.27 nm, and the surface has abundant amino, hydroxyl, and carbonyl groups. In addition, in vitro antibacterial experiments were conducted to study the antibacterial activity of CDs against Staphylococcus aureus (S. aureus). As shown in Figure 4, the antibacterial effect of CDs on S. aureus. The CDs exhibited inhibition of the growth of S. aureus, and their MICs were found to be 2.5 mg/mL. The MBC value was calculated by plating aliquots of the bacterial culture on agar plates that showed no growth of bacteria after 24 h. Following 24 h of incubation at 37 °C, the minimal bactericidal concentration (MBC) was determined as the lowest concentration at which no colony formation was observed and the MBC value is 5 mg/mL. Notably, the antibacterial effect was less pronounced over shorter incubation periods. This may be attributed to the time required for CDs to release antibacterial components and fully engage with the cellular structure of S. aureus.

Figure 4. Antibacterial activity of S. aureus at different incubation times and different concentration.

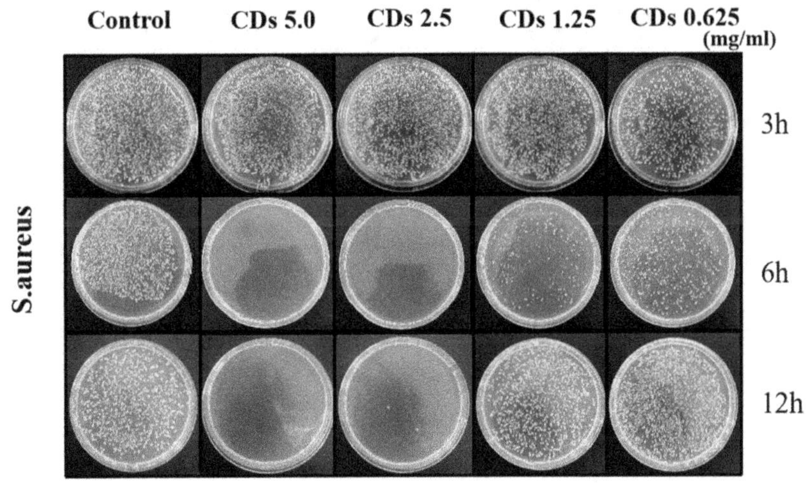

3 BIOFUELS

Biofuels are transportation fuels that are made from biomass materials, can be utilizing as fuel additives or used on their own. Using biofuels means less gasoline and diesel fuel is burned, which can reduce the amount of crude oil imported from other countries and environmentally friendly and sustainability. Owing to this advantages, world bioethanol and biodiesel production increased from 67 to 110.4 billion liters and from 12 to 41 billion liters respectively,over the decade of 2008-2018. Any hydrocarbon fuel produced by living, organic matter capable of growth within a short time frame is termed a biofuel. Biofuels can replace petroleum-derived gasoline and diesel to power transportation, as they are renewable fuels—unlike fossil fuels, which require millions of years to form. Obtainable through chemical reactions in laboratory or industrial settings, biofuels have become a significant area of research for renewable energy development. Further, biofuels are commonly classified into bioethanol and biodiesel.

Figure 5. Life cycle for biofuels.

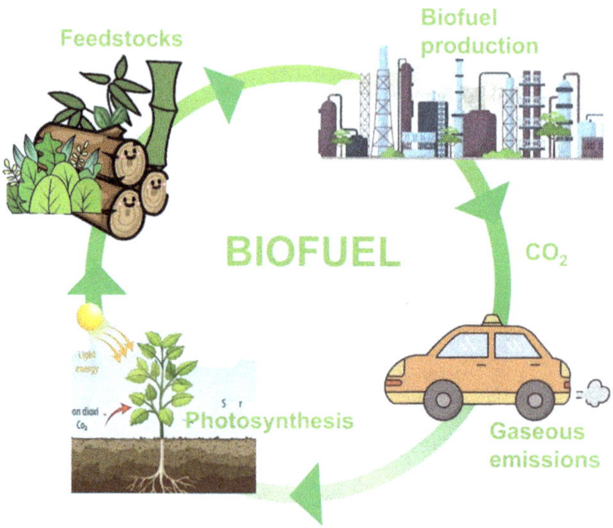

Bioethanol is a cleaner and first-generation biofuels derived from organic material such as plants and their residues, including corn, sorghum, and barley, as well as potato skins, rice, sugar cane, sugar beet, and yard clippings, etc. Ethanol has approximately one-third lower value than petrol, also had lower vapor pressure that enables it to store in pure fuel form, which is a challenge for gasoline. But, in a colder climate, the lower vapor pressure of pure ethanol creates a significant problem with starting the engine. Therefore, in cold climates, ethanol is mixed with gasoline by 40% to obtain a higher evaporative emission.

Biodiesel is another important first-generation biofuel that can be used to replace petroleum diesel fuel. Biodiesel development can be categorized into four generations based on feedstock sources. First-generation biodiesel is produced from edible oil crops (such as rapeseed, soybeans, and sunflowers) through transesterification to yield fatty acid methyl esters (FAME). Although this process is relatively mature, the use of high-quality edible oils results in elevated production costs and contributes to the "food vs. fuel" debate [3,4]. Additionally, FAME's high oxygen content leads to lower calorific value and poorer stability compared to conventional diesel, limiting its blending ratio to below 20%.

3.1 Biomass Conversion Technology

1. Bioethanol Production Technologies

(1) Advanced Pretreatment and Enzymatic Hydrolysis

Lignocellulosic biomass requires effective delignification to enhance cellulose accessibility. Conventional methods (e.g., steam explosion, acid/alkali treatment) face high energy consumption and inhibitor formation. Recent advances include:

a. Surfactant-assisted organosolv pretreatment: Adding PEG or Tween during glycerol organosolv pretreatment reduces lignin re-condensation by 35% and cuts energy use by 20% (Song et al., 2024). This improves enzymatic hydrolysis, slashing cellulase dosage by 50% and achieving a net energy gain of 1.4 MJ/kg.

b. Versatile peroxidase (VP) biocatalysis: Heterologously expressed in Pichia pastoris, VP enables mild delignification of rice straw at 30°C (Teo et al., 2024). This process increases glucose yield by 1.5-fold (to 4.5 mg/mL) and raises the bioethanol titer to 7.8 mg/mL, representing a 1.4-fold enhancement relative to untreated biomass.

(2) Detoxification and Fermentation Optimization

Pretreatment-derived inhibitors (furan derivatives, phenolics) impede microbial fermentation. Novel detoxification strategies involve:

a. Calcium carbonate-mediated detoxification: Treating corn stover hydrolysate with 0.4% $CaCO_3$ elevates fermentable solid loading from 10% to 21% (Shah et al., 2025). This non-adsorptive mechanism stabilizes cell membranes, enabling Rhodosporidium toruloides to produce 17 g/L biomass and 8 g/L lipids.

b. Consolidated bioprocessing (CBP): Engineered Clostridium thermocellum strains concurrently hydrolyze cellulose and ferment sugars, eliminating external enzymes and reducing costs by 30% (Bakali et al., 2024; de Araujo Guilherme et al., 2019).

2. Biodiesel Production Technologies

(1) Heterogeneous Catalysts for Transesterification

Conventional homogeneous catalysts (KOH, NaOH) suffer from saponification and purification issues. Calcium oxide (CaO)-based catalysts offer reusability but face glycerol-induced leaching. Recent solutions include:

a. Anti-leaching CaO/La_2O_3 catalysts: La_2O_3 competitively adsorbs glycerol via hydrogen bonding, shielding CaO active sites. At a Ca/La molar ratio of 0.12, the catalyst retains 96.7% activity after 10 cycles—surpassing pure CaO by >40% 3. Density functional theory (DFT) confirms La_2O_3 disrupts glycerol-CaO synergy, reducing Ca^{2+} leaching by 90% (Fu et al., 2025).

b. Waste-derived catalysts: Biosilica from rice husk enhances catalytic porosity, increasing fatty acid methyl ester (FAME) yield to 98% while enabling catalyst reuse.

(2) Non-esterification Pathways

To circumvent FAME's limitations (low blend ratios, oxidation instability), novel routes have emerged:

a. Photocatalytic conversion: Ru-doped $ZnIn_2S_4$ catalysts under visible light convert biomass-derived furans (e.g., 2-methylfuran) into C_{10}–C_{18} diesel alkanes with >86% selectivity, co-producing hydrogen .

b. High-biomass-content diesel: Blending bio-oils with petroleum fractions achieves 40% biomass content, meeting diesel standards without engine modifications.

3.2 Challenges and Future Directions

1. Bioethanol

(1) It is one of the widely used biofuels worldwide, occupying the dominant position in global biofuels. At present, the vast majority of fuel ethanol produced in industrial production is made from grain crops, which has scale limitations and unsustainability in the long run. The second-generation biofuel ethanol, which uses lignocellulose as raw material, is the key to replacing petroleum on a large scale in the future. However, China has not made fundamental breakthroughs in cellulase production technology, pentose fermentation strain construction, and other aspects, resulting in high production costs and unsuitability for industrial production. However, with the advancement of biorefinery technology, researchers are committed to developing new microbial strains and optimizing reaction conditions to improve yield and quality, and broaden raw material sources.

(2) Although there has been progress in the second-generation bioethanol production technology, there are still technical bottlenecks, such as low cellulose

hydrolysis efficiency, high production costs, and large initial investment, which restrict its large-scale commercial application.

(3) Enzyme cost: Cellulases still contribute 20–25% of ethanol production costs. CBP and VP biocatalysts show promise but lack industrial validation.

(4) AI-driven process optimization: Machine learning predicts optimal conditions for enzymatic hydrolysis (pH, temperature), boosting yields by 15–20%.

(5) In the future, more attention will be paid to environmentally friendly production processes, such as developing efficient cellulose hydrolysis technology and low-cost fermentation processes to reduce production costs. At the same time, explore diversified applications, strengthen international cooperation and technological exchanges, and jointly tackle key technological challenges. Establish a sound standard testing system to ensure product quality.

2. Biodiesel

(1) Due to its wide range of raw materials (such as waste oil and algae) and environmental characteristics (reducing CO_2 emissions by 78%), it is considered an ideal alternative energy source. The research focuses on the full chain optimization of production, covering raw material screening, catalyst innovation (chemical catalysts and lipases), process improvement (such as carbon capture and utilization CCU), and the introduction of artificial intelligence (AI) to achieve production automation. Homogeneous catalysts have high efficiency but are difficult to recover, while heterogeneous catalysts can be reused but have lower activity. Lipases exhibit excellent performance under mild conditions, but enzyme cost remains a bottleneck. Although the production cost of biodiesel is still higher than that of fossil diesel (about 20-30% higher), its full lifecycle carbon emissions are reduced by 50-80%.

(2) The high production cost is the main factor restricting its large-scale application. Currently, the production cost of biodiesel is 20-30% higher than that of fossil diesel (Akram et al., 2022). Although policy incentives and large-scale production can narrow the cost gap, further cost reduction is still needed. The stability of raw material supply is also a challenge, for example, there are certain difficulties in the collection and pretreatment of waste oil, and the high cost of microalgae cultivation limits its large-scale application as raw materials. At the same time, further innovation is needed in the production technology of biodiesel to improve the performance and reaction efficiency of catalysts, reduce equipment investment and operating costs.

(3) Hybrid solar-biofuel systems: Solar-powered pretreatment cuts energy use by 30% while enabling net-zero biorefineries.

(4) In the future, we can focus on the development of non grain raw materials and enzyme engineering improvement to promote energy transformation under the circular economy model. Through technological innovation, such as AI driven

smart factories, the optimization and automation of production processes can be achieved, improving production efficiency and reducing costs. Strengthen policy coordination, such as implementing carbon pricing policies to enhance the market competitiveness of biodiesel. In addition, it is necessary to improve the collection, storage, and transportation system of waste oil and other raw materials to ensure stable supply of raw materials.

Bioethanol and biodiesel production technologies are evolving toward integrated biorefineries that maximize resource efficiency and minimize waste. Key advances—surfactant-assisted pretreatment, VP delignification, CaO/La_2O_3 catalysts, and photocatalytic conversion—address historical limitations in yield, cost, and sustainability. Future progress hinges on scaling innovations (e.g., CBP, anti-leaching catalysts) and policy support for carbon-negative systems. With biomass offering 3–5× lower lifecycle emissions than fossil fuels, these technologies are indispensable for a sustainable energy transition.

4 BIOCHAR

Biochar is a type of environmentally friendly and novel material characterized by its high refractoriness, stability, and carbon-rich composition, produced through the pyrolysis of biomass under oxygen-limited conditions. Globally, the efficient conversion of biomass into biochar and bioenergy has become one of the key strategies for addressing climate change and reducing greenhouse gas emissions. Typical biomass feedstocks include forest residues from logging and processing, livestock manure, municipal solid waste, and agricultural by-products such as crop straw. These raw materials can be processed into biochar through pyrolysis and other techniques, and the conversion pathway is illustrated in Figure 6. Meanwhile, the rapid expansion of modern society, the continuous growth of the population, and the steady improvement of living standards have collectively driven a significant increase in meat consumption, thereby accelerating the development of intensive livestock farming. With the rapid development of livestock and poultry farming, the livestock and poultry farming model has gradually expanded, and the livestock and poultry manure produced in the farming process has also increased. Most of these manures are directly used as fertilizers or randomly piled up without being treated, causing serious pollution to the environment. A large number of surplus livestock and poultry manure far exceeds the soil carrying capacity. If it cannot be used reasonably, effectively and safely, it will cause a large waste of resources and a series of environmental pollution problems. In recent years, although the preparation of livestock and poultry manure biochar has attracted a lot of attention, there are few studies on livestock and poultry manure biochar briquette fuel. Compared with traditional composting methods, biochar conversion technology effectively alleviates the accumulation pollution problem faced by the livestock and poultry farming industry and also creates considerable economic benefits. Compared with biochar powder, shaped fuel has significant advantages in physical properties and application performance. First, its bulk density is 30%-50% higher than that of powder form, and its mechanical strength is 2-3 times higher, which is mainly due to the high-pressure densification effect during the forming process. Second, this form effectively solves the problem of dust pollution caused by carbon powder, and has better operability in storage and logistics. From the perspective of resource utilization efficiency, the forming process can reduce the loss of a small amount of carbon material during pyrolysis. The amount of agricultural and forestry waste biomass generated each year is large and comes from a wide range of sources. If it is prepared into biomass solid molded fuel, it has the advantages of being clean, environmentally friendly, renewable, easy to transport, and easy to use (Ravindiran et al., 2024). It can replace traditional fossil energy such as coal and oil and reduce dependence on non-renewable energy. This chapter provides a critical and comprehensive review the fabrication of biochar focused on livestock manure and forest waste.

4.1 Biochar Production Technologies

The yield and physicochemical properties of biochar are significantly dependent on the type of raw material and the parameters of the thermal conversion process. Currently, the main thermochemical conversion technologies for preparing biochar include pyrolysis, hydrothermal carbonization, and gasification.

Figure 6. Biomass waste can be used as raw material for biochar production through pyrolysis.

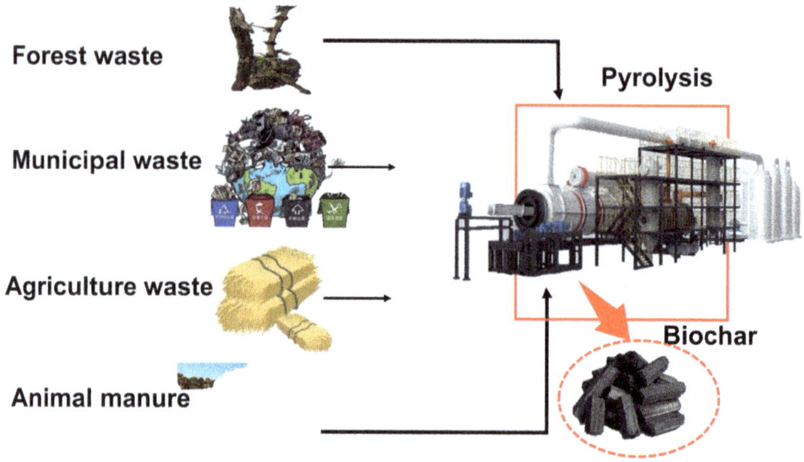

1. Pyrolysis method

Among the existing biochar production technologies, the pyrolysis conversion process has become the mainstream preparation method due to its ease of operation and wide adaptability to raw materials. The pyrolysis process refers to the process in which biomass is pyrolyzed under high temperature and oxygen-free conditions at 300-1000°C to generate solid, liquid and gaseous products. The solid substances generated include a mixture of elements such as C, O, N, Ca and S. In the actual preparation of biochar, the pyrolysis method can be divided into slow decomposition and fast decomposition. Slow pyrolysis is currently the most widely used method for preparing biochar. Its principle is to change the properties of biochar by changing the process conditions. During the slow pyrolysis process, biomass will degrade, decompose and transform to form carbon-rich solids and condensable and non-condensable volatile products. Ayaz et al. used cow dung as raw material and synthesized three forms of biochar by slow pyrolysis (Ayaz et al., 2021). They characterized the biochar and revealed their differences in morphology, specific surface area, pore structure, surface charge and oxygen-containing functional groups. Rapid high-temperature pyrolysis is a process in which biomass materials are rapidly heated from low-temperature, oxygen-deficient, normal pressure, ultra-high heating reaction speed, and ultra-short product residence time to a relatively high temperature, where macromolecules are decomposed, a large amount of small molecular gas products and a large number of condensable volatiles are generated, and a small amount of coke products are produced.

2. Hydrothermal carbonization

Hydrothermal carbonization technology is a breakthrough process in the field of biomass recycling, which has shown unique advantages in wet biomass conversion. This technology uses sub/supercritical water medium as the reaction environment, which can directly treat raw materials with high moisture content, eliminating the necessary drying pretreatment step in traditional pyrolysis process. Compared with other conversion methods, this technology has the following remarkable characteristics: simple and efficient process, the obtained hydrothermal carbon products generally contain more than 60% carbon, and have high energy density. It can not only be used as an independent process, but also can be coupled with other conversion technologies, showing good system compatibility. These characteristics make it an effective way to solve the problem of high-water content biomass treatment, and improve the rate of resource conversion while reducing energy consumption and cost. Hydrothermal carbonization has become the preferred way to prepare carbon materials from biomass raw materials and reduce carbon emissions.

3. Gasification

Gasification is a thermochemical conversion process that occurs at temperatures above 750 °C, using oxygen, air or steam as the gasifying agent. It simultaneously generates biochar, bio-oil and syngas by utilizing the heat released from the pyrolysis and oxidation reactions of biomass waste. As a multi-functional technology that combines carbon fixation and energy recovery, gasification has been widely regarded as one of the key pathways for the large-scale production of biochar. It refers to the thermochemical conversion of biomass raw materials under specific temperature conditions to generate combustible gas, solid charcoal and liquid extracts (Palaniappan, 2025). Gasification technology is divided into two process types: using gasifiers and not using gasifiers. Gasification pyrolysis has similar characteristics to fast pyrolysis. However, the main product of gasification pyrolysis is combustible gas, and the biochar yield is lower than that of slow pyrolysis.

4.2 Advanced Pyrolysis of Animal Manure

Pyrolysis is one of the most promising thermochemical conversion technologies for the large-scale production of biochar. Among them, co-pyrolysis and catalytic pyrolysis, as two mainstream advanced pyrolysis processes, have been widely applied because they can effectively enhance pyrolysis efficiency, regulate product distribution, and improve the physicochemical properties of biochar without significantly modifying the existing reactor system. The biochar yield from non-catalytic single pyrolysis of fecal-like raw materials ranges from 24.49 to 47.60 wt%. However, due to the extremely low inherent carbon content of organic feces, the solid residue rate of pyrolysis is significantly low, and the corresponding biochar yield is generally low. Therefore, the properties and yield of biochar depend on the type of co-feedstocks used for the biochar production. The addition of co-feedstocks has a huge effect on the quality of the final product, various types of materials with different characteristics have been chosen to be pyrolyzed with manure. Heating rate, retention time, and inert gas type play pivotal roles in the pyrolysis of animal manure (Su et al., 2022). Here, we mainly discuss the chicken litter co-pyrolysis with sawdust as an example to introduce the preparation process and biochar as results products.

Due to the complexity of the composition of livestock and poultry manure, the calorific value of biomass solid molded fuel prepared from it is generally low and the ash content is high. In addition, since livestock and poultry manure contain a certain number of heavy metals, the ash after the fuel is burned has the risk of contaminating the soil. In particular, chicken farming is particular. Chicken feed contains stone powder (the main component is $CaCO_3$). The stone powder content in the feed during the peak egg-laying period can be as high as 8.5% or more. Therefore, the ash content of biochar fuel prepared from chicken manure will be higher, resulting in low combustion efficiency. Finding a new type of environmentally friendly chicken litter biochar fuel with a simple preparation process, low cost, high calorific value and low ash content is of great significance to solving the environmental pollution problem of livestock and poultry manure and replacing fossil energy to reduce CO_2 emissions. In view of the problems and shortcomings of the above-mentioned prior art, a chicken litter biochar briquette fuel is prepared by pyrolyzing a mixture of fresh chicken litter (from laying hen farms) and wood processing waste, such as fir wood chips, as an auxiliary material. The fuel has the advantages of high calorific value and low ash content, low cost, easy storage and transportation, and ecological and environmental protection. While solving the problem of chicken litter environmental pollution, it also provides a direction for the energy utilization of poultry and livestock manure and other abundant agricultural and forestry wastes in mountainous areas.

1. Preparation process

First, chicken litter is selected as the raw material and sawdust is selected as the auxiliary material. The sawdust is dried in a blast drying oven at 150°C for 4 hours, and the fresh chicken litter is dried in a blast drying oven at 105°C for 4 hours. Then the dried chicken litter is crushed by a crusher and passed through a 20-mesh stainless steel sieve to pick out the feather impurities contained therein, and the dried sawdust is passed through a 7-mesh stainless steel sieve. Chicken litter and sawdust of different proportions (10-50wt%) are mixed in a mixer for 10-15 minutes, and the mixed materials are hot extruded into firewood rods in a rod making machine at a temperature range of 130-170°C. The fully mixed materials are made into firewood rods using a spiral rod making machine. The prepared firewood sticks were cut into small square pieces of about 3 cm and weighed to calculate the carbon yield after carbonization. During carbonization, chicken litter firewood sticks were pyrolyzed in an atmosphere tube furnace under the protection of inert gas argon. The pyrolysis temperature was set to 500°C and the pyrolysis time was set to 1 h, so as to obtain the required biochar solid shaped fuel. The scheme illustrated in Fig. 7.

Figure 7. Preparation process of biochar by co-pyrolysis of chicken litter and sawdust

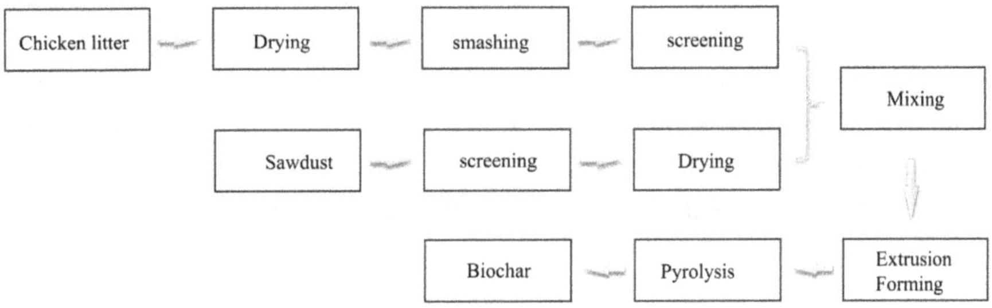

2. Yield and heating value

The yield and heating value of biochar prepared with different sawdust addition amounts are shown in Table 2. It can be seen that with the increase in sawdust addition, the yield of biochar gradually decreases, while the calorific value of biochar increases significantly.

Table 2. The yield and heating value of biochar as a function of sawdust addition rate

sawdust /%	**yield** /%	**heating value** /(MJ·Kg^{-1})
10	53.26	11.20
20	49.37	13.07
30	47.29	14.30
40	46.46	16.22
50	45.71	18.00

This is because sawdust contains more organic matter such as hemicellulose, cellulose and lignin compared to chicken manure. The more sawdust is added, the more organic matter decomposes and volatilizes at high temperature, so the biochar yield gradually decreases. As the amount of sawdust added increases, the less ash is contained in the biochar, and the more fixed carbon is produced, which is conducive to the improvement of the calorific value of biochar. However, when the amount of sawdust added increases, the difficulty of preparing firewood will increase, resulting in an increase in the cost of biochar preparation, and the amount of chicken manure processed will also decrease. The appropriate amount of sawdust added is 30%, and its calorific value is 14.30 kJ/kg, which is very close to the expected calorific value (15 kJ/kg). The calorific value of biochar can be further improved by optimizing other parameters.

4.3 Current Research Status, Challenges and Future Directions

(1) Biochar has demonstrated potential applications in various fields such as energy, water treatment, and environmental sustainability (Ravindiran et al., 2024). In the field of energy, it can be used for energy conversion and storage; In terms of wastewater treatment, as a catalyst and adsorbent, it can effectively remove heavy metals, organic pollutants, etc (Kaetzl et al., 2025; Nguyen et al., 2019); In the field of environment, it contributes to soil improvement and promotes carbon neutrality. Biochar can improve soil structure, increase soil porosity and water holding capacity, thereby increasing crop yields, such as increasing potato yields by up to 340% in some cases. In addition, biochar can also be combined with microorganisms to remediate heavy metal contaminated soil, reducing crop uptake of heavy metals. Biochar is a stable form of carbon with great potential for carbon sequestration, with the potential to remove up to 6% of global emissions annually. By 2023, biochar carbon removal accounted for over 90% of delivered carbon credits.

(2) Traditional biochar production has problems of high energy consumption and low efficiency. New production equipment such as microwave pyrolysis equipment has a high investment cost, which is 2.3 times that of traditional equipment. Moreover, the pyrolysis parameters of different raw materials vary greatly, requiring the development of intelligent control systems. Due to the uncertainty of biochar pyrolysis process and its environmental aging process, there may be potential environmental risks in its use, and the long-term safety and ecological health impacts are still unknown. The diverse raw materials and preparation methods of biochar result in its variable properties, making it challenging to establish a relationship between raw materials, production, characteristics, and application. There is no conclusive evidence to suggest that large-scale application of biochar has long-term environmental friendliness and cost-effectiveness.

(3) In the future, energy co production will be achieved, and the pyrolysis process will simultaneously generate bio oil and synthesis gas to meet some of the energy needs of the factory. Empowered by carbon trading, each ton of biochar can obtain carbon credits, incentivizing companies to participate in the carbon market; Developing smart agriculture, using IoT sensors to control the release of biochar fertilizers, accurately regulating nutrient release, and reducing non-point source pollution.

5 SUMMARY AND PROSPECT

This chapter systematically elaborates on the three major products of biomass resource conversion-biomass-derived carbon dots (BCDs), biofuels, and the preparation process, key properties, and application potential of biochar. BCDs are synthesized from green materials such as plant waste using hydrothermal or microwave methods. With their unique optical properties, efficient corrosion inhibition performance, and antibacterial activity, they have demonstrated significant value in multiple fields; Biofuels (bioethanol, biodiesel) achieve renewable energy substitution through pre-treatment, fermentation, and catalytic conversion technologies. Although facing raw material costs and technological bottlenecks, their low-carbon advantages are clear; Biochar is prepared from agricultural and forestry waste, as well as livestock and poultry manure, through processes such as pyrolysis. It plays a prominent role in soil improvement, carbon sequestration, and pollution control, and the use of formed fuel technology further enhances its practicality.

In the future, BCDs need to focus on material diversification and functional regulation, optimize doping processes to enhance performance stability, and expand their applications to high-end fields; Biofuels should break through the bottleneck of non grain raw material utilization, promote enzyme engineering and AI driven process optimization, reduce production costs, and enhance large-scale production capacity; Biochar needs to develop intelligent pyrolysis systems, combined with carbon trading mechanisms to achieve a win-win situation for economic and environmental benefits, while exploring its cross disciplinary applications in energy storage, deep wastewater treatment, and other fields. The coordinated development of the three will accelerate the efficient utilization of biomass resources throughout the entire chain, providing key technological support for global energy transformation and carbon neutrality goals.

ACKNOWLEDGEMENTS

This work was supported by the Guizhou Provincial Basic Research Program (No. QKHJC-ZK[2023]YB468), and the Doctoral Talent Project of the Tongren Municipal Bureau of Science and Technology (No. TSKY [2022]2).

REFERENCES

Akram, F., Haq, I., Raja, S. I., Mir, A. S., Qureshi, S. S., Aqeel, A., & Shah, F. I. (2022). Current trends in biodiesel production technologies and future progressions: A possible displacement of the petro-diesel. *Journal of Cleaner Production, 370,* 133479. DOI: 10.1016/j.jclepro.2022.133479

Ayaz, M., Feizienė, D., Tilvikienė, V., Akhtar, K., Stulpinaitė, U., & Iqbal, R. (2021). Biochar Role in the Sustainability of Agriculture and Environment. *Sustainability (Basel), 13*(3), 1330. DOI: 10.3390/su13031330

Bakali, K., Ampaire, W., & Extension, K. P. (2024). Bioethanol Production from Sugarcane Bagasse by Fermentation Process Using Saccharomyces cerevisiae as Yeast Species. *IAA Journal of Applied Sciences, 11*(3), 32–41. DOI: 10.59298/IAAJAS/2024/113.3241

de Araujo Guilherme, A., Dantas, P. V. F., Padilha, C. E. A., dos Santos, E. S., & de Macedo, G. R. (2019). Ethanol production from sugarcane bagasse: Use of different fermentation strategies to enhance an environmental-friendly process. *Journal of Environmental Management, 234,* 44–51. DOI: 10.1016/j.jenvman.2018.12.102 PMID: 30599329

Dong, S., Chen, J., Wang, K., Xiao, H., & Huang, T. (2025). Novel functional carbon dot based on Angelica residue and polyethyleneimine: A super-efficient corrosion and scale inhibitor. *Journal of Colloid and Interface Science, 679,* 64–74. DOI: 10.1016/j.jcis.2024.09.222 PMID: 39357227

E, S., He, C., Wang, J., Mao, Q., & Chen, X. (2021). Tunable organelle imaging by rational design of carbon dots and utilization of uptake pathways. *ACS Nano, 15*(9), 14465-14474. doi:https://doi.org/DOI: 10.1021/acsnano.1c04001

Edison, T. N. J. I., Atchudan, R., Sethuraman, M. G., Shim, J.-J., & Lee, Y. R. (2016). Microwave assisted green synthesis of fluorescent N-doped carbon dots: Cytotoxicity and bio-imaging applications. *Journal of Photochemistry and Photobiology. B, Biology, 161,* 154–161. DOI: 10.1016/j.jphotobiol.2016.05.017 PMID: 27236237

Fu, J., Yang, G., Lu, M., Fang, K., Lv, P., & Xiong, Z. (2025). Chemical simulation of high-performance CaO/La2O3 catalysts on its anti-leaching property during biodiesel production. *New Journal of Chemistry = Nouveau Journal de Chimie*, *49*(26), 11180–11190. DOI: 10.1039/D5NJ01085C

He, C., Xu, P., Zhang, X., & Long, W. (2022). The synthetic strategies, photoluminescence mechanisms and promising applications of carbon dots: Current state and future perspective. *Carbon*, *186*, 91–127. DOI: 10.1016/j.carbon.2021.10.002

Kaetzl, K., Riegel, M., Joseph, B., Ossenbrink, R., Gerber, H., Gwenzi, W., Morck, T., Laner, D., Heinrich, T., Kromrey, V., Friedrich, K., Wachendorf, M., & Stenchly, K. (2025). Biogenic activated carbons from conservation grassland biomass for organic micropollutants removal in municipal wastewater. *Environmental Science and Ecotechnology*, *26*, 100588. DOI: 10.1016/j.ese.2025.100588 PMID: 40599981

Kim, H. H., Lee, Y. J., Park, C., Yu, S., Won, S. O., Seo, W.-S., Park, C., & Choi, W. K. (2018). Bottom-Up Synthesis of Carbon Quantum Dots With High Performance Photo- and Electroluminescence. *Particle & Particle Systems Characterization*, *35*(7), 1800080. DOI: 10.1002/ppsc.201800080

Kuang, T., Jin, M., Lu, X., Liu, T., Vahabi, H., Gu, Z., & Gong, X. (2023). Functional carbon dots derived from biomass and plastic wastes. *Green Chemistry*, *25*(17), 6581–6602. DOI: 10.1039/D3GC01763J

Liu, J., Li, R., & Yang, B. (2020). Carbon dots: A new type of carbon-based nanomaterial with wide applications. *ACS Central Science*, *6*(12), 2179–2195. DOI: 10.1021/acscentsci.0c01306 PMID: 33376780

Nguyen, V.-T., Hung, C.-M., Nguyen, T.-B., Chang, J.-H., Wang, T.-H., Wu, C.-H., Lin, Y.-L., Chen, C.-W., & Dong, C.-D. (2019). Efficient Heterogeneous Activation of Persulfate by Iron-Modified Biochar for Removal of Antibiotic from Aqueous Solution: A Case Study of Tetracycline Removal. *Catalysts*, *9*(1), 49. DOI: 10.3390/catal9010049

Onyeachu, I. B., Chauhan, D. S., Quraishi, M. A., & Obot, I. B. (2021). Influence of hydrodynamic condition on 1,3,5-tris(4-methoxyphenyl)-1,3,5-triazinane as a novel corrosion inhibitor formulation for oil and gas industry. *Corrosion Engineering, Science and Technology*, *56*(2), 154–161. DOI: 10.1080/1478422X.2020.1827348

Palaniappan, M. (2025). Time-dependent properties of temperature aged vinyl–based biocomposites developed using waste reinforcements. *Biomass Conversion and Biorefinery*, *15*(4), 5339–5349. DOI: 10.1007/s13399-024-05489-2

Ravindiran, G., Rajamanickam, S., Janardhan, G., Hayder, G., Alagumalai, A., Mahian, O., Lam, S. S., & Sonne, C. (2024). Production and modifications of biochar to engineered materials and its application for environmental sustainability: A review. *Biochar*, *6*(1), 62. DOI: 10.1007/s42773-024-00350-1

Shah, A. M., Kamal, R., Wang, Q., Zhang, J., Lu, H., & Zhao, Z. K. (2025). Calcium carbonate-mediated detoxification of corn stover hydrolysate to facilitate the fermentation by Rhodosporidium toruloides. *Biomass and Bioenergy*, *200*, 108036. DOI: 10.1016/j.biombioe.2025.108036

Song, G., Sun, C., Madadi, M., Dou, S., Yan, J., Huan, H., Aghbashlo, M., Tabatabaei, M., Sun, F., & Ashori, A. (2024). Dual assistance of surfactants in glycerol organosolv pretreatment and enzymatic hydrolysis of lignocellulosic biomass for bioethanol production. *Bioresource Technology*, *395*, 130358. DOI: 10.1016/j.biortech.2024.130358 PMID: 38253243

Su, G., Ong, H. C., Mohd Zulkifli, N. W., Ibrahim, S., Chen, W. H., Chong, C. T., & Ok, Y. S. (2022). Valorization of animal manure via pyrolysis for bioenergy: A review. *Journal of Cleaner Production*, *343*, 130965. DOI: 10.1016/j.jclepro.2022.130965

Teo, K. S. K., Kondo, K., Khattab, S. M. R., Watanabe, T., Nagata, T., & Katahira, M. (2024). Enhancing Bioethanol Production from Rice Straw through Environmentally Friendly Delignification Using Versatile Peroxidase. *Journal of Agricultural and Food Chemistry*, *72*(5), 2657–2666. DOI: 10.1021/acs.jafc.3c07998 PMID: 38288662

Travlou, N. A., Giannakoudakis, D. A., Algarra, M., Labella, A. M., Rodríguez-Castellón, E., & Bandosz, T. J. (2018). S- and N-doped carbon quantum dots: Surface chemistry dependent antibacterial activity. *Carbon*, *135*, 104–111. DOI: 10.1016/j.carbon.2018.04.018

Wu, X., Li, J., Deng, C., Yang, L., Lv, J., & Fu, L. (2022). Novel carbon dots as effective corrosion inhibitor for N80 steel in 1 M HCl and CO_2-saturated 3.5 wt% NaCl solutions. *Journal of Molecular Structure*, *1250*, 131897. DOI: 10.1016/j.molstruc.2021.131897

Wu, Y., Guo, L., & She, Y. (2022). Insight on the corrosion inhibition performance of psidium guajava linn leaves extract. *Journal of Molecular Liquids*, *346*, 117858. DOI: 10.1016/j.molliq.2021.117858

Wu, Y., Zeng, L., Zheng, Y., You, F., & Liu, X. (2025). Pectin-derived nitrogen-doped carbon dots as green corrosion inhibitors for N80 steel in 1 M sulfuric acid solution. *Applied Surface Science*, *687*, 162245. DOI: 10.1016/j.apsusc.2024.162245

Zhao, D., Liu, X., Zhang, R., Xiao, X., & Li, J. (2021). Preparation of two types of silver-doped fluorescent carbon dots and determination of their antibacterial properties. *Journal of Inorganic Biochemistry*, *214*, 111306. DOI: 10.1016/j.jinorgbio.2020.111306 PMID: 33221600

Zheng, S., Feng, L., Hu, Z., Li, J., Zhu, H., & Ma, X. (2023). Study on the corrosion inhibition of biomass carbon quantum dot self- aggregation on Q235 steel in hydrochloric acid. *Arabian Journal of Chemistry*, *16*(4), 104605. DOI: 10.1016/j.arabjc.2023.104605

Chapter 11
Current Status and Future Potential of Specialty Chemicals

Yuru Shen
http://orcid.org/0000-0001-5253-7393
School of Materials and Chemical Engineering, Tongren University, Tongren, China

Yun Du
Changsha Center for Disease Prevention and Control, China

Wu Liu
School of Medical Laboratory, Hunan University of Medicine, China

Chen Hao
http://orcid.org/0000-0001-9849-2418
School of Chemistry and Chemical Engineering, Jiangsu University, Zhenjiang, China

ABSTRACT

This chapter outlines the current state and future prospects of the specialty chemicals sector, defined as high-value, performance-driven compounds characterized by molecular specificity and tailored applications. It begins by contrasting these with commodity chemicals, focusing on attributes like molecular precision, application-specific optimization, and regulatory complexity. The review traces the field's evolution from early polymer innovations to modern advancements in green chemistry, AI, and nanotechnology. Interdisciplinary advances in pharmaceuticals, agriculture, electronics, and environmental remediation are explored, highlighting roles for synthetic biology and advanced materials. Key challenges addressed include scalability,

DOI: 10.4018/979-8-3373-4402-7.ch011

regulatory fragmentation, and sustainability trade-offs. Finally, future opportunities driven by quantum computing, bio-inspired design, and circular economy principles are discussed, emphasizing the need for interdisciplinary collaboration and global governance to promote equitable, sustainable growth.

1. ADVANCED CONCEPTS IN SPECIALTY CHEMICALS

Specialty chemicals, commonly referred to as "functional chemicals," constitute a category of high-value substances defined by their molecular specificity, performance-oriented design, and tailored applications. In contrast to commodity chemicals, which emphasize large-volume production and cost efficiency, specialty chemicals are formulated to meet exact functional demands within niche markets. This differentiation stems from three core characteristics: molecular precision, application-focused optimization, and market exclusivity. For example, ionic liquids such as 1-ethyl-3-methylimidazolium bis(trifluoromethanesulfonyl)imide ([EMIM][NTf_2]) demonstrate adjustable thermal stability (with decomposition temperatures exceeding 300 °C) and tunable solubility through cation–anion pairing, making them suitable for use as nonflammable electrolytes in lithium–sulfur batteries (Majidi et al., 2020). Nevertheless, the distinction between specialty and commodity chemicals remains debated. Consider zeolite catalysts employed in petrochemical cracking: while their engineered pore structures qualify them as specialty materials in certain settings, their widespread use in ethylene production obscures this classification, moving them toward commodity status. Regulatory frameworks further complicate definitions. The European Union's REACH regulation, for instance, designates some specialty chemicals as "substances of very high concern" (SVHC) based on hazard potential rather than molecular distinctiveness, whereas the U.S. Environmental Protection Agency highlights molecular-level customization within its Green Chemistry metrics (Hayler et al., 2019, Wissinger et al., 2021). Such ambiguities present challenges for both academic discourse and industrial strategy.

The debate on lignin valorization illustrates this point (Zakzeski et al., 2010). Derived as a byproduct of cellulose extraction, lignin is typically regarded as a commodity chemical due to its abundance. Yet, when depolymerized via enzymatic hydrolysis—a process requiring stereochemical control and nanoscale catalysts—into compounds like vanillin or syringaldehyde, it functions as a specialty chemical used in fragrances and bioplastics. This duality underscores the need for a functional definition: specialty chemicals are defined not by production volume alone, but by performance in specific applications and high R&D intensity. Economically, although commodity chemicals account for the majority of global output by volume, specialty chemicals generate the greater share of industry profits, driven by high margins in fields such as pharmaceuticals and advanced materials (Katsumori Matsuoka, 2025).

The evolution of specialty chemicals reflects over two centuries of technological advancement and regulatory change. The 1960s to 1970s saw the emergence of high-performance polymers such as Kevlar®, produced via interfacial polymerization of terephthaloyl chloride and p-phenylenediamine. This period was defined by breakthroughs in material science and large-scale industrial trials, which established the foundation for modern composite materials used in aerospace and ballistic protection. In the following decades, the industry underwent a paradigm shift toward sustainability. Catalyzed by the U.S. Pollution Prevention Act of 1990, green chemistry initiatives gained momentum—exemplified by the development of biodegradable surfactants derived from sugar-based feedstocks (Sheldon, 2018). Advances such as lipase-catalyzed kinetic resolution for ibuprofen synthesis reduced waste generation by 90% compared to conventional stoichiometric methods (W. Yang et al., 2025). This progress was driven largely by interdisciplinary collaboration between biochemists and process engineers.

The 21st century has ushered in a third wave driven by artificial intelligence (AI) and nanotechnology. Machine learning models now predict perovskite solar cell efficiencies with 95% accuracy, bypassing years of trial-and-error experimentation (R. Zhang et al., 2025). Concurrently, nanostructured materials such as graphene oxide aerogels are transforming water desalination by enabling high salt rejection under low pressure. However, these advances face limitations. The scalability of electrochemical CO_2 reduction remains constrained by the "overpotential wall," as even advanced copper catalysts require current densities that exceed industrial feasibility for ethylene production. Such challenges underscore the enduring gap between laboratory innovation and commercial application—a recurring theme in specialty chemical development (Y. Yang et al., 2025).

The next frontier for specialty chemicals lies at the intersection of quantum simulation, biohybrid systems, and dynamic materials. Quantum chemistry tools like density functional theory (DFT) are now standard in catalyst design, with projects like MIT's "MolGAN" generating numerous candidate metal-organic frameworks (MOFs) hourly for methane storage (Tom et al., 2024). However, the veracity of these models remains debated. A study revealed that DFT-predicted IrO_2 catalysts for water splitting achieved high Faradaic efficiency in simulations but significantly lower under real-world conditions—a disparity attributed to surface reconstruction effects neglected in atomic-scale models (S. Wang et al., 2023). Nature-inspired design offers parallel promise. Mussels' adhesive proteins, rich in 3,4-dihydroxyphenylalanine (DOPA), have inspired underwater adhesives with lap shear strengths surpassing commercial products such as "Krazy Glue" (Matos-Pérez et al., 2012). Similarly, spider silk's hierarchical nanofibrils, reinforced through β-sheet stacking, have guided the synthesis of synthetic fibers with tensile strengths rivaling Kevlar (Guessous et al., 2024). Yet, replicating biological self-assembly fidelity remains elusive. DNA origami, for instance, suffers from high error rates per fold, limiting its utility in nanoelectronics (Saleem et al., 2025).

Dynamic covalent chemistry introduces another paradigm: materials capable of reversible bond reorganization (Shahi et al., 2022). Vitrimer polymers, featuring dynamic imine bonds, enable reshaping at 180°C while retaining mechanical strength—ideal for 3D-printed aerospace components (Sanchez-Rexach et al., 2020). However, their industrial adoption is stymied by trade-offs: increasing crosslinking density enhances thermal resilience but reduces fracture toughness. Such trade-offs underscore the need for multi-objective optimization frameworks, integrating life cycle assessment (LCA) metrics to balance performance, cost, and sustainability.

The trajectory of specialty chemicals reveals a field oscillating between specialization and convergence. While molecular precision defines its identity, interdisciplinary fluidity—blending AI, biotechnology, and materials science—will dictate its future (Lu et al., 2025). Policymakers and researchers alike must navigate tensions between innovation velocity and ethical accountability (Du et al., 2025). For instance, CRISPR-edited microbes engineered to produce taxadiene (a paclitaxel precursor) promise sustainable oncology solutions but risk unintended ecological impacts if released into ecosystems. Likewise, the EU's recent mandate for circularity may accelerate bio-based R&D but could disproportionately burden SMEs lacking computational resources for LCA modeling (Du et al., 2025). As the specialty chemical sector approaches a substantial valuation in the coming years, its evolution will hinge on harmonizing atomic-scale innovation with systemic socio-economic resilience.

2. HISTORICAL MILESTONES AND PARADIGM SHIFTS

Specialty chemicals have evolved through transformative discoveries and regulatory interventions that redefined industrial paradigms, oscillating between scientific curiosity and societal needs. The 20th century's first wave of innovation was driven by wartime demands, exemplified by Wallace Carothers' development of nylon-6,6 at DuPont in 1935 (Seavey et al., 2003). This step-growth polymerization breakthrough revolutionized textiles and later found applications in bulletproof vests, yet its success relied on trial-and-error experimentation, with efficiency metrics like yield and purity secondary to scalability. Such limitations became apparent in the penicillin scaling effort during WWII, where biochemical engineer Howard Florey's deep-tank fermentation and solvent extraction systems transformed Alexander

Fleming's serendipitous discovery into a viable antibiotic—though its instability necessitated lyophilization, a technique borrowed from food science.

The environmental movements of the 1970s catalyzed a paradigm shift toward sustainability, formalized by Paul Anastas' 12 Principles of Green Chemistry in 1998 (Martinengo et al., 2024). This era prioritized atom economy and waste minimization, exemplified by replacing tetraethyl lead in gasoline with methylcyclopentadienyl manganese tricarbonyl (MMT). However, progress faced industrial resistance; General Motors lobbied against the EPA's 1975 lead phase-out, citing engine redesign costs (Harris et al., 2005). Enzymatic catalysis emerged as a cornerstone of this shift, with immobilized lipase (e.g., Novozym®435) enabling enantiomer-specific syntheses of ibuprofen at 99% enantiomeric excess (ee), reducing waste by 90% compared to stoichiometric methods (Simić et al., 2022). Yet enzyme stability remained a bottleneck until directed evolution of Thermomyces lanuginosus lipase achieved significant thermostability, a feat pioneered by Codexis (Simić et al., 2022). Policy frameworks like the EU's REACH regulation (2006) further shaped this landscape, mandating registration of 30,000+ chemicals and accelerating bio-based alternatives like citric acid plasticizers—though SMEs faced R&D cost surges of 22% post-implementation (Short, 2007).

The 21st century's AI-driven era has dramatically accelerated R&D timelines, exemplified by Düren and colleagues, who directly employed a density functional theory/coupled-cluster (DFT/CC)-based potential energy surface in GCMC simulations to predict methane adsorption in CuBTC (copper 1,3,5-benzenetricarboxylate) (Prakash et al., 2013). Generative adversarial networks (GANs) later designed non-toxic battery solvents with ionic conductivity rivaling conventional options but with significantly lower toxicity, validated via standardized skin assays. China's "Made in China 2025" strategy exemplified national pivots, channeling substantial investments into semiconductor-grade germanium, reducing reliance on key foreign suppliers. This strategic push triggered trade disputes, including the addition of numerous Chinese firms to trade restriction lists over dual-use polymer risks. The COVID-19 pandemic underscored supply chain fragility, with nitrile glove shortages prompting a biotech firm to engineer a microorganism for nitrile precursor fermentation, achieving yields multiple times higher than traditional routes.

Persistent challenges, however, complicate this progress. Bio-based polylactic acid (PLA) bioplastics exhibit a significantly higher fossil energy demand compared to PET when accounting for corn cultivation. Similarly, amine scrubbers for carbon capture incur a substantial energy penalty, driving interest in ionic liquids that can considerably reduce steam requirements. Ethical debates intensify with frontier technologies: CRISPR-edited Yarrowia lipolytica strains avoid harvesting threatened yew trees for paclitaxel but pose unquantified gene-drive risks in open fermenters. Quantum dot displays have relied on cadmium-based compounds restricted under REACH, forcing manufacturers to adopt cadmium-free alternatives such as InP, despite higher costs. Even advanced AI tools like AlphaFold2, while predicting protein structures with high accuracy, can falter in modeling solvent effects (Brown et al., 2024).

The asbestos paradox epitomizes these tensions: chrysotile's replacement by aramid fibers (e.g., Nomex®) improved safety but tripled costs and demanded brake disc redesigns. Likewise, PFAS restrictions in the EU's 2023 proposal, targeting 10,000+ compounds critical for semiconductors and firefighting foams, pit thermal stability against environmental risk (Phelps et al., 2024, Siegel et al., 2023). These historical threads reveal specialty chemicals as a field perpetually balancing innovation velocity with ethical accountability, where interdisciplinary collaboration—spanning materials science, computational modeling, and regulatory science.

Figure 1. Schematic diagram of carbon capture by nanomaterials

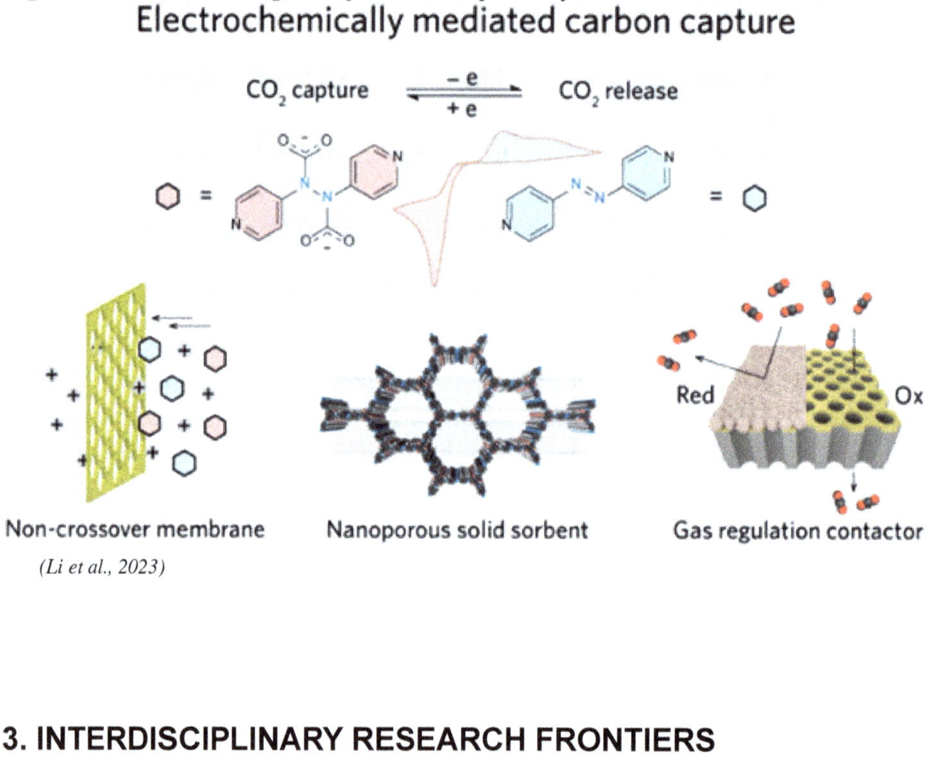

(Li et al., 2023)

3. INTERDISCIPLINARY RESEARCH FRONTIERS

Specialty chemicals drive innovation by providing tailored molecular solutions, yet each industry faces distinct technical and socio-economic challenges that require interdisciplinary collaboration (Wang et al., 2025). In pharmaceuticals, the advancement of precision medicine depends on ultra-high-purity active pharmaceutical ingredients (APIs) and novel delivery systems. For example, mRNA vaccine stabilizers such as SM-102 demonstrate this precision through ionizable lipid tails engineered to destabilize endosomes at pH 5.5, achieving up to 95% mRNA release efficiency. However, batch variability remains a production challenge, as illustrated by Moderna's 2021 COVID-19 vaccine recall, which was linked to peroxide residues in the synthesis of DLin-MC3-DMA (Shi et al., 2024). CRISPR therapies introduce further complexity, requiring Cas9 nuclease cultures with extremely low endotoxin levels (<0.1 EU/mL) and driving production costs up to ten times higher than those of small-molecule drugs. Startups like Intellia Therapeutics are working to bypass ex vivo manipulation through in vivo base editing, yet off-target risks persist. An AstraZeneca study published in Nature Biotechnology in 2023 revealed that dCas9-DNA mismatches can induce unintended mutations in up to 12% of hepatocytes. Personalized medicine also intensifies supply chain vulnerabilities. Monoclonal antibodies such as pembrolizumab rely on ultra-pure glucose (≥99.9% purity) for CHO cell cultures. During Ukraine's 2022 glucose shortages, API prices surged by 300%. Although biosynthetic alternatives—such as Pichia pastoris engineered via CRISPR to secrete monoclonal antibodies—have achieved titers of 8 g/L in fed-batch cultures (a 40% improvement over CHO systems), scalability remains limited by regulatory and economic barriers (Casteleijn et al., 2025, Shirk et al., 2024).

Agricultural innovation strikes a balance between productivity and sustainability through the adoption of biopesticides and precision farming. Engineered Bacillus thuringiensis Cry toxins, for example, demonstrate targeted efficacy against resistant pests, achieving 90% mortality at an application rate of 10 g/ha - comparable to spinosad but with 60% lower mammalian toxicity (LD50 >5,000 mg/kg vs. 2,000 mg/kg) (Faheem et al., 2021). Nonetheless, regulatory challenges persist, as seen in the EPA deregulation of RNAi-based dsRNA sprays, which drew criticism over potential risks to non-target arthropods and exposed ongoing ambiguities in governance frameworks (Zarrabian et al., 2025). Precision agriculture further enhances sustainability through smart delivery systems. Silicon nanocapsules release imidacloprid in response to elevated soil moisture, significantly reducing groundwater contamination. However, the application of atomic layer deposition (ALD) to coat particles with TiO_2 results in prohibitively high production costs, limiting accessibility for small-scale farmers in Sub-Saharan Africa. In the realm of biofertilizers, CRISPR-edited Azotobacter chroococcum demonstrates substantial nitrogen-fixing capacity, potentially offsetting a large portion of conventional urea use. Its performance, however, remains constrained under highly acidic soil conditions. This has spurred the development of integrated solutions such as Pseudomonas fluorescens–nanoclay composites, which buffer soil pH while delivering iron chelates to crops (Oluwole et al., 2025).

The electronics industry's relentless pursuit of miniaturization continues to push the boundaries of material science (Bo et al., 2023). In extreme ultraviolet (EUV) lithography, high-purity germanium (Ge) undergoes advanced purification processes to meet exceptionally stringent purity standards that surpass those of conventional silicon. Developments in epitaxial growth techniques, such as Ge-on-Si integration, have enabled large-diameter wafer uniformity with atomic-scale surface smoothness. Meanwhile, flexible electronics commonly utilize conductive polymers like PEDOT:PSS, though their mechanical fragility has historically constrained wearable applications. This limitation is now being mitigated through the integration of graphene quantum dots (GQDs), which significantly enhance strain tolerance while preserving high electrical performance. Traditional chemical synthesis of GQDs generates hazardous byproducts, prompting growing interest in environmentally benign biosynthetic routes. Microbial platforms have demonstrated efficient production of reduced graphene oxide with high conversion rates. In quantum computing hardware, ultra-pure cryogenic materials are critical: superconducting thin films such as NbN require deposition in ultra-high vacuum environments to ensure qubit coherence (Hadke et al., 2025, Paudel et al., 2022). The operational reliability of cryogenic systems is often impacted by pump-related downtime, driving innovation toward cryogen-free cooling architectures.

Environmental solutions rely on closing material loops through sustainable chemistry. Iron nanoparticle-impregnated granular activated carbon (GAC) enables high removal efficiency for perchlorate (ClO_4^-); however, citrate-capped Fe@C composites are now essential to prevent leaching and ensure long-term stability (Xie et al., 2018, Xu et al., 2013, Xu et al., 2016). Electrocoagulation with aluminum electrodes facilitates substantial phosphate recovery from wastewater, though the resulting byproducts require stabilization to mitigate potential soil degradation. Carbon capture systems are increasingly transitioning from conventional amine scrubbers to metal–organic frameworks (MOFs), which offer significantly higher adsorption capacity under ambient conditions—though hydrophobic surface modifications are critical to sustain performance in humid environments (Zhao et al., 2025). Enzymatic recycling of PET using CRISPR-engineered Ideonella sakaiensis PETase demonstrates enhanced monomer recovery compared to traditional glycolytic methods, which yield high-purity outputs at considerable energy demand. Advanced enzyme variants, refined through multiple genetic modifications, achieve rapid polymer degradation; however, immobilization on magnetic supports introduces substantially increased capital costs for reactor systems (Chen et al., 2025, Ma et al., 2025).

Cross-cutting challenges reveal inherent systemic trade-offs. Breakthroughs in laboratory settings often fail to scale linearly: for example, while electrocatalytic ammonia synthesis demonstrates high Faradaic efficiency using RuBisCo enzymes in vitro, pilot plants report significantly reduced efficiencies due to mass transfer limitations (Appel et al., 2013, Brachi et al., 2024). Regulatory disparities further complicate market entry: Cyrene™, a bio-based solvent compliant in Europe, faces restrictions in the U.S. due to residual solvents in its distillation byproducts. Meanwhile, ethical concerns regarding material access exacerbate global inequities. Gold nanoparticles used in AuroLase® cancer therapy come at a prohibitively high cost for low-income countries, whereas iron oxide alternatives, though more affordable, show substantially lower tumor-targeting efficacy (Bauri et al., 2023).

Emerging solutions integrate biology with data science. Machine learning accelerates enzyme discovery for bioremediation, reducing R&D time from several months to just a few weeks. Blockchain-based material passports enable traceability of rare earth elements from mining to final application, as demonstrated in BMW's pilot tracking neodymium in EV motors. These advances highlight a crucial reality: the industrialization of specialty chemicals requires not only molecular innovation but also resilient supply chains, regulatory foresight, and ethical accountability. With the UN's 2030 Sustainable Development Goals (SDGs) increasing urgency, the convergence of synthetic biology, AI-driven design, and circular economy principles will determine which technologies move beyond the lab to transform global markets (Choudhary et al., 2025).

Figure 2. Schematic of bioorthogonal POLY-PROTAC nanoparticles for tumour-specific protein degradation and precision cancer therapy.

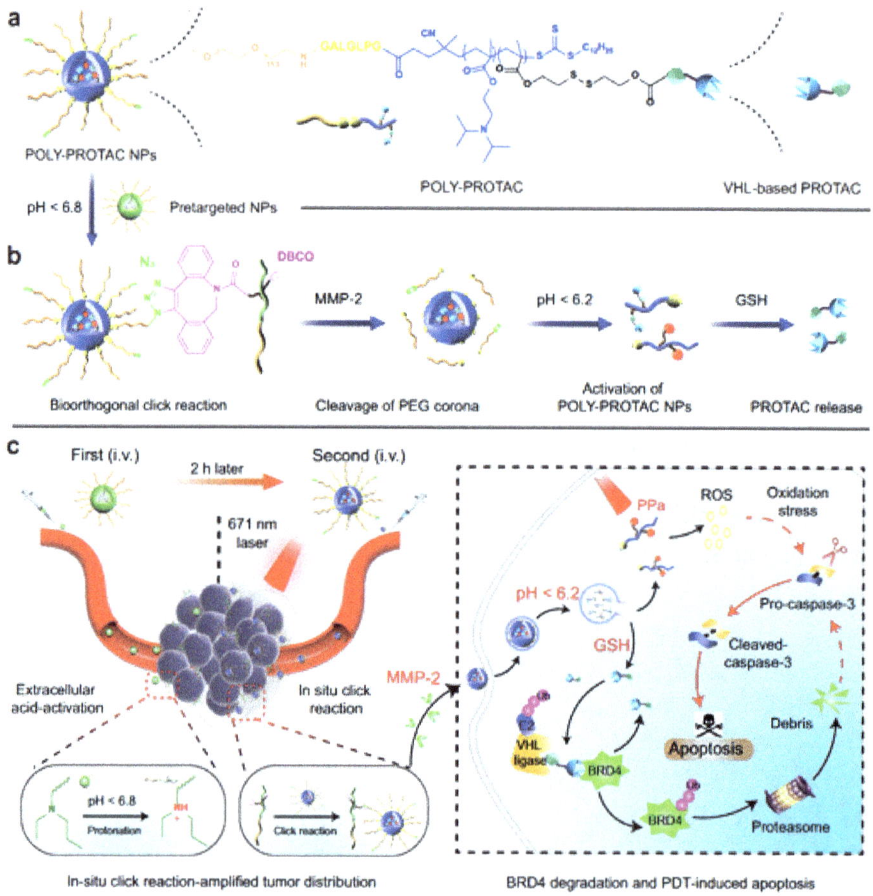

(Gao et al., 2022)

4. CHALLENGES AND CRITICAL ISSUES

The specialty chemical industry, though often characterized as innovation-led, confronts multidimensional challenges that reveal persistent tensions between scientific aspirations and industrial constraints. At the molecular scale, the demand for precision clashes with the requirements of scalable production. For example, in CRISPR-engineered Yarrowia lipolytica strains designed to synthesize paclitaxel precursors, laboratory-scale bioreactors attain high yields, yet scaling to larger fermenters introduces operational heterogeneity (W. Li et al., 2022). According to

Novartis' bioprocess audit, pH gradients and shear stress under these conditions reduce cell viability. Similarly, enzyme immobilization—a key strategy for industrial biocatalysis—faces comparable scalability dilemmas. Novozymes' Eversa® lipase maintains stability in controlled lab assays but experiences activity loss in continuous-flow reactors due to biofilm accumulation, an issue that has remained unresolved since the product's commercialization (Casteleijn et al., 2025). Nanotechnology applications also encounter trade-offs between performance and scale: graphene oxide aerogels demonstrate effective salt rejection in laboratory desalination trials, but industrial production via ambient-pressure drying compromises porosity, leading to a decline in water permeability. These technical hurdles reflect wider systemic barriers: a significant proportion of nanomaterial innovations fail to progress beyond technology readiness level (TRL).

Economic and supply chain fragilities compound these technical hurdles. The sector's reliance on rare inputs exacerbates geopolitical risks. Lanthanum oxide, essential for catalytic converters and battery cathodes, is sourced 85% from China's Bayan Obo mine (Arshi et al., 2018). The price surge of the early 2020s prompted EU automakers to lobby for REACH exemptions related to lanthanum recycling quotas. In contrast, the 2023 U.S. Inflation Reduction Act linked green hydrogen subsidies to domestic mineral sourcing, sparking trade disputes over vanadium pentoxide exports (Abedi et al., 2026, Field et al., 2017). Circular economy models face economic challenges: chemically recycled PET via glycolysis achieves high purity but requires substantially higher energy consumption than virgin PET production. Enzymatic recycling using Ideonella sakaiensis PETase significantly reduces energy use but yields limited monomer recovery per cycle. Startups like Carbios are piloting enzyme-membrane hybrid systems; however, high capital costs per ton of capacity hinder widespread adoption beyond niche applications (Ghosh et al., 2025).

Regulatory fragmentation further complicates market dynamics. Divergent policies encourage compliance arbitrage: for instance, the EU's REACH regulation mandates the substitution of substances of very high concern (SVHCs) such as bisphenol A with alternatives like bio-based citric acid, whereas the U.S. EPA's TSCA adopts a cost-benefit approach, permitting the use of BPA in medical devices until 2025 (Khan et al., 2025). This regulatory divergence has enabled East Asian manufacturers to dominate the global BPA market, despite EU import restrictions. Carbon pricing mechanisms further entrench this fragmentation: the EU Emissions Trading System (ETS) imposes a significant price on CO_2 allowances, incentivizing a shift toward methanol-to-olefins processes that utilize green hydrogen (Jiang et al., 2024). In contrast, the U.S. 45Q tax credit supports amine scrubber technology, leaving metal-organic framework (MOF)-based carbon capture systems without comparable subsidies and therefore commercially at a disadvantage (Teel et al., 2026).

Ethical and societal analyses reveal systemic inequities through critical trade-offs. Emerging technologies, for instance, pose dual-use dilemmas: while CRISPR-edited microbes may be designed to degrade herbicides such as 2,4-D, they could also be repurposed for bioweapon production—a risk that has spurred the development of biosafety containment systems funded by agencies like DARPA (da Silva et al., 2024). Similarly, quantum dots based on cadmium selenide, which are banned under EU regulations, remain vital for U.S. defense display technologies, underscoring a persistent tension between environmental governance and national security priorities. Equitable access to technological innovation remains elusive, as illustrated by gold nanoparticles used in photothermal cancer therapy (e.g., AuroLase®), whose high cost places them out of reach for low-income countries. In such settings, iron oxide alternatives, though more affordable, still entail substantial treatment costs, particularly for rare diseases (Debnath et al., 2024).

Sustainability metrics frequently obscure underlying compromises. Life cycle assessments (LCAs), for example, often omit upstream land-use impacts: polylactic acid (PLA) bioplastics consume substantially more fossil energy than PET when corn cultivation is accounted for. Moreover, a significant proportion of products marketed as "carbon-neutral" specialty chemicals depend on unverified offset mechanisms, such as biodiversity credits in the Amazon rainforest that lack demonstrated protection outcomes. Industrial symbiosis models also face practical behavioral barriers. While the Kalundborg Symbiosis achieved substantial energy savings over a long period through trust-based collaboration among multiple industries, attempts to replicate this model in fragmented industrial clusters in Asia—such as India's Vapi Industrial Park—have stalled due to antitrust concerns that hinder cross-firm data sharing (Subbiah et al., 2025, Tabone et al., 2010).

Critical debates highlight the strategic crossroads facing the sector, particularly whether research and development should prioritize laboratory breakthroughs like room-temperature ionic liquids or focus on incremental advances in catalyst durability. The awarding of the 2020 Nobel Prize in Chemistry for CRISPR technology underscores the risks of prioritizing "eureka moments" over sustained process engineering, while in carbon capture, economic tensions persist between established amine scrubbers—which incur a significant energy penalty—and emerging metal-organic frameworks that remain largely unproven (Cao et al., 2025). The scientific community is divided on this front, as reflected in reports of a split among experts, with some advocating for hybrid systems and others calling for mandates on direct air capture. Further contention arises from bioengineering ethics, exemplified by a temporary moratorium on gene drives for ocean cleanup under the UN Convention on Biological Diversity due to potential risks to marine ecosystems—a decision actively contested by NGOs such as The Ocean Cleanup.

Addressing these multifaceted challenges requires adaptive governance frameworks. Competing approaches are exemplified by the EU's Carbon Border Adjustment Mechanism (CBAM), which taxes carbon-intensive imports, and the U.S. Inflation Reduction Act (IRA), which subsidizes domestic green technology (Peng et al., 2024, F. Zhang et al., 2025). Yet neither fully addresses supply chain vulnerabilities, as illustrated by Tesla's shift to Indonesian nickel sources despite associated deforestation concerns. Interdisciplinary solutions are emerging: machine learning now helps predict regulatory hurdles for novel chemicals, reducing time-to-market, while BMW utilizes blockchain-based material passports to trace neodymium from extraction to electric motor. As the sector expands, its resilience will depend on balancing atomic-scale precision with broad socioeconomic adaptability—a delicate tightrope walk that demands unprecedented collaboration among academia, industry, and policymakers (Lombardo et al., 2022).

Figure 3. Nanoparticle-plant interactions: uptake, translocation, oxidative stress, cellular and physio-logical responses of plants to nanoparticles exposure: positive and negative impacts

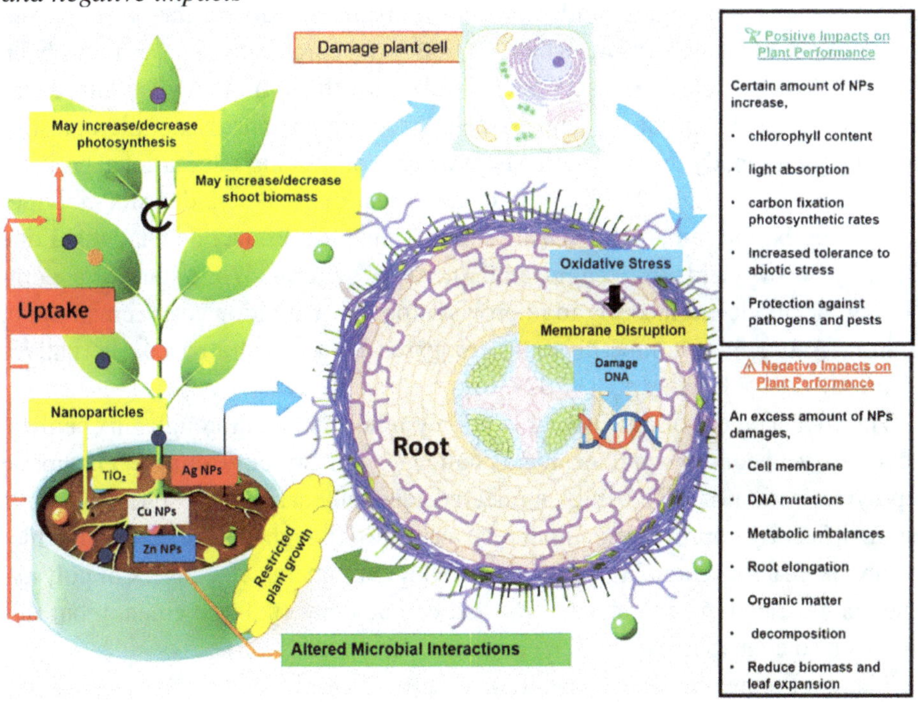

(Bouhadi et al., 2025)

5. FUTURE TRENDS AND OPPORTUNITIES

The specialty chemical sector stands at a pivotal crossroads of disruptive innovation, where emerging technologies and shifting societal values are reshaping value creation. Quantum computing and artificial intelligence are transforming molecular discovery, though hardware limitations such as high error rates in current quantum devices necessitate hybrid workflows (Paudel et al., 2022). Meanwhile, AI-driven molecule discovery is expediting drug development but raising ethical concerns, as many computationally generated molecules encounter practical setbacks in preclinical trials due to unmodeled properties (Lu et al., 2025). This trajectory mirrors challenges observed in advanced fields like CRISPR editing, where predictive accuracy does not always account for complex real-world interactions (Cao et al., 2025, Hartley et al., 2022).

Nature-inspired design continues to drive the discovery of novel bioactive compounds, though the scalability of synthetic biology platforms remains a topic of debate. For example, researchers at Caltech have engineered E. coli to produce taxadiene, a key precursor to paclitaxel, by using CRISPR to modify mevalonate pathways, achieving significantly elevated titers. This marks a major shift from conventional plant extraction, which is highly resource-intensive and has been linked to environmental degradation. However, synthetic biology also introduces ecological risks; for instance, gene drives intended to degrade ocean microplastics could unintentionally disrupt marine ecosystems, prompting international calls for a proposed moratorium. Similarly, biohybrid materials like spider silk–polymer composites exhibit remarkable tensile strength but face production challenges due to relatively low yields from engineered microbial systems. While startups are employing metabolic engineering to improve output, fermentation costs remain high, limiting commercial competitiveness compared to petroleum-based alternatives (Cao et al., 2025, Hartley et al., 2022, Simić et al., 2022).

The advancement of the circular economy relies on breakthroughs in chemical recycling, yet economic and technical challenges remain. For example, enzymatic depolymerization of PET using engineered Ideonella sakaiensis PETase variants has significantly improved monomer recovery rates, though the associated enzyme-membrane reactors entail high costs. Meanwhile, the chemical recovery of polystyrene via gasification generates syngas that can be converted into ethanol, but this process also contributes to carbon emissions (Liu et al., 2025).

The environmental impact per ton of styrene continues to challenge carbon neutrality claims. Meanwhile, industrial symbiosis models are gaining traction—exemplified by the Kalundborg Symbiosis, which reuses substantial by-products—yet many systems still lose an estimated $985 million annually due to siloed information. In contrast, researchers examined the industrial symbiosis network in Finland's

Sodankylä region, uncovering a compelling balance between investment and return. While the system's full lifecycle carries a €115.20 million price tag, the collaborative model unlocks €6.42 million in waste management savings for businesses — and simultaneously curbs greenhouse gas emissions by nearly a million tonnes of CO_2, hundreds of tonnes of CH_4 and N_2O. When optimized, the architecture reveals a sustainable annual surplus of €0.63 million in operational efficiency (Haq et al., 2021).

Green chemistry metrics are evolving beyond traditional life-cycle assessments, yet policy arbitrage undermines global climate efforts. The EU's Product Environmental Footprint (PEF) now requires real-time monitoring of microplastic emissions during polymerization, while the U.S. EPA promotes circularity-by-design mandates. However, some Chinese manufacturers exploit regulatory gaps by exporting goods containing Substances of Very High Concern (SVHCs) to the EU, prompting proposals for substantial fines on non-compliant imports. Carbon pricing mechanisms are reshaping R&D priorities: the EU Emissions Trading System (ETS) drives methanol-to-olefins (MTO) processes using green hydrogen, whereas the U.S. 45Q tax credit supports amine scrubber deployment. Startups like LanzaTech capitalize on this divide by converting captured CO_2 into ethanol and qualifying for both incentives. Yet such regulatory shopping risks undermining net-zero commitments, especially as global CO_2 emissions continued to rise despite numerous climate pledges (Solomon et al., 2024, Teel et al., 2026, Woeldgen et al., 2025).

Ethical and societal frontiers ignite vigorous debates regarding accessibility and equity. While open-source platforms such as PubChem democratize access to molecular data, the pharmaceutical industry's patent thickets often hinder generic competition, as seen with cancer drugs approved since 2010 that remain without generic alternatives due to evergreening strategies (Coca-Lopez et al., 2025). Decentralized laboratories, including BioFoundries, empower DIYbio communities to innovate in areas like biofertilizer engineering; however, such projects frequently fail because of insufficient or non-standardized biosafety protocols. Meanwhile, greenwashing continues to erode consumer confidence—for example, Unilever's "plastic neutral" initiative lost credibility in 2023 after audits revealed that collected plastics were ultimately incinerated (Lange, 2021). To counter such practices, the EU's Digital Product Passport (DPP) mandates comprehensive tracking of material origins (Wiesinger et al., 2024).

Strategic pathways for the 2030s require systemic coordination. Governments must invest in the development of error-corrected qubits to reduce quantum simulation errors to a threshold that could unlock room-temperature superconductors. The WHO's proposed Global Biofoundry Accord would establish standardized safety protocols for gene editing, thereby preventing the misuse of synthetic biology. Meanwhile, the World Bank's "Green Circular Finance" initiative could de-risk industrial symbiosis projects through blended finance models. UNESCO's

Open Science Framework advocates for a tiered patent system, which would allow generic production of life-saving drugs while still protecting upstream innovations (Bauer et al., 2020).

The future of the specialty chemical sector depends on aligning exponential technologies with planetary boundaries. While the convergence of quantum computing and AI promises autonomous catalyst design, it also brings societal costs—such as job displacement and biodiversity risks—that demand proactive governance. The sector's trajectory will hinge on whether innovators embrace inclusive futurism, where profit coexists with ecological stewardship and equitable access, with decisions made now shaping humanity's chemical legacy for generations (Bollini et al., 2023).

Figure 4. The interaction between specialty chemicals and artificial intelligence

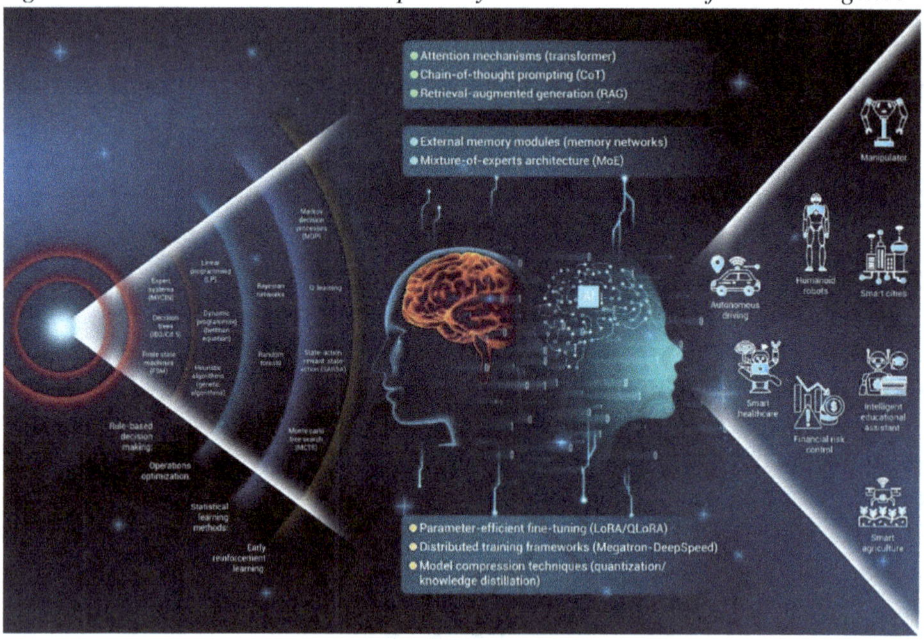

(Huang et al., 2025)

6. INTERDISCIPLINARY COLLABORATION

The most transformative innovations in the specialty chemical sector arise not from isolated expertise, but from the convergence and integration of diverse disciplines, a synergy that is reshaping molecular discovery, industrial strategy, and global governance despite persistent systemic barriers. Using MIT's "Self-Healing Catalyst" platform as an example, the intersection of materials science and computational chemistry demonstrates how machine learning models trained on density functional theory reaction pathways can predict novel catalysts for CO_2 hydrogenation, with validation through high-throughput experimentation significantly reducing development time while achieving high selectivity—a breakthrough unattainable through traditional trial-and-error methods. However, limitations persist; quantum chemistry simulations often overlook solvent effects, as exemplified by inaccuracies in predicting protein-ligand binding in aqueous environments. Nanotechnology further bridges disciplinary divides, as seen in the synthesis of graphene oxide aerogels enhanced by Bayesian optimization algorithms to improve desalination performance, though scaling these advances requires deeper collaboration with chemical engineers to adapt laboratory techniques to industrial-scale processes (Huang et al., 2025, Shahi et al., 2022, Zhou et al., 2023).

Data science, while democratizing innovation, simultaneously amplifies global inequities. Open-source platforms such as the Materials Project allow startups to identify promising battery cathodes without relying on expensive synchrotron facilities, yet most of their contributors are affiliated with U.S. or EU institutions, marginalizing researchers from Asia and Africa. A similar imbalance manifests in intellectual property, where multinational corporations hold a majority of AI-generated molecule patents, hindering open innovation. Blockchain technology is redefining trust in collaborative R&D; for instance, the EU's Horizon Europe program employs smart contracts to automate royalty distribution for cross-border patent collaborations, reducing related legal disputes, while in Asia, China's "Chemical Blockchain Alliance" traces rare earth elements from mine to magnet, ensuring conflict-free sourcing—a model adopted by BMW for its electric vehicle motors. Despite these advances, data silos remain entrenched: industrial laboratories often withhold proprietary datasets, primarily due to fears of corporate espionage (Champion et al., 2022, Wei et al., 2026).

The promise of synthetic biology hinges on interdisciplinary risk governance, as seen in the development of engineered Yarrowia lipolytica strains for pharmaceutical precursors, which requires collaboration among geneticists, metabolic engineers, and ethicists to address concerns like gene drives. Tensions between bioprospecting and ecological preservation were further highlighted when the UN Convention on Biological Diversity imposed a moratorium on marine synthetic biology over fears of uncontrolled plankton blooms. In materials science, biohybrid innovations such as spider silk–polymer composites exemplify the convergence of biology and polymer science, though scalability remains a challenge; yeast-expressed silk genes initially yielded low production levels, necessitating metabolic engineering expertise to increase titers. Such partnerships often falter over intellectual property conflicts, with a significant portion of academic–industry collaborations failing to commercialize innovations (Enkavi et al., 2019, K. Kim et al., 2025).

Policy frameworks for cross-disciplinary collaboration exhibit significant divergence across regions. In the United States, the National Nanotechnology Initiative channels substantial annual funding into integrated laboratory facilities, resulting in transformative advances such as CRISPR-Cas9 gene-editing technology. Conversely, the European Union's Horizon Europe program adopts a "mission-oriented" approach, exemplified by its Green Deal funding initiative aimed at developing carbon-neutral specialty chemicals. Across Asia, the emphasis leans toward corporate-led research and development. For instance, Japan's RIKEN Institute partners with Mitsubishi Chemical to engineer enzymatic PET recycling systems, achieving high monomer recovery—an accomplishment that necessitates coordinated efforts among bioengineers, chemists, and policy-makers to align regulatory and economic incentives. Addressing global challenges increasingly requires transnational coordination. The OECD's Global Science Forum, for example, mandates interdisciplinary review panels for AI-driven chemical discovery to ensure ethical and safety considerations are systematically integrated. Similarly, the WHO's Global Biofoundry Accord establishes standardized gene-editing protocols, designed to prevent misuse while facilitating international knowledge exchange. Nevertheless, geopolitical tensions continue to hinder such cooperation. Recent decoupling between the U.S. and China, for instance, has delayed joint initiatives in quantum catalysis, resulting in a significant reduction in cross-border research grants (Rabiee et al., 2025, Scala et al., 2026).

Systemic barriers to effective collaboration include misaligned incentives, communication gaps, and regulatory friction. Academia prioritizes high-impact publications, while industry focuses on patents and return on investment, with research indicating that many academic projects lack immediate industrial applicability. Communication challenges arise from disparate terminologies, such as "throughput" in biology versus "training epochs" in AI modeling. Additionally, divergent regulatory frameworks across regions impose inconsistent compliance burdens, discouraging cross-border R&D initiatives. Emerging strategies address these issues: Stanford's Bio-X program cultivates T-shaped researchers proficient in biology, engineering, and entrepreneurship; the Novo Nordisk Foundation's Convergence Science Grants mandate interdisciplinary teams including chemists, ethicists, and policymakers; and MIT's Climate Grand Challenge integrates climate science, materials engineering, and behavioral economics to advance carbon-negative chemical design (Chowdhury et al., 2024, Ismail et al., 2025).

The future of the specialty chemical sector depends on its ability to transcend disciplinary boundaries in order to tackle grand global challenges. The convergence of quantum computing and AI, for example, will require physicists to work alongside synthetic biologists to design light-harvesting proteins; similarly, circular economy models necessitate collaboration between chemists and blockchain developers to track material lifecycles. As highlighted in the IPCC's report, climate change cannot be solved by any single field—only through interdisciplinary dialogue and creative friction can we develop solutions that balance innovation with equity. The sector's anticipated valuation will rely not on isolated breakthroughs, but on integrated ecosystems where biologists brainstorm with software engineers, and policymakers engage directly with laboratory scientists. This reflects the unspoken truth of specialty chemicals: its most transformative advances will emerge not from molecules alone, but from the intersections between disciplines (Artz et al., 2018, Curutchet & Mennucci, 2017).

7. EDUCATION AND CAREER PATHWAYS

The evolution of the specialty chemical sector demands a paradigm shift in education and workforce development, as automation and artificial intelligence reshape traditional roles. The next generation of professionals must navigate interdisciplinary skill sets, ethical challenges, and global market dynamics, often finding misalignment with outdated academic structures and persistent industry silos. Central to this transformation is the growing need for hybrid competencies, such as process analytical technology (PAT) and real-time quality control, which are now highly sought after by European chemical companies for R&D roles. Yet academia has been slow to adapt, with few chemistry programs incorporating PAT into core curricula. Equally critical is expertise in regulatory affairs, particularly under frameworks like EU REACH 2.0, which require proficiency in tools such as ToxTree software and QSAR modeling for interpreting complex dossiers—a skill gap reflected in the limited formal training received by recent graduates (Barbon et al., 2026, E. J. Kim et al., 2021).

Nanotechnology and computational chemistry are redefining the boundaries of expertise. Quantum chemistry simulations—enabled by tools such as Materials Studio—now rival traditional wet-lab methods in drug discovery. Yet, few academic programs systematically incorporate training in quantum algorithm optimization or GPU-accelerated molecular dynamics simulations. This disconnect is reflected in industry hiring trends: startups like Schrödinger increasingly seek candidates with hybrid skills in molecular dynamics and cloud computing. A LinkedIn report further identified "quantum-ready" chemists as the fastest-growing job category within specialty chemicals. This emerging skills gap highlights a broader tension between industry needs and current training. While Industry 4.0 demands fluency in AI-driven molecular design, an NSF survey revealed that U.S. chemical engineering graduates lack proficiency in coding (Carnero Canales et al., Muntean et al., 2025).

Global education systems continue to grapple with curricular inertia, illustrated by divergent reform approaches. The EU's Erasmus Mundus Joint Masters in Green Chemistry integrates life cycle assessment with industrial symbiosis case studies, yet a significant proportion of its graduates struggle to secure sustainability-related roles, exposing a persistent skills-to-market gap. In contrast, the National University of Singapore offers an industry-sponsored "Smart Chemicals" track, co-developed with Bayer and Siemens, where students design AI-driven catalysts for CO_2 hydrogenation, achieving strong placement in petrochemical R&D. Meanwhile, the U.S. system reflects credentialing fragmentation: while ACS Green Chemistry certification is often required for EPA-funded roles, many employers prefer proprietary credentials like MITx's "Chemical Engineering for Energy Systems," which emphasizes reactor scalability over sustainability metrics. This divergence fuels an ongoing debate over whether accrediting bodies should mandate circular economy modules or whether market-driven certifications are more responsive to evolving industry needs (Ruini et al., 2025, Sbaiti et al., 2024)?

Corporate partnerships frequently undermine academic integrity, as illustrated by the "Chemical Valley" scandal in which Dow Chemical funded researchers at the University of Texas to downplay the risks of PFAS in firefighting foams—a stark illustration of the conflict between intellectual property commercialization and public health imperatives. In response to such controversies, "ethics-by-design" reforms are gaining traction; for instance, ETH Zurich now mandates that industry-sponsored projects incorporate ethics review boards with student representation, a model that has been adopted by numerous universities across Europe. Meanwhile, open-source education platforms democratize access while simultaneously exacerbating inequities (Leyfman et al., 2023). Although Coursera's "AI for Drug Discovery" course, developed in partnership with Insilico Medicine, attracted a large number of global enrollments, only a small fraction of participants completed the capstone project due to insufficient computational resources. This disparity is particularly acute in low-income nations, where cloud computing expenses can surpass average monthly incomes (Pley et al., 2020).

The emergence of "chempreneurship" is redrawing traditional career pathways, with accelerators such as IndieBio fostering startups focused on biohybrid materials, a significant portion of which are launching ventures in enzyme recycling (Ouyang et al., 2024). Nevertheless, substantial funding gaps persist, as seed-stage investors tend to prioritize nanotechnology over sustainability-focused ventures. Platforms like Kolabtree allow specialists to monetize niche expertise, such as CRISPR assay design; however, many freelancers report irregular income, highlighting the precarity inherent in the gig economy. Concurrently, global mobility trends are reshaping professional geographies, exemplified by initiatives like the UAE's "Chemical Innovation Hub," which attracts talent with tax-free salaries and fast-track visas, though cultural isolation remains a major barrier to long-term retention. Conversely, while Germany's "Blue Card" program targets skilled migrants, its stringent language requirements exclude a considerable number of otherwise qualified applicants from non-EU countries (Giovanis et al., 2021).

Three strategies are gaining traction to address the growing skills gap. First, microcredentialing: systems such as the Royal Society of Chemistry's "Digital Badges" certify modular competencies—for example, "MOF Synthesis" or "Blockchain Traceability"—which can be accumulated and applied toward full degree programs. Second, hybrid learning ecosystems: initiatives like MIT's Climate Grand Challenge bring together climate science, materials engineering, and behavioral economics, enabling student cohorts to collaboratively develop carbon-negative chemical solutions for partners such as Shell and Unilever. Third, policy advocacy: efforts including UNESCO's Open Educational Resources (OER) mandate free access to computational tools like Avogadro, thereby democratizing skill acquisition in developing nations (Cumberland et al., 2023, Otto, 2019).

The future workforce in the specialty chemical sector must combine technical expertise with ethical responsibility. While AI and automation may replace some traditional roles, they also open doors for hybrid professionals skilled in regulatory compliance, sustainable design, and cross-cultural collaboration. As highlighted in recent IPCC findings, the sector's value depends not solely on specialized knowledge but on ecosystems where educators, policymakers, and industry leaders collaboratively design curricula that prioritize planetary health alongside profitability. Ultimately, the success of educational reform should be measured not by patents or publications, but by graduates' capacity to engineer solutions that bridge disciplinary and geopolitical divides—transforming resources into equity (Pathak et al., 2021).

8. CONCLUSION AND CALL TO ACTION

The specialty chemical industry stands at a crossroads, where converging technologies offer unprecedented opportunities but are shadowed by critical paradoxes, revealing that innovation cannot be separated from its socio-ecological consequences amidst a fragmented global governance framework. To move forward, the sector must redefine its social contract by fundamentally rethinking education through interdisciplinary competencies, implementing policy reforms for international cooperation, and adopting circularity-by-design principles that require blended finance to scale. Urgently addressing ethical shortcomings is imperative, as past collusion and greenwashing have undermined trust, necessitating countermeasures like mandatory ethics reviews and transparent lifecycle assessments. With three imperatives demanding immediate action—educational equity, policy convergence through a global treaty, and corporate accountability via independent ecological risk assessments—the industry's legacy will be shaped not by molecular breakthroughs but by its capacity to align human innovation with planetary boundaries through transformative collaboration.

REFERENCES

Abedi, M., Chang, A., Hassan, H. I., Yang, A., Hickethier, M. L., Bushagour, A. H., Aghajari, M., Stoerzinger, K. A., Layton, A., Abbasi, B., Chang, C.-H., & Feng, Z. (2026). Recovering Critical Metals from Brines: Technology Landscape, Circular Economy, and Future Trends. *ACS Applied Energy Materials*, *9*(4), 2005–2034. Advance online publication. DOI: 10.1021/acsaem.5c03407

Appel, A. M., Bercaw, J. E., Bocarsly, A. B., Dobbek, H., DuBois, D. L., Dupuis, M., Ferry, J. G., Fujita, E., Hille, R., Kenis, P. J. A., Kerfeld, C. A., Morris, R. H., Peden, C. H. F., Portis, A. R., Ragsdale, S. W., Rauchfuss, T. B., Reek, J. N. H., Seefeldt, L. C., Thauer, R. K., & Waldrop, G. L. (2013). Frontiers, Opportunities, and Challenges in Biochemical and Chemical Catalysis of CO_2 Fixation. *Chemical Reviews*, *113*(8), 6621–6658. DOI: 10.1021/cr300463y

Arshi, P. S., Vahidi, E., & Zhao, F. (2018). Behind the Scenes of Clean Energy: The Environmental Footprint of Rare Earth Products. *ACS Sustainable Chemistry & Engineering*, *6*(3), 3311–3320. DOI: 10.1021/acssuschemeng.7b03484

Artz, J., Müller, T. E., Thenert, K., Kleinekorte, J., Meys, R., Sternberg, A., Bardow, A., & Leitner, W. (2018). Sustainable Conversion of Carbon Dioxide: An Integrated Review of Catalysis and Life Cycle Assessment. *Chemical Reviews*, *118*(2), 434–504. DOI: 10.1021/acs.chemrev.7b00435

Barbon, A. P. A. C., Nolasco-Perez, I. M., Barbon, S.Jr, Máquina, A. D. V., Su, W.-H., & Barbin, D. F. (2026). Process analytical technologies applied to quality control of emerging alternative protein food products: Challenges and future trends. *Journal of Pharmaceutical and Biomedical Analysis Open*, *7*, 100105. DOI: 10.1016/j.jpbao.2026.100105

Bauer, B., Bravyi, S., Motta, M., & Chan, G. K.-L. (2020). Quantum Algorithms for Quantum Chemistry and Quantum Materials Science. *Chemical Reviews*, *120*(22), 12685–12717. DOI: 10.1021/acs.chemrev.9b00829

Bauri, S., Tripathi, S., Choudhury, A. M., Mandal, S. S., Raj, H., & Maiti, P. (2023). Nanomaterials as Theranostic Agents for Cancer Therapy. *ACS Applied Nano Materials*, *6*(23), 21462–21495. DOI: 10.1021/acsanm.3c04235

Bo, R., Xu, S., Yang, Y., & Zhang, Y. (2023). Mechanically-Guided 3D Assembly for Architected Flexible Electronics. *Chemical Reviews*, *123*(18), 11137–11189. DOI: 10.1021/acs.chemrev.3c00335

Bollini, P., Diwan, M., Gautam, P., Hartman, R. L., Hickman, D. A., Johnson, M., Kawase, M., Neurock, M., Patience, G. S., Stottlemyer, A., Vlachos, D. G., & Wilhite, B. (2023). Vision 2050: Reaction Engineering Roadmap. *ACS Engineering Au, 3*(6), 364–390. DOI: 10.1021/acsengineeringau.3c00023

Bouhadi, M., Javed, Q., Jakubus, M., Elkouali, M., Fougrach, H., Ansar, A., Ban, S. G., Ban, D., Heath, D., & Cerne, M. (2025). Nanoparticles for Sustainable Agriculture: Assessment of Benefits and Risks. *Agronomy (Basel), 15*(5), 1131. Advance online publication. DOI: 10.3390/agronomy15051131

Brachi, M., El Housseini, W., Beaver, K., Jadhav, R., Dantanarayana, A., Boucher, D. G., & Minteer, S. D. (2024). Advanced Electroanalysis for Electrosynthesis. *ACS Organic & Inorganic Au, 4*(2), 141–187. DOI: 10.1021/acsorginorgau.3c00051

Brown, B. P., Stein, R. A., Meiler, J., & McHaourab, H. S. (2024). Approximating Projections of Conformational Boltzmann Distributions with AlphaFold2 Predictions: Opportunities and Limitations. *Journal of Chemical Theory and Computation, 20*(3), 1434–1447. DOI: 10.1021/acs.jctc.3c01081

Cao, X., Wang, X., Chen, R., Chen, L., Liu, Y., & Wang, M. (2025). Improving Bacillus subtilis as Biological Chassis Performance by the CRISPR Genetic Toolkit. *ACS Synthetic Biology, 14*(3), 677–688. DOI: 10.1021/acssynbio.4c00844

Carnero Canales, C. S., Marquez Cazorla, J. I., Marquez Cazorla, R. M., Santos, A. M. d., Lobato Duarte, J., Oliveira Catarin Nunes, L., Reis, T. C., Cerazi Salvador, L., Santos-Filho, N. A., Sábio, R. M., Santos, H. A., & Pavan, F. R. Redefining Therapies for Drug-Resistant Tuberculosis: Synergistic Effects of Antimicrobial Peptides, Nanotechnology, and Computational Design. *Advanced Healthcare Materials, n/a*(n/a), e03964. DOI: 10.1002/adhm.202503964

Casteleijn, M. G., Abendroth, U., Zemella, A., Walter, R., Rashmi, R., Haag, R., & Kubick, S. (2025). Beyond In Vivo, Pharmaceutical Molecule Production in Cell-Free Systems and the Use of Noncanonical Amino Acids Therein. *Chemical Reviews, 125*(3), 1303–1331. DOI: 10.1021/acs.chemrev.4c00126

Champion, W. M., Khaliq, M., & Mihelcic, J. R. (2022). Advancing Knowledge to Reduce Lead Exposure of Children in Data-Poor Low- and Middle-Income Countries. *Environmental Science & Technology Letters, 9*(11), 879–888. DOI: 10.1021/acs.estlett.2c00656

Chen, Y., Xu, Y., Zhou, Y., Deng, Z., Feng, Y., Zhang, R., Yang, C., Li, H., Yu, H., & Zhang, G. (2025). Tailored magnetic ZIF-67-derived FeCo-LDH for PETase immobilization toward effective removal and biodegradation of high-crystallinity micro- and nano-PET. *Journal of Hazardous Materials*, *500*, 140419. DOI: 10.1016/j.jhazmat.2025.140419

Choudhary, S., Venkatrayalu, G., Ravi Kiran, B., & Poluri, K. M. (2025). Advances and Perspectives of Microalgal Biorefineries for Bioenergy. *Energy & Fuels*, *39*(26), 12367–12406. DOI: 10.1021/acs.energyfuels.5c01676

Chowdhury, S., Westenberg, R., Wennerholm, K., Cardiff, R. A. L., Beliaev, A. S., Noireaux, V., Carothers, J. M., & Peralta-Yahya, P. (2024). Carbon Negative Synthesis of Amino Acids Using a Cell-Free-Based Biocatalyst. *ACS Synthetic Biology*, *13*(12), 3961–3975. DOI: 10.1021/acssynbio.4c00359

Coca-Lopez, N., Alcolea-Rodriguez, V., Bañares, M. A., Brockhauser, S., Gorenflot, J., Henderson, A., Hildebrandt, R., Jeliazkova, N., Kochev, N., Lozano Diz, E., Pilat, Z., Polli, D., Strömert, P., Sturm, C., Vanna, R., & Portela, R. (2025). Artificial Intelligence-Powered Raman Spectroscopy through Open Science and FAIR Principles. *ACS Nano*, *19*(44), 38189–38218. DOI: 10.1021/acsnano.5c09165

Cumberland, D. M., Deckard, T. G., Kahle-Piasecki, L., Kerrick, S. A., & Ellinger, A. D. (2023). Making sense of the digital badging landscape in education and workplace settings: A scoping review of the empirical literature. *European Journal of Training and Development*, *48*(1-2), 253–275. DOI: 10.1108/EJTD-06-2022-0067

Curutchet, C., & Mennucci, B. (2017). Quantum Chemical Studies of Light Harvesting. *Chemical Reviews*, *117*(2), 294–343. DOI: 10.1021/acs.chemrev.5b00700

da Silva, A. P., Poquioma Hernández, H. V., Comelli, C. L., Guillén Portugal, M. A., Moreira Delavy, F., de Souza, T. L., de Oliveira, E. C., de Oliveira-Ribeiro, C. A., Silva de Assis, H. C., & de Castilhos Ghisi, N. (2024). Meta-analytical review of antioxidant mechanisms responses in animals exposed to herbicide 2,4-D herbicide. *The Science of the Total Environment*, *924*, 171680. DOI: 10.1016/j.scitotenv.2024.171680

Debnath, M., Sarkar, S., Debnath, S. K., Dkhar, D. S., Kumari, R., Vaskuri, G. S. S. J., Srivastava, A., Chandra, P., Prasad, R., & Srivastava, R. (2024). Photothermally Active Quantum Dots in Cancer Imaging and Therapeutics: Nanotheranostics Perspective. *ACS Applied Bio Materials*, *7*(12), 8126–8148. DOI: 10.1021/acsabm.4c01190

Du, H., Liu, K., Xu, T., Xu, C., Lin, M., Fang, Z., Kim, S.-W., Seo, J.-Y., Chen, J., Ma, H., Hsiao, B. S., DeVetter, L. W., Piao, Z., Si, C., Chen, C., Yang, Q., Lee, S.-Y., Yao, Y., & Pan, X. (2025). Lignocellulosic Films: Preparation, Properties, and Applications. *Chemical Reviews*, *125*(24), 11666–11814. DOI: 10.1021/acs.chemrev.5c00267

Enkavi, G., Javanainen, M., Kulig, W., Róg, T., & Vattulainen, I. (2019). Multiscale Simulations of Biological Membranes: The Challenge To Understand Biological Phenomena in a Living Substance. *Chemical Reviews*, *119*(9), 5607–5774. DOI: 10.1021/acs.chemrev.8b00538

Faheem, A., Qin, Y., Nan, W., & Hu, Y. (2021). Advances in the Immunoassays for Detection of Bacillus thuringiensis Crystalline Toxins. *Journal of Agricultural and Food Chemistry*, *69*(36), 10407–10418. DOI: 10.1021/acs.jafc.1c02195

Field, F. R.III, Wallington, T. J., Everson, M., & Kirchain, R. E. (2017). Strategic Materials in the Automobile: A Comprehensive Assessment of Strategic and Minor Metals Use in Passenger Cars and Light Trucks. *Environmental Science & Technology*, *51*(24), 14436–14444. DOI: 10.1021/acs.est.6b06063

Gao, J., Hou, B., Zhu, Q., Yang, L., Jiang, X., Zou, Z., Li, X., Xu, T., Zheng, M., Chen, Y.-H., Xu, Z., Xu, H., & Yu, H. (2022). Engineered bioorthogonal POLY-PROTAC nanoparticles for tumour-specific protein degradation and precise cancer therapy. *Nature Communications*, *13*(1), 4318. Advance online publication. DOI: 10.1038/s41467-022-32050-4

Ghosh, J., Repon, M. R., Rupanty, N. S., Asif, T. R., Tamjid, M. I., & Reukov, V. (2025). Chemical Valorization of Textile Waste: Advancing Sustainable Recycling for a Circular Economy. *ACS Omega*, *10*(12), 11697–11722. DOI: 10.1021/acsomega.4c10616

Giovanis, E., Akdede, S. H., & Ozdamar, O. (2021). Impact of the EU Blue Card programme on cultural participation and subjective well-being of migrants in Germany. *PLoS One*, *16*(7), e0253952. DOI: 10.1371/journal.pone.0253952

Guessous, G., Blake, L., Bui, A., Woo, Y., & Manzanarez, G. (2024). Disentangling the Web: An Interdisciplinary Review on the Potential and Feasibility of Spider Silk Bioproduction. *ACS Biomaterials Science & Engineering*, *10*(9), 5412–5438. DOI: 10.1021/acsbiomaterials.4c00145

Hadke, S., Kang, M.-A., Sangwan, V. K., & Hersam, M. C. (2025). Two-Dimensional Materials for Brain-Inspired Computing Hardware. *Chemical Reviews*, *125*(2), 835–932. DOI: 10.1021/acs.chemrev.4c00631

Haq, H., Välisuo, P., & Niemi, S. (2021). Modelling Sustainable Industrial Symbiosis. *Energies*, *14*(4), 1172. DOI: 10.3390/en14041172

Harris, A. R., & Davidson, C. I. (2005). The Role of Resuspended Soil in Lead Flows in the California South Coast Air Basin. *Environmental Science & Technology*, *39*(19), 7410–7415. DOI: 10.1021/es050642s

Hartley, S., Taitingfong, R., & Fidelman, P. (2022). The principles driving gene drives for conservation. *Environmental Science & Policy*, *135*, 36–45. DOI: 10.1016/j.envsci.2022.04.021

Hayler, J. D., Leahy, D. K., & Simmons, E. M. (2019). A Pharmaceutical Industry Perspective on Sustainable Metal Catalysis. *Organometallics*, *38*(1), 36–46. DOI: 10.1021/acs.organomet.8b00566

Huang, J., Xu, Y., Wang, Q., Wang, Q., Liang, X., Wang, F., Zhang, Z., Wei, W., Zhang, B., Huang, L., Chang, J., Ma, L., Ma, T., Liang, Y., Zhang, J., Guo, J., Jiang, X., Fan, X., An, Z., & Fei, A. (2025). Foundation models and intelligent decision-making: Progress, challenges, and perspectives. *Innovation (Cambridge (Mass.))*, *6*(6), 100948. Advance online publication. DOI: 10.1016/j.xinn.2025.100948

Ismail, A., Tekli, J., & Wex, B. (2025). Automating Deep Learning-Based Generation and Evaluation of De Novo Chemical Reaction with ChemRxnSAGE. *Journal of Chemical Information and Modeling*, *65*(20), 10945–10960. DOI: 10.1021/acs.jcim.5c01692

Jiang, J., Ye, B., Zeng, Z., Yang, X., Sun, Z., Shao, S., Feng, K., & Tan, X. (2024). Carbon Abatement and Leakage in China's Regional Carbon Emission Trading. *Environmental Science & Technology*, *58*(40), 17661–17673. DOI: 10.1021/acs.est.4c04738

Khan, F. R., Molden, N., & Booth, A. M. (2025). As the "Wheels Come Off", Including Tire Emissions in the Plastics Treaty Would Show Drive. *Environmental Science & Technology*, *59*(50), 26914–26916. DOI: 10.1021/acs.est.5c13479

Kim, E. J., Kim, J. H., Kim, M.-S., Jeong, S. H., & Choi, D. H. (2021). Process Analytical Technology Tools for Monitoring Pharmaceutical Unit Operations: A Control Strategy for Continuous Process Verification. *Pharmaceutics*, *13*(6), 919. DOI: 10.3390/pharmaceutics13060919

Kim, K., Shim, K., Wang, Y. W., & Yang, D. (2025). Synthetic Biology Strategies for the Production of Natural Colorants and Their Non-Natural Derivatives. *ACS Synthetic Biology*, *14*(3), 662–676. DOI: 10.1021/acssynbio.4c00799

Lange, J.-P. (2021). Managing Plastic Waste—Sorting, Recycling, Disposal, and Product Redesign. *ACS Sustainable Chemistry & Engineering*, *9*(47), 15722–15738. DOI: 10.1021/acssuschemeng.1c05013

Leyfman, Y., Pawar, S., Joshi, M., Kieft, A. V., Menon, G. P., Wilkerson, W. B., Balasubramanian, M., Jackewicz, S., Azeez, A., Chorya, H., Pai, P., Coloma, H. S., Wilson, S., & Park, C. H. (2023). Impact of a digital education platform to combat global inequity in oncology education amongst healthcare professionals in lower socioeconomic areas. *JCO Global Oncology*, *9*(Supplement_1), 140–140. DOI: 10.1200/GO.2023.9.Supplement_1.140

Li, W., Cui, L., Mai, J., Shi, T.-Q., Lin, L., Zhang, Z.-G., Ledesma-Amaro, R., Dong, W., & Ji, X.-J. (2022). Advances in Metabolic Engineering Paving the Way for the Efficient Biosynthesis of Terpenes in Yeasts. *Journal of Agricultural and Food Chemistry*, *70*(30), 9246–9261. DOI: 10.1021/acs.jafc.2c03917

Li, X., Mathur, A., Liu, A., & Liu, Y. (2023). Electrifying Carbon Capture by Developing Nanomaterials at the Interface of Molecular and Process Engineering. *Accounts of Chemical Research*, *56*(20), 2763–2775. DOI: 10.1021/acs.accounts.3c00321

Liu, X., Park, H., Ackermann, Y. S., Averous, L., Ballerstedt, H., Besenmatter, W., Blazquez, B., Bornscheuer, U. T., Branson, Y., Casey, W., de Lorenzo, V., Dong, W., Floehr, T., Godoy, M. S., Ji, Y., Jupke, A., Klankermayer, J., San Leon, D., Liu, L., & Chen, G.-Q. (2025). Exploring biotechnology for plastic recycling, degradation and upcycling for a sustainable future. *Biotechnology Advances*, *81*, 108544. Advance online publication. DOI: 10.1016/j.biotechadv.2025.108544

Lombardo, T., Duquesnoy, M., El-Bouysidy, H., Årén, F., Gallo-Bueno, A., Jørgensen, P. B., Bhowmik, A., Demortière, A., Ayerbe, E., Alcaide, F., Reynaud, M., Carrasco, J., Grimaud, A., Zhang, C., Vegge, T., Johansson, P., & Franco, A. A. (2022). Artificial Intelligence Applied to Battery Research: Hype or Reality? *Chemical Reviews*, *122*(12), 10899–10969. DOI: 10.1021/acs.chemrev.1c00108

Lu, C., & Weusthuis, R. A. (2025). Microbial production of medium-chain-length diols: Current stage and perspectives. *Bioresource Technology*, *435*, 132932. Advance online publication. DOI: 10.1016/j.biortech.2025.132932

Ma, Y., Tayefi, S. H., Mogharabi-Manzari, M., & Luo, X. (2025). Advances in immobilized enzyme systems for enhanced microplastic biodegradation: A review. *International Journal of Biological Macromolecules*, *328*, 147656. DOI: 10.1016/j.ijbiomac.2025.147656

Majidi, L., Hemmat, Z., Warburton, R. E., Kumar, K., Ahmadiparidari, A., Hong, L., Guo, J., Zapol, P., Klie, R. F., Cabana, J., Greeley, J., Curtiss, L. A., & Salehi-Khojin, A. (2020). Highly Active Rhenium-, Ruthenium-, and Iridium-Based Dichalcogenide Electrocatalysts for Oxygen Reduction and Oxygen Evolution Reactions in Aprotic Media. *Chemistry of Materials*, *32*(7), 2764–2773. DOI: 10.1021/acs.chemmater.9b04117

Martinengo, B., Diamanti, E., Uliassi, E., & Bolognesi, M. L. (2024). Harnessing the 12 Green Chemistry Principles for Sustainable Antiparasitic Drugs: Toward the One Health Approach. *ACS Infectious Diseases*, *10*(6), 1856–1870. DOI: 10.1021/acsinfecdis.4c00172

Matos-Pérez, C. R., White, J. D., & Wilker, J. J. (2012). Polymer Composition and Substrate Influences on the Adhesive Bonding of a Biomimetic, Cross-Linking Polymer. *Journal of the American Chemical Society*, *134*(22), 9498–9505. DOI: 10.1021/ja303369p

Muntean, E., Leba, M., & Ionica, A. C. (2025). AI-Driven Arm Movement Estimation for Sustainable Wearable Systems in Industry 4.0. *Sustainability (Basel)*, *17*(14), 6372. DOI: 10.3390/su17146372

Oluwole, O. A., Campos, E. V. R., de Oliveira, J. L., & Fraceto, L. F. (2025). Integrating the Nano–Phyto–Micro Triad for Climate-Resilient and Sustainable Agriculture. *ACS Omega*, *10*(45), 53702–53721. DOI: 10.1021/acsomega.5c07151

Otto, D. (2019). Adoption and Diffusion of Open Educational Resources (OER) in Education: A Meta-Analysis of 25 OER-Projects. *International Review of Research in Open and Distance Learning*, *20*(5), 122–140. DOI: 10.19173/irrodl.v20i5.4472

Ouyang, J., Zhang, Z., Li, J., & Wu, C. (2024). Integrating Enzymes with Supramolecular Polymers for Recyclable Photobiocatalytic Catalysis. *Angewandte Chemie International Edition*, *63*(16), e202400105. DOI: 10.1002/anie.202400105

Pathak, M., Roy, J., Patel, S., Some, S., Vyas, P., Das, N., & Shukla, P. (2021). Communicating climate change findings from IPCC reports: Insights from outreach events in India. *Climatic Change*, *168*(3), 23. DOI: 10.1007/s10584-021-03224-8

Paudel, H. P., Syamlal, M., Crawford, S. E., Lee, Y.-L., Shugayev, R. A., Lu, P., Ohodnicki, P. R., Mollot, D., & Duan, Y. (2022). Quantum Computing and Simulations for Energy Applications: Review and Perspective. *ACS Engineering Au*, *2*(3), 151–196. DOI: 10.1021/acsengineeringau.1c00033

Peng, L., Guo, Y., Liu, S., He, G., & Mauzerall, D. L. (2024). Subsidizing Grid-Based Electrolytic Hydrogen Will Increase Greenhouse Gas Emissions in Coal Dominated Power Systems. *Environmental Science & Technology*, *58*(12), 5187–5195. DOI: 10.1021/acs.est.3c03045

Phelps, D. W., Parkinson, L. V., Boucher, J. M., Muncke, J., & Geueke, B. (2024). Per- and Polyfluoroalkyl Substances in Food Packaging: Migration, Toxicity, and Management Strategies. *Environmental Science & Technology*, *58*(13), 5670–5684. DOI: 10.1021/acs.est.3c03702

Pley, C., van der Heijden, M. R., & Tulinius, C. (2020). A health professional's guide to the intersection of public health with intellectual property rights in trade and investment: The case of tobacco plain packaging. *Journal of Public Health Policy*, *41*(1), 52–62. DOI: 10.1057/s41271-019-00195-0

Prakash, M., Sakhavand, N., & Shahsavari, R. (2013). H_2, N_2, and CH_4 Gas Adsorption in Zeolitic Imidazolate Framework-95 and -100: Ab Initio Based Grand Canonical Monte Carlo Simulations. *The Journal of Physical Chemistry. C, Nanomaterials and Interfaces*, *117*(46), 24407–24416. DOI: 10.1021/jp408075y

Rabiee, N., & Rabiee, M. (2025). Engineered Metal–Organic Frameworks for Targeted CRISPR/Cas9 Gene Editing. *ACS Pharmacology & Translational Science*, *8*(4), 1028–1049. DOI: 10.1021/acsptsci.5c00047

Ruini, A., Bundgaard, A. M., Mosgaard, M. A., Løkke, S., & Pizzol, M. (2025). Normative and empirical solutions to the upcycling dilemma. *Sustainable Production and Consumption*, *54*, 466–475. DOI: 10.1016/j.spc.2025.01.022

Saleem, S. S., Ravichandran, S., Ravichandran, S., Chinnadurai, K. R., Saravanan, A., & Vickram, S. (2025). A Comprehensive Review Molecular Origami Classical DNA Folding Principle Empowering Modern Nanobiotechnology. *ACS Applied Bio Materials*, *8*(10), 8467–8512. DOI: 10.1021/acsabm.5c01094

Sanchez-Rexach, E., Johnston, T. G., Jehanno, C., Sardon, H., & Nelson, A. (2020). Sustainable Materials and Chemical Processes for Additive Manufacturing. *Chemistry of Materials*, *32*(17), 7105–7119. DOI: 10.1021/acs.chemmater.0c02008

Sbaiti, M., Abbara, A., Al-Haffar, M., Budhathoki, S., Wihba, K., Cardoso Pinto, A., & Streule, M. (2024). Curricular innovations to address inequalities in global health education. *European Journal of Public Health*, *34*(Supplement_3), ckae144.1231. Advance online publication. DOI: 10.1093/eurpub/ckae144.1231

Scala, P., Serio, B., & Giudice, V. (2026). 3D In Vitro Models of the Bone Marrow Niche. *ACS Biomaterials Science & Engineering*, *12*(1), 110–127. DOI: 10.1021/acsbiomaterials.5c01421

Seavey, K. C., Khare, N. P., Liu, Y. A., Williams, T. N., & Chen, C.-C. (2003). A New Phase-Equilibrium Model for Simulating Industrial Nylon-6 Production Trains. *Industrial & Engineering Chemistry Research*, *42*(17), 3900–3913. DOI: 10.1021/ie030112+

Shahi, S., Roghani-Mamaqani, H., Hoogenboom, R., Talebi, S., & Mardani, H. (2022). Stimuli-Responsive Covalent Adaptable Hydrogels Based on Homolytic Bond Dissociation and Chain Transfer Reactions. *Chemistry of Materials*, *34*(2), 468–498. DOI: 10.1021/acs.chemmater.1c03678

Sheldon, R. A. (2018). Metrics of Green Chemistry and Sustainability: Past, Present, and Future. *ACS Sustainable Chemistry & Engineering*, *6*(1), 32–48. DOI: 10.1021/acssuschemeng.7b03505

Shi, Y., Zhen, X., Zhang, Y., Li, Y., Koo, S., Saiding, Q., Kong, N., Liu, G., Chen, W., & Tao, W. (2024). Chemically Modified Platforms for Better RNA Therapeutics. *Chemical Reviews*, *124*(3), 929–1033. DOI: 10.1021/acs.chemrev.3c00611

Shirk, B. D., Heichel, D. L., Eccles, L. E., Rodgers, L. I., Lateef, A. H., Burke, K. A., & Stoppel, W. L. (2024). Modifying Naturally Occurring, Nonmammalian-Sourced Biopolymers for Biomedical Applications. *ACS Biomaterials Science & Engineering*, *10*(10), 5915–5938. DOI: 10.1021/acsbiomaterials.4c00689

Short, P. L. (2007). New Regulatory Era in Europe Begins: European Chemicals Agency opens does as companies prepare for new regulations. *Chemical and Engineering News*, *85*(33), 33–35. DOI: 10.1021/cen-v085n033.p033

Siegel, H. G., Nason, S. L., Warren, J. L., Prunas, O., Deziel, N. C., & Saiers, J. E. (2023). Investigation of Sources of Fluorinated Compounds in Private Water Supplies in an Oil and Gas-Producing Region of Northern West Virginia. *Environmental Science & Technology*, *57*(45), 17452–17464. DOI: 10.1021/acs.est.3c05192

Simić, S., Zukić, E., Schmermund, L., Faber, K., Winkler, C. K., & Kroutil, W. (2022). Shortening Synthetic Routes to Small Molecule Active Pharmaceutical Ingredients Employing Biocatalytic Methods. *Chemical Reviews*, *122*(1), 1052–1126. DOI: 10.1021/acs.chemrev.1c00574

Solomon, M. B., Rabha, S. S., Fimbres-Weihs, G., Goyal, H., Taghikhah, F. R., Varghese, J. J., Wenger, S. R., Liang, W., Kearns, E. R., Huang, J., Kaisare, N. S., & D'Alessandro, D. M. (2024). Decarbonization in Australia and India: Bilateral Opportunities and Challenges for the Net Zero Transformation. *ACS Engineering Au*, *4*(3), 295–311. DOI: 10.1021/acsengineeringau.3c00060

Subbiah, G., Jena, A., Kalidhas, A. M., Kele, V., Aadiwal, V., Awasthi, K., & K, K. P. (2025). Sustainable Food Production and Valorization of Food and Agricultural Waste: Recent Advances and Future Prospects. *ACS Food Science & Technology*, *5*(7), 2549–2562. DOI: 10.1021/acsfoodscitech.5c00403

Tabone, M. D., Cregg, J. J., Beckman, E. J., & Landis, A. E. (2010). Sustainability Metrics: Life Cycle Assessment and Green Design in Polymers. *Environmental Science & Technology*, *44*(21), 8264–8269. DOI: 10.1021/es101640n

Teel, T., Zoelle, A., Williams, J. M., Gavvalapalli, N., Woods, M., Hancu, D., Fout, T., & Homsy, S. (2026). Insights from Initial Engineering Designs of Point Source Capture at Industrial Facilities. *Industrial & Engineering Chemistry Research*, *65*(3), 1751–1770. DOI: 10.1021/acs.iecr.5c03762

Tom, G., Schmid, S. P., Baird, S. G., Cao, Y., Darvish, K., Hao, H., Lo, S., Pablo-García, S., Rajaonson, E. M., Skreta, M., Yoshikawa, N., Corapi, S., Akkoc, G. D., Strieth-Kalthoff, F., Seifrid, M., & Aspuru-Guzik, A. (2024). Self-Driving Laboratories for Chemistry and Materials Science. *Chemical Reviews*, *124*(16), 9633–9732. DOI: 10.1021/acs.chemrev.4c00055

Under pressure, Japan's chemical giants pivot. (2025). *C&EN Global Enterprise*, *103*(10), 29-31. DOI: 10.1021/cen-10310-feature3

Wang, H., Ben, J., Zhang, S., Wang, K., & Xie, P. (2025). Green Methanol from CO_2 Hydrogenation at Industrial Scale: Progress, Challenges, and Perspectives. *Chem & Bio Engineering*. Advance online publication. DOI: 10.1021/cbe.5c00113

Wang, S., Shen, T., Yang, C., Luo, G., & Wang, D. (2023). Engineering Iridium-Based Oxygen Evolution Reaction Electrocatalysts for Proton Exchange Membrane Water Electrolyzers. *ACS Catalysis*, *13*(13), 8670–8691. DOI: 10.1021/acscatal.3c01511

Wei, D., Wang, Z., Lin, H., Yin, X. P., & Wang, Y. (2026). Research Progress in Artificial Intelligence-Assisted Preparation of High-Quality Biomaterials. *ACS Omega*, acsomega.5c09392. Advance online publication. DOI: 10.1021/acsomega.5c09392

Wiesinger, H., Shalin, A., Huang, X., Siegrist, A., Plinke, N., Hellweg, S., & Wang, Z. (2024). LitChemPlast: An Open Database of Chemicals Measured in Plastics. *Environmental Science & Technology Letters*, *11*(11), 1147–1160. DOI: 10.1021/acs.estlett.4c00355

Wissinger, J. E., Visa, A., Saha, B. B., Matlin, S. A., Mahaffy, P. G., Kummerer, K., & Cornell, S. (2021). Integrating Sustainability into Learning in Chemistry. *Journal of Chemical Education*, *98*(4), 1061–1063. DOI: 10.1021/acs.jchemed.1c00284

Woeldgen, E., Teoh, R., Stettler, M. E. J., & Malina, R. (2025). Sustainable Aviation Fuel Deployment Strategies in Europe: Supply Chain Implications and Climate Benefits. *Environmental Science & Technology*, *59*(25), 12447–12457. DOI: 10.1021/acs.est.5c02364

Xie, Y., Ren, L., Zhu, X., Gou, X., & Chen, S. (2018). Physical and chemical treatments for removal of perchlorate from water-A review. *Process Safety and Environmental Protection*, *116*, 180–198. DOI: 10.1016/j.psep.2018.02.009

Xu, J., Gao, N.-y., Deng, Y., & Xia, S. (2013). Nanoscale iron hydroxide-doped granular activated carbon (Fe-GAC) as a sorbent for perchlorate in water. *Chemical Engineering Journal*, *222*, 520–526. DOI: 10.1016/j.cej.2012.07.141

Xu, J., Gao, N.-y., Zhao, D.-y., Liu, Z., Tang, M., Chen, Y., & Zhu, Y. (2016). Different iron salt impregnated granular activated carbon (Fe-GAC) for perchlorate removal: Characterization, performance and mechanism. *Colloids and Surfaces. A, Physicochemical and Engineering Aspects*, *509*, 99–108. DOI: 10.1016/j.colsurfa.2016.08.039

Yang, W., Liu, Y., Li, R., Lv, J., Zhang, Y., Ren, Y., Yuan, Z., & Zhang, Z. (2025). Toward sustainable and scalable synthesis of ibuprofen: Integrative insights into batch and continuous flow strategies. *AIChE Journal. American Institute of Chemical Engineers*. Advance online publication. DOI: 10.1002/aic.70198

Yang, Y., Liu, G., Zhang, Y., Feng, S., Xu, D., & Chen, Y. M. (2025). Advances in Gel-Based Solar Interfacial Evaporators: Advantages, Performance and Applications. *ACS Energy Letters*, *10*(12), 6123–6150. DOI: 10.1021/acsenergylett.5c02413

Zakzeski, J., Bruijnincx, P. C. A., Jongerius, A. L., & Weckhuysen, B. M. (2010). The Catalytic Valorization of Lignin for the Production of Renewable Chemicals. *Chemical Reviews*, *110*(6), 3552–3599. DOI: 10.1021/cr900354u

Zarrabian, M., Adhikary, L., Nita, M., Sriyanka, L., & Sherif, S. M. (2025). Toxicological and Functional Assessment of Minicell-Encapsulated dsRNA on Biocontrol Agents in Agriculture. *ACS Environmental Au*, *5*(4), 427–441. DOI: 10.1021/acsenvironau.5c00067

Zhang, F., Gallagher, K. S., Deng, M., Liu, H., Orvis, R., & Xuan, X. (2025). Assessing the Policy Gaps for Achieving China's Carbon Neutrality Target. *Environmental Science & Technology*, *59*(34), 18124–18133. DOI: 10.1021/acs.est.4c12478

Zhang, R., Motes, B., Tan, S., Lu, Y., Shih, M.-C., Hao, Y., Yang, K., Srinivasan, S., Bawendi, M. G., & Bulović, V. (2025). Machine Learning Prediction of Organic–Inorganic Halide Perovskite Solar Cell Performance from Optical Properties. *ACS Energy Letters*, *10*(4), 1714–1724. DOI: 10.1021/acsenergylett.4c03592

Zhao, Z., Zuo, H., Fang, S., Duan, M., Zhao, D., & Chen, T. (2025). Hydrophobic metal–organic frameworks and membranes for enhanced carbon capture. *Chemical Engineering Journal*, *526*, 170657. DOI: 10.1016/j.cej.2025.170657

Zhou, T., Yuk, H., Hu, F., Wu, J., Tian, F., Roh, H., Shen, Z., Gu, G., Xu, J., Lu, B., & Zhao, X. (2023). 3D printable high-performance conducting polymer hydrogel for all-hydrogel bioelectronic interfaces. *Nature Materials, 22*(7), 895-+. DOI: 10.1038/s41563-023-01569-2

Chapter 12
Synthesis and Purification Technologies of Pharmaceutical Intermediates:
From Process Intensification to Quality Control

Kai Wang
http://orcid.org/0000-0001-7221-4575
Tongren University, China

Sujitraj Sheth
http://orcid.org/0000-0003-4082-6424
Guizhou University, China

ABSTRACT

This chapter analyzes pharmaceutical intermediates as the strategic link between basic feedstocks and APIs. It examines the evolution of synthetic methodologies, highlighting green synthesis, continuous flow chemistry, and biocatalysis as key approaches for process intensification. The text details purification strategies, contrasting traditional crystallization with advanced methods such as preparative chromatography and membrane separation. Furthermore, it emphasizes the "Quality by Design" (QbD) paradigm, integrating impurity profiling to ensure regulatory compliance. Exploring artificial intelligence-driven synthetic route planning and process automation, this work presents synthesis, purification, and quality control as an interdependent system essential for modern pharmaceutical engineering.

DOI: 10.4018/979-8-3373-4402-7.ch012

INTRODUCTION

This chapter presents a comprehensive analysis of pharmaceutical intermediates, elucidating their critical role within the modern pharmaceutical value chain. Far from being mere transient chemical entities, pharmaceutical intermediates serve as the strategic bridge connecting basic petrochemical feedstocks to complex active pharmaceutical ingredients (APIs). As a cornerstone of drug development, efficient intermediate management is instrumental in simplifying synthetic routes, reducing research and development (R&D) expenditures, and accelerating time-to-market for novel therapeutics. By systematically examining intermediates through the lenses of structural classification and therapeutic application, this analysis establishes the fundamental structure–property relationships that guide rational synthetic pathway design and align technical development with evolving market demands.

The evolution of synthetic technologies serves as the engine of molecular construction. While foundational organic reactions provide the essential methodology for carbon skeleton assembly, the focus of this chapter extends to the optimization of reaction kinetics and thermodynamics to maximize yield and cost-efficiency. Furthermore, the discussion pivots to modern process intensification as a key driver of industrial innovation. We explore how green synthesis, quantified by metrics such as Process Mass Intensity (PMI), mitigates the environmental footprint of traditional manufacturing; how continuous flow chemistry revolutionizes the safety profile of hazardous high-energy reactions; and how biocatalysis overcomes the stereoselectivity bottlenecks inherent in chiral synthesis.

Parallel to synthesis, Purification Technology is examined as the critical determinant of product safety and efficacy. The chapter delineates a tiered approach to separation: conventional methods (e.g., crystallization) serve as the primary stage for bulk impurity removal and cost control, while advanced technologies provide high-resolution solutions for stringent quality requirements. Specifically, the application of preparative chromatography for trace impurity control, membrane separation for continuous processing, and supercritical fluid extraction for thermosensitive compounds highlights the industry's shift toward precision purification.

Underpinning these technical operations is the quality control (QC) system. Moving beyond the traditional "testing into quality" approach, this chapter emphasizes the quality by design (QbD) paradigm, a cornerstone of modern regulatory frameworks such as ICH Q8 and ICH Q11. Analytical methodologies (e.g., HPLC and GC-MS) provide the empirical data necessary for rigorous impurity profiling, while QbD shifts the focus from post-production inspection to pre-emptive process design. This "prevention-over-detection" strategy serves as a robust risk management framework, coordinating synthesis and purification parameters to ensure process capability and regulatory compliance.

This chapter integrates these domains into a holistic technical framework. It posits that synthesis, purification, and quality control are not isolated silos but interdependent components of a closed-loop manufacturing system. These components are explicitly linked to the industrial imperatives of supply chain resilience, digitalization (e.g., AI-driven synthesis and process analytical technology), and quantified sustainability. This chapter contributes to the existing literature not only by reviewing technologies but also by creating an integrated framework that connects technical solutions to strategic business and regulatory drivers—a perspective that is often fragmented in specialized reviews.

1. INTRODUCTION TO DRUG INTERMEDIATES

1.1 Definition, Scope, and Classification of Pharmaceutical Intermediates

1.1.1 Definition of Drug Intermediates

Definition and Scope

A pharmaceutical intermediate is a defined chemical entity produced during the synthesis of an active pharmaceutical ingredient. Unlike raw materials, intermediates usually possess a significant portion of the final molecular skeleton; unlike the API, they are not intended for direct therapeutic use and require further chemical transformation. They act as molecular "building blocks," converting basic feedstocks—such as petrochemicals or fermentation products—into complex structures through a controlled sequence of reactions (e.g., halogenation, nitration, reduction). Each intermediate represents a stable, characterizable stage in the synthetic pathway, essential for modular synthesis strategies.

Industrial Case Study: 6-APA

The compound 6-aminopenicillanic acid (6-APA) serves as a classic platform intermediate. Its core β-lactam ring is the essential structural feature of penicillin-class antibiotics. By utilizing 6-APA as a versatile hub, chemists can efficiently introduce various side chains to create a portfolio of distinct APIs, such as penicillin G or ampicillin. This modular approach enables the targeted modification of antibacterial spectra and pharmacokinetics while streamlining industrial manufacturing (Liu et al., 2022). This aligns with the "divergent synthesis strategy" widely employed in natural product synthesis to access structural diversity from common intermediates (Li et al., 2018).

Strategic Value

The strategic importance of intermediates is multifaceted:

- **Process Efficiency:** They determine the robustness and feasibility of an API's synthetic route.
- **Economic Impact:** The cost of intermediate synthesis often accounts for over 60% of the total API manufacturing cost, making process optimization a key driver of affordability.
- **Quality Control:** The purity of an intermediate directly dictates the impurity profile of the final API. Controlling impurities at the intermediate stage—preventing their propagation downstream—is a fundamental tenet of modern process development (Farrant, 2020; Fathi et al., 2025; Weeranoppanant & Adamo, 2019).

1.1.2 Classification of Drug Intermediates

Drug intermediates are generally classified based on two primary dimensions: their molecular structure and their therapeutic application. This dual classification system facilitates strategic assessment regarding synthetic complexity and market demand drivers.

1. Classification by Molecular Structure

Based on the chemical skeleton, intermediates are systematically categorized into heterocyclic, aromatic, and aliphatic classes.

- **Heterocyclic Intermediates:** Containing heteroatoms (N, O, S) within the ring structure, these are the most prevalent class in modern medicinal chemistry, serving as the core pharmacophores.
    - o *Examples:* Nitrogen-containing heterocycles (pyridine, pyrimidine) are crucial precursors for antimetabolite antitumor drugs and kinase inhibitors. Emerging studies also demonstrate the utility of complex hybrid heterocycles, incorporating thiazole and phenol moieties, in the development of metal-complex based therapeutics (Alharbi et al., 2025).
- **Aromatic Intermediates:** Built upon a benzene ring skeleton, these intermediates provide structural rigidity and stability.
    - o *Examples:* *p*-Nitrophenol is a classic aromatic intermediate used in the industrial synthesis of acetaminophen (paracetamol).
- **Aliphatic Intermediates:** Consisting of open-chain or non-aromatic cyclic structures, often used as linkers or side chains to adjust lipophilicity.
    - o *Examples:* Key aliphatic chains are indispensable in the synthesis of steroid hormones and macrolide antibiotics.

2. CLASSIFICATION BY APPLICATION (THERAPEUTIC AREA)

From an industrial perspective, categorizing intermediates by therapeutic application aligns upstream chemical manufacturing with downstream market dynamics. This classification allows for the assessment of volume requirements, regulatory stringency regarding impurities, and specific synthetic challenges inherent to different pharmacophore classes.

- **Anti-infective Intermediates:**
  - *Market Characteristic:* Characterized by high-volume, cost-sensitive manufacturing. The synthesis often integrates fermentation-derived precursors with chemical semi-synthesis. Supply chain stability is critical due to the global magnitude of demand.
  - *Key Intermediates:* The industry relies heavily on β-lactam cores. For instance, 7-ACA (7-aminocephalosporanic acid) acts as the pivotal platform for cephalosporin production. Modern industrial routes have shifted from chemical deacylation to enzymatic processes to reduce waste and improve yield.
- **Oncology (Anti-tumor) Intermediates:**
  - *Market Characteristic:* Defined by high-potency active pharmaceutical ingredient (HPAPI) protocols. The synthesis demands strict containment measures and focuses on constructing complex heterocyclic scaffolds or chiral centers typically found in targeted therapies.
  - *Key Intermediates:*
    - *Natural Product Derivatives:* 10-DAB (10-deacetylbaccatin II) serves as the strategic starting material for the semi-synthesis of taxanes (paclitaxel, docetaxel).
    - *Kinase Inhibitor Scaffolds:* The synthesis of tyrosine kinase inhibitors necessitates highly functionalized pyrimidine, quinoline, or indazole intermediates, where controlling regio-isomerism is a primary synthetic hurdle.
- **Cardiovascular Intermediates:**
  - *Market Characteristic:* Driven by the manufacturing of anti-hypertensives (Sartans) and lipid-lowering agents (Statins). Given the generic nature of this market, process economics and enantiomeric purity are the dominant drivers.
  - *Key Intermediates:* The synthesis of Atorvastatin requires a specific chiral side chain (e.g., ATS-9). The industrial focus is on achieving >99.9% enantiomeric excess (ee) through asymmetric hydrogenation or biocatalysis to meet stringent API specifications (Campos et al., 2019).

- **Central Nervous System (CNS) Intermediates:**
    - o *Market Characteristic:* Intermediates in this class are designed to construct molecules with specific lipophilicity profiles to facilitate Blood-Brain Barrier (BBB) permeability. The chemistry often involves rigid bicyclic or tricyclic frameworks.
    - o *Key Intermediates:* The production of SSRIs (e.g., Sertraline) relies on tetralone derivatives. The critical synthetic step often involves asymmetric reductive amination or resolution processes to establish the cis-configuration essential for receptor binding affinity.
- **Metabolic and Endocrine Intermediates:**
    - o *Market Characteristic:* A rapidly expanding sector driven by chronic management therapies (e.g., GLP-1 agonists, DPP-4 inhibitors). This field is currently at the forefront of adopting Green Chemistry and Biocatalysis to handle complex chirality on a metric-ton scale.
    - o *Key Intermediates:* The synthesis of Gliptins (e.g., Sitagliptin) utilizes β-amino acid derivatives. The industrial synthesis of the sitagliptin intermediate represents a paradigm shift, utilizing engineered transaminases to install the chiral amine directly, replacing high-pressure hydrogenation steps.

2.2 The Strategic Significance of Synthesis and Purification Technologies

The advancement of synthesis and purification technologies for drug intermediates is not merely a technical exercise; it serves as the technological backbone of the pharmaceutical value chain. Its significance extends beyond the laboratory, acting as a decisive factor in industrial efficiency, regulatory compliance, supply chain resilience, and patient accessibility (Campos et al., 2019).

2.2.1 Driving Process Intensification and Sustainability

Research and development (R&D) in this field is the primary engine for Process Intensification—the drive to create manufacturing processes that are substantially smaller, cleaner, and more energy-efficient.

- **Synthesis-Purification Synergy:** The traditional boundary between synthesis and purification is blurring. Strategic innovation focuses on the synergy between these two stages. For instance, designing a highly stereoselective synthetic route can eliminate the need for costly downstream chiral resolution, thereby drastically reducing the Process Mass Intensity (PMI)—the total mass of materials used per kilogram of product—and the environmental footprint (E-factor).

- **Green Transformation:** Shifting from stoichiometric reagents to catalytic systems and adopting enabling technologies (e.g., continuous flow chemistry) not only enhances safety for hazardous reactions but also ensures long-term sustainability, addressing the industry's growing pressure to meet Environmental, Social, and Governance (ESG) targets.

2.2.2 Enabling Supply Chain Resilience and Economic Viability

Intermediates act as the critical "decoupling point" in the complex pharmaceutical supply chain, balancing cost, speed, and risk.

- **Convergent Synthesis & Risk Mitigation:** By utilizing advanced intermediates, complex API synthesis can be broken down into modular stages. This allows for **convergent synthesis strategies**, where molecular fragments are synthesized in parallel and assembled. This approach significantly reduces lead times compared to linear synthesis and mitigates the risk of batch failures in late-stage production, thereby enhancing supply chain robustness against disruptions.
- **Strategic Outsourcing (CDMOs):** The standardization of high-quality intermediates facilitates a robust global supply ecosystem. Pharmaceutical companies leverage Contract Development and Manufacturing Organizations (CDMOs) for "advanced intermediates." This strategy reduces internal Capital Expenditure (CAPEX) and allows innovators to focus resources on the final, high-value API conversion and Regulatory Starting Material (RSM) definition strategies.

2.2.3 Accelerating Drug Development and Market Access

The robustness of the intermediate supply chain is directly correlated with the speed of clinical development and commercial launch.

- **Addressing the "Chemistry Bottleneck":** Speed is paramount, especially in early-phase clinical trials (Phase I/II). A scalable intermediate synthesis pathway allows for the rapid delivery of kilogram-scale batches required for toxicology and efficacy studies, preventing chemistry issues from delaying promising therapies. Innovative approaches, such as remote-controlled adaptive laboratories, have demonstrated the potential to significantly accelerate this cycle by automating routine synthetic tasks (Godfrey, Masquelin, & Hemmerle, 2013).

- **Seamless Technology Transfer:** A well-characterized intermediate process—underpinned by Quality by Design (QbD) principles as outlined in ICH Q11—provides the foundation for seamless technology transfer from pilot plants to commercial facilities. This accelerates the timeline from regulatory approval to market launch.

3. SYNTHESIS PRINCIPLES AND MODERN TECHNOLOGIES

The synthesis of pharmaceutical intermediates is the core engine of API manufacturing. This section transitions from foundational synthetic strategies to advanced enabling technologies, highlighting the paradigm shift from simple "molecule making" to "precision engineering" characterized by sustainability, safety, and efficiency.

3.1 Industrial Synthetic Strategies: From Skeleton to Function

While the fundamental organic toolkit remains valid, industrial synthesis prioritizes selectivity, cost-efficiency, and scalability over mere conversion. Furthermore, the control of Critical Process Parameters (CPPs)—such as temperature, pressure, and solvent choice—is essential. Utilizing Design of Experiments (DoE) helps optimize these multivariate conditions to balance reaction kinetics against impurity formation (Krska et al., 2017; Schneider, 2017).

3.1.1 Skeleton Construction Strategies

- **Nucleophilic Substitution & Alkylation:** These reactions are pivotal for building the carbon-heteroatom framework.
    - *Industrial Application:* In the synthesis of the local anesthetic intermediate *p-nitroanisole*, chemoselective alkylation is key. *p*-Nitrophenol is treated with dimethyl sulfate under controlled alkaline conditions to selectively methylate the phenolic hydroxyl group. Modern improvements replace toxic methylating agents with greener alternatives like dimethyl carbonate.
    - *Advanced Application:* Recent developments in photo-induced copper catalysis have expanded the scope of alkylation, enabling the direct functionalization of C-H bonds under mild conditions.
- **Condensation & Cyclization:** Constructing heterocyclic rings—the pharmacophores of most drugs—relies heavily on condensation.

- *Industrial Application:* The Ugi multi-component reaction (MCR) exemplifies efficiency, allowing the one-pot assembly of complex peptide-like structures from simple starting materials.
- *Strategic Reversibility:* Conversely, hydrolysis is equally critical. For example, while amide bond formation constructs the penicillin backbone, the industrial production of the key intermediate 6-APA relies on the enzymatic hydrolysis of Penicillin G. This highlights the reversible interplay between condensation and hydrolysis in strategic process design.

3.1.2 Functional Group Transformations

- **Redox Manipulations:** Adjusting the oxidation state is essential for tuning reactivity.
  - *Industrial Application:* The synthesis of *p*-nitrobenzaldehyde involves precise oxidation, whereas reducing nitro groups to anilines is a standard step for sulfonamide intermediates.
  - *Innovation:* Novel photocatalytic systems now allow for "redox-neutral" transformations, avoiding harsh stoichiometric oxidants (like $KMnO_4$) or reductants (like Fe/HCl), thereby significantly reducing metal waste.

3.2 Advanced Enabling Technologies

To address the limitations of traditional batch chemistry—such as safety risks and high environmental footprints—the industry is adopting advanced manufacturing technologies.

3.2.1 Green Chemistry & Alternative Solvents

Green synthesis aims to minimize the E-factor (waste per kg product) by systematically adopting sustainable methodologies throughout the drug development lifecycle (Stefanache et al., 2025).

- **Ionic Liquids (ILs):** As non-volatile, recyclable solvents, ILs offer a solution to volatile organic compound (VOC) emissions. Studies have shown that ILs can enhance catalyst stability and stereoselectivity. For example, recent work utilizing water-compatible ionic liquid systems has enabled the one-pot asymmetric synthesis of chiral 1,3-diarylpropan-1-ols without external metal catalysts, adhering to atomic economy principles (Hoffmann, 2012; Yu et al., 2024).

3.2.2 Continuous Flow Chemistry

Flow chemistry represents a paradigm shift from "batch" to "continuous" manufacturing. This trend is now formally supported by regulatory frameworks like ICH Q13 (Continuous Manufacturing of Drug Substances and Drug Products), which provides guidance for its implementation. To frame its strategic advantages, a direct comparison with traditional batch synthesis is essential (Table 1).

Table 1. Comparative Analysis of Batch vs. Continuous Flow Synthesis

Feature	Batch Synthesis	Continuous Flow Synthesis
Safety	High risk for hazardous reactions (accumulation)	Inherently safer (small volumes, superior heat transfer)
Process Control	Difficult to control; temperature/ concentration gradients	Precise control over parameters (temperature, time, mixing)
Scalability	Challenging; requires re-engineering of equipment	Simple scale-up by running for longer ("scale-out")
Product Quality	Higher batch-to-batch variability	Consistent quality, tighter impurity profile
Capital Expenditure	High for large, specialized vessels	Lower for modular, smaller-footprint equipment
Adoption Barriers	Established infrastructure and know-how	Higher initial R&D investment, need for new skill sets

- **Safety & Efficiency:** It is particularly superior for hazardous reactions like diazotization. In traditional batch reactors, the accumulation of unstable diazonium salts poses an explosion risk. Flow reactors, with their minimal internal volume and excellent heat transfer, allow these intermediates to be generated and consumed *in situ* instantly. Furthermore, specialized enabling tools have been developed to enhance mass transfer in biphasic liquid-liquid systems, overcoming the mixing limitations often encountered in traditional multiphase synthesis (Weeranoppanant, 2019).
- **Case Study:** In the production of sulfamethoxazole intermediates, switching to a continuous flow process increased the yield by 10–15% while significantly improving the product purity profile compared to the batch process (Aroh & Jensen, 2018; Lei, Ang, & Wu, 2023). Similarly, the continuous synthesis of Celecoxib has demonstrated how flow reactors can efficiently manage fluorinated precursors to streamline the production of complex heterocyclic APIs (Ivanova et al., 2021). Furthermore, the synthesis of Diazepam has been successfully adapted to continuous flow, achieving high purity and efficiency, which underscores the versatility of this technology for CNS drug intermedi-

ates (Nicholas et al., 2022). Furthermore, the power of multistep telescoping was demonstrated in the seven-step continuous synthesis of Linezolid, which successfully eliminated the need for any intermediate purification, exemplifying the peak of process intensification (Russell & Jamison, 2019). This integrated approach, which combines multistep synthesis with in-line downstream processing, is best conceptualized as a holistic system with distinct operational advantages and technical challenges (**Fig. 1**).

3.2.3 Biocatalysis and Enzymatic Resolution

Enzymes offer unparalleled chemo- and stereoselectivity, solving the "chiral challenge" in intermediate synthesis (Sun et al., 2018). Beyond single-step replacements, the concept of biocatalytic retrosynthesis is reshaping how entire synthetic routes are designed for high-value intermediates (Green & Turner, 2016).

- **Chiral Resolution:** A classic application is the kinetic resolution of Ibuprofen intermediates. Lipases can selectively hydrolyze the (S)-ester while leaving the (R)-isomer untouched, yielding high-purity (S)-Ibuprofen under ambient conditions (neutral pH, room temperature). This bio-route eliminates the need for toxic chiral auxiliaries and heavy metal catalysts used in traditional asymmetric synthesis (Calvó-Tusell et al., 2023; Mathebula et al., 2022).

3.2.4 Microwave-Assisted Synthesis

Microwave irradiation offers rapid volumetric heating, often reducing reaction times from hours to minutes.

- **Application:** While primarily utilized in early-phase development and Lead Optimization for library synthesis, it is effective for synthesizing challenging heterocyclic intermediates, such as benzimidazoles. Recent studies have combined microwave technology with green solvents (e.g., ethanol) to achieve efficient tandem alkylation reactions, though scalability remains a consideration for large-scale manufacturing (Clarke et al., 2018).

Figure 1. Conceptual framework of Multistep Continuous Flow Synthesis (MCFS) and integrated downstream processing.

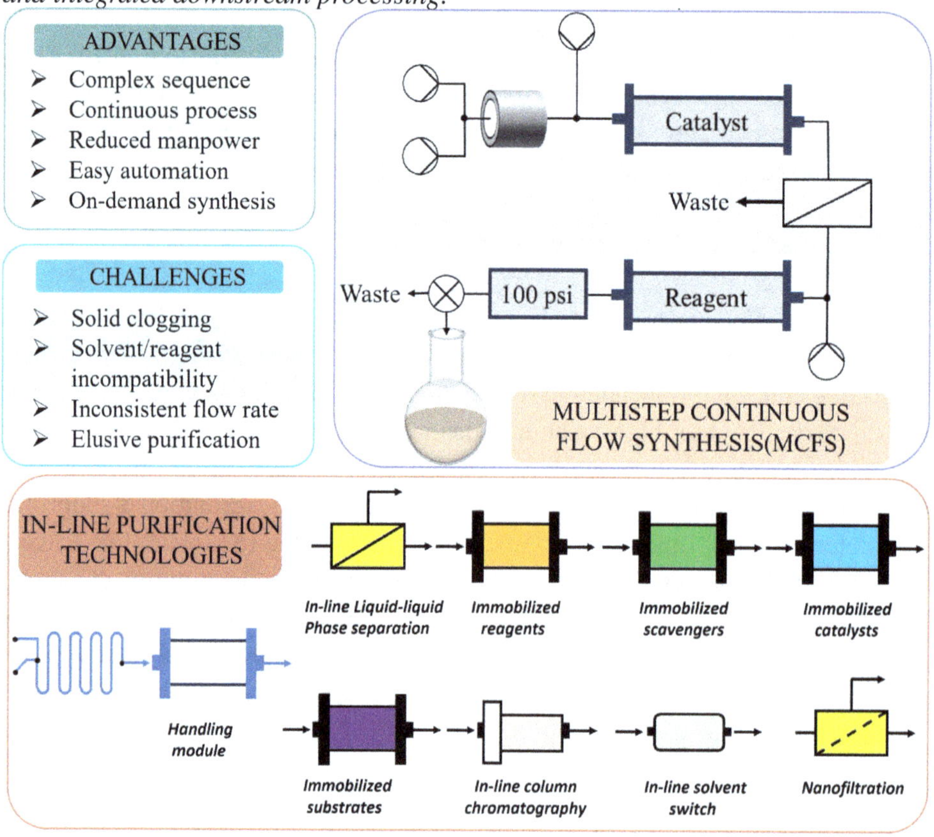

The diagram contrasts the operational advantages (e.g., automation, on-demand synthesis) with technical challenges (e.g., clogging, solvent incompatibility), while providing a taxonomy of enabling in-line purification technologies—ranging from liquid-liquid phase separation to nanofiltration—essential for closing the continuous manufacturing loop.

4. PURIFICATION TECHNOLOGY AND PROCESS DESIGN OF DRUG INTERMEDIATES

4.1 Conventional Purification Methodologies: Engineering and Optimization

Conventional purification techniques constitute the foundational unit operations in pharmaceutical intermediate manufacturing. Governed by fundamental physicochemical principles—specifically thermodynamic solubility differentials and vapor-liquid equilibrium (VLE)—methods such as recrystallization and distillation offer a balance between economic viability and operational robustness for bulk separation.

4.1.1 Recrystallization: Thermodynamics and Kinetics

Recrystallization is the predominant method for solid intermediates requiring purities in the range of 80%–99%. The efficacy of this process relies on the precise manipulation of temperature-dependent solubility profiles. Successful process design must address three critical engineering dimensions:

1. **Thermodynamic Compatibility:** Taking sulfonamide intermediates (e.g., sulfamethoxazole) as a case study, solvent selection is predicated on the "like-dissolves-like" principle modulated by temperature. An ethanol-water cosolvent system is preferred over pure polar or non-polar solvents because it offers a steep solubility gradient—sulfamethoxazole exhibits a >5-fold solubility increase from 20°C to 60°C, whereas impurities (e.g., p-aminobenzenesulfonamide) show only a 2-fold change. This differential is the driving force for selective precipitation.
2. **Chemical Inertness:** The solvent must be chemically orthogonal to the solute. For instance, acetic acid is contraindicated for sulfonamides to prevent acylation of the amine moiety.
3. **Kinetic Control of Nucleation:** The cooling profile dictates crystal habit and purity. Shock cooling (e.g., ice bath) induces rapid primary nucleation, leading to solvent occlusion and purities <95%. Conversely, controlled gradient cooling (e.g., 5°C/h) maintains metastable supersaturation, promoting the growth of uniform, high-purity crystals (>98%) with superior filtration characteristics. For thermolabile compounds, such as penicillin side chains, vacuum evaporative crystallization is employed to reduce thermal stress. Furthermore, recent advances have expanded these kinetic principles to chiral crystallization, providing a scalable strategy for enhancing enantiomeric purity without the need for chiral chromatography (Putman & Armstrong, 2022).

4.1.2 Distillation Strategies

Distillation separates liquid intermediates based on relative volatility (α).

- **Atmospheric Distillation:** Applicable to thermally stable, low-boiling compounds (e.g., ethyl acetate derivatives). Process efficiency is controlled by the reflux ratio and column hydrodynamics. Unlike laboratory settings where "drop counting" is used, industrial control relies on maintaining optimal vapor velocities to prevent entrainment, which can degrade purity from >99.5% to <98%.
- **Vacuum and Azeotropic Distillation:** For high-boiling (>200°C) or thermosensitive species like isophytol (a Vitamin E intermediate), vacuum distillation is imperative. Reducing system pressure to 1 kPa lowers the boiling point from 300°Cto 150°C, mitigating oxidative pyrolysis (Yield >90%).
- **Azeotropic Breaking:** In systems forming azeotropes (e.g., amine intermediates in ethanol-water), entrainers such as benzene modify the VLE. By forming a ternary low-boiling azeotrope (69°C), water is selectively removed, shifting the equilibrium to yield anhydrous amine purities of 99%, significantly surpassing the 92% limit of simple distillation.

4.2 Advanced Purification Technologies: High-Resolution Solutions

When conventional methods fail to meet stringent purity specifications (≥99.9%) or accommodate complex molecular architectures (e.g., chiral, peptidic, or viscous species), advanced separation technologies provide the necessary selectivity.

4.2.1 Preparative High-Performance Liquid Chromatography (Prep-HPLC)

Prep-HPLC represents the gold standard for isolating high-value species. Unlike analytical chromatography, the objective function in Prep-HPLC is mass throughput and recovery (Pichon et al., 2014).

- **Stationary Phase Engineering:** C18 reversed-phase silica (10 nm pore size) is standard for hydrophobic discrimination of peptides. For chiral ibuprofen precursors, polysaccharide-based chiral stationary phases (CSPs) are utilized to exploit steric hindrance for enantiomeric resolution (R_s>2.5).

- **Gradient Elution:** Linear gradients (e.g., 10%–50% acetonitrile/water) sharpen peak shapes and prevent tailing, improving purification efficiency by approximately 40% compared to isocratic methods.
- **Loading Capacity:** Process economics dictate operating near the saturation limit. A typical loading factor of 5% column capacity represents the trade-off point between throughput and the strict 99.9% purity threshold; overloading beyond this point causes isotherm non-linearity and peak overlap.

4.2.2 Centrifugal Partition Chromatography (CPC)

CPC utilizes a liquid-liquid partition mechanism, eliminating the solid stationary phase. This feature addresses the critical issue of irreversible adsorption found in traditional HPLC (Sánchez-Trasviña et al., 2021).

- **Mechanism:** Separation is driven by the partition coefficient (K) in a Biphasic Liquid System (BLS) held under centrifugal force. An ideal K range of 0.5–1.5 ensures optimal resolution.
- **Industrial Advantages:** CPC eliminates the cost of silica and allows for total sample recovery. Its ability to handle high solid loads makes it compatible with continuous flow reactors, facilitating a seamless transition from synthesis to purification (Örkényi et al., 2017) (**Fig. 2**). Recent innovations even integrate membrane-based phase separators to achieve continuous liquid-liquid partition without centrifugal forces, further broadening the applicability of this technique (Viano et al., 2023).

Figure 2. Schematic representation of the operational modes in Centrifugal Partition Chromatography (CPC).

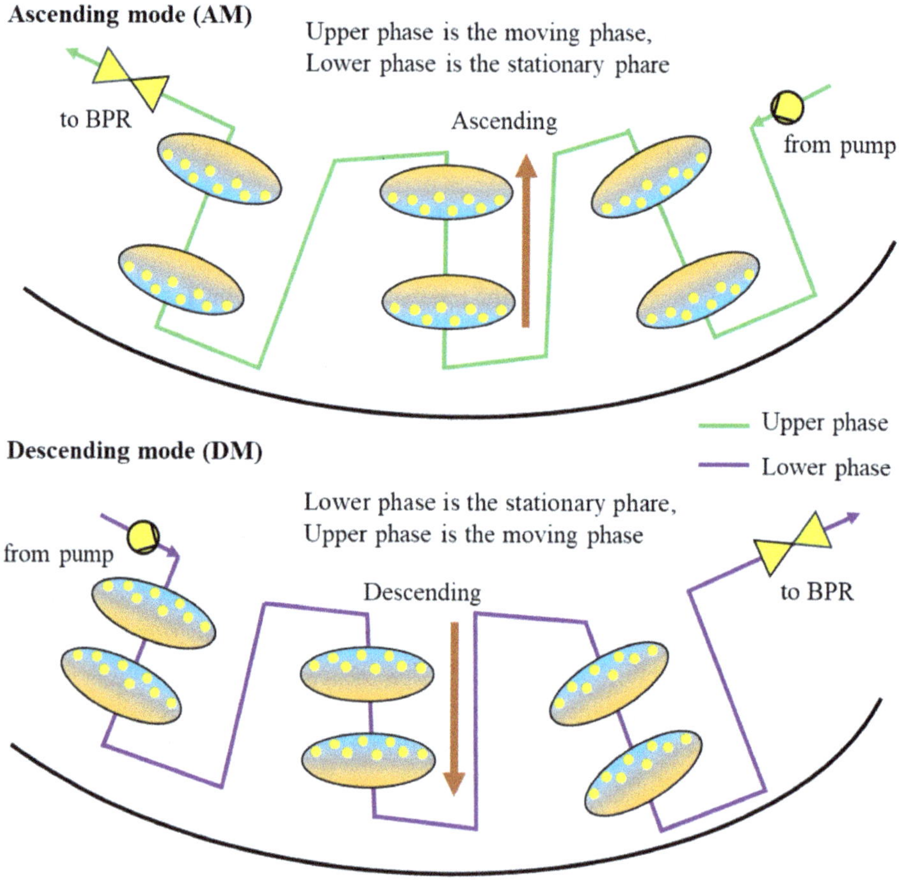

The diagram illustrates the hydrodynamic behavior within the rotor chambers: (Top) Ascending Mode (AM), utilized when the lighter (upper) phase acts as the mobile phase while the denser (lower) phase is stationary; (Bottom) Descending Mode (DM), utilized when the denser (lower) phase acts as the mobile phase while the lighter (upper) phase is stationary.

Abbreviation: BPR, Back Pressure Regulator.

4.2.3 Membrane Separation Technology

Membrane processes offer a non-thermal, phase-change-free separation alternative, crucial for energy efficiency and product stability (Bi et al., 2024; Ma et al., 2024).

- **Case Study: β-Lactam Antibiotics (6-APA):** A dual-stage cascade is employed:

  1. **Ultrafiltration (UF):** Membranes with a Molecular Weight Cut-Off (MWCO) of 3000 Da retain macromolecular impurities (e.g., penicillin acylase) via size exclusion, while the smaller 6-APA (339 Da) permeates, achieving >99% deproteination.
  2. **Nanofiltration (NF):** The permeate is subsequently concentrated using 200 Da NF membranes. This step concentrates the solution from 1% to 10% while removing inorganic salts. Compared to thermal evaporation, this process reduces energy consumption by 60% (Fig. 3). Beyond concentration, recent studies have also established rigorous kinetic models for the removal and recovery of sulfonamide intermediates (e.g., sulfadiazine) using polyamide nanofiltration membranes, further clarifying the separation mechanisms (Zhu, Hu, & Yang, 2021).

- **Maintenance:** Regular Clean-in-Place (CIP) protocols (e.g., 0.1 mol/L NaOH) are essential to mitigate fouling and restore flux.

Figure 3. Schematic representation and operational principles of membrane separation technology.

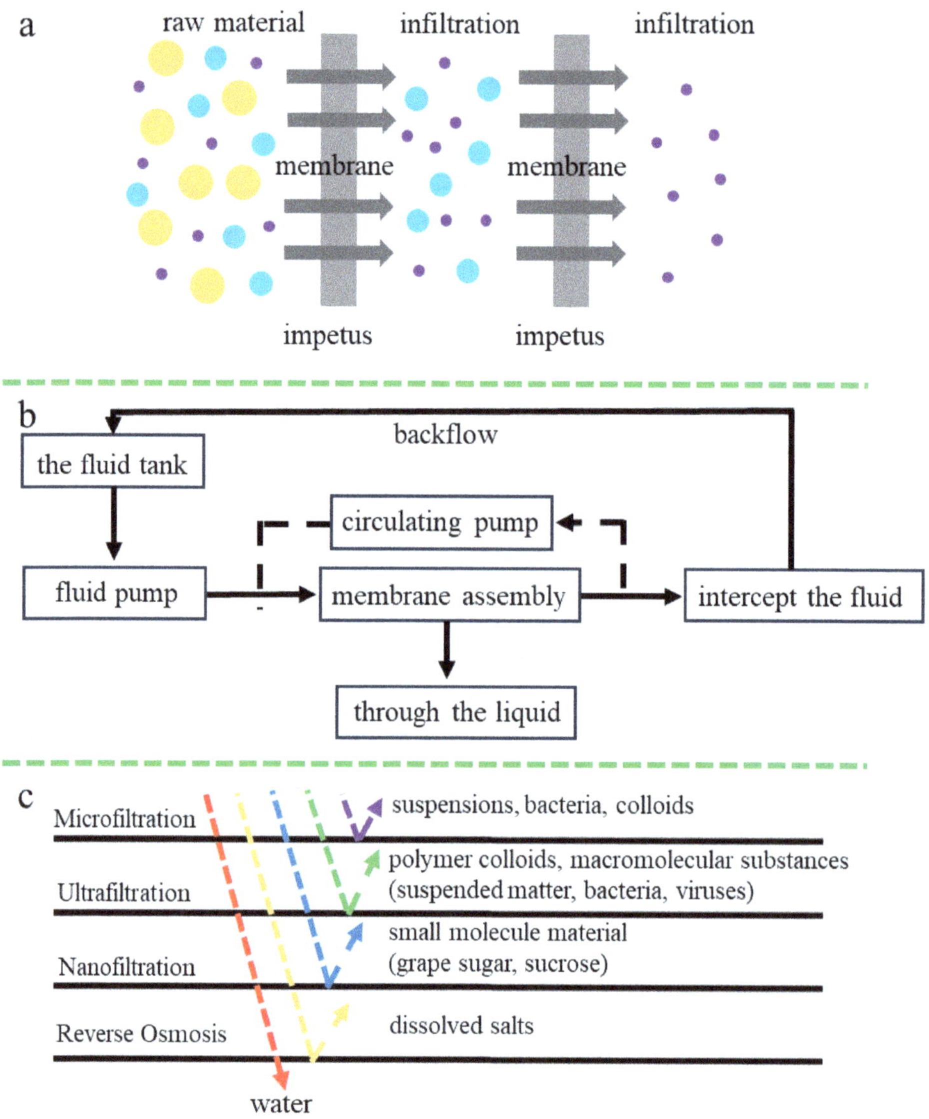

(a) Illustration of the size-exclusion separation mechanism: Under driving pressure ("impetus"), the feed stream undergoes stepwise filtration where components are fractionated based on molecular size differences, allowing selective permeation of active ingredients.

(b) Process flow diagram of a single-stage membrane separation system with partial recirculation (backflow) to enhance separation efficiency.

(c) Classification of the membrane filtration spectrum: A comparison of Microfiltration, Ultrafiltration, Nanofiltration, and Reverse Osmosis based on their specific rejection capabilities for suspensions, macromolecules, small molecules, and dissolved salts. (Source: Bi et al., 2024).

4.2.4 Supercritical Fluid Extraction (SFE)

Supercritical CO_2 ($scCO_2$) serves as a tunable "green solvent" for lipophilic, thermosensitive intermediates (e.g., carotenoids, Vitamin E).

- **Thermodynamic Advantages:** Operating above the critical point (31°C, 7.38 MPa), $scCO_2$ combines gas-like diffusivity with liquid-like solvating power. Low-temperature operation (35°C) increases intermediate retention to 95%, significantly outperforming thermal solvent extraction (80%).
- **Solvent Residue Elimination:** Depressurization leads to the immediate gasification of CO_2, reducing solvent residues to <1 ppm, far below the regulatory limits for Class 1 solvents like n-hexane.
- **Tunable Selectivity:** Solvating power is density-dependent. By modulating pressure (e.g., from 15 MPa to 25 MPa), fractionation can be achieved, selectively extracting the target intermediate while distinguishing it from impurities based on polarity.

4.3 Digitalization in Purification: Process Analytical Technology (PAT) and Real-Time Control

Modern purification processes rely on real-time monitoring to ensure quality attributes. PAT is not a purification method itself but an integrated system for process control, a key enabler of QbD and continuous manufacturing. These technologies are integral to process control (Fig. 4).

- **UV-Vis Spectroscopy:** Provides rapid, non-destructive in-line monitoring of conjugated systems, serving as the primary detector for fraction collection in chromatography.
- **Vibrational Spectroscopy (IR/Raman):** Raman spectroscopy is particularly effective for monitoring polymorph formation during crystallization in aqueous environments, where IR is limited by water absorption.
- **Nuclear Magnetic Resonance (NMR):** While traditionally an offline characterization tool, benchtop flow-NMR has emerged as a critical PAT tool

for structural elucidation and purity assessment during continuous flow manufacturing.

Figure 4. Comparative assessment of principal in-line spectroscopic techniques employed in Process Analytical Technology (PAT).

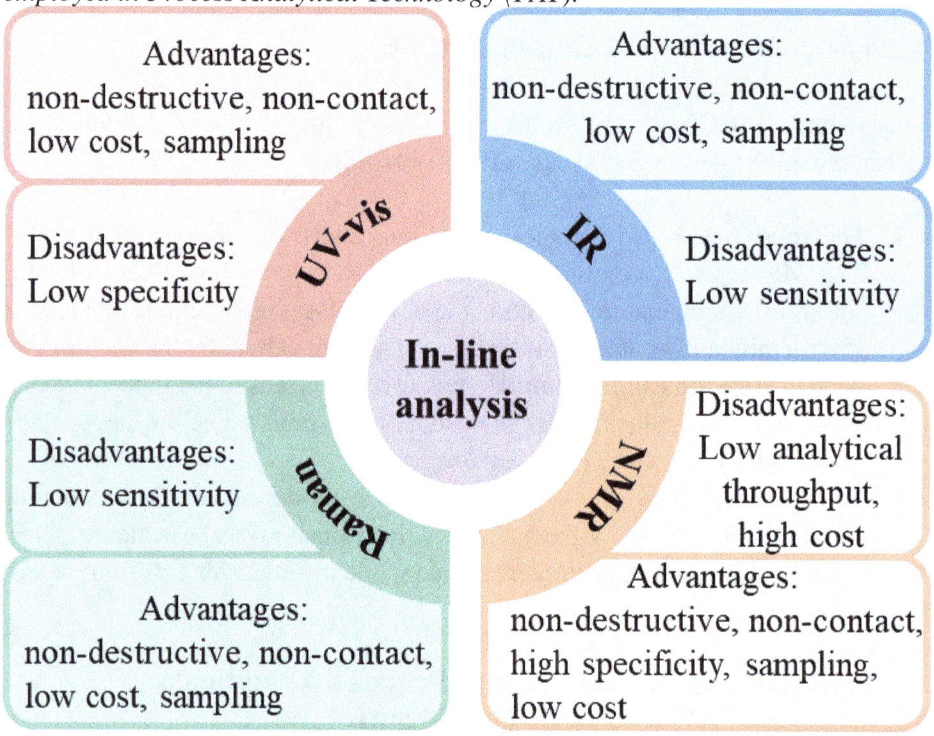

The diagram delineates the operational trade-offs, highlighting the specific advantages (e.g., non-destructive nature, sampling capabilities) and limitations (e.g., sensitivity, specificity, and cost) associated with UV-Vis, IR, Raman, and NMR methodologies.

Abbreviations: UV–Vis, Ultraviolet–visible spectroscopy; IR, Infrared spectroscopy; NMR, Nuclear magnetic resonance spectroscopy.

4.4 Strategic Selection Framework

The selection of a purification strategy requires a heuristic approach matching physicochemical properties to technical capabilities:

- **Thermal Stability:** Thermolabile compounds necessitate non-thermal methods (SFE, Membrane) or vacuum techniques.
- **Phase & Polarity:** Solid, stable compounds favor Recrystallization (Cost-optimal); Lipophilic liquids favor SFE or Distillation.
- **Purity Requirements:** High-value chiral or isomeric separations mandate Prep-HPLC or CPC despite higher operational costs.

This "fit-for-purpose" selection strategy ensures the alignment of technical feasibility with economic and regulatory requirements in intermediate manufacturing.

5. QUALITY CONTROL SYSTEMS FOR SYNTHESIS AND PURIFICATION

5.1 Critical Quality Attributes (CQAs) and Analytical Strategies

The quality assurance of pharmaceutical intermediates is the bedrock of final drug product efficacy and safety. The definition of Critical Quality Attributes (CQAs) extends beyond basic purity to a comprehensive impurity profiling strategy, strictly adhering to the frameworks established by the International Council for Harmonisation (ICH) guidelines (Q3A, Q3C, Q6A, and Q11).

5.1.1 Specification Establishment and Impurity Profiling

Specifications are not merely pass/fail criteria but a set of rigorous controls ensuring the intermediate fits its intended purpose in the synthetic pathway.

- **Purity and Assay:** While a purity threshold of ≥98.0% (via area normalization) is a standard baseline, assay determinations using external standards are preferred to preclude mass balance errors. For chiral intermediates, such as L-carnitine precursors, the Enantiomeric Excess (e.e.) is typically strictly controlled at ≥99.0% to prevent the propagation of isomeric impurities.
- **Organic Impurities (ICH Q3A):** A hierarchical control strategy is employed. Identified impurities are typically limited to ≤0.15%, while unspecified impurities are capped at 0.10% (identification threshold). Special attention is mandated for Potential Genotoxic Impurities (PGIs)—such as alkylating agents or hydrazines formed during synthesis—which require control at ppm levels complying with ICH M7 to ensure the cumulative risk is negligible (Schmidtsdorff & Schmidt, 2022).

- **Residual Solvents (ICH Q3C):** Control strategies differentiate solvents by toxicity class. Class I solvents (e.g., benzene) are generally avoided; if used, they must be controlled to typically ≤2 ppm. Class II solvents (e.g., acetonitrile, ≤410 ppm) require routine monitoring, whereas Class III solvents (e.g., ethanol) are controlled by nonspecific loss on drying (LOD) unless specific GMP requirements dictate otherwise.
- **Elemental Impurities (ICH Q3D):** The paradigm has shifted from nonspecific colorimetric limit tests to specific quantification using Inductively Coupled Plasma Mass Spectrometry (ICP-MS), ensuring toxic elements like Cd, Pb, As, and Hg are controlled within the Permitted Daily Exposure (PDE) limits.

5.1.2 Advanced Analytical Methodologies

- **High-Performance Liquid Chromatography (HPLC):** HPLC remains the workhorse for organic impurity profiling. Method development focuses on orthogonality—using complementary stationary phases (e.g., C18 vs. Phenyl-Hexyl) to uncover co-eluting peaks. System Suitability Tests (SST) are critical; for instance, in the analysis of cephalosporin intermediate 7-ACA, a resolution (Rs) of >2.0 between the active moiety and its deacetylated degradant is a mandatory acceptance criterion, alongside injection precision (RSD ≤1.0% for n=6).
- **Headspace Gas Chromatography (HS-GC):** For volatile impurities, static headspace sampling eliminates non-volatile matrix interference. Using a DB-624 column with Flame Ionization Detection (FID), this technique achieves limits of quantification (LOQ) sufficient to monitor Class I solvents at sub-ppm levels.
- **Hyphenated Techniques (HPLC-MS):** Mass spectrometry is indispensable for structural elucidation of unknown impurities appearing above the identification threshold (0.10%). For example, in paclitaxel side-chain synthesis, Electrospray Ionization (ESI-MS/MS) facilitates the identification of oxidative degradants (e.g., [M+H]+ m/z 328) by analyzing fragmentation patterns, thereby guiding process optimization to suppress specific degradation pathways.

5.2 Process Control via Quality by Design (QbD)

Modern pharmaceutical engineering transitions from "Quality by Testing" (QbT) to **Quality by Design (QbD)** (ICH Q8/Q11). This approach posits that quality cannot be tested into a product but must be built into the manufacturing process through a deep understanding of Critical Process Parameters (CPPs).

5.2.1 Design of Experiments (DoE) and Design Space

Process optimization utilizes statistical tools to define the "Design Space"—the multidimensional combination of input variables that assures quality.

- **Screening and Optimization:** Initial single-factor experiments identify key variables (e.g., reaction stoichiometry, temperature). Subsequently, Response Surface Methodology (RSM), such as a Box-Behnken design, models interactions between factors. In the crystallization of an antineoplastic intermediate, an interaction was revealed between solvent composition (ethanol:water ratio) and cooling rate; optimizing these concurrently raised purity from 97.5% to 99.2%, a result unattainable by optimizing factors individually.
- **Process Analytical Technology (PAT):** Real-time release testing is enabled by in-line sensors. Near-Infrared Spectroscopy (NIRS) can monitor reaction kinetics non-invasively, triggering automated quench sequences when conversion reaches >99%. Similarly, Focused Beam Reflectance Measurement (FBRM) monitors chord length distributions during crystallization, ensuring the median particle size (D50) remains within the 50–100 μmμm range to optimize filtration and drying rates.

5.2.2 Process Validation and Continuous Improvement

Process validation (ICH Q11) is a lifecycle approach comprising three stages: Process Design, Process Qualification (PPQ), and Continued Process Verification (CPV).

- **Robustness and Capability:** Validation goes beyond producing three conforming batches. It demonstrates process capability (CpkCpk) and robustness against variability. For an azithromycin intermediate, validation data must prove that the impurity profile remains compliant even when CPPs (e.g., pH ± 0.2, Temp ± 1°C) fluctuate within their proven acceptable ranges (PAR).
- **Deviation Management (CAPA):** When deviations occur (e.g., pp-nitroaniline purity drop to 96.8%), a Root Cause Analysis (RCA) is conducted. If the root cause is identified as delayed quenching leading to azo-impurity formation, Corrective and Preventive Actions (CAPA) are implemented—such as integrating in-line HPLC monitoring to objectively determine endpoints—thereby preventing recurrence.

5.3 Analytical Method Validation and Data Integrity

Analytical methods must be validated in accordance with **ICH Q2(R1)** to demonstrate they are suitable for their intended use.

- **Validation Parameters:** Key performance characteristics include Specificity (peak purity confirmation via DAD or MS), Linearity ($R^2>0.999$), Accuracy (Recovery 98%–102%), and Precision. Sensitivity is defined by the Limit of Quantification (LOQ), typically required to be ≤ 50% of the impurity reporting threshold (e.g., LOQ of 0.05% for a 0.10% limit).
- **Data Integrity and ALCOA+:** In the era of digital compliance, data management is governed by the ALCOA+ principles (Attributable, Legible, Contemporaneous, Original, Accurate). Chromatography Data Systems (CDS) like Empower™ must utilize audit trails to record all metadata, including integration changes and abort sequences. Raw data retention and security are paramount to ensure complete traceability and defensibility of the quality decisions throughout the product lifecycle.

6. CONCLUSION AND FUTURE OUTLOOK

The manufacturing of pharmaceutical intermediates has evolved from a series of discrete chemical operations into a highly integrated, cybernetic system where molecular design, process engineering, and quality assurance are inextricably coupled. This chapter has demonstrated this closed-loop knowledge network that governs the entire product lifecycle, establishing a nexus between synthesis, purification, and quality control.

6.1 Future Outlook: Emerging Paradigms and Strategic Directions

The trajectory of the pharmaceutical intermediate industry is shifting towards intelligence, sustainability, and molecular complexity. To advance, it is crucial to address the technical and economic hurdles associated with emerging paradigms.

6.2.1 Emerging Paradigms and Industrial Challenges

- **The Digitalization of Synthesis (AI, Machine Learning, and Digital Twins):** Artificial Intelligence (AI) is revolutionizing retrosynthetic planning... The next frontier is the development of digital twins—high-fidelity virtual models of a process—that can predict performance, optimize parameters in silico, and enable real-time autonomous control.

  - o Challenge & Adoption Barrier: The "Black Box" nature of some ML models, the scarcity of high-quality, structured data (especially failed reactions), and the high computational cost and expertise required to build and validate digital twins limit widespread adoption (Iglesias, Haddad, & Sans, 2022).

- **Sustainability and Green Manufacturing:** The industry is under pressure to transition from petrochemical feedstocks to bio-based alternatives and implement circular economy principles...
  - o Challenge & Adoption Barrier: The economic viability of green technologies remains a barrier. Robust Techno-Economic Analysis (TEA) and Life Cycle Assessments (LCA) are needed to justify the high CAPEX of green technologies, as the cost of downstream processing for bio-based materials and catalyst recycling often impede industrial-scale adoption (Sheldon & Woodley, 2018).
- **The Complexity Frontier (Peptides and Oligonucleotides):** As therapeutic modalities shift towards biologics and complex small molecules, their synthesis presents unique engineering challenges...
  - o Challenge & Adoption Barrier: There is a critical lag (knowledge gap) in the development of highly selective, robust catalysts and purification technologies capable of operating under mild conditions for these large, complex, and often fragile molecules.

6.2.2 Strategic Directions for Future Research

To address these challenges, academic and industrial research should converge on the following strategic vectors:

- **Process Intensification via Micro-reaction Technology:** Future research should prioritize the transition from batch to continuous flow manufacturing...
- **Advanced Separation Materials:** The bottleneck in downstream processing calls for the development of next-generation separation media...
- **Integration of Biocatalysis and Chemocatalysis:** Developing chemo-enzymatic cascade reactions—where enzymes and chemical catalysts operate in a single pot—represents a frontier in efficiency...

By addressing these challenges through interdisciplinary collaboration, the field can evolve to meet the demands of next-generation therapeutics, creating a more efficient, sustainable, and resilient pharmaceutical manufacturing ecosystem.

REFERENCES

Alharbi, A. M., Katary, M., Alshehri, K. M., Asghar, B. H., Omran, M. M., Elshaarawy, R. F. M., & Zakaria, R. (2025). Hybrid thiazolyl–benzylidene–phenol metal complexes as novel chemotherapeutic agents with anti-topoisomerase I activity in human breast carcinoma: Synthesis, in vitro and in silico studies. *RSC Advances*, *15*(30), 20552–20569. DOI: 10.1039/D5RA00889A PMID: 40530316

Aroh, K. C., & Jensen, K. F. (2018). Efficient kinetic experiments in continuous flow microreactors. *Reaction Chemistry & Engineering*, *3*(1), 94–101. DOI: 10.1039/C7RE00163K

Bi, Y., Dong, J., Zhou, Y., Zhang, M., Chen, X., & Zhang, Y. (2024). Application of membrane separation technology in the purification of pharmaceutical components. *Preparative Biochemistry & Biotechnology*, *54*(9), 1107–1115. DOI: 10.1080/10826068.2024.2328673 PMID: 38526323

Calvó-Tusell, C., Liu, Z., Chen, K., Arnold, F. H., & Garcia-Borràs, M. (2023). Reversing the enantioselectivity of enzymatic carbene N–H insertion through mechanism-guided protein engineering. *Angewandte Chemie International Edition*, *62*(35), e202303879. DOI: 10.1002/anie.202303879 PMID: 37260412

Campos, K. R., Coleman, P. J., Alvarez, J. C., Dreher, S. D., Garbaccio, R. M., Terrett, N. K., Tillyer, R. D., Truppo, M. D., & Parmee, E. R. (2019). The importance of synthetic chemistry in the pharmaceutical industry. *Science*, *363*(6424), eaat0805. Advance online publication. DOI: 10.1126/science.aat0805 PMID: 30655413

Clarke, C. J., Tu, W. C., Levers, O., Bröhl, A., & Hallett, J. P. (2018). Green and sustainable solvents in chemical processes. *Chemical Reviews*, *118*(2), 747–800. DOI: 10.1021/acs.chemrev.7b00571 PMID: 29300087

Farrant, E. (2020). Automation of synthesis in medicinal chemistry: Progress and challenges. *ACS Medicinal Chemistry Letters*, *11*(8), 1506–1513. DOI: 10.1021/acsmedchemlett.0c00292 PMID: 32832016

Fathi, E. M., Salem, M. E., Sroor, F. M., Mohamed, F. G., Elwahy, A. H. M., Abdel-Megid, M., & Emara, M. (2025). Synthesis and evaluation of novel bis-chalcone derivatives containing a thiophene moiety as potential anticancer agents: In vitro, in silico, and mechanistic studies. *ACS Omega*, *10*(24), 25921–25937. DOI: 10.1021/acsomega.5c02394 PMID: 40584358

Godfrey, A. G., Masquelin, T., & Hemmerle, H. (2013). A remote-controlled adaptive medchem lab: An innovative approach to enable drug discovery in the 21st century. *Drug Discovery Today*, *18*(17-18), 795–802. DOI: 10.1016/j.drudis.2013.03.001 PMID: 23523957

Green, A. P., & Turner, N. J. (2016). Biocatalytic retrosynthesis: Redesigning synthetic routes to high-value chemicals. *Perspectives in Science*, *9*, 42–48. DOI: 10.1016/j.pisc.2016.04.106

Hoffmann, R. W. (2012). Natural product synthesis: Changes over time. *Angewandte Chemie International Edition*, *52*(1), 123–130. DOI: 10.1002/anie.201203319 PMID: 23203402

Iglesias, D., Haddad, D., & Sans, V. (2022). Recent developments in process digitalisation for advanced nanomaterial syntheses. *Chemistry Methods*, *2*(12), e202200031. DOI: 10.1002/cmtd.202200031

Ivanova, M., Legros, J., Poisson, T., & Jubault, P. (2021). Continuous flow synthesis of Celecoxib from 2-bromo-3,3,3-trifluoropropene. *Journal of Flow Chemistry*, *12*(2), 147–151. DOI: 10.1007/s41981-021-00205-x PMID: 34631154

Krska, S. W., DiRocco, D. A., Dreher, S. D., & Shevlin, M. (2017). The evolution of chemical high-throughput experimentation to address challenging problems in pharmaceutical synthesis. *Accounts of Chemical Research*, *50*(12), 2976–2985. DOI: 10.1021/acs.accounts.7b00428 PMID: 29172435

Lei, Z., Ang, H. T., & Wu, J. (2023). Advanced in-line purification technologies in multistep continuous flow pharmaceutical synthesis. *Organic Process Research & Development*, *28*(5), 1355–1368. DOI: 10.1021/acs.oprd.2c00374

Li, L., Chen, Z., Zhang, X., & Jia, Y. (2018). Divergent strategy in natural product total synthesis. *Chemical Reviews*, *118*(7), 3752–3832. DOI: 10.1021/acs.chemrev.7b00653 PMID: 29516724

Liu, C., Zhou, Y., Wu, G., Gao, K., Li, L., Tu, H., & Chen, Z. (2022). Sandwich-likely structured, magnetically-driven recovery, biomimetic composite penicillin G acylase-based biocatalyst with excellent operation stability. *Colloids and Surfaces. A, Physicochemical and Engineering Aspects*, *639*, 128245. DOI: 10.1016/j.colsurfa.2021.128245

Ma, R., Li, J., Zeng, P., Duan, L., Dong, J., Ma, Y., & Yang, L. (2024). The application of membrane separation technology in the pharmaceutical industry. *Membranes (Basel)*, *14*(1), 24. DOI: 10.3390/membranes14010024 PMID: 38248714

Mathebula, N. P., Sheldon, R. A., & Bode, M. L. (2022). Lipase-catalysed enzymatic kinetic resolution of aromatic Morita-Baylis-Hillman derivatives by hydrolysis and transesterification. *Chembiochem : A European Journal of Chemical Biology*, *23*(21), e202200435. DOI: 10.1002/cbic.202200435 PMID: 36049111

Nicholas, R. J., McGuire, M. A., Hyun, S.-H., Cullison, M. N., & Thompson, D. H. (2022). Development of an efficient, high purity continuous flow synthesis of Diazepam. *Frontiers in Chemical Engineering*, *4*, 877498. DOI: 10.3389/fceng.2022.877498

Örkényi, R., Éles, J., Faigl, F., Vincze, P., Prechl, A., Szakács, Z., Kóti, J., & Greiner, I. (2017). Continuous synthesis and purification by coupling a multistep flow reaction with centrifugal partition chromatography. *Angewandte Chemie International Edition*, *56*(30), 8742–8745. DOI: 10.1002/anie.201703852 PMID: 28548374

Pichon, V., Brothier, F., & Combès, A. (2014). Aptamer-based-sorbents for sample treatment—A review. *Analytical and Bioanalytical Chemistry*, *407*(3), 681–698. DOI: 10.1007/s00216-014-8129-5 PMID: 25286873

Putman, J. I., & Armstrong, D. W. (2022). Recent advances in the field of chiral crystallization. *Chirality*, *34*(10), 1338–1354. DOI: 10.1002/chir.23492 PMID: 35904758

Russell, M. G., & Jamison, T. F. (2019). Seven-step continuous flow synthesis of Linezolid without intermediate purification. *Angewandte Chemie International Edition*, *58*(23), 7678–7681. DOI: 10.1002/anie.201901814 PMID: 30964955

Sánchez-Trasviña, C., Flores-Gatica, M., Enriquez-Ochoa, D., Rito-Palomares, M., & Mayolo-Deloisa, K. (2021). Purification of modified therapeutic proteins available on the market: An analysis of chromatography-based strategies. *Frontiers in Bioengineering and Biotechnology*, *9*, 717326. DOI: 10.3389/fbioe.2021.717326 PMID: 34490225

Schmidtsdorff, S., & Schmidt, A. H. (2022). Nitrosamine impurities in pharmaceuticals: A review on the current status of the regulatory environment and analytical methods. *Journal of Pharmaceutical and Biomedical Analysis*, *209*, 114526. DOI: 10.1016/j.jpba.2021.114526

Schneider, G. (2017). Automating drug discovery. *Nature Reviews. Drug Discovery*, *17*(2), 97–113. DOI: 10.1038/nrd.2017.232 PMID: 29242609

Sheldon, R. A., & Woodley, J. M. (2018). Role of biocatalysis in sustainable chemistry. *Chemical Reviews*, *118*(2), 801–838. DOI: 10.1021/acs.chemrev.7b00203 PMID: 28876904

Stefanache, A., Marcinschi, A., Marin, G.-A., Mitran, A.-M., Lungu, I. I., Miftode, A. M., & Hancianu, M. (2025). Green chemistry approaches in pharmaceutical synthesis: Sustainable methods for drug development. *AppliedChem*, *5*(2), 22. DOI: 10.3390/appliedchem5020013

Sun, H., Zhang, H., Ang, E. L., & Zhao, H. (2018). Biocatalysis for the synthesis of pharmaceuticals and pharmaceutical intermediates. *Bioorganic & Medicinal Chemistry*, *26*(7), 1275–1284. DOI: 10.1016/j.bmc.2017.06.043 PMID: 28709846

Viano, R., Milani, L., Orkenyi, R., & Adamo, A. (2023). Continuous separation and purification of chemical species by continuous liquid–liquid partition chromatography implemented with membrane-based phase separators. *Organic Process Research & Development*, *27*(12), 2296–2308. DOI: 10.1021/acs.oprd.3c00239

Weeranoppanant, N. (2019). Enabling tools for continuous-flow biphasic liquid–liquid reaction. *Reaction Chemistry & Engineering*, *4*(2), 235–243. DOI: 10.1039/C8RE00230D

Weeranoppanant, N., & Adamo, A. (2019). In-line purification: A key component to facilitate drug synthesis and process development in medicinal chemistry. *ACS Medicinal Chemistry Letters*, *11*(1), 9–15. DOI: 10.1021/acsmedchemlett.9b00491 PMID: 31938456

Yu, P., Liu, Q., Zuo, L., Zhang, X., Yin, C., & Zhou, H. (2024). Water-mediated one-pot multi-step synthesis of chiral 1,3-diarylpropan-1-ols by the asymmetric hydrofunctionalisation of simple alkynes. *Green Chemistry*, *26*(7), 4074–4078. DOI: 10.1039/D3GC04945K

Zhu, H., Hu, B., & Yang, F. (2021). Removal of sulfadiazine by polyamide nanofiltration membranes: Measurement, modeling, and mechanisms. *Membranes (Basel)*, *11*(2), 143. DOI: 10.3390/membranes11020104 PMID: 33540550

Chapter 13
Recent Advances in Anticancer Drug Screening

Wei Xiao
http://orcid.org/0009-0005-5797-2331
Tongren University, China

Xiao bin Wang
Jiangsu Ocean University, China

ABSTRACT

Cancer has become a major global public health threat. Although cancer diagnosis and treatment technologies continue to advance, the screening and development of effective anticancer drugs still face multiple bottlenecks. The technical level of anticancer drug screening directly determines the scientific rationality of subsequent directions and the efficacy of candidate drugs, serving as the core breakthrough point for enhancing overall success rates. Accordingly, this chapter systematically explores anticancer drug screening, focusing on three core dimensions: first, analyzing the technical pathways and inherent limitations of traditional screening methods; second, clarifying the innovative directions and practical application values of the latest screening technologies; and third, summarizing solutions to the core technical challenges in the screening process and the key barriers in clinical application translation.

DOI: 10.4018/979-8-3373-4402-7.ch013

INTRODUCTION

Cancer has emerged as a major global public health issue, posing a serious threat to human life and safety. According to data from the World Health Organization, millions of new cancer cases are reported annually worldwide, with projections exceeding 35 million by 2050 (Kamita et al., 2024). Cancer not only inflicts physical and psychological damage on patients, leading to a decline in quality of life and survival, but its high treatment costs also exacerbate family financial pressures and societal healthcare burdens. Significant breakthroughs have been achieved in cancer research regarding pathogenesis, diagnostic techniques, and treatment modalities (Al-Hussaniy et al., 2025; Cai et al., 2025; Pradhan et al., 2025; Lin, 2025). However, the development of effective anticancer drugs still faces immense challenges.

Anticancer drug discovery is a complex, interdisciplinary system engineering project involving molecular biology, pharmacology, and other fields. It requires stages such as elucidating molecular mechanisms, screening drug targets, and designing and synthesizing drug candidates. The journey from research and development to market approval often spans decades. Within this process, screening potential anticancer compounds represents the critical starting point, directly determining the direction and success rate of the entire endeavor. This screening must accurately identify active substances from a vast pool of natural products, chemically synthesized compounds, and biologics. Ideal drug candidates must to possess specificity, while balancing therapeutic efficacy with safety to minimize damage to normal cells.

Current anticancer drug screening methods are broadly categorized into traditional approaches and emerging technologies. Traditional methods, such as in vitro cell-based assays and animal models, exhibit significant limitations. New technologies encompass three-dimensional (3D) cell culture, advanced high-throughput screening (HTS), virtual screening, the integration of targeted therapy principles, and the application of nanotechnology (Zhang et al., 2016; Hassan et al., 2022; Ciftci, et al., 2025). Furthermore, the convergence of multi-omics technologies, medical imaging, and 3D bioprinting holds promise for addressing the shortcomings of traditional methods.

Core challenges persist in current drug screening: Technologically, there is a disconnect between in vitro models and the complex in vivo tumor microenvironment (TME), making it difficult to mimic tumor heterogeneity and dynamic characteristics. The precision of high-throughput screening remains insufficient, and screening technologies for rare targets and personalized tumor models lag behind. In terms of the pipeline, candidate drugs must undergo preclinical studies, multi-phase clinical trials, and regulatory approval, resulting in prolonged translation timelines and high costs that hinder the delivery of innovative therapies to patients.

The future of anticancer drug screening must evolve toward greater efficiency and precision. Accelerating the application of computational biology and artificial intelligence in drug design and virtual screening can significantly enhance the efficiency and accuracy of candidate drug discovery. With ongoing technological breakthroughs, the challenges in anticancer drug development will be progressively overcome, bringing new hope to cancer patients.

TRADITIONAL ANTICANCER DRUG SCREENING METHODS

Cell-Based Assays

Cell-based assays have long been a core and relied-upon technical approach in anticancer drug screening. Valued for their strong operational controllability and high reproducibility of results, they serve as crucial tools in the initial stages of drug discovery. The standardized procedure for these assays primarily involves three steps: First, researchers establish a standardized cultivation system for cancer cell lines in cell culture dishes/flasks. By controlling conditions such as culture medium composition, temperature (37 °C), and CO_2 concentration (5%), they ensure stable proliferation of cancer cells to the logarithmic growth phase, providing a uniform cellular model for subsequent drug testing. Second, potential drug compounds are prepared at different concentrations using serial dilution methods and then co-incubated with cancer cells. The incubation duration, typically ranging from 24 to 72 hours, is set based on the cell type and the drug's mechanism of action to ensure sufficient interaction. Finally, the impact of the drug on the physiological functions of cancer cells is analyzed using specific detection methods, such as assessing cell growth rate, proliferative activity, apoptosis rate, and the activation status of key signaling pathways.

The 3-(4,5-dimethylthiazol-2-yl)-2,5-diphenyltetrazolium bromide (MTT) assay is a commonly used technique for detecting cell viability. Its principle relies on the reduction of yellow MTT to purple formazan crystals by mitochondrial succinate dehydrogenase in living cells, an activity absent in dead cells (Surin et al., 2017). When a drug inhibits cancer cells, the decrease in viable cells leads to reduced formazan production. By measuring the absorbance at 570 nm using a microplate reader, the number of living cells can be quantified, allowing for the calculation of the drug's half-maximal inhibitory concentration (IC_{50}), which provides a quantitative basis for evaluating drug activity. The colony formation assay, on the other hand, focuses on assessing the long-term proliferative capacity of cancer cells after drug treatment (Wan et al., 2025). Following drug exposure, a low number of cells (100-500 cells/dish) are seeded into culture dishes and continuously cultivated until surviving cells form colonies containing 50 or more cells. After staining with crystal violet, an image analysis system is used to count the number and measure the area of colonies, enabling the calculation of the colony formation rate. This metric objectively reflects the inhibitory effect of the drug on cancer cell survival and proliferation and is significant for screening drugs with long-term antitumor potential.

Traditional two-dimensional (2D) cell culture systems possess significant limitations and struggle to adequately simulate the complexity of the in vivo tumor microenvironment, representing a core bottleneck constraining the accuracy of drug screening. In these systems, cancer cells grow only as a monolayer attached to the dish surface. This setup lacks the three-dimensional tissue structure and intricate surrounding microenvironment characteristic of in vivo tumors. Consequently, cancer cells tend to lose in vivo-specific functional phenotypes, such as epithelial-mesenchymal transition traits, the self-renewal capacity of cancer stem cells, and drug resistance-associated phenotypes. These defects often lead to a loss of efficacy for drugs identified as effective in 2D screens when they progress to animal studies or clinical trials. Statistics indicate that approximately 60% of candidate drugs showing positive results in 2D screens fail in later stages due to significant discrepancies between in vitro and in vivo outcomes, severely reducing the efficiency of drug development.

To overcome these limitations, the integration of multi-omics technologies and medical imaging is driving the evolution of anticancer drug screening toward simulation of the in vivo microenvironment (Figure 1) (Long et al., 2025). At the molecular mechanism level, genomics, utilizing techniques like whole-exome sequencing and single-cell sequencing, can precisely identify gene mutation sites and copy number variations in cancer cells, clarifying potential drug targets. Transcriptomics dynamically analyzes changes in gene expression within cancer cells following drug treatment, revealing regulatory mechanisms of signaling pathways. Epigenomics investigate the drug's impact on epigenetic modifications in cancer cells, thereby refining the understanding of its mechanism of action. At the functional and metabolic levels, proteomics employ mass spectrometry to quantitatively analyze drug-induced differences in protein expression, while metabolomics detect changes in intracellular metabolites. These approaches identify key regulatory proteins and elucidate the drug's effects on energy metabolism pathways. Furthermore, radiomics quantifies the three dimensional structural features of tumors using imaging techniques like CT and MRI, and pathomics analyzes histopathological morphological changes in tumor tissue via digital pathology slides. This extends the screening dimension from the cellular level to the tissue/individual level. The integration of multi-omics data with imaging information facilitates the construction of multi-dimensional screening models that more closely approximate real in vivo conditions. This approach not only enhances the accuracy of drug activity assessment but also aids in predicting potential toxic side effects, providing comprehensive and reliable technological support for anticancer drug screening.

Figure 1. This document outlines the sources and processing procedures for omics data, encompassing genomic, transcriptomic, epigenomic, proteomic, metabolomic, radiomics, and pathomics datasets.

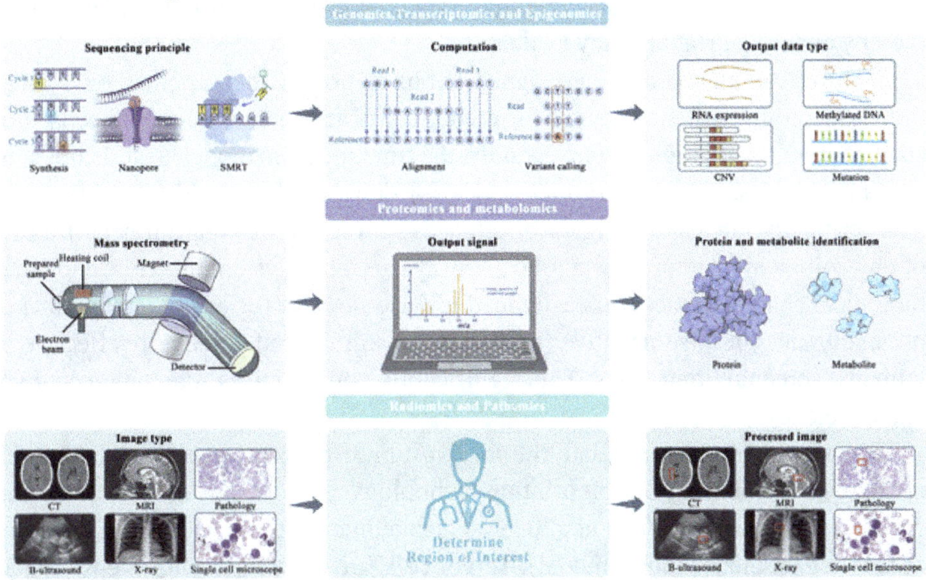

Source: Adapted from Xinrui Long et al. (2025)

Animal Model-Based Detection Methods

In the field of anticancer drug screening, animal models (particularly murine models) hold an indispensable and central position. Among these, genetically engineered mouse models (GEMMs) are constructed using precise gene-editing technologies with the core objective of mimicking specific types of human cancer and harboring the characteristic gene mutations associated with those cancers (Jean & Jaurand, 2018). Specifically, researchers can genetically modify the mouse genome to carry genes frequently mutated in human cancers. For instance, engineering mice to carry mutations in the tumor suppressor gene p53 (a gene often mutated in various cancers leading to uncontrolled cell proliferation) enables the creation of disease models that correspond to specific human cancers.

Compared to cell-based assays, GEMMs provide a research environment that more closely approximates human physiological conditions. They can realistically reflect a drug's effects on tumor growth rate, metastatic patterns, and the overall survival of the experimental animals, yielding results more aligned with the drug's actual in vivo performance. Furthermore, xenograft models are commonly used animal models for anticancer drug screening. These models are established by transplanting human cancer cells or tumor tissue fragments into immunodeficient mice. Leveraging the inability of the mouse immune system to reject the transplanted cells, human cancer cells survive, proliferate, and form human-like tumor tissues, thereby enabling drug efficacy evaluation.

However, animal models have significant limitations in their application. Firstly, the financial investment required for model construction and maintenance is substantial, resulting in high costs. Secondly, the timeline from model establishment to drug efficacy assessment is protracted, slowing down the drug development process. More critically, intrinsic biological differences between animals and humans in terms of physiological structure, metabolic pathways, immune system function, and other biological characteristics can easily lead to false-positive (effective in animals but not in humans) or false-negative (ineffective in animals but potentially effective in humans) screening outcomes. This significantly compromises screening accuracy and efficiency.

To more accurately simulate the in vivo tumor microenvironment, integrated approaches combining 3D bioprinting technology with animal models are being developed (Figure 2) (Wu et al., 2023). This method first utilizes an in vivo tumor model (e.g., a mouse tumor model). Tumor cells are isolated through steps such as tumor resection, mincing, and enzymatic digestion to prepare a bioink. Subsequently, digital light processing (DLP) bioprinting is employed to construct a tumor model with a structure closely resembling the in vivo architecture for ex vivo drug screening. Concurrently, in vivo screening can be performed using the original animal model. This integrated approach offers dual advantages: On the one hand, it leverages the precision of 3D bioprinting to construct tumor models with complex structures and microenvironments, enhancing the accuracy of ex vivo screening. On the other hand, it relies on the holistic in vivo environment of animal models for comprehensive drug efficacy assessment. This can compensate for the shortcomings of both traditional animal models and simple cell-based assays, providing a more reliable platform for anticancer drug screening and helping to address the drug development challenges arising from biological differences between species.

Traditional screening technologies (2D cell culture, animal models) offer the advantages of low short-term initial screening costs and simple operation. 2D cell culture allows for rapid preliminary screening of drug activity with a short cycle and minimal investment. However, the validation phase using animal models is costly, plagued by significant species differences, long experimental cycles, and numerous ethical restrictions, resulting in an overall low development efficiency. Furthermore,

2D cell culture suffers from a high false-positive rate due to the lack of an in vivo microenvironment simulation, whereas animal models have limited physiological relevance to the human body. Consequently, the success rate of translating screening results into clinical applications remains low at only 5%-10%. In personalized anticancer drug development scenarios, the difficulty in reusing models further increases costs and reduces efficiency.

Figure 2. This diagram illustrates the workflow for obtaining tumor cells from an in vivo tumor model, processing them, and subsequently employing bioink and DLP bioprinting technology for in vitro drug screening, while also facilitating in vivo drug screening.

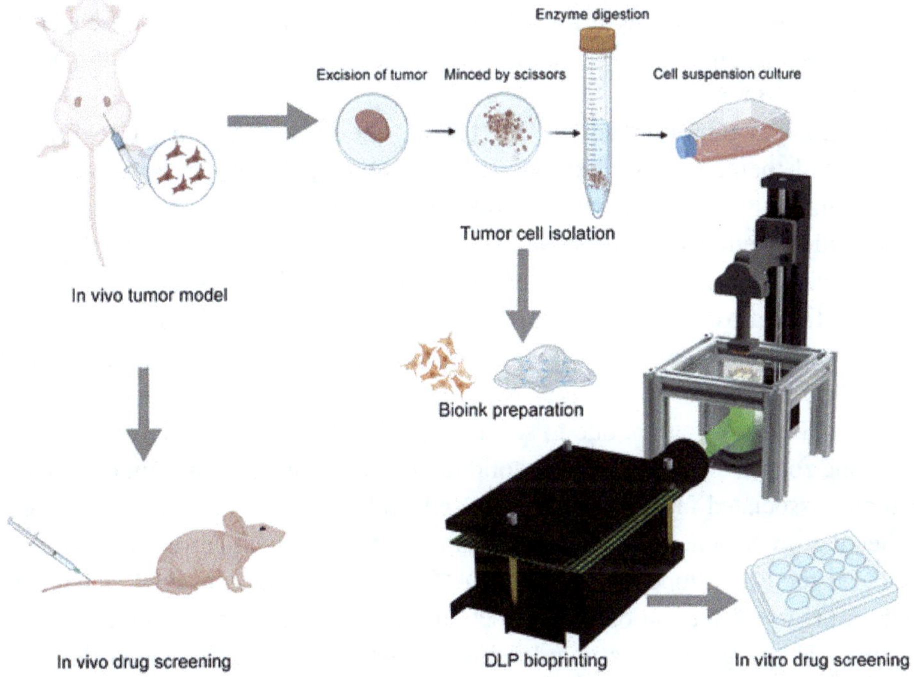

Source: Adapted from Bing-Xuan Wu et al. (2025)

RECENT ADVANCES IN ANTICANCER DRUG SCREENING

3D Cell Culture Technology

In the field of anticancer drug screening, three-dimensional (3D) cell culture technology has emerged as a core alternative to traditional two-dimensional (2D) culture, effectively overcoming key bottlenecks, such as distorted cellular phenotypes and biased drug evaluation inherent in 2D systems, owing to its high fidelity in simulating the in vivo tumor microenvironment (TME) (Sun & Ma, 2023). This system constructs a three-dimensional growth environment containing extracellular matrix components (e.g., collagen, laminin, and Matrigel), which not only provides physical support to cancer cells akin to the in vivo setting but also regulates cell adhesion, proliferation, and differentiation via signaling molecules. Consequently, cell morphology and gene expression patterns more closely resemble the actual state of solid tumors.

Multicellular tumor spheroids (MCTS) are the most widely used 3D culture models, with their core advantage lying in recapitulating the gradient microenvironment characteristics of solid tumors (Kochanek et al., 2019; Bowers et al., 2020). Construction methods can be selected as needed, primarily including the hanging drop method, microwell plate method, and bioreactor culture. A critical feature is the formation of oxygen-nutrient-metabolic waste gradients: Cells on the spheroid surface receive ample oxygen and nutrients, leading to active proliferation. Oxygen tension and nutrient concentrations decrease toward the interior, slowing cell proliferation. The core region, due to severe hypoxia and nutrient deprivation, forms a necrotic zone analogous to that in solid tumors, accompanied by high expression of tumor-associated factors like hypoxia-inducible factors (HIF) and vascular endothelial growth factor (VEGF). This characteristic closely aligns with the clinical TME. Currently developed integrated "3D-MCTS + impedance sensing" technology enables precise detection of drug effects on inhibiting tumor cell migration and invasion. This approach boasts a detection sensitivity 3 - 5 times higher than 2D culture and shortens the screening cycle to within 72 hours, making it suitable for high-throughput screening of anti-metastatic drugs.

Cancer organoids are a novel 3D culture model that has risen to prominence recently. Their core strengths of being patient-derived and organ-mimicking are driving a shift toward personalized anticancer drug screening, with their key value concentrated on clinical translatability (Selvin et al., 2023). These models are derived from patient tumor tissues or biopsy samples, cultured over 3-4 weeks to form structures that maximally retain the tumor's genetic mutations, epigenetic features, and phenotypic characteristics. They can be used for individual patient drug sensitivity testing. For example, screening a panel of 10-15 chemotherapy or targeted drugs on colorectal cancer patient-derived organoids can yield drug sensitivity data within 7-10 days, providing direct evidence for formulating personalized treatment plans and reducing the risk of trial-and-error therapy. Furthermore, cancer organoids can be used to construct tumor evolution models, simulating the development of drug resistance under long-term drug exposure, offering a research platform for developing novel anti-resistance therapies.

High-Throughput Screening (HTS) Technology

High-throughput screening (HTS) technology has overcome the bottleneck of low efficiency and low throughput in traditional anticancer drug screening. Through the synergy of automation, miniaturization, and multi-dimensional detection, it enables the rapid screening of vast compound libraries, shortening the initial drug discovery screening phase from years to months. Consequently, HTS has become a cornerstone technology supporting anticancer drug discovery. Its primary objective is the systematic testing of thousands to millions of compounds within a short timeframe to identify potential active compounds that specifically act on cancer cells or tumor-associated targets.

The efficient operation of an HTS system relies on the coordinated work of three key modules: automated liquid handling systems, automated sample processing robots, and high-content imaging and detection systems, ensuring both the precision and high throughput of the process. The standardized workflow is as follows: First, a suitable assay model (e.g., cancer cell line, molecular target protein) is selected based on the screening objective and seeded into multi-well plates (96, 384, or 1536 wells). The plates are then incubated in a fully automated incubator to allow sufficient interaction between the compounds and the assay model. Post-incubation, an automated transport system transfers the plates to the detection module, where indicators such as absorbance or fluorescence intensity for each well are rapidly measured. Simultaneously, high-content imaging systems perform cellular imaging and analysis (Akhtari et al., 2021). Finally, specialized software is used for data normalization, calculating compound inhibition or activation rates, thereby facilitating the identification of active compounds.

With technological iteration, next-generation HTS platforms have transcended the limitations of 2D cell culture. By integrating 3D cell culture technology and functional genomics detection, they significantly enhance the clinical translatability of screening outcomes. The incorporation of 3D culture allows drug screening to be conducted under conditions that better mimic the in vivo TME. Meanwhile, functional genomics detection employs a combined compound screening + genetic perturbation assay mode to elucidate drug activity and mechanisms of action, effectively reducing interference from false-positive compounds.

The integration of 3D-compatible well plates with HTS technology achieves a synergistic effect of "3D microenvironment simulation + high-throughput screening", balancing in vivo relevance with screening efficiency and precision (Figure 3) (Malakpour-Permlid et al., 2021). Taking the preparation of heterogeneous and homogeneous 3D 96-well plates as an example: heterogeneous 3D 96-well plates are fabricated by preparing nanofiber membranes via electrospinning. These membranes undergo cryogenic freezing, mechanical shearing, and homogenized dispersion before being thermally crosslinked to the plate substrate. Subsequent air-drying and oxygen plasma treatment create a growth microenvironment for cells that closely resembles the in vivo conditions. Such optimized compatible substrates further enhance the stability and reproducibility of screening models, providing robust support for the rapid identification of clinically promising anticancer drugs from massive compound libraries and propelling anticancer drug discovery steadily toward greater efficiency, precision, and personalization.

Figure 3. Schematic illustration of the fabrication process for high-throughput non-homogeneous (top) and homogeneous (bottom) 3D PCL-based 96-well plates.

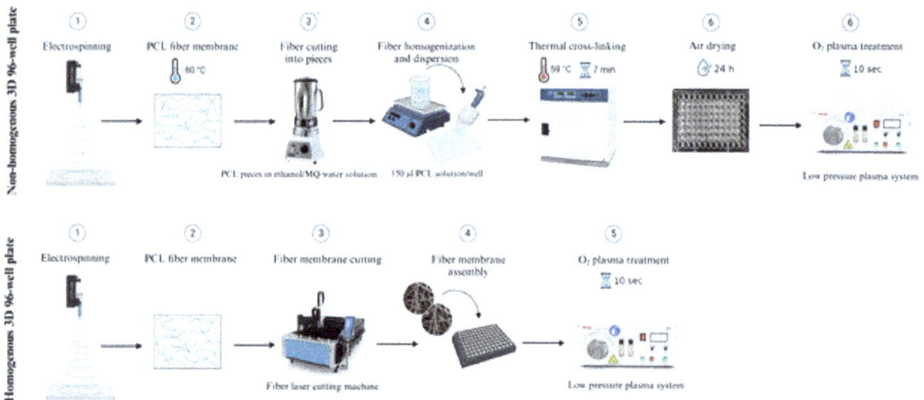

Source: Adapted from Atena Malakpour-Permlid et al. (2025)

Computational and Virtual Screening Technologies

In the process of cancer drug discovery, computational and virtual screening methods have become pivotal supporting technologies for anticancer drug research and development, leveraging their core advantages of high efficiency and low cost. These approaches enable the screening of potential active compounds from vast chemical libraries without the need for large-scale physical experiments, significantly shortening the screening timeline and reducing expenditures.

Structure-based drug design (SBDD) is a central method within computational and virtual screening. It uses the three-dimensional structure of a target protein to perform virtual screening of small-molecule libraries (Lam et al., 2024). Such target proteins are often key therapeutic targets, like abnormally active protein kinases in cancer cells, which play crucial roles in carcinogenesis and progression. Researchers determine the spatial conformation and chemical properties of the target protein's active site and key functional regions by resolving its 3D structure. Molecular docking algorithms are then employed to simulate the binding process between small-molecule compounds and the target protein. By calculating parameters such as binding energy, hydrogen bonds, and hydrophobic interactions, the binding potential of compounds is quantitatively assessed and ranked. Compounds demonstrating strong binding affinity and high specificity are prioritized for subsequent experimental validation.

Ligand-based drug design (LBDD) represents another important approach. It relies on the chemical structures of known active compounds (ligands) to construct predictive models for novel compound activity, typically by analyzing ligand structural features and pharmacophore models (Szwabowski et al., 2023). Such models can capture key structural information related to compound activity, rapidly predict the potential anticancer activity of new compounds, and provide a core reference for screening efforts.

With technological advancements, machine learning and artificial intelligence (AI) algorithms have been widely integrated into the field of anticancer drug screening, further enhancing both efficiency and accuracy. For instance, the deep learning framework “Shennong” integrates single-cell RNA sequencing (scRNA-seq) data with perturbation data to accurately predict cellular drug responses at the single-cell level. This capability not only aids in the precise identification of drug targets and elucidation of molecular mechanisms but also facilitates the assessment of potential drug-induced tissue damage, providing strong support for drug safety evaluation and propelling toward greater efficiency and precision (Zhang et al., 2025).

As illustrated in Figure 4, AI-driven drug discovery establishes a systematic and highly efficient pipeline (Paithankar et al., 2025). The process begins with a data preparation phase, where data cleaning and preprocessing lay the foundation for high-quality data. The target prediction stage then employs AI to locate potential drug targets, defining the direction. This is followed by a hit identification phase, screening compounds that interact with the targets, and a lead optimization phase, where AI assists in enhancing compound properties. After in silico validation of the compound efficacy, experimental testing confirms its therapeutic effect and safety. Furthermore, AI can effectively address traditional drug development challenges such as broad-spectrum effects, significant side effects, and narrow therapeutic windows. It plays a vital role in personalized medicine, toxicity prediction, and targeted drug design, thereby enhancing the precision and safety of cancer treatment. This approach aligns with individual patient variability and clinical needs, accelerating the translation of drugs from the research bench to clinical application.

Figure 4. Applications of AI in the development of drugs.

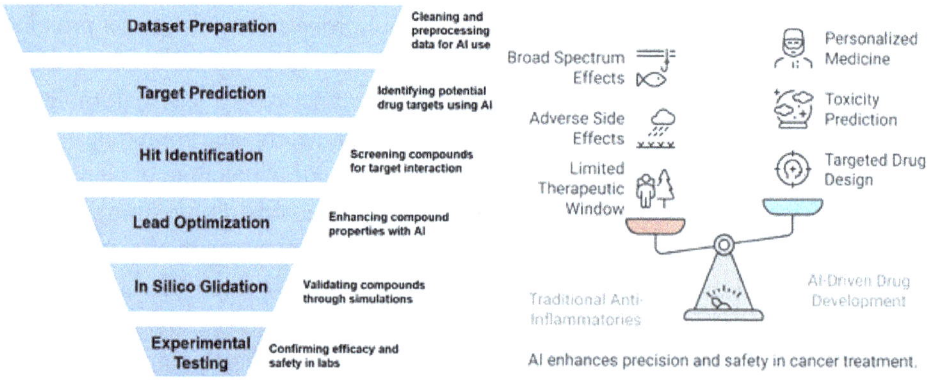

Source: Adapted from Vivek Paithankar et al. (2025)

Integration of Targeted Therapy and Precision Medicine

The advent of targeted therapy and precision medicine has fundamentally reshaped the core philosophy and developmental trajectory of anticancer drug screening, providing a novel research perspective for cancer treatment. Distinct from traditional anticancer therapies reliant on broad-spectrum cytotoxic agents, targeted therapy is centered on precise localization. It specifically inhibits key molecules or signaling pathways that regulate cancer cell growth, survival, and metastasis. This approach breaks through the limitation of the "indiscriminate attack" on cancer cells characteristic of conventional treatments, enabling precise intervention against tumor targets (Vandghanooni, 2018).

As specific molecules (including proteins, gene fragments, etc.) reflect the biological characteristics of tumor cells, their presence and changes in expression levels provide important evidence for cancer diagnosis, efficacy evaluation, and drug screening (Lin, 2023). For example, in the clinical diagnosis and treatment of breast cancer, human epidermal growth factor receptor 2 (HER2)-positive tumors represent a typical cancer type characterized by specific biomarkers. These tumor cells overexpress the HER2 protein, which persistently drives abnormal proliferation. Trastuzumab, developed for this target, specifically binds to the HER2 protein and blocks its mediated signaling pathways, effectively inhibiting cancer cell growth and metastasis, thereby providing a precise treatment strategy for this subtype of breast cancer.

The principles of targeted therapy and precision medicine are now deeply integrated into the anticancer drug screening workflow. Researchers first conduct a systematic analysis of patient tumor samples to identify the presence, expression levels, and mutational characteristics of specific biomarkers. Based on these analytical results, they then screen for compounds that can target and modulate the molecular pathways associated with those biomarkers. This transforms the screening process from the search for broad-spectrum cytotoxic compounds to target-specific screening based on the molecular profile of the tumor. This personalized screening paradigm revolutionizes traditional anticancer drug screening systems. It significantly increases the probability of matching individual patients with effective drugs. Concurrently, because the drugs act precisely on specific targets within cancer cells, nonspecific damage to normal cells is reduced. This lowers the risk of adverse drug reactions and improves patients' quality of life during treatment.

As illustrated in Figure 5, nanoparticles offer unique technological support for targeted therapy (Maleki et al., 2025). The endothelial cells in normal tissues are densely packed, making it difficult for nanoparticles to penetrate. In contrast, tumor tissues exhibit endothelial leakage (enhanced permeability and retention effect), and tumor cell surfaces often express specific receptors. Nanoparticles can thus efficiently accumulate in cancer cells through mechanisms like transcytosis. Using nanoparticles as drug carriers to encapsulate agents targeting specific biomarkers or signaling pathways leverages the characteristics of the tumor microenvironment and cancer cell surface receptors for precise drug delivery. This further enhances drug specificity, minimizes impact on normal cells, provides a powerful tool for both anticancer drug screening and the implementation of targeted therapy, and propels precise anticancer treatment toward greater efficacy and safety.

Figure 5. The schematic illustrates two targeting mechanisms for tumor-directed nanocarriers: passive (EPR effect) and active (ligand conjugation).

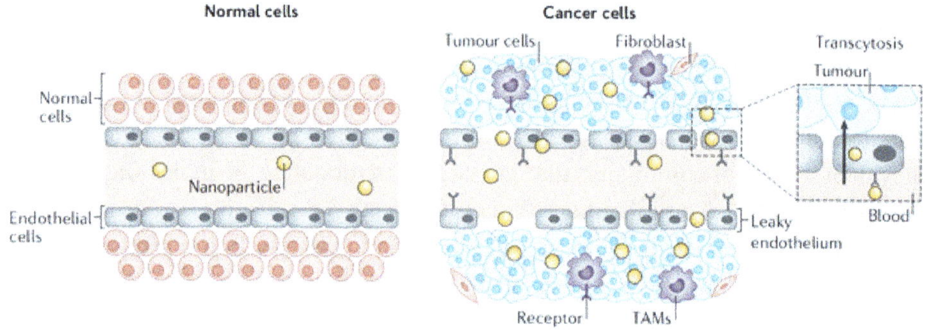

Source: Adapted from Hassan Maleki et al. (2025)

Although initial platform construction and model development for such new technologies require substantial investment, they offer irreplaceable core value in the field of anticancer drug screening: 3D organoids can highly replicate the human physiological microenvironment and tumor heterogeneity, addressing the low physiological relevance of traditional models. High-throughput screening enables the large-scale testing of millions of compounds, significantly increasing the screening throughput. Computational and virtual screening technologies leverage AI algorithms to accurately predict drug activity and toxicity, substantially narrowing the range of candidate compounds. The integration of targeted therapy and precision medicine anchors the process to individual patient characteristics, enabling "tailored" drug development. Compared to traditional methods, the comprehensive advantages of new technologies are significant (see **Table 1**). The synergistic application of multiple technologies can shorten the drug development cycle by 50%-80%, reduce long-term R&D costs by 30%-40%, effectively circumvent the high false-positive rates associated with traditional techniques, and increase the clinical translation success rate to 20%-35%. These technologies are well suited for precision medicine applications such as personalized anticancer drug development and the elucidation of drug resistance mechanisms.

Table 1. Comparison of traditional and novel anticancer drug screening technologies

Technology type	Technology principle	Advantages	Disadvantages	Screening efficiency	Clinical relevance	Application scenarios
2D Cell culture	Monolayer culture of tumor cell lines, detection of cell viability/proliferation	Low cost, simple operation, high reproducibility	Inability to simulate TME, loss of cellular phenotype	High	Low	Primary screening of large compound libraries
Animal models	Transplantation of patient tumor tissue into immunodeficient mice	Simulates in vivo tumor growth and metastasis	High cost, long cycle, and species differences	Low	High	Preclinical efficacy validation of drug candidates
3D cell culture	Culture of patient-derived tumor cells as 3D organoids with tissue-like structures	Retains tumor heterogeneity and simulates the microenvironment	High cost, longer culture period (3-4 weeks)	Medium	High	Personalized drug sensitivity testing, drug resistance mechanism studies
High-throughput screening	Automated microplate-based detection of compound activity	Extremely high throughput (millions of compounds per week)	Single-target focus, prone to false positives	Very High	Medium	Primary screening of large compound libraries
Computational and virtual screening	Uses algorithms such as molecular docking, machine learning, deep learning to predict compound-target binding affinity	Low cost, short cycle, and no physical experiments required	Relies on high-quality data, requires experimental validation	Very High	Medium	Target-directed drug design, drug repurposing studies
Integration of targeted therapy and precision medicine	Combines patient multi-omics data (genomics, transcriptomics) to identify specific molecular targets, design targeted drugs, and make formulate personalized treatment plans	Enables precision treatment, strong drug targeting, low side effects	Difficult for some targets, limited by target mutation frequency, relatively high treatment cost	Medium	High	Personalized anticancer drug development, treatment of rare target tumors

KEY POINTS IN ANTICANCER DRUG SCREENING AND CLINICAL TRANSLATION

Core Technical Challenges and Solutions in Anticancer Drug Screening

During the initial stages of anticancer drug development, the reliability and efficiency of screening technologies directly determine the subsequent research direction and success rate. Although current anticancer drug screening has evolved from traditional cell-based models to a multidimensional, high-precision technological system, three major core technical challenges persist, forming key bottlenecks that constrain drug development efficiency.

The first challenge is the disconnect between in vitro screening models and the in vivo tumor microenvironment (TME), leading to the failure of numerous candidate drugs in clinical phases. Current screening predominantly relies on two-dimensional (2D) cell culture models and simple animal models. The former fails to replicate the three-dimensional (3D) growth state of tumors, cell-cell interactions, and stromal microenvironment components, such as angiogenesis and immune cell infiltration. The latter, exemplified by xenograft models in immunodeficient mice, struggles to recapitulate the genetic heterogeneity and complex pathological features of human tumors. For instance, some small-molecule drugs demonstrating potent anticancer activity in 2D cell models have been eliminated in Phase I clinical trials due to insufficient efficacy, often because they cannot penetrate the dense tumor stroma or are cleared by immune cells. It is estimated that approximately 60% of anticancer drug candidates are halted in the preclinical stage due to a lack of in vivo efficacy despite promising in vitro results, resulting in significant wastage of research resources (Kunnumakkara et al., 2019).

The second challenge is the insufficient precision of high-throughput screening (HTS) technologies, which constrains the quality of candidate drugs. While mainstream HTS technologies (e.g., fluorescence-based cell viability and protein binding assays) enable the rapid screening of thousands to tens of thousands of compounds, they often focus on a single target or a singular biological effect (such as proliferation inhibition). This approach overlooks the synergistic mechanisms involving multiple pathways and targets in tumorigenesis and progression. Taking drug screening for EGFR-mutant lung cancer as an example, traditional techniques often focus solely on a drug's inhibitory effect on EGFR kinase activity, neglecting its potential activating effects on downstream pathways like PI3K/AKT, which can lead to drug resistance in clinical application (Xu et al., 2024). Furthermore, screening data noise (e.g., compound non-specific binding, and detection signal interference) reduces screening accuracy. Approximately 20% of positive candidate compounds are later identified as false positives in subsequent validation, significantly increasing development costs and timelines.

The third challenge is the lag in screening technologies for rare targets and personalized tumors, hindering the ability to meet the demands of precision oncology (Drost & Clevers, 2018). With the deepening of research into tumor molecular subtyping, rare driver mutations such as ALK fusions and ROS1 rearrangements are being progressively identified. Patient populations corresponding to these targets often constitute less than 5% of cases. However, traditional screening technologies rely on "Universal" models, making it difficult to establish dedicated screening systems. Concurrently, personalized tumor heterogeneity imposes high requirements for "customized" screening technologies. Current personalized screening approaches based on organoids or circulating tumor cells (CTCs) still face challenges, including difficult sample acquisition, long culture periods, and high costs, preventing their routine clinical application.

As illustrated in Figure 6, technologies related to extracellular vesicles (EVs) offer a novel pathway to overcome the aforementioned bottlenecks (Assunçao, 2025). At the diagnostic level, liquid biopsy for analyzing molecular features (e.g., DNA, RNA, proteins) carried by EVs (such as exosomes) enables precise assessment of tumor status. This provides a basis for identifying rare targets, analyzing personalized tumor molecular characteristics, and facilitating customized screening. For prognosis, molecular markers within EVs can help predict events like tumor metastasis, aiding in the selection of drugs with greater therapeutic potential. At the treatment level, engineered natural or artificially synthesized EVs can serve as drug delivery vehicles, achieving targeted drug delivery to tumor sites. This approach addresses the pain point of the disconnect between in vitro models and the in vivo microenvironment, thereby enhancing the drug's in vivo efficacy. Moreover, EVs derived from engineered cells can simulate signal transduction relevant to the in vivo TME, enhancing the authenticity of in vitro screening models and the precision of HTS. This opens up an efficient new avenue for anticancer drug development.

Figure 6. Extracellular vesicles hold significant potential in precision oncology, as the tumor-associated molecules they carry can serve as biomarkers for diagnosis, prognosis, and treatment response via liquid biopsy. Following nanoengineering modifications, they can also optimize targeted delivery and function as adjuvants in cancer therapy.

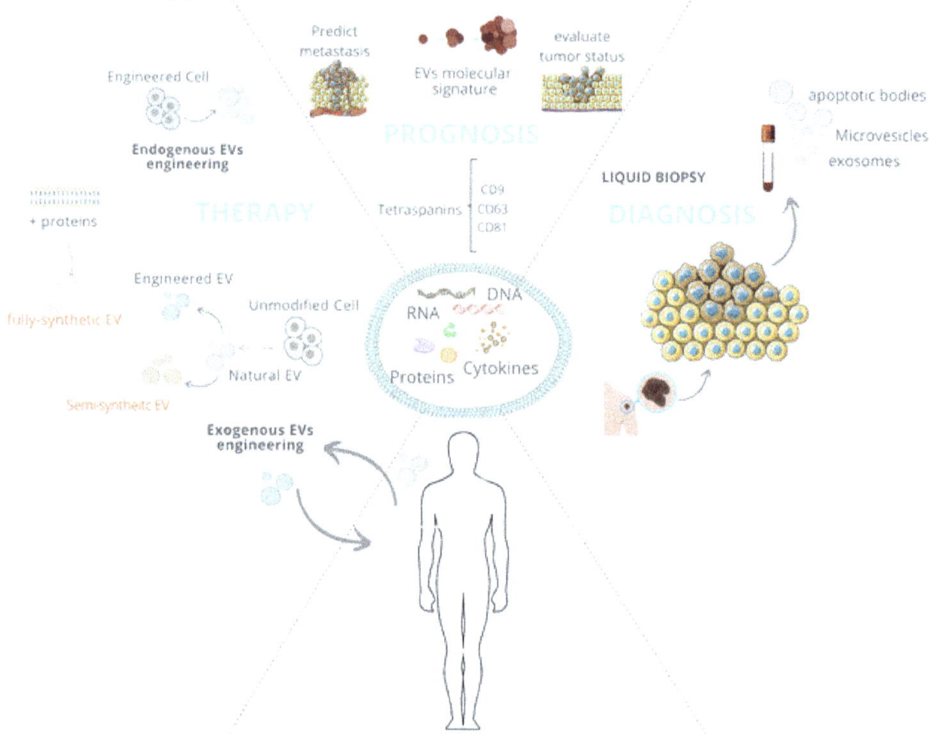

Source: Adapted from Raphaela Rebeca Silveira Assunç ao et al. (2025)

Clinical translation pathway and associated roles of anticancer drugs

The clinical translation of anticancer drugs is a systematic endeavor that bridges laboratory candidate molecules with clinically approved therapeutics. It encompasses four core stages: preclinical research, clinical trials, regulatory approval, and post-marketing surveillance. This process relies on the collaborative advancement of multiple disciplines, including pharmacology, medicine, and statistics. Technical deficiencies or insufficient data at any stage can lead to translational failure (Al-Karmalawy, 2025). Globally, only approximately 5% of anticancer candidate drugs achieve marketing approval. The core challenges lie in the consistency between preclinical data and clinical reality, the confirmation of efficacy in large populations, and the long-term management of safety. For translating drugs targeting specific molecules, such as those related to the BEX family of proteins, strategies across all stages must be optimized according to the target's characteristics to enhance translation efficiency.

Preclinical research serves as an essential entry point for drug translation. Its primary objectives are to validate the efficacy, pharmacokinetic properties, and safety of candidate drugs through in vitro and in vivo experiments, providing a scientific basis for clinical trials. This phase typically spans 2 - 4 years (Cully, 2019) and involves the following key components: (1) Pharmacodynamic validation: A multi-tiered model system is employed. Foundational models (e.g., two-dimensional cell lines such as A549 and MCF-7 cells, three-dimensional patient-derived organoids, and PDOs) are used for preliminary assessment of inhibitory activity and mechanism of action. Advanced models (e.g., cell-derived xenografts (CDX), patient-derived xenografts (PDX) and humanized immune system mouse models) are used for in vivo validation. Notably, PDX models, which preserve tumor genetic heterogeneity and microenvironment features, demonstrate a concordance rate of over 70% between preclinical efficacy results and clinical response, far exceeding that of CDX models (~40%). For instance, the PARP inhibitor olaparib was rapidly approved following successful screening in PDX models. The specific characteristics of various models are detailed in Table 2. (2) Pharmacokinetic (PK) Studies: These studies elucidate the drug's absorption, distribution, metabolism, and excretion (ADME) profiles in vivo. Parameters such as oral bioavailability and peak plasma concentration are used to evaluate the suitability of administration routes. They clarify the drug's enrichment capability in tumor tissues and the risk of accumulation in vital organs, and define metabolic characteristics and excretion pathways to mitigate clinical risks. (3) Toxicological Assessment: This involves determining the median lethal dose (LD50) via acute toxicity tests and establishing the No Observed Adverse Effect Level (NOAEL) through long-term toxicity studies. Combined with special toxicity tests (e.g., genotoxicity and reproductive toxicity), a safe dosage range is defined, providing the basis for clinical dose design.

Table 2. Multidimensional comparison of preclinical models for anticancer drugs

Model type	Genetic heterogeneity retention	Microenvironment simulation	Consistency with clinical trial drug response	Cost	Timeline
2D cell Lines	Low, genomic stability decreases after passage, losing patient tumor characteristics	Very low, monolayer cells, no cell interaction or TME components	Low (20%-30%)	Very Low	3-7 days
3D organoids	High, derived from patient primary cells, retaining genetic diversity and mutations	Relatively high, 3D structure simulating cell connections, lacks vascular/immune components	Relatively High (60%-70%)	Medium	2-4 weeks
Cell Line-derived xenograft	Low, single genetic background, loses patient tumor heterogeneity	Relatively low, tumor formed in mice, and the microenvironment differs significantly from humans	Medium (approx. 40%);	Medium	4-8 weeks
Patient-derived xenograft	High; Transplants patient tumor tissue, fully retains individual specificity	High; Carries patient stromal cells, closely mimics human TME	High (≥70%)	High	8-16 weeks
Humanized immune system mouse model	High, combines PDX features, retains tumor genetic heterogeneity	High, reconstitutes human immune system, simulates tumor-immune interaction	High (75%-85%)	Very High	12-24 weeks

Clinical trials are the critical phase for verifying drug safety, efficacy, and appropriate patient populations in humans. They are conducted in Phases **I**, **II**, and **III**, spanning 5 - 7 years (Dowden, 2019).

Phase **I** trials typically enroll 20-30 patients with advanced cancer. The primary goals are to explore the safe dosage range and administration schedule. This is achieved through dose-escalation studies to determine the maximum tolerated dose (MTD), assessing the safety and PK properties of different regimens, and early efficacy observations.

Phase **II** trials enroll 100-300 target patients, focusing on preliminary efficacy validation in specific tumor types. Key endpoints include objective response rate (ORR) and progression-free survival (PFS). This phase aims to identify efficacy-related biomarkers and explore potential combination therapies.

Phase **III** trials enroll over 1,000 patients using a randomized, controlled, multi-center design. Primary endpoints are typically overall survival (OS) and PFS. This large-scale study monitors rare adverse reactions, and subgroup analyses help clarify the drug's applicable population. A single-agent Phase III trial can cost over $1 billion and last 2-5 years.

Regulatory approval constitutes the final gate, where agencies critically assess the drug's benefit-risk profile (Moore et al., 2020). Sponsors submit a new drug application (NDA) or biologics license application (BLA), containing comprehensive data from preclinical studies, clinical trials, and manufacturing processes. Expert review teams evaluate the veracity of efficacy data, the controllability of safety issues, and the overall benefit-risk ratio. Furthermore, special regulatory pathways exist to expedite the approval of urgently needed drugs, such as the Breakthrough Therapy designation by the US FDA and the Breakthrough Therapy Drug program by China's NMPA.

Conducted after drug approval, post-marketing surveillance aims to evaluate long-term safety and effectiveness in large, real-world populations, addressing limitations of clinical trials, such as limited sample size and follow-up duration. Safety monitoring utilizes spontaneous reporting systems and active surveillance to detect rare and delayed adverse events. Efficacy monitoring employs real-world evidence (RWE) generated from observational studies and other real-world data (RWD) analyses to assess the drug's actual performance in complex, heterogeneous clinical settings.

Figure 7. Cancersclosely related to the BEX family proteins.

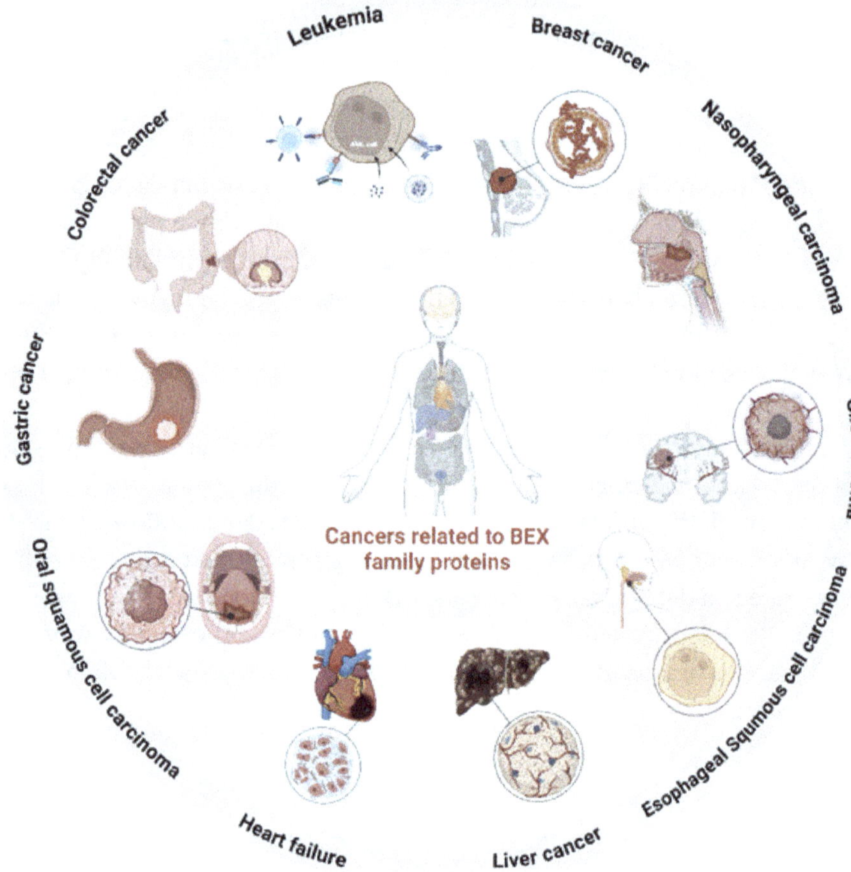

Source: Adapted from Pingping Wang et al. (2025)

Multidimensional Strategies for Overcoming Bottlenecks in Anticancer Drug Screening and Translation

To address the core bottlenecks in anticancer drug screening and translation, breakthrough strategies must be developed from three dimensions: technological innovation, system optimization, and policy support, thereby connecting the "screening-clinical-market" translation chain and enhancing drug development efficiency and success rates (Motohashi, 2025).

First, technological innovation is needed to overcome the disconnect between models and the challenge of screening accuracy. On the one hand, tumor organoid models should be promoted. By culturing patient-derived tumor tissues in vitro, these models preserve the three-dimensional tumor architecture, cellular heterogeneity, and microenvironment features. For example, the colorectal cancer organoid model developed by the Hubrecht Institute in the Netherlands increased the clinical treatment response rate by 25% (Sasaki, 2018). On the other hand, microfluidic chip (organ-on-a-chip) models should be developed to simulate dynamic interactions among blood vessels, immune cells, and tumor cells in vivo. The lung-on-a-chip model by Emulate in the U.S. has been adopted by the FDA as an auxiliary tool for drug screening to assess pulmonary toxicity. Furthermore, artificial intelligence (AI) technologies should be integrated to analyze tumor multi-omics data, construct tumor signaling network models, and identify candidate drugs that regulate multiple pathways. For instance, Zhejiang University's deep learning framework "Shennong" successfully predicted the potential efficacy of exemestane in lung cancer, offering a new approach for cross-cancer screening. Additionally, high-content screening technology should be introduced, which uses fluorescence imaging to simultaneously detect drug effects on multiple indicators such as cell morphology, apoptosis, and signaling pathways, thereby avoiding bias from single-parameter screening.

Second, system optimization is required to break down translational barriers and reduce trial risks. This can be achieved by establishing industry-academia-research-hospital collaborative platforms that promote deep cooperation among pharmaceutical companies, universities, research institutions, and hospitals. Screening phases should be guided by clinical needs; for example, the lung cancer drug translation platform jointly operated by the Shanghai Institute of Materia Medica (CAS) and Shanghai Chest Hospital uses patient-derived circulating tumor cells (CTCs) to construct models, shortening the clinical translation cycle by 1-2 years. Moreover, a preclinical-clinical data-sharing system should be established to integrate pharmacokinetic, toxicological, and clinical trial data for timely optimization of strategies (Yin et al., 2024). Furthermore, adaptive clinical trial designs should be promoted, such as the FDA-approved "basket trial", which enrolls patients with different tumor types sharing the same target, thereby accelerating drug development for rare targets. Real-world data (RWD) should also be utilized to assist trial design; for instance, China's National Medical Products Administration (NMPA) already permits the use of RWD for post-market drug reevaluation, which can shorten phase IV trial cycles (Psioda et al., 2021).

Third, policy support and full-cycle regulation should be strengthened to boost motivation and ensure quality. To address insufficient incentives for drugs targeting rare targets and personalized therapies, policy measures must be enhanced. Examples include expanding the scope of orphan drug designation and providing support, such as subsidies, fast-track reviews, and market exclusivity. In China, the number of orphan drug designations increased by 45% from 2020 to 2023, with 58 new designations in 2023 alone (Zhao et al., 2023). Additionally, personalized drug approval pathways should be established; for instance, the FDA's Breakthrough Therapy designation has shortened the approval cycle for personalized anticancer drugs to 6 months, which is more than one year faster than conventional approval. Meanwhile, full life-cycle regulation must be improved. During the screening stage, standardization of technologies such as organoids and high-throughput screening should be promoted to ensure reproducibility of results (Xiang et al., 2024). In the clinical trial phase, blockchain technology can be adopted to enable data traceability and prevent fraud (Yun et al., 2025). After marketing, a risk-benefit reevaluation mechanism should be implemented (Wang, 2023). For example, China's NMPA issued the Post-Market Surveillance Management Measures for Anticancer Drugs in 2024, requiring companies to establish active adverse reaction monitoring systems and submit quarterly safety reports for high-risk drugs.

CHALLENGES AND FUTURE DEVELOPMENT STRATEGIES IN ANTICANCER DRUG SCREENING

Cancer drug screening technology has evolved from traditional cell models to an integrated system of 3D biomimetic models + AI screening. However, two core challenges remain to be overcome. First, there is a biological gap in preclinical-to-clinical translation. Current models cannot fully simulate the human immune system, tumor microenvironment, and individual variations, leading to approximately 60% of candidate drugs failing due to inconsistencies between in vitro and in vivo efficacy. Second, there are technological limitations in screening for metastatic cancers. Most existing models focus on primary tumors and struggle to simulate key steps in metastasis, resulting in slow progress in developing effective drugs for metastatic cancer. To address these challenges, future anticancer drug screening should advance in three directions:

First, develop multidimensional biomimetic models: construct humanized organ-immune system dual-reconstituted mouse models and tumor metastasis microfluidic chips, and integrate patient gut microbiome and metabolomics data to narrow the preclinical-to-clinical gap and enhance the accuracy of individualized screening.

Second, strengthen multi-omics integration and AI prediction: Correlate multi-omics data with drug screening data to build efficacy prediction models; develop multimodal AI models that integrate various drug-related data to optimize the entire process, potentially increasing the clinical translation success rate of candidate drugs to over 20%.

Third, focus on metastatic cancers and rare targets: Identify metastasis-specific targets through single-cell sequencing; establish cross-border and cross-institutional rare-target drug alliances to integrate global resources, reduce development costs, and address research gaps.

Innovation in anticancer drug screening technology is a key factor in breaking through current development bottlenecks. Future drug screening will rely on the construction of multidimensional biomimetic models and the deep integration of multi-omics and AI technologies to effectively bridge the biological gap between preclinical research and clinical application, significantly improving the clinical translation efficiency of drug candidates. At the same time, this innovative approach can address the shortcomings in screening technologies for metastatic cancers and rare targets, advancing the drug screening model from traditional "effective compound screening" to precise "individual patient efficacy prediction." Ultimately, this will help cancer treatment achieve a fundamental leap from broad-spectrum therapy to personalized cures.

REFERENCES

Akhtari, F. S., Green, A. J., Small, G. W., Havener, T. M., House, J. S., Roell, K. R., Reif, D. M., McLeod, H. L., Wiltshire, T., & Motsinger-Reif, A. A. (2021). High-throughput screening and genome-wide analyses of 44 anticancer drugs in the 1000 Genomes cell lines reveals an association of the NQO1 gene with the response of multiple anticancer drugs. *PLOS Genetics*, *17*(8), e1009732. DOI: 10.1371/journal.pgen.1009732 PMID: 34437536

Al-Hussaniy, H. A., Oraibi, A. I., Al-Tameemi, Z. S., Alburghaif, A. H., Naji, M. A., & Naji, F. A. (2025). The association of KRAS and P53 gene mutations and MDM2 expression with the occurrence of colorectal cancer. *Current Cancer Drug Targets*, *25*, e15680096349198. DOI: 10.2174/0115680096349198250407100724 PMID: 40357780

Al-Karmalawy, A. A., Eissa, M. E., Ashour, N. A., Yousef, T. A., Khatib, A. O. A., & Hawas, S. S. (2025). Medicinal chemistry perspectives on anticancer drug design based on clinical applications (2015–2025). *RSC Advances*, *15*(43), 36441–36471. DOI: 10.1039/D5RA05472A PMID: 41041285

Bowers, H. J., Fannin, E. E., Thomas, A., & Weis, J. A. (2020). Characterization of multicellular breast tumor spheroids using image data-driven biophysical mathematical modeling. *Scientific Reports*, *10*(1), 11583. DOI: 10.1038/s41598-020-68324-4 PMID: 32665565

Cai, Q. Y., Cai, W. Q., Liu, Y. F., Li, M. L., Xiang, S. E., Ji, H., Zhang, Z. S., Yu, N., Hu, Y. Q., Sun, Y., Yin, K. Y., Wu, H. M., & Li, M. W. (2025). Advancing lymph node metastasis in head and neck squamous cell carcinoma: Integrative mechanisms, emerging diagnostics, and translational therapies. *Critical Reviews in Oncology/Hematology*, *215*, 104854. DOI: 10.1016/j.critrevonc.2025.104854 PMID: 40681008

Ciftci, F., Özarslan, A. C., Kantarci, İ. C., Yelkenci, A., Tavukcuoglu, O., & Ghorbanpour, M. (2025). Advances in drug targeting, drug delivery, and nanotechnology applications: Therapeutic significance in cancer treatment. *Pharmaceutics*, *17*(1), 121. DOI: 10.3390/pharmaceutics17010121 PMID: 39861768

Cully, M. (2019). CRISPR restores expression in models of vision loss. *Nature Reviews. Drug Discovery*, *18*(3), 174. DOI: 10.1038/d41573-019-00016-9 PMID: 30824889

Drost, J., & Clevers, H. (2018). Organoids in cancer research. *Nature Reviews. Cancer*, *18*(7), 407–418. DOI: 10.1038/s41568-018-0007-6 PMID: 29692415

Hassan, M. I., Anjum, D., Mohammad, T., Alam, M., Khan, M. S., Shahwan, M., Shamsi, A., & Yadav, D. K. (2022). Integrated virtual screening and MD simulation study to discover potential inhibitors of Lyn-kinase: Targeting cancer therapy. *Journal of Biomolecular Structure & Dynamics*, *41*(20), 10558–10568. DOI: 10.1080/07391102.2022.2154849 PMID: 36495308

Jean, D., & Jaurand, M.-C. (2018). Mesotheliomas in genetically engineered mice unravel mechanism of mesothelial carcinogenesis. *International Journal of Molecular Sciences*, *19*(8), 2191. DOI: 10.3390/ijms19082191 PMID: 30060470

Kamita, M., Waweru, H., Githinji, M., Kibiro, E., & Makokha, F. (2024). Establishing a cancer registry and baseline data for Nyandarua County, Kenya: A step towards establishing a central cancer registry. *Heliyon*, *10*(20), e39056. https://doi .org/. DOI: 10.1016/j.heliyon.2024.e39056 PMID: 39492911

Kochanek, S. J., Close, D. A., & Johnston, P. A. (2019). High content screening characterization of head and neck squamous cell carcinoma multicellular tumor spheroid cultures generated in 384-well ultra-low attachment plates to screen for better cancer drug leads. *Assay and Drug Development Technologies*, *17*(1), 17–36. DOI: 10.1089/adt.2018.896 PMID: 30592624

Kunnumakkara, A. B., Bordoloi, D., Sailo, B. L., Roy, N. K., Thakur, K. K., Banik, K., Shakibaei, M., Gupta, S. C., & Aggarwal, B. B. (2019). Cancer drug development: The missing links. *Experimental Biology and Medicine (Maywood, N.J.)*, *244*(8), 663–689. DOI: 10.1177/1535370219839163 PMID: 30961357

Lam, H. Y. I., Guan, J. S., Ong, X. E., Pincket, R., & Mu, Y. G. (2024). Protein language models are performant in structure-free virtual screening. *Briefings in Bioinformatics*, *25*(6), bbae480. DOI: 10.1093/bib/bbae480 PMID: 39327890

Lin, S. H., Zhou, C. X., Chen, H. L., Zhou, X. L., Yang, H. J., Sun, L. T., Zhang, L. Y., & Zhang, Y. X. (2025). Advances in hereditary colorectal cancer: How precision medicine is changing the game. *Cancers (Basel)*, *17*(21), 3461. DOI: 10.3390/cancers17213461 PMID: 41228253

Lin, Y., & Cai, H. (2023). Biological functions and therapeutic potential of SHCBP1 in human cancer. *Biomedicine & Pharmacotherapy = Biomédecine & Pharmacothérapie*, *160*, 114362. DOI: 10.1016/j.biopha.2023.114362 PMID: 36739763

Long, X. R., Sun, K., Lai, S., Liu, Y. C., Sun, J., Chen, W. Q., Liu, R. H., He, X. Y., Zhao, S., & Huang, K. (2025). Artificial intelligence and anti-cancer drug response. *Acta Pharmaceutica Sinica. B*, *15*(7), 3355–3371. DOI: 10.1016/j.apsb.2025.05.009 PMID: 40698144

Malakpour-Permlid, A., Rodriguez, M. M., Untracht, G. R., Andersen, P. E., Oredsson, S., Boisen, A., & Zor, K. (2025). High-throughput non-homogenous 3D polycaprolactone scaffold for cancer cell and cancer-associated fibroblast mini-tumors to evaluate drug treatment response. *Toxicology Reports*, *14*, 101863. DOI: 10.1016/j.toxrep.2024.101863 PMID: 39758801

Maleki, H., Aiyelabegan, H. T., Javadi, P., Abdi, F., Mirzavi, F., Behjani, Z. Z., Rizvanov, A. A., Takallu, S., Kumar, R., Barhaghtalab, R. H., Hosseinpouri, A., & Vakili-Ghartavol, R. (2025). Nanotechnology-mediated precision drug delivery strategies for breast cancer treatment. *Biomedicine & Pharmacotherapy = Biomédecine & Pharmacothérapie*, *188*, 118224. DOI: 10.1016/j.biopha.2025.118224 PMID: 40456194

Moore, A. R., Rosenberg, S. C., & Malek, F. M. S. (2020). RAS-targeted therapies: Is the undruggable drugged? *Nature Reviews. Drug Discovery*, *19*(8), 533–552. DOI: 10.1038/s41573-020-0068-6 PMID: 32528145

Motohashi, S., Katsuta, E., & Ban, D. (2025). Advances and challenges in drug screening for cancer therapy: A comprehensive review. *Bioengineering (Basel, Switzerland)*, *12*(12), 1315. DOI: 10.3390/bioengineering12121315 PMID: 41463612

Munro, D. H. (2019). Estimating clinical trial success rates from small molecule drug discovery and development. *Nature Reviews. Drug Discovery*, *18*, 495–496.

Paithankar, V., Devnania, D., & Nimburkar, T. A. (2025). AI-guided discovery of novel anti-inflammatory agents for cancer therapy: A new era in drug development. *Intelligent Hospitals*, *1*(1), 100007. DOI: 10.1016/j.inhs.2025.100007

Pradhan, A., Pattnaik, G., Das, S., Acharya, B., & Patra, C. N. (2025). Advancements in lung cancer: Molecular insights, innovative therapies, and future prospects. *Medical Oncology*, *42*(9), 383. DOI: 10.1007/s12032-025-02725-1 PMID: 40719980

Psioda, M. A., Xu, J. W., Jiang, Q., Ke, C. L., Yang, Z., & Ibrahim, J. G. (2021). Bayesian adaptive basket trial design using model averaging. *Biostatistics (Oxford, England)*, *22*(1), 19–34. DOI: 10.1093/biostatistics/kxz014 PMID: 31107534

Sasaki, N., & Clevers, H. (2018). Studying cellular heterogeneity and drug sensitivity in colorectal cancer using organoid technology. *Current Opinion in Genetics & Development*, *52*, 117–122. DOI: 10.1016/j.gde.2018.09.001 PMID: 30261425

Selvin, T., Berglund, M., Lenhammar, L., Jarvius, M., Fryknäs, M., Larsson, R., & Andersson, C. R. (2023). Phenotypic screening platform identifies statins as enhancers of immune cell-induced cancer cell death. *BMC Cancer*, *23*(1), 164. DOI: 10.1186/s12885-023-10645-4 PMID: 36803614

Sun, Y. Q., & Ma, H. Y. (2023). Application of three-dimensional cell culture technology in screening anticancer drugs. *Biotechnology Letters, 45*(9), 1073–1092. DOI: 10.1007/s10529-023-03410-x PMID: 37421554

Surin, A. M., Sharipov, R. R., Krasil'nikova, I. A., Boyarkin, D. P., Lisina, O. Y., Gorbacheva, L. R., Avetisyan, A. V., & Pinelis, V. G. (2017). Disruption of functional activity of mitochondria during MTT assay of viability of cultured neurons. *Biochemistry. Biokhimiia, 82*(6), 737–749. DOI: 10.1134/S0006297917060104 PMID: 28601083

Szwabowski, G. L., Cole, J. A., Baker, D. L., & Parrill, A. L. (2023). Structure-based pharmacophore modeling 1: Automated random pharmacophore model generation. *Journal of Molecular Graphics & Modelling, 121*, 108429. DOI: 10.1016/j.jmgm.2023.108429 PMID: 36804368

Vandghanooni, S., Eskandani, M., Barar, J., & Omidi, Y. (2018). Bispecific therapeutic aptamers for targeted therapy of cancer: A review on cellular perspective. *Journal of Molecular Medicine : Official Organ of the "Gesellschaft Deutscher Naturforscher und Ärzte", 96*(9), 885–902. DOI: 10.1007/s00109-018-1669-y PMID: 30056527

Wan, C. P., Hu, X. M., Yang, P. X., Li, M. X., Nisar, M. F., Zhou, Z. H., Gan, Y. D., Cai, Y., & Liu, Z. H. (2025). Screening of theaflavins that inhibit proliferation of nasopharyngeal carcinoma cells through metabolomics approaches. *Food Science & Nutrition, 13*(8), e70642. DOI: 10.1002/fsn3.70642 PMID: 40755495

Wang, H. (2023). Quantitative drug safety and benefit-risk evaluation: Practical and cross-disciplinary approaches. *The American Statistician, 77*(1), 111–112. DOI: 10.1080/00031305.2022.2160592

Wu, B. X., Wu, Z., Hou, Y.-Y., Fang, Z.-X., Deng, Y., Wu, H.-T., & Liu, J. (2023). Application of three-dimensional (3D) bioprinting in anti-cancer therapy. *Heliyon, 9*(10), e20475. DOI: 10.1016/j.heliyon.2023.e20475 PMID: 37800075

Xiang, D. X., He, A. N., Zhou, R., Wang, Y. G., Xiao, X. Y., Gong, T., Kang, W. Y., Lin, X. L., Wang, X. C., Liu, L. X., Chen, Y. G., Gao, S. R., & Liu, Y. B. (2024). Building consensus on the application of organoid-based drug sensitivity testing in cancer precision medicine and drug development. *Theranostics, 14*(8), 3300–3316. DOI: 10.7150/thno.96027 PMID: 38855182

Xu, X. P., Liu, Y., Gong, Q., Ma, L., Wei, W., Zhao, L. Q., & Luo, Z. B. (2024). PARP1 promotes EGFR-TKI drug resistance via PI3K/AKT pathway in non-small-cell lung cancer. *Cancer Chemotherapy and Pharmacology, 94*(2), 209–221. DOI: 10.1007/s00280-024-04668-2 PMID: 38609654

Yin, S. Y., Yu, Y., Wu, N., Zhuo, M. L., Wang, Y. M., Niu, Y. J., Ni, Y. Q., Hu, F., Ding, C. M., Liu, H. S., Cheng, X. H., Peng, J., Li, J., He, Y., Li, J. X., Wang, J. Y., Zhang, H. S., Zhai, X. Y., Liu, B., & Xi, J. Z. J. (2024). Patient-derived tumor-like cell clusters for personalized chemo- and immunotherapies in non-small cell lung cancer. *Cell Stem Cell*, *31*(5), 717–733. DOI: 10.1016/j.stem.2024.03.008 PMID: 38593797

Yun, D. Y., Wu, X. H., Chen, X., Yang, Y. H., Shang, Y. J., Liu, S. P., Gunasekeran, D. V., Lin, D. R., Liu, L. X., Zhao, L. Q., Chen, W. B., Chen, J. J., Jin, L., Wu, Y. X., Hong, J. M., Hu, W. L., Lin, Z. Z., Cheung, C. Y., Xie, P. C., & Lin, H. T. (2025). An artificial intelligence and blockchain technology-based data management framework for multicenter randomized controlled trials. *Science Bulletin*, *70*(6), 856–860. DOI: 10.1016/j.scib.2025.01.041 PMID: 39933985

Zhang, P. J., Wang, X. Y., Cen, X. F., Zhang, Q., Fu, Y. Y., Mei, Y. Q., Wang, X. R., Wang, R. Y., Wang, J. J., Ouyang, H. W., Liang, T. B., Xia, H. G., Han, X. P., & Guo, G. J. (2025). A deep learning framework for in silico screening of anticancer drugs at the single-cell level. *National Science Review*, *12*(2), nwae451. DOI: 10.1093/nsr/nwae451 PMID: 39872221

Zhang, W. J., Zhuang, A., Gu, P., Zhou, H. F., & Fan, X. Q. (2016). A review of the three-dimensional cell culture technique: Approaches, advantages and applications. *Current Stem Cell Research & Therapy*, *11*(4), 370–380. DOI: 10.2174/1574888X11666160217154007 PMID: 26899396

Zhao, Z. Y., Pei, Z. Y., Hu, A. X., Zhang, Y. H., & Chen, J. (2023). Analysis of incentive policies and initiatives on orphan drug development in China: Challenges, reforms and implications. *Orphanet Journal of Rare Diseases*, *18*(1), 220. DOI: 10.1186/s13023-023-02684-8 PMID: 37501126

Chapter 14
Research on Development and Performance Evaluation of Traditional Chinese Medicine

Hai Yang
http://orcid.org/0000-0002-6155-5603
Tongren University, China

Li Chen
http://orcid.org/0009-0007-1442-8226
Tongren Hospital of Traditional Chinese Medicine, China

Luo Hui Yong
Tongren Hospital of Traditional Chinese Medicine, China

Yang Xian
Tongren Hospital of Traditional Chinese Medicine, China

Chen Shi Xue
Tongren University, China

ABSTRACT

This chapter provides a systematic overview of the historical evolution, modern innovations, performance evaluation systems, and expanding applications of traditional Chinese medicine (TCM). It offers an in-depth examination of TCM's theoretical

DOI: 10.4018/979-8-3373-4402-7.ch014

foundations, its integration with modern technologies such as molecular biology and evidence-based medicine, its global development trajectory, and a multidimensional evaluation system encompassing efficacy, safety, and quality. The study also analyzes the challenges facing the development of TCM and its future prospects, emphasizing its unique value in global health governance.

INTRODUCTION

Traditional Chinese Medicine (TCM) possesses profound historical roots and enduring cultural significance, representing a unique fusion of philosophical wisdom, empirical practice, and ecological holism. Over millennia, TCM has evolved into a dynamic system of health preservation, diagnosis, and treatment, deeply intertwined with cosmological principles such as Yin-Yang, Five Elements, and meridian theory. From its ancient origins documented in classics such as *The Yellow Emperor's Canon of Internal Medicine* (Huangdi Neijing) and *Shennong's Classic of Materia Medica* to its modern-day innovations, TCM has continuously adapted to societal changes while preserving its core tenets. This chapter explores the historical trajectory, contemporary advancements, and scientific evaluation of TCM, emphasizing its role as a bridge between traditional knowledge and modern medical paradigms.

The first section traces the historical development of Traditional Chinese Medicine (TCM). In ancient times, theories such as Yin-Yang and the Five Elements, along with the diagnostic and therapeutic systems of acupuncture and herbal medicine, were established with classical texts as their core. During the dynastic periods, institutional improvements, academic specialization, and innovations in the theory of warm diseases took place, gradually leading to cross-cultural medical exchanges. In modern times, amidst the convergence of Chinese and Western medicine, TCM has explored a development path that integrates tradition with modern science, achieving a systematic revival.

The second section introduces contemporary innovations in TCM. On the scientific front, technologies such as chromatography and molecular biology have been used to elucidate mechanisms of action, establishing a comprehensive research chain. Clinically, TCM has integrated with evidence-based medicine to validate efficacy across multiple fields, forming a distinctive evidence system. Globally, TCM has expanded to 183 countries and regions and gained policy recognition from the WHO, though it faces challenges regarding standards and public perception.

The third section introduces the three-dimensional evaluation system for TCM. Efficacy evaluation balances traditional experience with modern quantitative methods, combining holistic efficacy assessment with precise validation. Safety evaluation establishes a full-process monitoring system, reducing the overestimation of risks by 40% compared to traditional methods. Quality evaluation forms a "chemical + bioactivity" dual-dimension paradigm, covering quality control needs across the entire industrial chain.

The fourth section briefly introduces the further expansion of traditional Chinese medicine (TCM), the accelerating integration of TCM wisdom with modern science, the broadening of TCM's application across multiple fields, and the continuous emergence of innovative medical service models.

The fifth section describes the current state of TCM culture, noting that through "diverse" and integrated development, TCM culture has gained widespread recognition.

Finally, the section outlines opportunities for the development of TCM. TCM must focus on achieving systematic breakthroughs in key areas to overcome development bottlenecks. It is essential to modernize TCM as soon as possible, establishing a complementary and coordinated development framework with Western medicine, thereby providing core support for building a global community of health for humanity.

Through a systematic review and discussion of TCM from all dimensions, this chapter aims to elucidate the precious cultural and practical heritage of TCM's millennia-long tradition, demonstrate its groundbreaking innovative value through deep integration with modern cutting-edge science, and highlight its pivotal role in improving global healthcare systems and advancing the construction of a global health community.

I. HISTORICAL DEVELOPMENT OF TRADITIONAL CHINESE MEDICINE

1. Ancient Origins

The foundation of the ancient Chinese traditional medical system is derived from the integration of philosophical wisdom and medical practice. The Yin and Yang theory represented a significant advancement in the field of dialectics, a framework that established itself as the core paradigm for explaining the law of life through *The Yellow Emperor's Canon of Internal Medicine* (Huangdi Neijing). The theory of Yin and Yang not only describes the dynamic equilibrium of qi and blood flow in the human body as "Yin is tranquil and Yang is firm", but also develops the spatiotemporal diagnostic method: "diagnose the ascent and descent of Yang Qi by day, assess Yin Essence via tongue coating observation by night." The Five elements theory evolved from the material element theory recorded in *the Hongfan chapter of The Book of Documents* into a dynamic model revealing the zang-fu viscera-state relationship in TCM: the Liver, corresponding to the element Wood, governs the free flow of qi and resonates with the vital qi of spring; the Heart, corresponding to Fire, dominates blood circulation and aligns with the seasonal rhythm of summer. This philosophical thought of human-nature isomorphism gave rise to unique technical systems such as the "Midday-Midnight Flow Doctrine" time-acupuncture therapy, which integrates circadian rhythms into acupuncture practice. The compilation of *Shennong's Classic of Materia Medica* marked the systematic establishment of the TCM pharmacology system. Its three-tiered classification framework demonstrates a holistic, ecological approach. The 120 upper-grade herbs, including Ganoderma lucidum (lingzhi) and Panax ginseng, are defined as 'herbs for nourishing life, lightening the body, and prolonging lifespan' and correspond to the cosmic-human cyclic order. The 120 middle-grade tonifying herbs, such as Angelica sinensis (danggui) and Paeonia veitchii (chishao), conform to human physiological rhythms for regulatory treatment. The 125 lower-grade pathogen-expelling herbs, such as Aconitum carmichaelii (fuzi) and Aconitum kusnezoffii (wutou), follow geographical and phenological characteristics for treating severe diseases. This "Three Categories Classification Method" embodies a more comprehensive ecological perspective compared to the plant-centered classification system in ancient Western Galenic medicine.

Archaeological and textual research confirms that the diagnostic and therapeutic system of traditional Chinese medicine had already taken shape during the Warring States period; by the Qin and Han dynasties, with the publication of the Four Classic Texts, it had evolved into a comprehensive theoretical and practical system. This is evidenced by the "Five-Toxin Formula" recorded in the *Liye Qin Slips* (Qin Dynasty), which employed mercury-based preparations to treat sores and ulcers, and the *Mawangdui Silk Manuscripts*' "*Treatise on Pregnancy and Childbirth*" (early Western Han Dynasty), which utilized moxibustion to correct fetal malposition. This theoretical paradigm has maintained innovative vitality for over a millennium: Zhang Zhongjing transformed the Yin-Yang theory into the Six Meridian System to interpret exogenous febrile diseases. Ge Hong's Emergency Prescriptions Kept in the Sleeve recorded a therapeutic method for opisthotonus, pioneering the practice of physical therapy in TCM. The use of datura (mandrake) for anesthesia, as recorded in Li Shizhen's *Compendium of Materia Medica*, predates the clinical use of ether anesthesia in the West by nearly three centuries. This medical tradition, which integrates philosophical speculation with empirical experience, still exhibits unique value in modern times. The translational application of arsenic trioxide for leukemia treatment and the revelation of the neurobiological mechanisms underlying acupuncture analgesia confirm the in-depth dialogue between ancient wisdom and modern science, offering Eastern solutions for the construction of a Global Community of Shared Health for Humanity.

The concept of a global community of health—centered on the principles of "health knows no borders, shared responsibility, and shared benefits"—serves as a vital guide for reforming the global health governance system. As a treasure of traditional Chinese medicine, its holistic view of "harmony between heaven and humanity," its preventive philosophy of "treating disease before it occurs," and its core value of "harmony in diversity" align closely with the underlying logic of the global community of health. It not only provides a systematic approach to addressing global health challenges such as chronic diseases and population aging but also aligns with the global shift in public health from a "disease-centered" to a "health-centered" approach. Its scientific validity has been validated by evidence-based research on Bai Zi Bu Shen Capsules and other studies. By intervening on 10 key biomarkers of aging, it increased telomerase activity in subjects by 76.7%, establishing a systematic intervention paradigm for global anti-aging research (Mei et al., 2024).

The diagnosis of Traditional Chinese Medicine (TCM) is rooted in five primary methods (**Figure 1**): Inspection (observation of physical signs), Olfaction (detection of odors), Auscultation (listening to sounds), Inquiry (assessment of symptoms through questioning), and Palpation (examination through tactile means). The objective of these techniques is to evaluate patients' health holistically. By analyzing bodily cues, sensory information, and subjective experiences. Practitioners integrate findings from these approaches to identify imbalances and develop personalized treatment plans in accordance with TCM principles.

Figure 1. Various types of diagnostic methods present in traditional Chinese medicine.

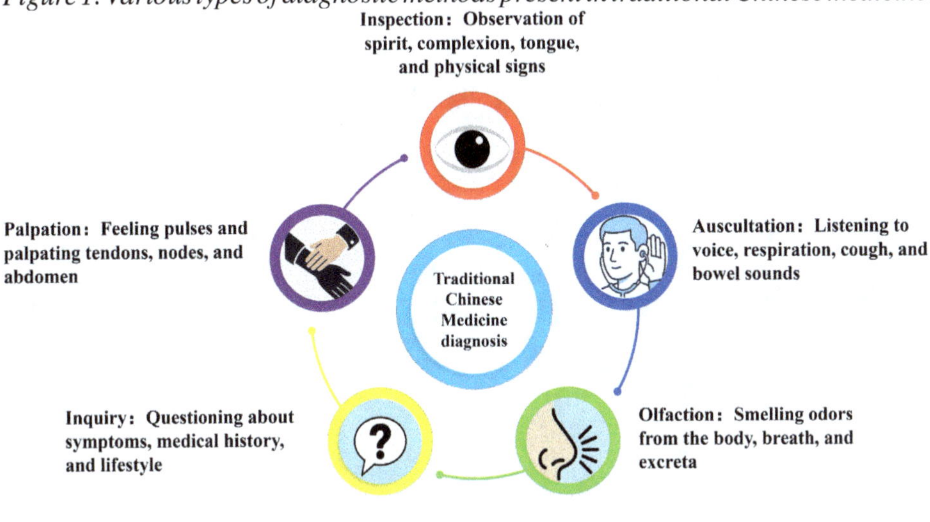

The Traditional Chinese Medicine (TCM) system is a comprehensive medical system developed over thousands of years of clinical practice. All TCM therapies are centered on the holistic view of TCM and the principle of syndrome differentiation and treatment, aiming to achieve the goals of disease prevention, diagnosis, and treatment by regulating the balance of qi and blood, as well as yin and yang, within the human body. It primarily includes external treatment techniques (acupuncture, moxibustion, tuina, and cupping), which regulate bodily functions by stimulating acupoints along the meridians on the body's surface. Modern research has confirmed that these techniques exert their effects through mechanisms such as neural modulation and improved microcirculation, and they are widely applied in pain management, sports injury rehabilitation, and the treatment of gynecological disorders; Herbal therapies (traditional Chinese medicine) follow the principles of "sovereign, minister, assistant, and messenger" in formula composition to achieve personalized treatment. A large number of modern drugs (such as artemisinin) originate from traditional herbal medicines and currently demonstrate unique value in the management of chronic diseases and adjunctive cancer therapy; Lifestyle Intervention Methods (Dietary Therapy, Lifestyle Guidance, Tai Chi, Qigong, Meditation) These embody the TCM philosophy of "preventing disease before it occurs" and are deeply integrated with modern health management. They can effectively improve population health and reduce the risk of chronic diseases. As modern medical research advances, the mechanisms of action for these traditional techniques are gradually being scientifically validated, demonstrating unique practical value in contemporary clinical practice.

Table 1. Traditional Chinese Medicine therapy

Treatment Technique Categories	Key Advantages	Primary Applications	Ref.
External treatment techniques (acupuncture, moxibustion, tuina, cupping)	Fast-acting, with significant localized effects and minimal side effects	Pain-related conditions, musculoskeletal disorders, relief of acute symptoms	Singh, K., et. al. 2024 Deng et al. 2025 Guo, J., et. al. 2024
Medication Therapy (Herbal Medicine)	Holistic approach, personalized treatment, suitable for chronic disease management	Chronic diseases, intractable conditions, and adjunctive cancer therapy	Wang, M., al. 2025 Lee, B. et. al. 2023 Wang, Y., al. 2025
Lifestyle interventions (dietary therapy, lifestyle guidance, tai chi, qigong, meditation)	Prevention-oriented, proactive health, suitable for everyone	Health management, disease prevention, and post-recovery care	Zhu, S. J., et. al. 2026 Siu, P. M., al. 2025 Xu, J., et. al. 2025

The historical roots of Traditional Chinese Medicine (TCM) are deeply connected to the *I Ching* (Yijing, or Book of Changes), a cosmological classic from the Zhou Dynasty (1046–256 BCE) that laid the foundation for TCM's core philosophy. The philosophical foundation of Traditional Chinese Medicine (TCM) theory lies in the theory of yin and yang, which originated in the *I Ching* (Book of Changes), and the theory of the Five Elements, as recorded in the *Book of Documents* (Shang Shu); these two systems were fully integrated during the Warring States period. As early as the early Shang Dynasty (1600–1046 BCE), there was empirical understanding linking illness to an imbalance between heaven and humanity; these insights were gradually systematized during the Western Han Dynasty (206–220 CE), culminating in the compilation of the *Yellow Emperor's Inner Canon* (Huangdi Neijing), which marked the formation of a complete theoretical system of Traditional Chinese Medicine. This text established correspondences between the Five Elements and human physiological, emotional and environmental factors, forming TCM's enduring focus on preventive care and bodily balance. Thus, the *I Ching's* dialectical thinking about universal order, combined with the Five Elements theory, bridges ancient Chinese cosmological cognition and practical medical experience, forming the philosophical basis of TCM's holistic health approach.

I Ching-TCM Integration: The *I Ching* is one of the pivotal ideological sources of the philosophical system of TCM. Its core follows the logical evolution of image-number thought: where Liangyi (Two Poles) generates Sixiang (Four Symbols), which in turn evolves into Bagua (Eight Trigrams). From a modern mathematical perspective, the Eight Trigrams can embody the law of dynamic balance through binary encoding (1 for Yang Yao “ ”, 0 for Yin Yao “ ”) (**Figure 2**). The image-number thinking paradigm of the I Ching has been inherited and developed by Traditional Chinese Medicine (TCM) within the three dimensions of philosophical guidance, cultural inheritance, and clinical practice. Integration of the qualitative cognition of “analogy through observing phenomena” with the logical analysis of “numerical deduction” has enabled the construction of core theoretical frameworks within TCM, including the Five Movements and Six Qi (a Heavenly Stems and Earthly Branches deduction system) and the Zangxiang Theory (a Five Elements attribution model).

Figure 2. The theoretical origin of TCM Xiangshu thinking. Source: Adapted from Qi WANG. (2025)

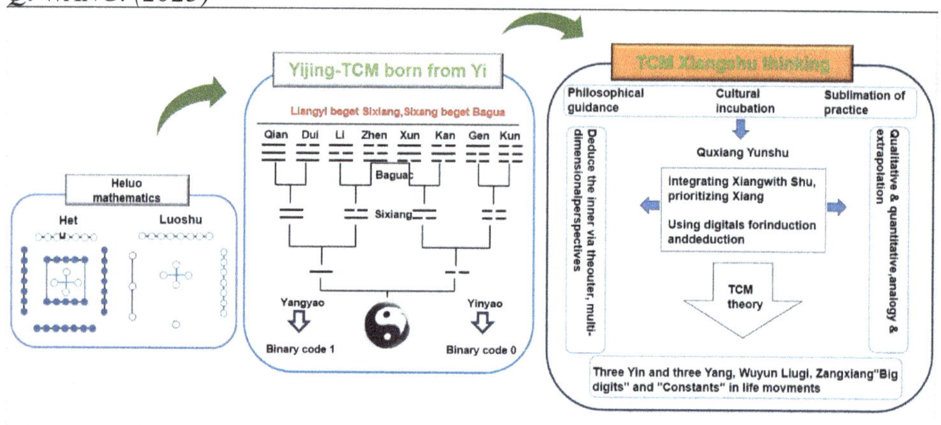

2. Traditional Chinese Medicine in the Dynastic Periods (Medieval)

The systematization, institutionalization, and theoretical refinement of TCM occurred during the medieval and dynastic periods in China (roughly spanning the 6th to 19th centuries). Under imperial patronage and through the efforts of medical scholars, TCM evolved into a sophisticated medical system with enduring philosophical foundations, clinical practices, and pharmacological innovations. Landmark medical texts occurred during the Sui-Tang, Song-Jin-Yuan, and Ming-Qing dynasties, each contributing distinct advancements (**Figure 3**).

Figure 3. The development of traditional Chinese Medicine During the Imperial Period.

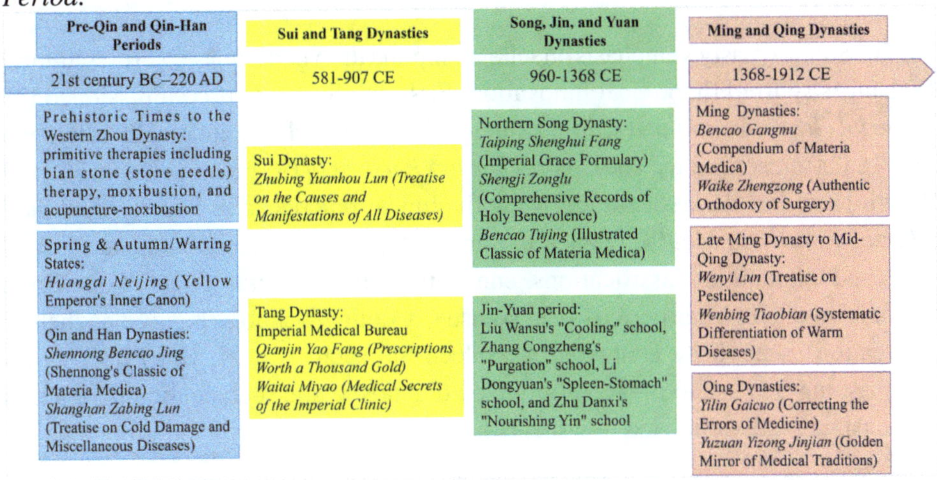

Sui and Tang Dynasties (581-907 CE): Integration and Institutionalization.

The Sui and Tang dynasties marked a golden age of comprehensive medical development, during which imperial support for medical education and standardization propelled advancements in medical practice. China's first state-sponsored medical college (Imperial Medical Bureau) was established during the Tang Dynasty. It standardized curricula for acupuncture, herbal medicine, and diagnosis, training physicians to serve the government. Concurrently, medical texts emerged, such as *Treatise on the Causes and Manifestations of All Diseases* (Zhubing Yuanhou Lun), compiled by Chao Yuanfang, is the first medical monograph to systematically expound the "etiology and syndrome" framework, laying the foundation for the etiology and syndrome differentiation in traditional Chinese medicine. Sun Simiao's *Prescriptions Worth a Thousand Gold* (Qianjin Yao Fang), which became a cornerstone text, emphasizing ethical medical practice, preventive healthcare, and the integration of Daoist and Buddhist healing principles. Wang Tao compiled the *Medical Secrets of the Imperial Clinic* (Waitai Miyao), a text that cataloged over 6,000 formulas, reflecting advancements in pharmacology and disease classification. Furthermore, the Silk Road facilitated exchanges with Persian, Indian, and Central Asian medicine, enriching TCM with new herbs (e.g., saffron) and therapeutic concepts.

Song, Jin, and Yuan Dynasties (960–1368 CE): Scholarly Refinement and Specialization.

The Song Dynasty's emphasis on scholarship and the Yuan Dynasty's cross-cultural exchanges spurred theoretical innovation. First, the Song court commissioned the *Imperial Grace Formulary* (Taiping Shenghui Fang) which systematically cataloged over 16,000 prescriptions. The *Comprehensive Records of Holy Benevolence* (Shengji Zonglu) integrated classical theories with clinical practices. Su Song's *Illustrated Classic of Materia Medica* (Bencao Tujing) introduced detailed botanical illustrations and taxonomic classifications, enhancing herbal accuracy. The "Four Great Schools" of the Jin-Yuan period—Liu Wansu's "Cooling" school, Zhang Congzheng's "Purgation" school, Li Dongyuan's "Spleen-Stomach" school, and Zhu Danxi's "Nourishing Yin" school—revolutionized diagnostic and therapeutic approaches, emphasizing individualized treatments based on constitution and environment.

Ming and Qing Dynasties (1368-1912 CE): Consolidation and Global Exchange.

The late imperial era saw TCM's codification and engagement with global medicine. Li Shizhen's *Compendium of Materia Medic*a (Bencao Gangmu) became a global benchmark, cataloging 1,892 substances with meticulous descriptions of their properties, preparations, and clinical applications. Chen Shigong's *Authentic Orthodoxy of Surgery* (Waike Zhengzong) systematically expounded the etiology, pathogenesis, diagnosis, and treatment of surgical diseases. It holds pioneering significance in the field of traditional Chinese surgical medicine and laid the theoretical foundation for surgical medicine. Additionally, Wu Youxing and later scholars developed the Wenbing school, addressing epidemic febrile diseases through novel diagnostic frameworks (e.g., differentiating "heat" and "dampness") and treatments like *Decoction to Clear Epidemic Toxin* (Qingwen Baidu Yin). *Treatise on Pestilenc* (Wenyi Lune), the first monograph systematically studying acute infectious diseases in the history of Chinese medicine, constitutes the core theoretical foundation of the warm disease doctrine system. *Systematic Differentiation of Warm Diseases* (Wenbing Tiaobian), a comprehensive masterpiece of warm disease studies, has systematically constructed the warm disease diagnosis and treatment system integrating "triple energizer syndrome differentiation" and "defensive-qi-nutrient-blood syndrome differentiation", established standardized therapeutic principles and prescriptions for different stages of warm diseases, serves as the core guiding classic for clinical treatment of exogenous febrile diseases in traditional Chinese medicine since the Ming and Qing dynasties, and remains a compulsory core classic in higher education of Chinese medicine to this day. Anatomical knowledge introduced by European

Jesuit missionaries sparked debates among Qing physicians like Wang Qingren, who sought to reconcile TCM theories with empirical observations in *Correcting the Errors of Medicine* (Yilin Gaicuo). The Qing court compiled the *Golden Mirror of Medical Traditions* (Yuzuan Yizong Jinjian), which became the state-sanctioned manual for acupuncture, pediatric care, and other medical disciplines.

Traditional Chinese Medicine (TCM) is a medical system with a history spanning thousands of years, whose core theoretical framework was first systematically expounded in the classic text *The Yellow Emperor's Inner Canon*. Through centuries of clinical practice and ongoing contemporary research, scholars have developed an increasingly profound understanding of core TCM concepts such as the complementary balance of yin and yang, the laws governing the circulation of qi and blood, and the interrelationships among the Five Elements. This evolution has preserved the essence of traditional theory while continuously incorporating new academic insights. TCM has had a profound influence on East Asian medicine, including Japan's Kampo and Korea's traditional medicine. The impact of TCM on Western medicine began in the 17th century, when Jesuit missionaries translated key texts such as The Compendium of Materia Medica, thereby initiating cross-cultural dialogue between TCM and modern medicine.

3. Transitional Period to Modern Times

The late 19th to mid-20th century marked a pivotal era in the evolution of Traditional Chinese Medicine (TCM), characterized by profound sociopolitical upheavals and the clash of Eastern and Western medical paradigms. During the late Qing Dynasty and the Republic of China period (1912–1949, a historical regime in Chinese history), the introduction of Western medicine by missionaries and colonial powers sparked debates over the legitimacy of TCM. The Republican government, influenced by modernization ideologies, sought to marginalize TCM through policies such as the 1929 proposal to abolish traditional practices, with its proponents dismissing them as "unscientific." However, grassroots resistance emerged as TCM practitioners organized associations, published journals, and emphasized its cultural and clinical value. Simultaneously, early attempts at modernization took shape: the establishment of TCM schools (e.g., the Shanghai Specialized TCM School in 1916) to formalize education, and the adaptation of classical texts like Zhao Xuemin's *Bencao Gangmu Shiyi* to incorporate anatomical terminology and systematic herbal classification. These efforts reflected a dual struggle: preserving TCM's core identity while cautiously embracing scientific methodologies—such as microscopy and chemical analysis—to enhance herbal quality control and win the trust and support of a skeptical public.

The post-1949 era under Communist rule ushered in a state-sponsored revival of TCM, driven by policies like the "Uniting Chinese and Western Medicine" campaign. The government institutionalized TCM through state-run hospitals, universities (e.g., Beijing University of Chinese Medicine, founded in 1956), and research centers, framing it as a national heritage and a complementary system to biomedicine. Scientific validation became a priority, with studies on herbal pharmacology (e.g., Artemisia annua's antimalarial properties) and acupuncture's neurophysiological mechanisms gaining global traction. Yet, globalization posed new challenges. TCM faced international skepticism due to its theoretical divergence from biomedicine, concepts like qi and meridians lacked standardized empirical validation and regulatory barriers in standardizing herbal safety for global markets. Despite achieving milestones like the WHO's inclusion of TCM in ICD-11 (2019), tensions persisted between modernization and cultural authenticity. Advocates stressed adherence to classical texts (e.g., Huangdi Neijing) and holistic philosophies, while critics demanded evidence-based reforms. This transitional period thus encapsulates TCM's resilience and adaptability, navigating both historical continuity and the demands of a globalized scientific era.

Figure 4 illustrates the distribution of academic research during the transitional period from Traditional Chinese Medicine to modernity. Research in this era focused heavily on validating clinical efficacy: 30% was dedicated to the efficacy of Chinese patent medicines, 9% to acupuncture efficacy, and 4% to other therapies, 13% explored the therapeutic mechanisms of TCM, and 9% interpreted the theoretical connotations of TCM, and 3% investigated the biological basis of acupuncture. Additionally, 32% of research efforts were dedicated to scientifically interpreting TCM principles through modern science——leveraging TCM's unique strengths in disease prevention——and treatment, and enhancing innovation across the entire TCM development chain. Despite progress, deeper exploration remains necessary in areas such as holistic theory interpretation (9%) and the biological basis of acupuncture (3%) to align ancient practices with evidence-based medical frameworks.

Figure 4. The distribution of various types of candidate academic progress. Source: Adapted from Xiaoxiao Zhang. (2025)

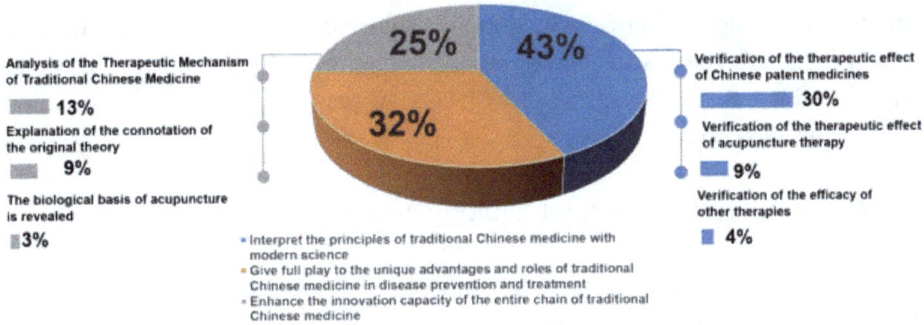

II. CONTEMPORARY DEVELOPMENT OF TRADITIONAL CHINESE MEDICINE

1. Integration with Modern Science and Technology

The Use of Modern Analytical Techniques in TCM Research

Modern analytical techniques, such as chromatography, have enabled the precise identification and quantification of the chemical constituents of traditional Chinese medicine, thereby driving innovation in research methods for traditional Chinese medicine. High-performance liquid chromatography (HPLC) and gas chromatography (GC), as the most widely used chromatographic techniques today, enable the efficient separation, qualitative identification, and quantitative analysis of various bioactive components in the complex matrices of traditional Chinese medicines. These methods are particularly well-suited for analyzing complex mixtures found in traditional Chinese medicine formulations, which often contain multiple components, low concentrations, and significant differences in polarity. For instance, HPLC has been instrumental in standardizing Panax ginseng extracts by quantifying ginsenosides, the primary active constituents linked to immunomodulatory and anti-fatigue effects. In addition, ultra-high-performance liquid chromatography–mass spectrometry (UPLC-MS), with its higher chromatographic resolution, lower detection limits, and faster analysis speed, enables the efficient screening and structural identification of trace active compounds in complex traditional Chinese medicine systems. It is currently widely used in research on the discovery of unknown components in commonly used traditional Chinese medicinal materials such as Salvia miltiorrhiza. Modern pharmacological studies have confirmed that the lipophilic tanshinone compounds found in Salvia miltiorrhiza possess distinct cardiovascular activities, including cardiomyoprotection, anti-myocardial ischemia, and improvement of microcirculation. These technologies not only ensure batch-to-batch consistency but also validate the holistic principles of TCM by mapping synergistic interactions among multiple components.

The Application of Molecular Biology Methods

The rapid advancement of modern molecular biology technologies has provided strong support for the scientific interpretation and innovative development of traditional Chinese medicine, enabling a deeper understanding of the therapeutic mechanisms of traditional Chinese herbal formulas at the molecular and cellular levels. A systematic application framework has now been established, covering efficacy validation, mechanism-of-action analysis, quality control, and safety assessment. By employing cutting-edge technologies such as transcriptomics, proteomics, and CRISPR gene editing, researchers can precisely elucidate how bioactive components in Chinese herbal medicines regulate key cellular signaling pathways.

Multi-omics integration technology (genomics, transcriptomics, proteomics, metabolomics) combined with mass spectrometry imaging (MSI) and bioinformatics analysis has enabled systematic decoding of the synergistic mechanisms of TCM, characterized by "multiple components - multiple targets - multiple pathways." Taking Lianhua Qingwen as an example, multi-omics analysis identified core components such as chlorogenic acid and baicalin, confirming that they combat pneumonia by regulating tumor necrosis factor-α (TNF-α), interleukin-6 (IL-6), and the TLR4/NF-κB signaling pathway, thereby providing multidimensional evidence for the screening of quality markers (Wang et al., 2024; Chen et al., 2025). Metabolomics studies of Ge Gen Qin Lian Tang demonstrated that compounds such as puerarin lower blood glucose levels by correcting glucose metabolism disorders (Wang et al., 2024). TCM metabolomics constructs a "syndrome-metabolism-drug" network, enabling the molecular quantification of "pattern differentiation and treatment" and the objective evaluation of therapeutic efficacy for Jinkui Shenqi Wan (Wang et al., 2024). Single-cell omics reveals the mechanism by which astragalus polysaccharides target macrophage and Treg cell subsets to regulate the immune microenvironment (Wu et al., 2025).

Molecular recognition technologies (including affinity ligand screening, molecular docking, surface plasmon resonance (SPR), biosensor interference (BLI), ultrafiltration, etc.) effectively address core challenges in screening active components and validating targets for traditional Chinese medicine. Compared to traditional separation and screening methods, these technologies significantly enhance efficiency and simplify operational workflows. In a study investigating the mechanism of action of the traditional Chinese medicine Lycium barbarum oil injection on acute lung injury, 11 key sesquiterpene bioactive compounds were first identified using ultra-high-performance liquid chromatography–mass spectrometry (UHPLC-MS); subsequently, network pharmacology analysis identified target molecules such as Usp18 and Map3k7; Finally, molecular docking and surface plasmon resonance (SPR) confirmed that the sesquiterpene bioactive components can stably bind to target proteins such as MAPK8, clarifying the regulatory relationship between "bioactive components - target - anti-inflammatory effect." The active compounds were shown to stably bind to targets such as MAPK8, establishing a "component-target-anti-inflammatory" relationship and revealing their mechanism of action in improving acute lung injury via Toll-like receptor (TLR) and NF-κB pathways. In the field of screening for active components in traditional Chinese medicine, affinity-based screening technologies have successfully identified representative bioactive compounds, such as berberine and baicalin, from commonly used herbal medicines like Coptis chinensis and Scutellaria baicalensis. Interactions are validated using SPR/BLI and molecular docking, providing evidence for the therapeutic mechanisms underlying diseases such as ulcerative colitis and diabetes (Wu et al., 2022; Li et al., 2025). Network pharmacology translates the traditional Chinese medicine theory of "sovereign, minister, assistant, and messenger" herbs into quantifiable molecular networks. For example, in a study of Astragalus and Cinnamon Twig Decoction (Wuwu Tang), eight core blood-penetrating components, including paeoniflorin, were identified via UPLC-HRMS. Network pharmacology was used to identify STAT3 as a key target, and experiments confirmed that it treats autoimmune arthritis by regulating the JAK/STAT pathway and balancing Th17 and Treg cells (Zhang et al., 2025).

Core technologies such as gene editing, molecular identification, and omics analysis provide foundational support for the precise construction, functional validation, and scenario-based application of organoids and organ-on-a-chip systems. Conversely, these systems offer molecular biology research validation scenarios that more closely mimic the human physiological environment. Together, they drive the application of fundamental research findings in clinical diagnosis, treatment, and new drug development. Organoid and organ-on-a-chip technologies provide highly realistic in vitro models for evaluating the efficacy and safety of traditional Chinese medicine, significantly enhancing the clinical translational value of research. Experiments using intestinal organoids demonstrated that Shenling Baizhu San can repair the intestinal epithelial barrier by the tight junction proteins occludin and ZO-1. Studies on liver organoids showed that compounds such as emodin in Epimedium regulate liver toxicity via the Nrf2/ARE pathway, providing molecular evidence to support risk reduction in clinical settings (Li et al., 2025).

DNA barcoding molecular identification technology has established a "molecular identification" system for traditional Chinese medicines based on specific gene sequences such as ITS2, rbcL, and matK. It can accurately identify medicinal materials such as ginseng, American ginseng, and notoginseng, as well as easily confused plants like plantain and echinacea. By comparing differences in gene sequences, it effectively distinguishes genuine products from adulterated, substituted, or hybrid varieties, fundamentally resolving issues of source confusion and eliminating intentional or unintentional contamination of low-quality raw materials in the TCM supply chain. Combined with high-throughput sequencing technology, this method has evolved into DNA metabolic barcoding technology, overcoming the limitation of traditional DNA barcoding that could only identify single components. This technology is used to verify the consistency of raw material sources in multi-component formulations, ensuring the stability of active ingredient content and therapeutic efficacy (Raclariu-Manolică & de Boer, 2022). Proteomics technologies provide crucial technical support for elucidating the multi-target regulatory properties of traditional Chinese medicine and revealing the mechanisms underlying its organ-protective effects. Proteomic analysis has shown that Qishen Yqi Pills improve ischemic heart failure by inhibiting mitochondrial fission (Li et al., 2025).

In summary, molecular biology techniques have established a comprehensive research framework, transforming traditional Chinese medical theory into scientific evidence through triple verification. The deep integration of molecular biology techniques with traditional Chinese medicine (TCM) research has not only elucidated the multi-component, multi-target mechanisms of action of TCM at the molecular level, but has also advanced TCM quality control from traditional identification methods to precise molecular-level quality control. This shift has driven a transition in research paradigms from empirical summarization to a comprehensive "material basis–mechanism of action–clinical validation" model, thereby laying the technical foundation for the establishment of international standards for TCM and the promotion of evidence-based practices.

Figure 5 presents a research design approach for evaluating the therapeutic effects of TCM combined with modern technology and databases.

Herb-Symptom Network Analysis: TCM herbs function by interacting with symptom-associated genes within the human protein interactome network (18,505 proteins, 327,924 interactions), where herb-target proximity predicts therapeutic effectiveness, a relationship validated through 12,474 herb-symptom indications and 1,936 hospital inpatient records.

Data-Driven Validation: By integrating symptom-gene associations (174 symptoms, 110,407 links) and herb-target databases (798 herbs, 2,270 targets), researchers identified 50 novel network-proximal herbs with clinically validated efficacy, bridging molecular patterns and real-world patient outcomes.

Predictive Framework: Symptom localization and shorter symptom-gene distances enable co-occurrence predictions, offering a systematic approach to optimize TCM herb prescriptions and uncover new therapeutic applications through computational and clinical synergy.

Figure 5. Study design. Source: Adapted from Xiao Gan et al. (2023).

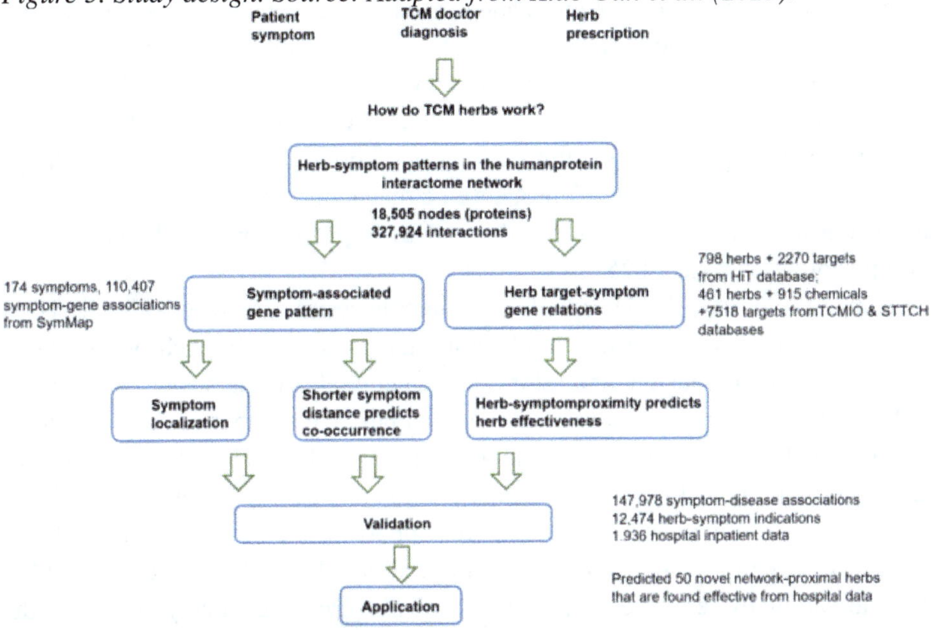

2. Clinical Practice and Evidence-Based Medicine

In the modern era, the clinical practice of Traditional Chinese Medicine (TCM) is undergoing profound transformations, with its integration with Evidence-Based Medicine (EBM) emerging as a crucial development trend. This integration not only enriches the clinical application of TCM but also offers scientific validation for its global recognition and dissemination.

Traditional Chinese Medicine (TCM) is a traditional medical system with a history spanning thousands of years. Its theoretical and practical frameworks have always been guided by clinical needs, giving rise to a distinctive diagnostic and therapeutic paradigm. In clinical practice, TCM practitioners conduct a systematic assessment of a patient's physiological condition and disease characteristics by integrating the five diagnostic methods: observation, auscultation, olfaction, inquiry, and palpation. Subsequently, guided by fundamental TCM theories such as the theory of yin-yang balance, the Five Elements theory, and the theory of meridians, they perform pattern differentiation and treatment to develop personalized intervention plans for patients. Common therapies include compound herbal medicine, acupuncture, moxibustion, and tuina (a distinctive TCM therapeutic technique). Evidence-based research confirms that, through cross-cultural adaptation, Chinese diagnostic techniques have evolved into a global system. Pulse diagnosis, tongue diagnosis, and meridian diagnosis have been refined in various regions and standardized through evidence-based validation within the context of traditional East Asian medicine, thereby providing support for clinical prescriptions and pattern differentiation (Kadier et al., 2025). Traditional Chinese medicine intervention systems, which utilize methods such as herbal medicine and acupuncture, exhibit regional variations: Japan's Kampo medicine focuses on health maintenance, Europe's electroacupuncture is used for pain relief, and Southeast Asia's moxibustion and tuina are employed for the rehabilitation of chronic conditions (Lu, 2025).

In modern society, TCM hospitals and clinics have proliferated worldwide. TCM clinical practice has expanded beyond traditional boundaries, from chronic diseases like diabetes, hypertension, and arthritis to functional disorders such as insomnia and anxiety. Over the past two decades, the cumulative number of clinical studies on traditional Chinese medicine formulas published in JCR Q1-ranked journals has increased more than 40-fold. The standardization and quality of research evidence have continued to improve, providing core support for the development of a high-quality evidence-based medicine system for traditional Chinese medicine. In 2024, the internationally renowned medical journal *The Lancet* officially published the results of a multicenter, large-scale, randomized, double-blind, controlled clinical trial evaluating the traditional Chinese medicine compound FYTF-919 for the treatment of acute intracerebral hemorrhage, marking a significant milestone in the field of evidence-based research in traditional Chinese medicine. This marks the first time the journal has published a multicenter study on a Chinese herbal formula. There is now clear evidence supporting the application of Traditional Chinese Medicine (TCM) in the treatment of acute and chronic diseases as well as functional disorders, and some of these research findings have been incorporated into international guidelines (Nielsen et al., 2024). For example, in the treatment of digestive system diseases, TCM herbal formulas demonstrate remarkable efficacy in regulating gastrointestinal function and improving symptoms, gaining the trust of numerous patients. Qing Dai's active component indigo treats ulcerative colitis through a microbiota-metabolite-immune pathway: it promotes Roseburia intestinalis, elevates butyrate, regulates T cells via mTORC1/HIF-1α to restore immune balance, with clinical studies verifying their negative correlation with UC (Ulcerative colitis) severity (Xiao et al., 2024).

Evidence-based medicine (EBM) advocates that clinical treatment decisions should not rely solely on a physician's experience, but rather combine the most reliable current clinical research evidence, the physician's professional expertise, and the patient's own needs and preferences to develop the most appropriate treatment plan for the patient. In recent years, integrating the principles and methods of evidence-based medicine into the field of traditional Chinese medicine has become a clear direction for the industry's development, helping to make TCM diagnostic and treatment protocols more standardized and scientific, and thereby more easily gain international recognition.

Firstly, in terms of clinical research, more and more TCM studies are employing scientific methodologies aligned with EBM requirements. Randomized controlled trials (RCTs) are widely recognized in the field of evidence-based medicine as the "gold standard" for evaluating the efficacy and safety of treatment regimens. In recent years, an increasing number of clinical studies on traditional Chinese medicine have begun to adopt this methodology to validate the practical value of TCM treatment regimens with more scientific and compelling evidence. For instance, numerous RCTs have been conducted on the effectiveness of acupuncture in pain relief. ACUITY is the first multicenter feasibility study to evaluate adherence to acupuncture treatment. The study recruited 79 patients with acute pain from three emergency departments in the United States and achieved a 98.1% adherence rate, thereby validating the study's feasibility and advocating for the assessment of adherence. Its limitations include a small sample size and the lack of cross-center quality control (Xing et al., 2025). These studies not only provide reliable evidence for the analgesic effect of acupuncture but also contribute to standardizing the operation procedures and treatment parameters of acupuncture.

Second, the development of evidence-based clinical practice guidelines for traditional Chinese medicine has advanced steadily. In 2011, with funding from the World Health Organization's Western Pacific Region, the China Academy of Chinese Medical Sciences, in collaboration with professional academic organizations such as the Chinese Association of Traditional Chinese Medicine and the Chinese Association of Acupuncture and Moxibustion, compiled and published China's first monograph, *Evidence-Based Clinical Practice Guidelines for Traditional Chinese Medicine*. This marks the first clinical practice guideline in the field of traditional Chinese medicine to be developed through international cooperation and based on international evidence-based methodology standards. These guidelines adhere to the tripartite principle of "evidence + experience + patient values," utilizing evidence-based medicine methodologies as a framework. By integrating modern randomized controlled trials, systematic reviews, and other evidence-based evidence, while also incorporating traditional resources such as ancient Chinese medical texts and the experience of renowned senior practitioners, an exclusive evidence-grading system tailored to traditional Chinese medicine has been established. These guidelines standardize clinical practices through unified diagnostic and therapeutic frameworks, reducing treatment variability; integrate traditional TCM experience with modern evidence-based data to provide personalized precision treatment plans for different diseases; guide TCM research toward evidence-based transformation, advancing the discipline's modernization; and support the international dissemination of TCM through a scientific system. For example, in treating coronary heart disease, TCM clinical guidelines combine traditional syndrome differentiation and treatment principles with modern scientific evidence to deliver more comprehensive and effective therapeutic approaches.

The integration of EBM into traditional Chinese medicine (TCM) clinical practice faces multiple practical challenges. The personalized nature of Traditional Chinese Medicines "differential diagnosis and treatment" approach differs fundamentally from the standardized clinical trial paradigm relied upon by evidence-based medicine, presenting a core methodological challenge for current evidence-based research in Traditional Chinese Medicine. TCM emphasizes dynamic treatment tailored to the individual, timing, and environment, making it difficult to comprehensively validate its clinical value through trial designs with uniform inclusion criteria and fixed intervention protocols. Furthermore, the complex composition of traditional Chinese herbal formulas and their multi-targeted mechanisms of action remain incompletely elucidated. Meanwhile, EBM's efficacy evaluation systems primarily depend on singular quantitative indicators (such as biochemical markers or mortality rates). This makes it difficult to precisely quantify and evaluate TCM's strengths in holistic regulation, symptom improvement, and quality-of-life enhancement, further complicating the integration of the two approaches.

To overcome the challenges of integration, interdisciplinary collaboration and innovation are essential. By integrating the clinical experience of TCM practitioners, the methodological strengths of evidence-based medicine researchers, the mechanistic research capabilities of pharmacologists, and the data analysis expertise of statisticians through interdisciplinary collaboration, it is possible to develop innovative research methods tailored to the unique characteristics of TCM—such as its personalized "differential diagnosis and treatment" approach and the dynamic evolution of syndromes. Chengdu University of Traditional Chinese Medicine (Sichuan Province, China) has taken the lead in developing the TCMKD platform to address common challenges in the field of Traditional Chinese Medicine research, such as the long-standing issues of scattered clinical resources, high data heterogeneity, and inconsistent data quality. This platform integrates nearly 10,000 proprietary Chinese medicines, over 3,800 herbal medicines, 36,000 components, 18,000 targets, and

23,000 diseases, constructing a knowledge graph comprising 143,000 entities and 3.6 million relationships. Its reliability has been validated through the analysis of lung cancer treatment formulas, leading to the identification of core drug pairs and the elucidation of their mechanisms of action. Moreover, leveraging network meta-analysis techniques enables systematic comparisons of efficacy differences among multiple intervention strategies, such as TCM compound formulas, acupuncture, and tuina massage, across various disease stages, thereby breaking through the limitations of traditional randomized controlled trials (RCTs) with single controls. At the same time, we should conduct in-depth analyses of the pharmacologically active components and core mechanisms of action in traditional Chinese medicine formulas, elucidating their holistic regulatory characteristics—characterized by "multiple components, multiple targets, and multiple pathways"—at the molecular level, thereby providing precise and reliable scientific evidence for the development and refinement of an evidence-based evaluation system for traditional Chinese medicine.

The deep integration of traditional Chinese medicine and evidence-based medicine holds immense potential for development and practical application. With the introduction of cutting-edge methods such as omics technologies and artificial intelligence, the clinical evidence base for TCM will continue to improve. The combination of traditional experience and modern science will provide a more robust quantitative basis for the efficacy of TCM, thereby significantly enhancing its academic credibility and influence within the global medical community. This will also promote the deep integration of traditional wisdom and modern science, accelerating the modernization and internationalization of TCM. Integrating evidence-based medicine into TCM clinical practice represents a significant advancement in the modernization of TCM. Although numerous challenges remain, through continuous exploration and innovation, this integration is bound to inject new vitality into the development of TCM and provide more effective treatment options for patients worldwide.

3. Global Expansion

Against the backdrop of globalization, the global influence of Traditional Chinese Medicine (TCM) continues to expand; it has evolved from a regional traditional medical system into an integral part of the international healthcare system. This global expansion not only showcases the unique charm and effectiveness of TCM but also promotes cultural exchanges and cooperation within the medical field.

The global spread of TCM began with the establishment of medical institutions abroad. Currently, over 183 countries and regions worldwide have established TCM clinics and hospitals. In the United States, for example, there are more than 50,000 licensed acupuncturists, and acupuncture has been widely covered by health insurance in many states. These medical institutions offer a variety of traditional Chinese medicine services, including acupuncture, herbal medicine, and tuina massage, to meet the healthcare needs of local residents and gradually reshape their perceptions of traditional Chinese medicine. Educational cooperation is another important driving force for the global expansion of TCM. Many universities and educational institutions around the world have introduced TCM-related majors and courses. In the United Kingdom, the University of Westminster offers undergraduate programs in Chinese acupuncture, integrating traditional TCM theory with modern scientific research methods. Through these educational programs, more and more foreign students have the opportunity to systematically learn TCM knowledge, fostering a growing pool of TCM professionals overseas. This not only promotes the inheritance and development of TCM but also helps TCM gain a firm foothold in the international academic community.

In addition to conventional pathways including the cross-border establishment of TCM medical facilities and international collaborative initiatives in TCM education, high-quality scientific research on traditional Chinese medicine (TCM) has become a critical determinant facilitating its global diffusion and the elevation of its international acceptance. As depicted in the accompanying diagram (**Figure 6**), research on TCM formulas like Xiaoyao San (XYS) is exploring the underlying mechanisms of its efficacy. Such studies explore how Traditional Chinese Medicine interacts with the body's physiological systems (such as the hypothalamic-pituitary-adrenal [HPA] axis), specifically through the interactions between various hormones and their receptors (corticotropin-releasing hormone [CRH], adrenocorticotropic hormone [ACTH], cortisol [CORT], glucocorticoid receptor [GR], corticotropin-releasing hormone receptor 2 [CRHR2]), as well as the involvement of neuropeptides such as apelin and its receptor APJ. These scientific explorations have not only deepened our understanding of the efficacy of TCM but have also provide a scientific basis for its integration into the global mainstream medicine system. When international medical researchers and the public see the scientific evidence behind TCM, their trust in and acceptance of it will increase significantly. For instance, research on how TCM modulates the blood-brain barrier (as shown in the figure, involving tight junction proteins such as Occludin, Claudin, and ZO-1) has opened up new avenues for the treatment of neurological disorders.

Moreover, international cooperation in TCM research is on the rise. Scientists from different countries are collaborating to conduct large-scale randomized controlled clinical trials of TCM therapies. These trials follow international-standard research protocols and can further verify the safety and effectiveness of TCM. Once the results are recognized by the international medical community, TCM can be more widely applied in global healthcare. This not only benefits patients worldwide but also consolidate TCM's position in the international medical arena, making it an indispensable part of the global healthcare system in the era of globalization.

Figure 6. Potential relationship between Xiao yao san, depression and HPAaxis. Source: Adapted from Jianbei Chen et al. (2022).

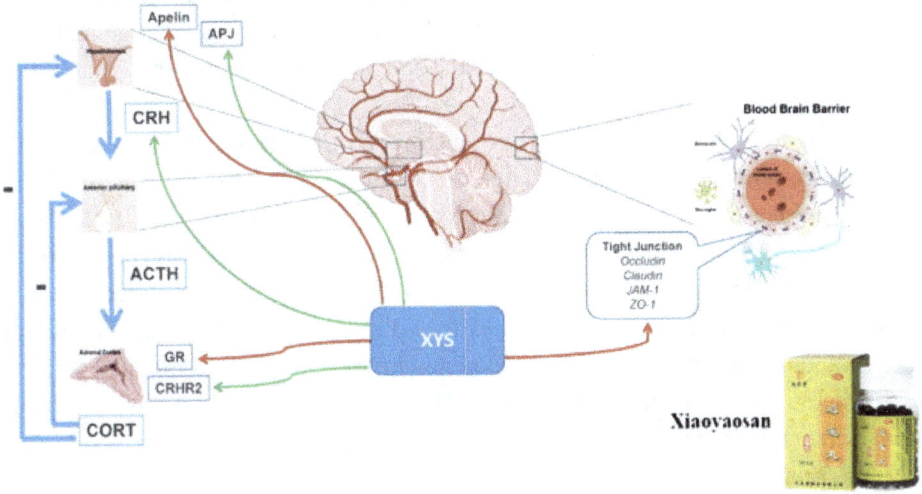

Top-level institutional design and policy coordination provide core support for the global promotion of traditional Chinese medicine. The World Health Organization has officially included traditional medical systems, including traditional Chinese medicine, in the International Classification of Diseases, 11th Revision (ICD-11). This landmark development has laid a solid institutional foundation for the international standardization and global development of traditional Chinese medicine. At the same time, many countries have actively promoted cooperation and exchange in TCM through government-led initiatives. China has established a series of TCM cooperation centers worldwide, such as the China-Hungary Traditional Chinese Medicine Center, which advances the clinical application, scientific research, and talent development of TCM through in-depth bilateral cooperation.

However, the global expansion of TCM also faces multiple persistent challenges. One of the main difficulties is the absence of unified international standards. The complexity of the ingredients in traditional Chinese herbal preparations and the multidimensional nature of traditional Chinese medical treatments pose the core challenges currently facing the establishment of a unified international quality control system and clinical efficacy evaluation standards. In addition, cultural differences hinder the promotion of TCM. Concepts such as Yin-yang balance and the Five-element theory in TCM are deeply rooted in traditional Chinese culture, are often difficult for Western societies to understand and accept in the short term. To address these systemic challenges, strengthening international collaborative governance in the field of standard-setting has become a strategic priority for the global development of traditional Chinese medicine. International organizations, governments, and TCM research institutions should cooperate to develop scientific, evidence-based international standards for TCM. In terms of cultural communication, more vivid and easy-to-understand popular science materials and cultural exchange activities should be organized to help international audiences better grasp the connotation and value of TCM.

Looking to the future, the global expansion of TCM has broad prospects. With the growing demand for natural and holistic medical approaches worldwide, TCM, with its unique theoretical system and diverse treatment modalities, will attract more attention. Through continuous scientific research innovation and cross-border academic collaboration, traditional Chinese medicine (TCM) is expected to play a more prominent role in improving the health and well-being of the global population and advancing the interdisciplinary development of international medical science.

In conclusion, the global expansion of TCM is an inevitable trend in the modern advancement of traditional Chinese medicine. Although it faces many challenges, with the joint efforts of all parties, TCM is bound to play an increasingly important role on the global healthcare stage and bring greater health benefits to people around the world.

III. PERFORMANCE EVALUATION OF TRADITIONAL CHINESE MEDICINE

1. Efficacy Evaluation

The evaluation of traditional Chinese medicine (TCM) efficacy is a crucial aspect of its research and development, serving as a key determinant of its clinical application and global recognition. As TCM continues to gain traction in modern healthcare, establishing a scientific and comprehensive efficacy evaluation system has become an imperative task.

Historically, TCM efficacy evaluation was deeply rooted in traditional practices. Traditional Chinese medicine practitioners use the Five Diagnostic Methods (observation, auscultation and olfaction, inquiry, and palpation) to assess a patient's condition before and after treatment. Changes in symptoms, such as reduced pain, improved appetite, or restored sleep quality, served as key indicators of treatment efficacy. The evaluation was also closely tied to TCM's concepts of syndrome differentiation. For example, in treating a patient with "spleen-stomach deficiency cold" syndrome, the resolution of symptoms like abdominal distension, loose stools, and cold intolerance, along with the restoration of the body's yin-yang balance and qi-blood harmony, was considered a sign of effective treatment. This traditional evaluation method, although rich in experience, lacked standardization and objectivity, making it difficult to communicate TCM's efficacy in the international medical community.

With the development of modern medicine and the widespread adoption of Evidence-Based Medicine (EBM) principles, the evaluation of TCM efficacy is gradually transitioning toward a standardized and scientific paradigm. Randomized controlled trials (RCTs), as the gold standard of modern clinical research, have become a key method for validating the efficacy of traditional Chinese medicine, and their scope of application continues to expand. For instance, in studies investigating TCM for mild to moderate depression, multiple RCTs employed head-to-head designs to directly compare the clinical efficacy of TCM formulas against placebos and first-line antidepressants. These studies not only used standardized tools such as the Hamilton Depression Rating Scale (HAMD) and the Montgomery-Åsberg Depression Rating Scale (MADRS) to quantitatively assess the degree of symptom improvement, but also revealed TCM's physiological mechanisms by measuring serum neurotransmitter levels (e.g., serotonin, dopamine) and incorporating brain functional imaging metrics. Through rigorous randomization, double-blinding, and variable control, the RCT effectively minimizes bias risks. It provided reproducible, verifiable evidence-based medical proof of TCM efficacy, advancing broader international recognition of TCM research outcomes within the medical community. A team led by Professor Cheng Lu from the China Academy of Chinese Medical Sciences and Professor Zhongde Zhang from Guangdong Provincial Hospital of Chinese Medicine conducted a systematic review and meta-analysis of 18 high-quality randomized controlled trials (RCTs) involving a total of 1,353 patients with rheumatoid arthritis-associated interstitial lung disease (RA-ILD). The results demonstrated that the remission rate of integrated traditional Chinese and Western medicine therapy for RA-ILD was 3.69 times higher than that of monotherapy, providing evidence-based medical support for this clinical regimen (Lu et al., 2024).

In addition to RCTs, multicenter clinical trials and real-world studies (RWS) have become crucial extensions of the efficacy evaluation system for Traditional Chinese Medicine (TCM), addressing the limitations of single-study designs from multiple dimensions. Multicenter clinical trials, by coordinating research across multiple medical institutions, are able to recruit larger patient groups with more diverse clinical characteristics. This effectively avoids geographical bias and sample homogeneity issues inherent in single-center studies, significantly enhancing the generalizability of research conclusions. Such trials provide more representative evidence to support the widespread application of TCM therapies. Real-world studies, conducted in routine clinical practice, observe the actual application patterns of TCM within real healthcare environments. They comprehensively incorporate complex clinical variables such as patient adherence to treatment, underlying comorbidities, and the combined use of different TCM therapies, thereby authentically reflecting efficacy and safety under natural clinical conditions. For instance, in acupuncture studies for chronic pain, integrating multi-center electronic health records, insurance payment data, and long-term patient follow-up outcomes enables systematic analysis of treatment efficacy differences across distinct patient subgroups (e.g., elderly patients and those with comorbidities), while also monitoring rare adverse events to provide an evidence-based foundation for developing personalized clinical treatment plans. In the field of idiopathic pulmonary fibrosis (IPF) treatment, a team led by Professor Jiansheng Li from Henan University of Chinese Medicine conducted a 26-week multicenter, randomized, double-blind trial that enrolled 80 patients with stable IPF. The study results not only verified the therapeutic effect of traditional Chinese medicine (TCM) on stable-stage IPF, but also established a standardized research paradigm for evaluating the efficacy of TCM in the treatment of rare pulmonary diseases (Li et al., 2022).

However, evaluating the efficacy of traditional Chinese medicine (TCM) poses numerous challenges due to its complexity. For instance, the use of multi-component, multi-target therapies makes it difficult to identify key active ingredients. Furthermore, the holistic nature of TCM efficacy assessment makes it challenging to quantify efficacy indicators using traditional methods. To address these challenges, innovative evaluation methods are urgently needed. For example, integrating omics and network pharmacology to obtain objective evidence, and developing new evaluation indicators that combine TCM theory with modern medical metrics.

Figure 7. Chest computed tomographic manifestations at enrolment and after treatment. Source: Adapted from Ke Hu et al. (2021).

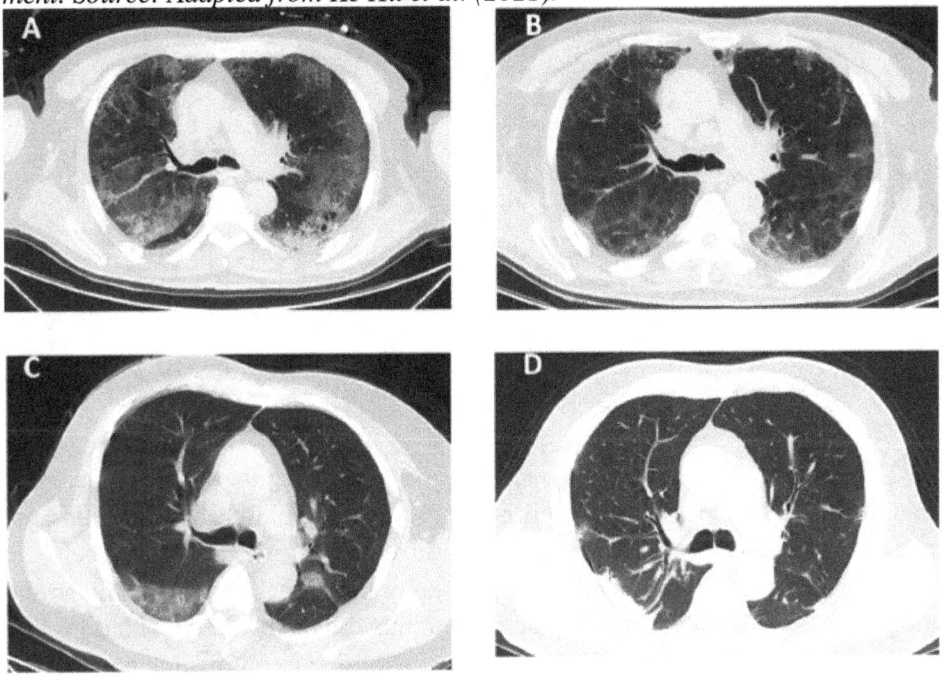

With the advancement of bioimaging technology, the efficacy of traditional Chinese medicine can now be visually demonstrated, enhancing its persuasiveness. For instance, based on data from this multicenter randomized controlled trial, combining Lianhua Qingwen capsules with standard treatment significantly improved clinical symptoms in COVID-19 patients. Lung CT images clearly demonstrate that patients treated with Lianhua Qingwen Capsules in combination showed greater improvement in lung condition compared to the control group (as defined by the "Diagnosis and Treatment Protocol for Novel Coronavirus Pneumonia (4th Edition)") (**Figure 7**). Furthermore, statistical analysis of CT improvement data and cure rates indicates that the treatment group demonstrated significantly higher rates of lung recovery than the control group in both the full analysis set (FAS) and the per-protocol set (PPS) (**Figure 8**).

Figure 8. The rate of clinical cure and the improvement of chest computed tomographic imaging. Source: Adapted from Ke Hu et al. (2021).

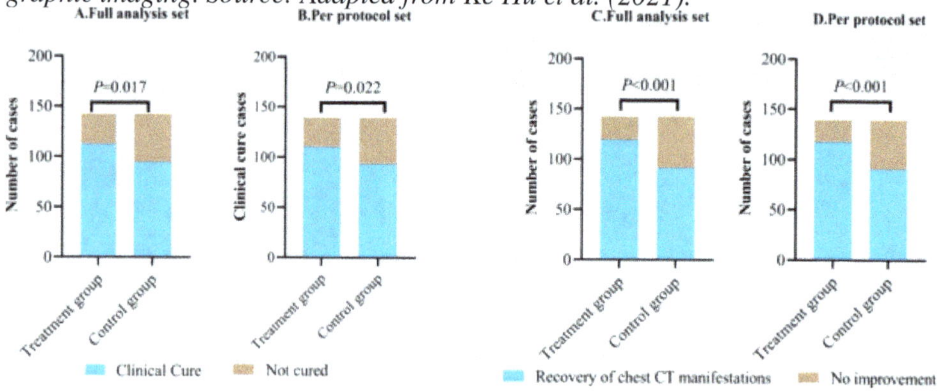

The application of bioimaging technology in evaluating the clinical efficacy of traditional Chinese medicine not only provides a quantifiable and objective measure of therapeutic outcomes but also effectively bridges the gap in understanding between the holistic and empirical diagnostic principles of traditional Chinese medicine and the precise, standardized technical systems of modern medicine. TCM has historically emphasized the function of lung qi circulation and its correlation with overall health status. By documenting physical changes in the lungs via CT scans, we can better connect these alterations with TCM theoretical understandings of pulmonary diseases and their treatment.

In conclusion, the efficacy evaluation of traditional Chinese Medicine is in a stage of continuous development and improvement. By combining traditional experience with modern scientific methods and constantly exploring and innovating evaluation systems, we can more accurately assess the efficacy of TCM, promote its scientific development, and enhance its status in the global medical field.

2. Safety Evaluation

The safety evaluation of traditional Chinese Medicine (TCM) is of paramount importance in its modern development. As TCM gains broader acceptance globally, ensuring the safety of its use becomes a critical factor in maintaining public health trust and promoting its sustainable integration into mainstream healthcare. This evaluation encompasses understanding potential risks, adverse reactions, and the safe application of TCM therapies and herbal medicines.

Historically, TCM's safety was evaluated through long-term clinical experience and traditional knowledge passed down through generations. Practitioners relied on classical TCM texts and their own observations to identify herbs with potential toxicity and understand appropriate dosages and contraindications. For example, herbs like Aconitum carmichaelii, known for its strong medicinal properties, were carefully processed and used in strict accordance with traditional prescriptions to minimize risks. However, this traditional approach lacked systematic and standardized assessment methods. With limited understanding of the chemical composition of herbal medicines and the complex interactions between different herbs in formulas, some potential safety issues might have gone unnoticed, such as cumulative toxicity from long-term use or adverse reactions due to improper combination with other medications.

In the modern era, TCM safety evaluation has evolved significantly with the integration of advanced scientific techniques and international safety assessment standards. Pharmacovigilance, a key concept in modern medicine, has been applied to TCM, establishing systems for the monitoring, detection, assessment, and prevention of adverse drug reactions. Pre-clinical studies using animal models and in-vitro experiments are now common to explore the toxicity of TCM herbs and formulas at the cellular and molecular levels. For instance, researchers use hepatocyte cell lines to study the potential hepatotoxicity of herbal extracts. Clinical trials also play a crucial role, closely monitoring patients' physiological parameters and reporting any adverse events during TCM treatment. Additionally, regulatory authorities around the world have been working on formulating safety guidelines for TCM products. For example, the United States Food and Drug Administration (FDA) has specific regulations regarding the labeling and safety evaluation of dietary supplements containing TCM ingredients, aiming to ensure consumer safety.

The team from Guang'anmen Hospital (China Academy of Chinese Medical Sciences) employed a large-scale, evidence-based research methodology, selecting 16 high-quality randomized controlled trials (RCTs) involving 1,898 patients with metabolic syndrome. Through a systematic meta-analysis integrating adverse event incidence rates and monitoring data on liver and kidney function indicators, the study demonstrated that adverse events in the traditional Chinese medicine (TCM) group were similar to those in the control group, with only mild diarrhea reported and no cases of severe organ damage. This provides an evidence-based safety assessment for TCM interventions in metabolic syndrome, confirming that their overall clinical safety is acceptable (Wu et al., 2020). Additionally, a team from the China National Medical Products Administration (NMPA) integrated bioavailability and bioaccessibility indicators to establish a comprehensive risk assessment system for exogenous and endogenous toxic substances in traditional Chinese medicine. This method reduces the overestimation of risk caused by traditional total content assessment methods by more than 40%, providing a more scientifically sound and clinically relevant evaluation strategy for TCM safety assessment and establishing a reference technical paradigm for the formulation of TCM safety limits (Zuo et al., 2024).

Despite these advancements, TCM safety evaluation still faces several challenges. The complexity of TCM products, with their multi-ingredient herbal formulas and diverse processing methods, makes it difficult to precisely identify the sources of potential risks. Moreover, the lack of a unified international standard for TCM safety evaluation leads to inconsistencies in regulatory requirements across different regions. Cultural differences also contribute to misunderstandings about TCM safety; for example, some traditional TCM practices that have been used safely in China for centuries may be misinterpreted as risky in other cultural contexts. To address these issues, international cooperation is essential. Scientists, regulatory agencies, and TCM experts should collaborate to develop standardized safety evaluation protocols that consider the unique characteristics of TCM. Advanced analytical techniques, such as metabolomics and genomics, can be further utilized to comprehensively analyze the chemical profiles and biological effects of TCM products, enabling more accurate risk assessment. Additionally, strengthening public education about TCM safety can help patients and healthcare providers make informed decisions, promoting the safe and rational use of TCM.

TCM often aims to regulate the body's internal balance, and its effects may interact with the functions of these complex physiological axes. For example, some TCM herbs or formulas may influence the HPA axis, which is involved in the body's stress response. If a TCM treatment affects the HPA axis in an unregulated way, it could potentially lead to imbalances in the body's stress-hormone levels, such as cortisol. This, in turn, might have implications for overall health and could potentially pose safety risks, especially if prolonged or excessive. Similarly, the HPG axis, which is related to the reproductive system, can also be influenced by TCM. Certain TCM therapies used for menopausal symptoms or other reproductive-related disorders need to be carefully evaluated for their impact on this axis. An improper influence could disrupt hormonal balance, leading to adverse effects like irregular menstruation or changes in libido. The brain-gut axis, as depicted in the diagram (**Figure 9**), has a significant role in neurodegenerative diseases like Alzheimer's. TCM treatments that target the gut microbiota or the communication between the gut and the brain need to be closely monitored for safety. Some herbal medicines might alter the gut microbiota composition, which could have both positive and negative consequences. For instance, an over-alteration might disrupt normal gut function, leading to digestive problems, and this could further impact the brain-gut axis communication and potentially affect the course of neurodegenerative diseases.

Moreover, considering the processes such as mitochondrial dysfunction, autophagy, apoptosis, and endoplasmic reticulum stress shown in the diagram, TCM safety evaluation should also assess how TCM interventions affect these cellular processes.

If a TCM formula inadvertently promotes excessive autophagy or apoptosis in brain cells, it could exacerbate neurodegenerative conditions rather than alleviate them.

In conclusion, by integrating the understanding of these complex physiological pathways with TCM safety evaluation, we can develop more comprehensive and accurate safety assessment models. This will not only help in preventing potential adverse effects but also in optimizing TCM treatments for better therapeutic outcomes while ensuring patient safety in the context of modern healthcare.

Figure 9. The relationship between neuroendocrine and Alzheimer's disease. Source: Adapted from Chujun Deng et al. (2022).

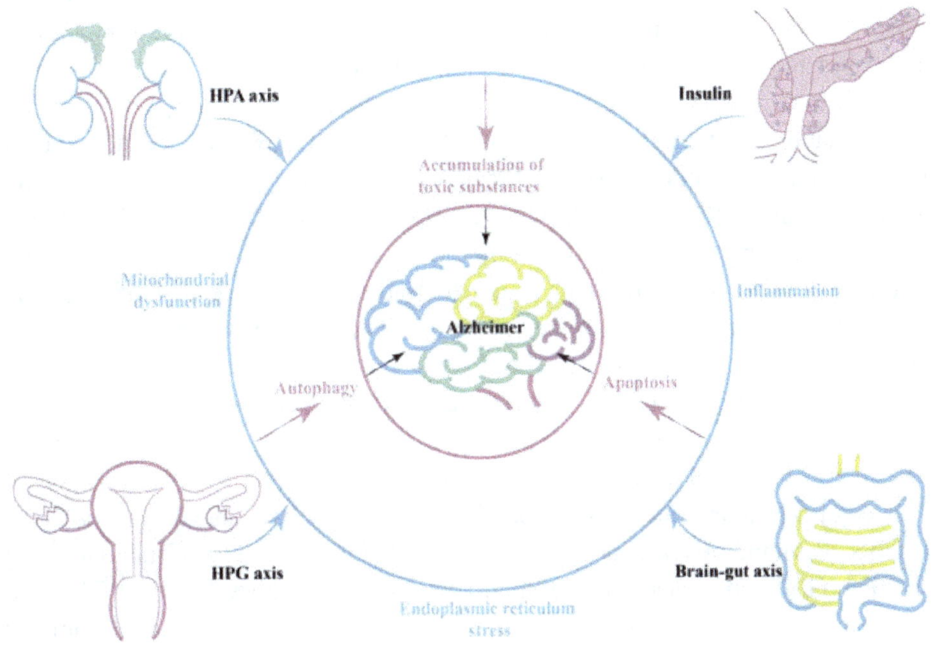

3. Quality Evaluation

Quality evaluation of traditional Chinese Medicine (TCM) is crucial for ensuring its clinical efficacy, safety, and sustainable development. As TCM expands globally and integrates with modern healthcare systems, establishing a robust quality evaluation system is not only a technical necessity but also a regulatory requirement. This evaluation aims to guarantee the consistency, purity, and potency of TCM products, ranging from herbal medicines to complex therapeutic formulations.

Historically, Chinese medicine quality evaluation primarily relied on traditional practices and empirical knowledge. Practitioners judged quality based on macroscopic characteristics such as appearance, color, aroma, and texture. For example, high-quality ginseng was identified by its dense texture, distinctive aroma, and characteristic shape. Traditional methods also emphasized the origin and harvesting time of herbs; Daodi herbs, which were cultivated in specific geographical regions with optimal environmental conditions, were prized for their superior quality. However, this approach was subjective and lacked the precision to detect microscopic variations, chemical impurities, or adulterations. Moreover, without standardized processing and storage procedures, the quality of TCM products could vary significantly, potentially affecting their therapeutic efficacy.

In the modern era, Chinese medicine quality evaluation has undergone a profound transformation with the adoption of advanced scientific technologies and international quality standards. Modern analytical techniques such as high-performance liquid chromatography (HPLC), gas chromatography–mass spectrometry (GC-MS), and Fourier-transform infrared spectroscopy (FTIR) have become indispensable core tools in the quality control research system for traditional Chinese medicine. Among these, HPLC technology enables the precise qualitative and quantitative detection of various active components in herbal medicines, providing critical technical support for ensuring the consistency of quality across different batches of herbal materials and preparations. For instance, when evaluating the quality of Salvia miltiorrhiza (Danshen), HPLC can measure the content of tanshinones, a key active component with cardiovascular effects. Additionally, modern quality control systems, such as Good Agricultural Practice (GAP) for herbal cultivation, Good Manufacturing Practice (GMP) for product processing, and Good Laboratory Practice (GLP) for research, have been introduced. GAP standardizes the cultivation process from soil quality and fertilization to harvesting and post-harvest processing, while GMP ensures that TCM products are manufactured under controlled conditions to minimize contamination risks and ensure product quality. GLP provides a standardized experimental framework for evaluating the safety and efficacy of TCM. It strictly defines requirements across the entire process, from laboratory environmental management and experimental protocol design to data collection, analysis, and report writing, ensuring the scientific rigor, accuracy, and reproducibility of research outcomes in pharmacology, toxicology, and component analysis. This establishes the core scientific basis for the clinical application, regulatory approval, and international recognition of TCM.

A research team from the University of Macau has successfully established a multidimensional comprehensive quality evaluation system for Huangqi Guizhi Wuweizi Tang (a classic traditional Chinese medicine formula consisting of Astragali Radix, Cinnamomi Ramulus, Schisandrae Chinensis Fructus and other herbal components), providing a standardized technical framework for analyzing the chemical composition of this classic formula, controlling the quality of its preparations, and evaluating the consistency of its clinical efficacy. Through fingerprint analysis, mass spectrometry, and anti-inflammatory assays, they identified five quality markers associated with therapeutic efficacy and developed a 7-minute simultaneous quantitative method, providing a comprehensive chemobiological strategy for traditional Chinese medicine formulations (Jiang et al., 2024). Additionally, a research team at Chengdu University of Traditional Chinese Medicine established a precision quality system for Carthamus tinctorius based on target-associated evaluation. Through multi-omics analysis, the team identified PTGS2 as a core analgesic target and established a validated bioassay whose results correlate positively with in vivo efficacy, thereby providing an efficacy-oriented paradigm for the study of traditional Chinese medicinal materials (Zhou et al., 2025).

Notwithstanding these advancements, Chinese medicine quality evaluation still faces multiple persistent challenges. The complex chemical composition of TCM products——with hundreds of potentially active compounds in a single herbal formula——makes it difficult to comprehensively assess quality using a limited number of markers. The lack of unified international quality standards remains a significant barrier; different countries may have divergent requirements for TCM product quality, creating regulatory confusion in international trade and compliance. Additionally, issues such as herb adulteration (where inferior or substitute herbs are used) and improper processing due to insufficient technical expertise in some regions undermine quality assurance initiatives. To address these challenges, international cooperation in standard-setting is crucial. Harmonizing quality evaluation criteria across countries can facilitate global trade and regulatory approval of TCM products. Continued research into advanced analytical techniques, such as multi-omics approaches that integrate genomics, proteomics, and metabolomics, can enable a more precise and comprehensive understanding of TCM quality. Strengthening regulatory oversight and enforcement and promoting targeted education on quality control among TCM producers and practitioners will also contribute to enhancing the overall quality of TCM products.

IV. THE EXPANSION OF TRADITIONAL CHINESE MEDICINE

Against the backdrop of a deep integration between the theoretical wisdom of traditional Chinese medicine and modern scientific research paradigms, the field of traditional Chinese medicine is gradually transcending the limitations of traditional applications and exploring innovative development pathways in emerging interdisciplinary fields such as bioinformatics, precision medicine, and systems biology. These expansions not only inject new momentum into the inheritance and development of TCM but also provide new solutions to global public health challenges. TCM is integrating with cutting-edge biomedical technologies such as synthetic biology, neuroscience, and stem cell technology, driving the efficient biosynthesis of bioactive substances and offering new strategies for the treatment of neurological disorders and regenerative medicine.

TCM's preventive philosophy of "preventing disease before it occurs" plays a pivotal role in public health, effectively controlling chronic and infectious diseases while offering a comprehensive, lifecycle-based health management model. Furthermore, interdisciplinary innovation is extending TCM's applications to aerospace, sports medicine, and geriatrics, while its integration with artificial intelligence and the Internet of Things is giving rise to smart diagnostic and telemedicine models, making TCM services more accessible and standardized.

V. TRADITIONAL CHINESE MEDICINE CULTURE INDUSTRY

As a vital vehicle for preserving a civilization that spans millennia, the traditional Chinese medicine (TCM) cultural industry is now reaping the dual benefits of policy support and rising consumer demand. By 2025, the industry's market size had surpassed 980 billion yuan, giving rise to diverse integrated business models such as "Traditional Chinese Medicine (TCM) + Cultural Tourism," "TCM + Wellness," and "TCM + Cultural Creativity." Cultural landmarks like the Beijing Barefoot Doctors Museum attract over 60,000 visitors annually, while new product categories—including food-medicine products and smart wellness devices—have gained significant popularity among younger demographics. This not only facilitates the contemporary adaptation of traditional wisdom but also injects new momentum into economic growth.

As the country enters a new stage of development, the TCM cultural industry will enter a critical period of "upholding tradition while innovating, and improving quality and efficiency." With the widespread adoption of technologies such as AI-based pattern identification and blockchain traceability, as well as the advancement of the "TCM Going Global" strategy, the industry will further integrate the entire chain of "cultural dissemination—service implementation—product export." It is projected that by 2030, the market size will exceed 2 trillion yuan. This will not only provide the public with a richer array of TCM solutions for health but will also serve as a vital window for showcasing the soft power of Chinese culture.

VI. FUTURE OUTLOOK

The development of traditional Chinese medicine faces both opportunities and challenges, and breakthroughs must be achieved in five key areas. First, we must promote integration with cutting-edge technologies such as artificial intelligence and synthetic biology, elucidate the scientific basis of the core theories of traditional Chinese medicine, and overcome key technical challenges—such as the material basis and mechanisms of action of traditional Chinese medicine formulas—to achieve digital and intelligent transformation; Second, we must improve the evidence-based clinical evidence system, conduct large-scale clinical trials and real-world studies, align with international standards to establish universal norms for TCM, and break down barriers to its international application; third, we must innovate talent cultivation models, nurture interdisciplinary professionals with both a foundation in traditional theory and modern scientific literacy, and build an academic system that integrates across disciplines. At the same time, we must promote the development of high-quality industries, establish a comprehensive quality supervision system covering the entire supply chain, foster new business models such as "Traditional Chinese Medicine (TCM) + Health" and "TCM + Cultural Tourism," and strengthen the protection and sustainable use of resources; we must also deepen international regulatory cooperation and joint research to promote the integration of TCM into the global healthcare system. It is projected that by 2035, TCM will have essentially achieved modernization, forming a complementary and coordinated relationship with Western medicine, and contributing core strength to the building of a global community of health.

REFERENCES

Chen, J., Lei, C., Li, X., Wu, Q., Liu, C., Ma, Q., & Chen, J. (2022). Research progress on classical traditional chinese medicine formula xiaoyaosan in the treatment of depression. *Frontiers in Pharmacology*, *13*, 925514. Advance online publication. DOI: 10.3389/fphar.2022.925514 PMID: 35991880

Chen, S., Qi, H., Zhu, C., Zhao, Y., Jiao, B., Tan, Y., &Wu, C. (2025). Quality and composition control of complex TCM preparations through a novel "Herbs-in vivo Compounds-Targets-Pathways" network methodology: The case of Lianhuaqingwen capsules, 213. DOI: 10.1016/j.phrs.2025.107650

Deng, C., Chen, H., Meng, Z., & Meng, S. (2022). Roles of traditional chinese medicine regulating neuroendocrinology on AD treatment. *Frontiers in Endocrinology*, *13*, 955618. Advance online publication. DOI: 10.3389/fendo.2022.955618 PMID: 36213283

Deng, Z., Li, H., Zhou, Q. and Zheng, H. (2025). Progress of Clinical Research on the Treatment of Neurogenic Lower Urinary Tract Dysfunction after Stroke with External Treatment of Chinese Medicine, 14. DOI: 10.12677/tcm.2025.148504

Guo, J., Chen, X., Wu, M., Wang, D., Zhao, Y., Li, Q., &Li, M. (2024). Traditional Chinese medicine FYTF-919 (Zhongfeng Xingnao oral prescription) for the treatment of acute intracerebral haemorrhage: a multicentre, randomised, placebo-controlled, double-blind, clinical trial, 404. DOI: 10.1016/S0140-6736(24)02261-X

Hu, K., Guan, W., Bi, Y., Zhang, W., Li, L., Zhang, B., & Zhong, N. (2021). Efficacy and safety of Lianhuaqingwen capsules, a repurposed Chinese herb, in patients with coronavirus disease 2019: A multicenter, prospective, randomized controlled trial. *Phytomedicine : International Journal of Phytotherapy and Phytopharmacology*, *85*, 153242. Advance online publication. DOI: 10.1016/j.phymed.2020.153242 PMID: 33867046

Jiang, M., Yang, L., Zou, L., Zhang, L., Wang, S., Zhong, Z., &Li, P. (2024). A comprehensive quality evaluation for Huangqi Guizhi Wuwu decoction by integrating UPLC-DAD/MS chemical profile and pharmacodynamics combined with chemometric analysis, 319. DOI: 10.1016/j.jep.2023.117325

Kadier, A., Ablimit, M., Tursun, H. and Ablat, N. (2025). The western transmission of traditional Chinese medicine: an investigation of the cultural elements of traditional Chinese medicine in biomedical systems of cross-Asia countries, 16. DOI: 10.3389/fphar.2025.1589275

Lee, B., Kwon, C.-Y., Suh, H.-W., Kim, Y. J., Kim, K.-I., Lee, B.-J. and Lee, J.-H. (2023). Herbal medicine for the treatment of chronic cough: a systematic review and meta-analysis, 14. DOI: 10.3389/fphar.2023.1230604

Li, J., Yu, X.-q., Xie, Y., Yang, S.-g., Zhao, L., Zhou, M. and Meng, Y. (2022). Efficacy and safety of traditional Chinese medicine treatment for idiopathic pulmonary fibrosis: An exploratory, randomized, double-blinded and placebo controlled trial, 13. DOI: 10.3389/fphar.2022.1053356

Li, L., Li, Q., Li, Y., Gong, D., & Zhao, B. (2025). Research progress and challenges of molecular recognition techniques in the screening of active ingredients in traditional. *Journal of Pharmaceutical Analysis*, *15*(9), 101243. Advance online publication. DOI: 10.1016/j.jpha.2025.101243 PMID: 41069865

Li, T., Yang, Y., Yang, F., Li, Y., Qiao, H., Yan, Y., & Ding, X. (2025). Organoids and Organoids-on-Chip in Traditional Chinese Medicine Research: Applications, Advantages, and Future Prospects. *Cell Biology International*, *49*(10), 1233–1244. Advance online publication. DOI: 10.1002/cbin.70067 PMID: 40772545

Li, J., Zhang, X., Hou, L., Liu, B.-Y., Fan, Y.-M., Zhang, Y., & Yin, X. (2025). Proteomic analysis reveals QiShenYiQi Pills ameliorates ischemia-induced heart failure through inhibition of mitochondrial fission, 138. DOI: 10.1016/j.phymed.2025.156435

Lu, A. (2025). Elevating traditional Chinese medicine in global health research: The case for Chinese herbal medicine formulas in mainstream therapeutics, 6. DOI: 10.1016/j.xinn.2025.100811

Lu, P., Li, L., Liu, B., Cao, Z., Geng, Q., Ji, X., &Lu, C. (2024). Efficacy and safety of integrated traditional Chinese and Western medicine for rheumatoid arthritis-interstitial lung disease: A systematic review and meta-analysis, 10. DOI: 10.1016/j.heliyon.2024.e38771

Mei, J., Ju, C., Wang, B., Gao, R., Zhang, Y., Zhou, S., &Xu, F. (2024). The efficacy and safety of Bazi Bushen Capsule in treating premature aging: A randomized, double blind, multicenter, placebo-controlled clinical trial, 130. DOI: 10.1016/j.phymed.2024.155742

Nielsen, A., Dyer, N. L., Lechuga, C., McKee, M. D. and Dusek, J. A. (2024). Fidelity to the acupuncture intervention protocol in the ACUpuncture In The EmergencY department for pain management (ACUITY) trial: Expanding the gold standard of STRICTA and CONSORT guidelines, 13. DOI: 10.1016/j.imr.2024.101048

Raclariu-Manolică, A. C. and de Boer, H. J. (2022). DNA barcoding and metabarcoding for quality control of botanicals and derived herbal products, DOI: 10.1016/b978-0-323-85542-6.00004-410.1016/b978-0-323-85542-6.00004-4

Singh, K., Gupta, J. K., Jain, D., Kumar, S., Singh, T. and Saha, S. (2024). Exploring the ancient wisdom and modern relevance of Chinese medicine: A comprehensive review, 11. DOI: 10.1016/j.prmcm.2024.100448

Siu, P. M., Yu, D. J., Yu, A. P., Recchia, F., Li, S. X., Chan, R. N. Y., &Irwin, M. R. (2025). Tai chi or cognitive behavioural therapy for treating insomnia in middle aged and older adults: randomised non-inferiority trial, 391. DOI: 10.1136/bmj-2025-084320

Wang, J., & Qi, F. (2020). Traditional Chinese medicine to treat COVID-19: The importance of evidence-based research. *Drug Discoveries & Therapeutics*, *14*(3), 149–150. DOI: 10.5582/ddt.2020.03054 PMID: 32669523

Wang, M., Yin, F., Kong, L., Yang, L., Sun, H., Sun, Y., &Wang, X. (2024). Chinmedomics: a potent tool for the evaluation of traditional Chinese medicine efficacy and identification of its active components, 19. DOI: 10.1186/s13020-024-00917-x

Wang, M., Yin, F., Kong, L., Yang, L., Sun, H., Sun, Y., &Wang, X. (2024). Chinmedomics: a potent tool for the evaluation of traditional Chinese medicine efficacy and identification of its active components, 19. DOI: 10.1186/s13020-024-00917-x

Wang, Y., Jiang, Y., Li, M., Xiao, Y., Zhao, Q., Zeng, J., &Wu, X. (2025). Rosavin derived from Rhodiola alleviates colitis in mice through modulation of Th17 differentiation, 136. DOI: 10.1016/j.phymed.2024.156318

Wang, M., Yang, Q., Li, Y., Zhao, Y., Zou, J., Luan, F., &Liu, F. (2025). Therapeutic potential of traditional Chinese medicine and mechanisms for the treatment of type 2 diabetes mellitus, 20. DOI: 10.1186/s13020-025-01222-x

Wu, H., Tian, J., Dai, D., Liao, J., Wang, X., Wei, X., &Tong, X. (2020). Efficacy and safety assessment of traditional Chinese medicine for metabolic syndrome, 8. DOI: 10.1136/bmjdrc-2020-001181

Wu, L., Wu, Q., Du, X., Ling, M. and Tong, H. (2025). Revolutionizing pharmacological research of traditional Chinese medicine with single-cell omics technologies, 186. DOI: 10.1016/j.fitote.2025.106846

Wu, Y.-Z., Zhang, Q., Wei, X.-H., Jiang, C.-X., Li, X.-K., Shang, H.-C. and Lin, S. (2022). Multiple anti-inflammatory mechanisms of Zedoary Turmeric Oil Injection against lipopolysaccharides-induced acute lung injury in rats elucidated by network pharmacology combined with transcriptomics, 106. DOI: 10.1016/j.phymed.2022.154418

Xi, Z., Dai, R., Ze, Y., Jiang, X., Liu, M., & Xu, H. (2025). Traditional Chinese medicine in lung cancer treatment. *Molecular Cancer*, *24*(1), 57. Advance online publication. DOI: 10.1186/s12943-025-02245-6 PMID: 40001110

Xiang, L., Chen, Z., Wei, S., & Zhou, H. (2022). Global Trade Pattern of Traditional Chinese Medicines and China's Trade Position. *Frontiers in Public Health*, *10*, 865887. Advance online publication. DOI: 10.3389/fpubh.2022.865887 PMID: 35602159

Xiao, W., Zhang, M., Zhao, D., Meng, F., Tang, Q., Hu, L., &Chen, W. (2025). TC-MKD: From ancient wisdom to modern insights-A comprehensive platform for traditional Chinese medicine knowledge discovery, 15. DOI: 10.1016/j.jpha.2025.101297

Xing, Y., Wang, M., Yuan, Y., Hu, J., Wang, Z., Sun, Z., &Mao, T. (2025). Gut microbiota-derived butyrate mediates the anticolitic effect of indigo supplementation through regulating CD4+ T cell differentiation, 4. DOI: 10.1002/imt2.70040

Xu, J., Li, H., Sze, D. M.-y., Chan, V. W. S. and Yang, A. W. H. (2025). Effectiveness of qigong and Tai Chi for quality of life in patients with cancer: an umbrella review and meta-analysis, 25. DOI: 10.1186/s12906-025-04875-1

Yue, J., Hao, D., Liu, S., Yu, J., Meng, L., Lv, J., & Guo, J. (2024). Research progress of traditional Chinese medicine in the treatment of allergic rhinitis. *Heliyon*, *10*(7), e29262. Advance online publication. DOI: 10.1016/j.heliyon.2024.e29262 PMID: 38617960

Zhang, S. Q., & Li, J. C. (2021). An introduction to traditional Chinese medicine, including acupuncture. *The Anatomical Record : Advances in Integrative Anatomy and Evolutionary Biology*, *304*(11), 2359–2364. DOI: 10.1002/ar.24782 PMID: 34626156

Zhang, X., He, X., Fang, Z., Yu, S., You, L., Zhu, S., & Zhang, B. (2025). Analyzing the highlights and trends of traditional Chinese medicine academic development based on the “Top 10 Academic Advances in TCM from 2020 to 2023&rdquo. *Zhongguo Ke Xue : Sheng Ming Ke Xue = Scientia Sinica Vitae*, *55*(2), 267–281. DOI: 10.1360/SSV-2024-0293

Zhang, Y., Lu, P., Qin, H., Zhang, Y., Sun, X., Song, X., Liu, Z., Peng, H., Liu, Y., Nwafor, E. O., Li, J., & Liu, Z. (2021). Traditional Chinese medicine combined with pulmonary drug delivery system and idiopathic pulmonary fibrosis: Rationale and therapeutic potential. *Biomedicine & Pharmacotherapy = Biomédecine & Pharmacothérapie, 133*, 111072. Advance online publication. DOI: 10.1016/j.biopha.2020.111072 PMID: 33378971

Zhang, W., Wu, P., Song, Y., Liang, D., Meng, G. and Zeng, H. (2025). Immunomodulatory mechanism of Huangqi-Guizhi-Wuwu Decoction in alleviating autoimmune arthritis based on network pharmacology and experimental validation, 143. DOI: 10.1016/j.phymed.2025.156644

Zhao, M., Che, Y., Gao, Y., & Zhang, X. (2024). Application of multi-omics in the study of traditional. *Frontiers in Pharmacology, 15*, 1431862. Advance online publication. DOI: 10.3389/fphar.2024.1431862 PMID: 39309011

Zhao, Y.-Z., Dai, Y.-Z., & Nie, K. (2022). Research Progress on the Antiemetic Effect of Traditional Chinese Medicine Against Chemotherapy-Induced Nausea and Vomiting: A Review. *Frontiers in Pharmacology, 12*, 790784. Advance online publication. DOI: 10.3389/fphar.2021.790784 PMID: 35222008

Zhou, Y., Wu, Q., Ren, C., Jiang, H., Huang, X., Chen, J., &Pei, J. (2025). Construction and application of a precise evaluation method for the quality of traditional Chinese medicine based on "target-combined quality evaluation" using safflower as an example, 16. DOI: 10.3389/fphar.2025.1554895

Zhu, S. J., Hinman, R. S., Nelligan, R. K., Li, P., De Silva, A. P., Harrison, J., & Bennell, K. L. (2026). Online Unsupervised Tai Chi Intervention for Knee Pain and Function in People With Knee Osteoarthritis. *JAMA Internal Medicine, 186*(1), 15. Advance online publication. DOI: 10.1001/jamainternmed.2025.5723 PMID: 41143827

Zuo, T.-T., Liu, J., Zan, K., Liu, L.-N., Wang, Q., Wang, Z., &Ma, S.-C. (2024). Bioaccessibility and bioavailability of exogenous and endogenous toxic substances in traditional Chinese medicine and their significance in risk assessment, 208. DOI: 10.1016/j.phrs.2024.107388

Compilation of References

Abdullahi, T., Harun, Z., Hafiz, M., & Othman, D. (2017). A review on sustainable synthesis of zeolite from kaolinite resources via hydrothermal process. *Advanced Powder Technology*, *28*(8), 1827–1840. DOI: 10.1016/j.apt.2017.04.028

Abedi, M., Chang, A., Hassan, H. I., Yang, A., Hickethier, M. L., Bushagour, A. H., Aghajari, M., Stoerzinger, K. A., Layton, A., Abbasi, B., Chang, C.-H., & Feng, Z. (2026). Recovering Critical Metals from Brines: Technology Landscape, Circular Economy, and Future Trends. *ACS Applied Energy Materials*, *9*(4), 2005–2034. Advance online publication. DOI: 10.1021/acsaem.5c03407

Aftab, W., Usman, A., Shi, J., Yuan, K., Qin, M., & Zou, R. (2021). Phase change material-integrated latent heat storage systems for sustainable energy solutions. *Energy & Environmental Science*, *14*(8), 4268–4291. DOI: 10.1039/D1EE00527H

Aggarwal, P., Rana, M., & Akhai, S. (2022). Briefings on e-waste hazard until COVID era in India. *Materials Today: Proceedings*, *71*, 389–393. DOI: 10.1016/j.matpr.2022.09.507

Ahmad, M., Rajapaksha, A. U., Lim, J. E., Zhang, M., Bolan, N., Mohan, D., Vithanage, M., Lee, S. S., & Ok, Y. S. (2014). Biochar as a sorbent for contaminant management in soil and water: A review. *Chemosphere*, *99*, 19–33. DOI: 10.1016/j.chemosphere.2013.10.071 PMID: 24289982

Akhtari, F. S., Green, A. J., Small, G. W., Havener, T. M., House, J. S., Roell, K. R., Reif, D. M., McLeod, H. L., Wiltshire, T., & Motsinger-Reif, A. A. (2021). High-throughput screening and genome-wide analyses of 44 anticancer drugs in the 1000 Genomes cell lines reveals an association of the NQO1 gene with the response of multiple anticancer drugs. *PLOS Genetics*, *17*(8), e1009732. DOI: 10.1371/journal.pgen.1009732 PMID: 34437536

Akolekar, D., Chaffee, A., & Howe, R. F. (1997). The transformation of kaolin to low-silica X zeolite. *Zeolites*, *19*(5-6), 359–365. DOI: 10.1016/S0144-2449(97)00132-2

Akram, F., Haq, I., Raja, S. I., Mir, A. S., Qureshi, S. S., Aqeel, A., & Shah, F. I. (2022). Current trends in biodiesel production technologies and future progressions: A possible displacement of the petro-diesel. *Journal of Cleaner Production*, *370*, 133479. DOI: 10.1016/j.jclepro.2022.133479

Alallaq, A. A., & Kashan, J. S. (2023). A review: In vivo studies of bioceramics as bone substitute materials. *Nano Select*, *4*(2), 123–144. https://doi.org/. DOI: 10.1002/nano.202200222

Alharbi, A. M., Katary, M., Alshehri, K. M., Asghar, B. H., Omran, M. M., Elshaarawy, R. F. M., & Zakaria, R. (2025). Hybrid thiazolyl–benzylidene–phenol metal complexes as novel chemotherapeutic agents with anti-topoisomerase I activity in human breast carcinoma: Synthesis, in vitro and in silico studies. *RSC Advances*, *15*(30), 20552–20569. DOI: 10.1039/D5RA00889A PMID: 40530316

Al-Hussaniy, H. A., Oraibi, A. I., Al-Tameemi, Z. S., Alburghaif, A. H., Naji, M. A., & Naji, F. A. (2025). The association of KRAS and P53 gene mutations and MDM2 expression with the occurrence of colorectal cancer. *Current Cancer Drug Targets*, *25*, e15680096349198. DOI: 10.2174/0115680096349198250407100724 PMID: 40357780

Ali, G., Afraz, M., Muhammad, F., Nisar, J., Shah, A., Munir, S., & Tasleem Hussain, S. (2022). Production of fuel range hydrocarbons from pyrolysis of lignin over zeolite Y, hydrogen. *Energies*, *16*(1), 215. DOI: 10.3390/en16010215

Ali, S., Ahsan Farooq Qaisar, M., Sufyan Javed, M., Umer, K., Alarfaji, S. S., Mateen, M., Chhattal, M., Ali, S., Parkash, A., Lama Tamang, T., & Qi, J. (2024). Two-dimensional MXene based innovative electrode materials for supercapacitors: Recent advances and prospects. *Fuel*, *377*, 132783. DOI: 10.1016/j.fuel.2024.132783

Aliyu, M., & AlZahrani, A. A. (2024). Integrated SMR-SOFC system for efficient hydrogen and power production. *International Journal of Hydrogen Energy*. Advance online publication. DOI: 10.1016/j.ijhydene.2024.12.354

Al-Karmalawy, A. A., Eissa, M. E., Ashour, N. A., Yousef, T. A., Khatib, A. O. A., & Hawas, S. S. (2025). Medicinal chemistry perspectives on anticancer drug design based on clinical applications (2015–2025). *RSC Advances*, *15*(43), 36441–36471. DOI: 10.1039/D5RA05472A PMID: 41041285

Amaral, M. M., Real, C. G., Yukuhiro, V. Y., Doubek, G., Fernandez, P. S., Singh, G., & Zanin, H. (2023). In situ and operando infrared spectroscopy of battery systems: Progress and opportunities. *Journal of Energy Chemistry*, *81*, 472–491. DOI: 10.1016/j.jechem.2023.02.036

Amri, I., Ouakouak, A., Hamdi, W., Srasra, E., & Hamdi, N. (2022). Removal of non-steroidal drug from waste water using synthetic zeolites from illito-kaolinitic clays. *Water, Air, and Soil Pollution*, *233*(9), 369. DOI: 10.1007/s11270-022-05846-9

Anderson, K. P., Wollmershauser, J. A., Ryou, H., Goswami, R., Rommel, S., Aindow, M., Gorzkowski, E. P., & Feigelson, B. N. (2024). Nanostructural effects beyond Hall-Petch: Towards superhard tungsten carbide. *Acta Materialia*, *275*, 120004. DOI: 10.1016/j.actamat.2024.120004

Andrea, D. (2010). *Battery management systems for large lithium-ion battery packs*. Artech house.

Appel, A. M., Bercaw, J. E., Bocarsly, A. B., Dobbek, H., DuBois, D. L., Dupuis, M., Ferry, J. G., Fujita, E., Hille, R., Kenis, P. J. A., Kerfeld, C. A., Morris, R. H., Peden, C. H. F., Portis, A. R., Ragsdale, S. W., Rauchfuss, T. B., Reek, J. N. H., Seefeldt, L. C., Thauer, R. K., & Waldrop, G. L. (2013). Frontiers, Opportunities, and Challenges in Biochemical and Chemical Catalysis of CO_2 Fixation. *Chemical Reviews*, *113*(8), 6621–6658. DOI: 10.1021/cr300463y

Argirusis, C., Angelara, C., Argirusis, N., Karantonis, A., Pandis, P. P., & Sourkouni, G. (2024). Preparation and characterization of supercapacitor cells using modified CNTs and bimetallic MOFs. *Processes (Basel, Switzerland)*, *12*(12), 2778. DOI: 10.3390/pr12122778

Armstrong McKay, D. I., Staal, A., Abrams, J. F., Winkelmann, R., Sakschewski, B., Loriani, S., Fetzer, I., Cornell, S. E., Rockström, J., & Lenton, T. M. (2022). Exceeding 1.5 °C global warming could trigger multiple climate tipping points. *Science*, *377*(6611), eabn7950. DOI: 10.1126/science.abn7950 PMID: 36074831

Aroh, K. C., & Jensen, K. F. (2018). Efficient kinetic experiments in continuous flow microreactors. *Reaction Chemistry & Engineering*, *3*(1), 94–101. DOI: 10.1039/C7RE00163K

Arrinda, M., Vertiz, G., Sanchéz, D., Makibar, A., & Macicior, H. (2022). Surrogate Model of the Optimum Global Battery Pack Thermal Management System Control. *Energies*, *15*(5), 1695. DOI: 10.3390/en15051695

Arshi, P. S., Vahidi, E., & Zhao, F. (2018). Behind the Scenes of Clean Energy: The Environmental Footprint of Rare Earth Products. *ACS Sustainable Chemistry & Engineering*, *6*(3), 3311–3320. DOI: 10.1021/acssuschemeng.7b03484

Artz, J., Müller, T. E., Thenert, K., Kleinekorte, J., Meys, R., Sternberg, A., Bardow, A., & Leitner, W. (2018). Sustainable Conversion of Carbon Dioxide: An Integrated Review of Catalysis and Life Cycle Assessment. *Chemical Reviews*, *118*(2), 434–504. DOI: 10.1021/acs.chemrev.7b00435

Ayalew, A. A. (2020). Development of kaolin clay as a cost-effective technology for defluoridation of groundwater. *International Journal of Chemical Engineering*, *2020*(1), 8820727. DOI: 10.1155/2020/8820727

Ayaz, M., Feizienė, D., Tilvikienė, V., Akhtar, K., Stulpinaitė, U., & Iqbal, R. (2021). Biochar Role in the Sustainability of Agriculture and Environment. *Sustainability (Basel)*, *13*(3), 1330. DOI: 10.3390/su13031330

Baerlocher, L. B. M., & Ch, D. H. O. (2007). *Atlas of zeolite framework types.*

Bai, H., Qing, S. L., Yang, D. P., Zhang, Y. X., Fan, X., & Tao, C. Y. (2017). Periodic potential oscillation during oxygen evolution catalyzed by manganese oxide at constant current. *Journal of the Electrochemical Society*, *164*(4), E78–E83. DOI: 10.1149/2.1241704jes

Bakali, K., Ampaire, W., & Extension, K. P. (2024). Bioethanol Production from Sugarcane Bagasse by Fermentation Process Using Saccharomyces cerevisiae as Yeast Species. *IAA Journal of Applied Sciences*, *11*(3), 32–41. DOI: 10.59298/IAAJAS/2024/113.3241

Baniasadi, M., Vakilchap, F., Bahaloo-Horeh, N., Mousavi, S. M., & Farnaud, S. (2019). Advances in bioleaching as a sustainable method for metal recovery from e-waste: A review. *Journal of Industrial and Engineering Chemistry*, *76*, 75–90. DOI: 10.1016/j.jiec.2019.03.047

Barbon, A. P. A. C., Nolasco-Perez, I. M., Barbon, S.Jr, Máquina, A. D. V., Su, W.-H., & Barbin, D. F. (2026). Process analytical technologies applied to quality control of emerging alternative protein food products: Challenges and future trends. *Journal of Pharmaceutical and Biomedical Analysis Open*, *7*, 100105. DOI: 10.1016/j.jpbao.2026.100105

Barton, A. D. K. C. D. (2000). Clay minerals. In *Encyclopedia of Soil Science*. https://doi.org/DOI: 10.2118/59292-MS

Batzner, S., Musaelian, A., Sun, L., Geiger, M., Mailoa, J. P., Kornbluth, M., Molinari, N., Smidt, T. E., & Kozinsky, B. (2022). E(3)-equivariant graph neural networks for data-efficient and accurate interatomic potentials. *Nature Communications*, *13*(1), 2453. DOI: 10.1038/s41467-022-29939-5 PMID: 35508450

Bauer, B., Bravyi, S., Motta, M., & Chan, G. K.-L. (2020). Quantum Algorithms for Quantum Chemistry and Quantum Materials Science. *Chemical Reviews*, *120*(22), 12685–12717. DOI: 10.1021/acs.chemrev.9b00829

Bauri, S., Tripathi, S., Choudhury, A. M., Mandal, S. S., Raj, H., & Maiti, P. (2023). Nanomaterials as Theranostic Agents for Cancer Therapy. *ACS Applied Nano Materials*, *6*(23), 21462–21495. DOI: 10.1021/acsanm.3c04235

Bazant, M. Z. (2013). Theory of chemical kinetics and charge transfer based on nonequilibrium thermodynamics. *Accounts of Chemical Research*, *46*(5), 1144–1160. DOI: 10.1021/ar300145c PMID: 23520980

Bedrov, D., Piquemal, J.-P., Borodin, O., MacKerell, A. D.Jr, Roux, B., & Schröder, C. (2019). Molecular Dynamics Simulations of Ionic Liquids and Electrolytes Using Polarizable Force Fields. *Chemical Reviews*, *119*(13), 7940–7995. DOI: 10.1021/acs.chemrev.8b00763 PMID: 31141351

Belluati, M., Tabasso, S., Gaudino, E. C., Cravotto, G., & Manzoli, M. (2024). Biomass-derived carbon-based catalysts for lignocellulosic biomass and waste valorisation: A circular approach. *Green Chemistry*, *26*(15), 8642–8668. DOI: 10.1039/D4GC00606B

Belviso, C., Cavalcante, F., Lettino, A., & Fiore, S. (2013). A and X-type zeolites synthesised from kaolinite at low temperature. *Applied Clay Science*, *80–81*, 162–168. DOI: 10.1016/j.clay.2013.02.003

Benghanem, M., Mellit, A., Almohamadi, H., Haddad, S., Chettibi, N., Alanazi, A. M., Dasalla, D., & Alzahrani, A. (2023). Hydrogen production methods based on solar and wind energy: A review. *Energies*, *16*(2), 757. DOI: 10.3390/en16020757

Best, S. M., Porter, A. E., Thian, E. S., & Huang, J. (2008). Bioceramics: Past, present and for the future. *Journal of the European Ceramic Society*, *28*(7), 1319–1327. DOI: 10.1016/j.jeurceramsoc.2007.12.001

Bhadrakumari, S., & Predeep, P. (2011). Performance of high temperature superconducting ceramic/thermoplastic polymer composites. *Journal of Polymer Engineering*, *31*(2-3), 205–207. DOI: 10.1515/polyeng.2011.043

Bhuvaneswari, V., Priyadharshini, M., Deepa, C., Balaji, D., Rajeshkumar, L., & Ramesh, M. (2021). Deep learning for material synthesis and manufacturing systems: A review. *Materials Today: Proceedings*, *46*, 3263–3269. DOI: 10.1016/j.matpr.2020.11.351

Bing, D., Qinglin, C., Yuji, W., Yijia, S., Rukang, Y., & Qiyuan, L. (2025). Selectively and deeply extracting manganese from modified manganese-rich slag to prepare lithium manganese oxide battery materials. *Separation and Purification Technology, 377*, 134374. Advance online publication. DOI: 10.1016/j.seppur.2025.134374

Bi, Y., Dong, J., Zhou, Y., Zhang, M., Chen, X., & Zhang, Y. (2024). Application of membrane separation technology in the purification of pharmaceutical components. *Preparative Biochemistry & Biotechnology, 54*(9), 1107–1115. DOI: 10.1080/10826068.2024.2328673 PMID: 38526323

Bochenek, D. (2022). Advanced ceramic materials with functional properties. *Materials (Basel), 15*(18), 6439–6440. DOI: 10.3390/ma15186439 PMID: 36143751

Bollini, P., Diwan, M., Gautam, P., Hartman, R. L., Hickman, D. A., Johnson, M., Kawase, M., Neurock, M., Patience, G. S., Stottlemyer, A., Vlachos, D. G., & Wilhite, B. (2023). Vision 2050: Reaction Engineering Roadmap. *ACS Engineering Au, 3*(6), 364–390. DOI: 10.1021/acsengineeringau.3c00023

Bo, R., Xu, S., Yang, Y., & Zhang, Y. (2023). Mechanically-Guided 3D Assembly for Architected Flexible Electronics. *Chemical Reviews, 123*(18), 11137–11189. DOI: 10.1021/acs.chemrev.3c00335

Bouhadi, M., Javed, Q., Jakubus, M., Elkouali, M., Fougrach, H., Ansar, A., Ban, S. G., Ban, D., Heath, D., & Cerne, M. (2025). Nanoparticles for Sustainable Agriculture: Assessment of Benefits and Risks. *Agronomy (Basel), 15*(5), 1131. Advance online publication. DOI: 10.3390/agronomy15051131

Bowers, H. J., Fannin, E. E., Thomas, A., & Weis, J. A. (2020). Characterization of multicellular breast tumor spheroids using image data-driven biophysical mathematical modeling. *Scientific Reports, 10*(1), 11583. DOI: 10.1038/s41598-020-68324-4 PMID: 32665565

Brachi, M., El Housseini, W., Beaver, K., Jadhav, R., Dantanarayana, A., Boucher, D. G., & Minteer, S. D. (2024). Advanced Electroanalysis for Electrosynthesis. *ACS Organic & Inorganic Au, 4*(2), 141–187. DOI: 10.1021/acsorginorgau.3c00051

Brown, B. P., Stein, R. A., Meiler, J., & McHaourab, H. S. (2024). Approximating Projections of Conformational Boltzmann Distributions with AlphaFold2 Predictions: Opportunities and Limitations. *Journal of Chemical Theory and Computation, 20*(3), 1434–1447. DOI: 10.1021/acs.jctc.3c01081

Cai, Q. Y., Cai, W. Q., Liu, Y. F., Li, M. L., Xiang, S. E., Ji, H., Zhang, Z. S., Yu, N., Hu, Y. Q., Sun, Y., Yin, K. Y., Wu, H. M., & Li, M. W. (2025). Advancing lymph node metastasis in head and neck squamous cell carcinoma: Integrative mechanisms, emerging diagnostics, and translational therapies. *Critical Reviews in Oncology/Hematology, 215*, 104854. DOI: 10.1016/j.critrevonc.2025.104854 PMID: 40681008

Calvó-Tusell, C., Liu, Z., Chen, K., Arnold, F. H., & Garcia-Borràs, M. (2023). Reversing the enantioselectivity of enzymatic carbene N–H insertion through mechanism-guided protein engineering. *Angewandte Chemie International Edition, 62*(35), e202303879. DOI: 10.1002/anie.202303879 PMID: 37260412

Campos, K. R., Coleman, P. J., Alvarez, J. C., Dreher, S. D., Garbaccio, R. M., Terrett, N. K., Tillyer, R. D., Truppo, M. D., & Parmee, E. R. (2019). The importance of synthetic chemistry in the pharmaceutical industry. *Science, 363*(6424), eaat0805. Advance online publication. DOI: 10.1126/science.aat0805 PMID: 30655413

Caoping, N., Jing, Z., Wenming, X., Guo, C., Wei, L., Jie, Z., Rajeev, A., & Xian-long, W. (2023). Configuration stability and physical properties of new diamondene structure. *Materials Today. Communications, 36*, 106465. DOI: 10.1016/j.mtcomm.2023.106465

Cao, X., Wang, X., Chen, R., Chen, L., Liu, Y., & Wang, M. (2025). Improving Bacillus subtilis as Biological Chassis Performance by the CRISPR Genetic Toolkit. *ACS Synthetic Biology, 14*(3), 677–688. DOI: 10.1021/acssynbio.4c00844

Cardenas-Sierra, I., Vijay, U., Aguesse, F., Antuñano, N., Ayerbe, E., Gold, L., & Stier, S. (2025). Integration of lithium-ion battery recycling into manufacturing through digitalization: A perspective. *Journal of Power Sources, 631*, 236158. DOI: 10.1016/j.jpowsour.2024.236158

Cardoso, A. M., Paprocki, A., Ferret, L. S., Azevedo, C. M. N., & Pires, M. (2015). Synthesis of zeolite Na-P1 under mild conditions using Brazilian coal fly ash and its application in wastewater treatment. *Fuel, 139*, 59–67. DOI: 10.1016/j.fuel.2014.08.016

Carnero Canales, C. S., Marquez Cazorla, J. I., Marquez Cazorla, R. M., Santos, A. M. d., Lobato Duarte, J., Oliveira Catarin Nunes, L., Reis, T. C., Cerazi Salvador, L., Santos-Filho, N. A., Sábio, R. M., Santos, H. A., & Pavan, F. R. Redefining Therapies for Drug-Resistant Tuberculosis: Synergistic Effects of Antimicrobial Peptides, Nanotechnology, and Computational Design. *Advanced Healthcare Materials, n/a*(n/a), e03964. DOI: 10.1002/adhm.202503964

Casteleijn, M. G., Abendroth, U., Zemella, A., Walter, R., Rashmi, R., Haag, R., & Kubick, S. (2025). Beyond In Vivo, Pharmaceutical Molecule Production in Cell-Free Systems and the Use of Noncanonical Amino Acids Therein. *Chemical Reviews*, *125*(3), 1303–1331. DOI: 10.1021/acs.chemrev.4c00126

Castro, C., Urbieta, M. S., Plaza Cazón, J., & Donati, E. R. (2019). Metal biorecovery and bioremediation: Whether or not thermophilic are better than mesophilic microorganisms. *Bioresource Technology*, *279*, 317–326. DOI: 10.1016/j.biortech.2019.02.028 PMID: 30755320

Cavelier, S., Mirmohammadi, S. A., & Barthelat, F. (2021). Titanium mesh-reinforced calcium sulfate for structural bone grafts. *Journal of the Mechanical Behavior of Biomedical Materials*, *118*, 10446–10450. DOI: 10.1016/j.jmbbm.2021.104461 PMID: 33756418

Cecilia, J. A., Vilarrasa-García, E., Morales-Ospino, R., Finocchio, E., Busca, G., Sapag, K., Villarroel-Rocha, J., Bastos-Neto, M., Azevedo, D. C. S., & Rodríguez-Castellón, E. (2022). Kaolinite-based zeolites synthesis and their application in CO_2 capture processes. *Fuel*, *320*, 123953. DOI: 10.1016/j.fuel.2022.123953

Champion, W. M., Khaliq, M., & Mihelcic, J. R. (2022). Advancing Knowledge to Reduce Lead Exposure of Children in Data-Poor Low- and Middle-Income Countries. *Environmental Science & Technology Letters*, *9*(11), 879–888. DOI: 10.1021/acs.estlett.2c00656

Chandel, S. S., Singh, P. K., Katiyar, P. K., & Randhawa, N. S. (2023). A review on environmental concerns and technological innovations for the valorization of steel industry slag. *Mining, Metallurgy & Exploration*, *40*(6), 2059–2086. DOI: 10.1007/s42461-023-00886-z

Chauhan, G., Jadhao, P. R., Pant, K. K., & Nigam, K. D. P. (2018). Novel technologies and conventional processes for recovery of metals from waste electrical and electronic equipment: Challenges & opportunities - A review. *Journal of Environmental Chemical Engineering*, *6*(1), 1288–1304. DOI: 10.1016/j.jece.2018.01.032

Chen, S., Qi, H., Zhu, C., Zhao, Y., Jiao, B., Tan, Y., &Wu, C. (2025). Quality and composition control of complex TCM preparations through a novel "Herbs-in vivo Compounds-Targets-Pathways" network methodology: The case of Lianhuaqingwen capsules, 213. DOI: 10.1016/j.phrs.2025.107650

Cheng, D., Sha, W., Xu, Z., Huang, L., Du, Y., Tang, S., Guo, Y., Cao, Y.-C., & Cheng, S. (2022). Computer vision analysis on material characterization images. *Advanced Intelligent Systems*, *4*(3), 2100158. DOI: 10.1002/aisy.202100158

Cheng, Q., Wang, Z., Wang, Y., Li, J.-T., & Fu, H. (2024). Recent advances in preferentially selective Li recovery from spent lithium-ion batteries: A review. *Journal of Environmental Chemical Engineering*, *12*(3), 112903. DOI: 10.1016/j.jece.2024.112903

Cheng, X.-Q., Li, H., Zhao, Z.-X., Wang, Y.-Z., & Wang, X.-M. (2021). The use of in-situ Raman spectroscopy in investigating carbon materials as anodes of alkali metal-ion batteries. *New Carbon Materials*, *36*(1), 93–105. DOI: 10.1016/S1872-5805(21)60007-0

Chen, J., Lei, C., Li, X., Wu, Q., Liu, C., Ma, Q., & Chen, J. (2022). Research progress on classical traditional chinese medicine formula xiaoyaosan in the treatment of depression. *Frontiers in Pharmacology*, *13*, 925514. Advance online publication. DOI: 10.3389/fphar.2022.925514 PMID: 35991880

Chen, L. H., Sun, M. H., Wang, Z., Yang, W., Xie, Z., & Su, B. L. (2020). Hierarchically structured zeolites: From design to application. *Chemical Reviews*, *120*(20), 11194–11294. DOI: 10.1021/acs.chemrev.0c00016

Chen, Y., Freunberger, S. A., Peng, Z., Fontaine, O., & Bruce, P. G. (2013). Charging a Li–O2 battery using a redox mediator. *Nature Chemistry*, *5*(6), 489–494. DOI: 10.1038/nchem.1646 PMID: 23695630

Chen, Y., Li, J., Li, L., Han, M., & He, J. (2024). In Situ Synthesis and Tribological Characterization of TiC–Diamond Composites: Effect of the Counterface Material on Wear Rate and Mechanism. *Coatings*, *14*(6), 735–755. DOI: 10.3390/coatings14060735

Chen, Y., Xu, Y., Zhou, Y., Deng, Z., Feng, Y., Zhang, R., Yang, C., Li, H., Yu, H., & Zhang, G. (2025). Tailored magnetic ZIF-67-derived FeCo-LDH for PETase immobilization toward effective removal and biodegradation of high-crystallinity micro- and nano-PET. *Journal of Hazardous Materials*, *500*, 140419. DOI: 10.1016/j.jhazmat.2025.140419

Chen, Z., Li, Z., Li, J., Liu, C., Lao, C., Fu, Y., & He, Y. (2019). 3D printing of ceramics: A review. *Journal of the European Ceramic Society*, *39*(4), 661–687. https://doi.org/. DOI: 10.1016/j.jeurceramsoc.2018.11.013

Chica, A. (2013). Zeolites: Promised materials for the sustainable production of hydrogen. *International Scholarly Research Notices*, *2013*(1), 907425. DOI: 10.1155/2013/907425

Choi, H. J., & Hong, S. B. (2022). Effect of framework Si/Al ratio on the mechanism of CO_2 adsorption on the small-pore zeolite gismondine. *Chemical Engineering Journal, 433*(P3), 133800. DOI: 10.1016/j.cej.2021.133800

Choudhary, S., Venkatrayalu, G., Ravi Kiran, B., & Poluri, K. M. (2025). Advances and Perspectives of Microalgal Biorefineries for Bioenergy. *Energy & Fuels, 39*(26), 12367–12406. DOI: 10.1021/acs.energyfuels.5c01676

Chowdhury, S., Westenberg, R., Wennerholm, K., Cardiff, R. A. L., Beliaev, A. S., Noireaux, V., Carothers, J. M., & Peralta-Yahya, P. (2024). Carbon Negative Synthesis of Amino Acids Using a Cell-Free-Based Biocatalyst. *ACS Synthetic Biology, 13*(12), 3961–3975. DOI: 10.1021/acssynbio.4c00359

Ciftci, F., Özarslan, A. C., Kantarci, İ. C., Yelkenci, A., Tavukcuoglu, O., & Ghorbanpour, M. (2025). Advances in drug targeting, drug delivery, and nanotechnology applications: Therapeutic significance in cancer treatment. *Pharmaceutics, 17*(1), 121. DOI: 10.3390/pharmaceutics17010121 PMID: 39861768

Clarke, C. J., Tu, W. C., Levers, O., Bröhl, A., & Hallett, J. P. (2018). Green and sustainable solvents in chemical processes. *Chemical Reviews, 118*(2), 747–800. DOI: 10.1021/acs.chemrev.7b00571 PMID: 29300087

Clifton, R. A. (1987). *Natural and synthetic zeolites*. U.S. Department of the Interior, Bureau of Mines.

Coca-Lopez, N., Alcolea-Rodriguez, V., Bañares, M. A., Brockhauser, S., Gorenflot, J., Henderson, A., Hildebrandt, R., Jeliazkova, N., Kochev, N., Lozano Diz, E., Pilat, Z., Polli, D., Strömert, P., Sturm, C., Vanna, R., & Portela, R. (2025). Artificial Intelligence-Powered Raman Spectroscopy through Open Science and FAIR Principles. *ACS Nano, 19*(44), 38189–38218. DOI: 10.1021/acsnano.5c09165

Cong, X., Pierce, R., & Liu, X. L. (2021). Development of recycled polypropylene-based sustainable composites with recycled carbon fibre/Kenaf fibre hybrid reinforcements. *Journal of Physics: Conference Series, 1765*(1), 012013. DOI: 10.1088/1742-6596/1765/1/012013

Cui, W., Cao, L., & Zhang, X. (2022). Bioceramics: Materials, properties, and applications. *Ceramic Science and Engineering, 65*, 110. DOI: 10.1016/B978-0-323-89956-7.00017-6

Cully, M. (2019). CRISPR restores expression in models of vision loss. *Nature Reviews. Drug Discovery, 18*(3), 174. DOI: 10.1038/d41573-019-00016-9 PMID: 30824889

Cumberland, D. M., Deckard, T. G., Kahle-Piasecki, L., Kerrick, S. A., & Ellinger, A. D. (2023). Making sense of the digital badging landscape in education and workplace settings: A scoping review of the empirical literature. *European Journal of Training and Development*, *48*(1-2), 253–275. DOI: 10.1108/EJTD-06-2022-0067

Cundy, C. S., & Cox, P. A. (2003). The hydrothermal synthesis of zeolites: History and development from the earliest days to the present time. *Chemical Reviews*, *103*(3), 663–702. Advance online publication. DOI: 10.1021/cr020060i

Curutchet, C., & Mennucci, B. (2017). Quantum Chemical Studies of Light Harvesting. *Chemical Reviews*, *117*(2), 294–343. DOI: 10.1021/acs.chemrev.5b00700

da Silva, A. P., Poquioma Hernández, H. V., Comelli, C. L., Guillén Portugal, M. A., Moreira Delavy, F., de Souza, T. L., de Oliveira, E. C., de Oliveira-Ribeiro, C. A., Silva de Assis, H. C., & de Castilhos Ghisi, N. (2024). Meta-analytical review of antioxidant mechanisms responses in animals exposed to herbicide 2,4-D herbicide. *The Science of the Total Environment*, *924*, 171680. DOI: 10.1016/j.scitotenv.2024.171680

Daochun, H., Lei, W., Minghe, C., Shaohui, F., Ning, W., & Ying, Z. (2023). Thermal conduction and strength of diamond-copper composite sandwich obtained by SPS diffusion bonding with Ti interlayer. *Journal of Materials Research and Technology*, *26*, 8806–8812. DOI: 10.1016/j.jmrt.2023.09.188

de Araujo Guilherme, A., Dantas, P. V. F., Padilha, C. E. A., dos Santos, E. S., & de Macedo, G. R. (2019). Ethanol production from sugarcane bagasse: Use of different fermentation strategies to enhance an environmental-friendly process. *Journal of Environmental Management*, *234*, 44–51. DOI: 10.1016/j.jenvman.2018.12.102 PMID: 30599329

Debnath, M., Sarkar, S., Debnath, S. K., Dkhar, D. S., Kumari, R., Vaskuri, G. S. S. J., Srivastava, A., Chandra, P., Prasad, R., & Srivastava, R. (2024). Photothermally Active Quantum Dots in Cancer Imaging and Therapeutics: Nanotheranostics Perspective. *ACS Applied Bio Materials*, *7*(12), 8126–8148. DOI: 10.1021/acsabm.4c01190

Dembele, S., Akcil, A., & Panda, S. (2022). Technological trends, emerging applications and metallurgical strategies in antimony recovery from stibnite. *Minerals Engineering*, *175*, 107304. DOI: 10.1016/j.mineng.2021.107304

Demidenko, M., Adamchuk, D., Liubimau, A., Uglov, V., Ishchenko, A., Chekan, M., Khama, M., & Maksimenko, S. (2024). High temperature synthesis and material properties of boron-enriched balk pyrolytic carbon. *Materials Science and Engineering B*, *307*, 117491. DOI: 10.1016/j.mseb.2024.117491

Demirocak, D. E., & Bhushan, B. (2014). In situ atomic force microscopy analysis of morphology and particle size changes in lithium iron phosphate cathode during discharge. *Journal of Colloid and Interface Science*, *423*, 151–157. DOI: 10.1016/j.jcis.2014.02.035 PMID: 24703680

Deng, Z., Li, H., Zhou, Q. and Zheng, H. (2025). Progress of Clinical Research on the Treatment of Neurogenic Lower Urinary Tract Dysfunction after Stroke with External Treatment of Chinese Medicine, 14. DOI: 10.12677/tcm.2025.148504

Deng, C., Chen, H., Meng, Z., & Meng, S. (2022). Roles of traditional chinese medicine regulating neuroendocrinology on AD treatment. *Frontiers in Endocrinology*, *13*, 955618. Advance online publication. DOI: 10.3389/fendo.2022.955618 PMID: 36213283

Deng, D., Kim, M. G., Lee, J. Y., & Cho, J. (2009). Green energy storage materials: Nanostructured TiO2 and Sn-based anodes for lithium-ion batteries. *Energy & Environmental Science*, *2*(8), 818–837. DOI: 10.1039/b823474d

Deng, J., Yang, X., & Zhang, G. (2022). Simulation study on internal short circuit of lithium ion battery caused by lithium dendrite. *Materials Today. Communications*, *31*, 103570. DOI: 10.1016/j.mtcomm.2022.103570

Denisart, L., Troncoso, J. F., Loup-Escande, E., & Franco, A. A. (2024). Breaking down the Barriers between the Digital and the Real: Mixed Reality applied to Battery Manufacturing R&D and Training. *Batteries & Supercaps*, *7*(7), e202400042. DOI: 10.1002/batt.202400042

Denisart, L., Zapata-Dominguez, D., David, X., Leclere, A., Lelong, R., Liu, C., Xu, J., Loup-Escande, E., & Franco, A. A. (2024). Combining virtual reality with mixed reality for efficient training in battery manufacturing. *Batteries & Supercaps*, *7*(1), e202300268. DOI: 10.1002/batt.202300268

Devi, L., & Kavitha, R. (2013). A review on non metal ion doped titania for the photocatalytic degradation of organic pollutants under UV/solar light: Role of photogenerated charge carrier dynamics in enhancing the activity. *Applied Catalysis B: Environment and Energy*, *140-141*, 559–587. DOI: 10.1016/j.apcatb.2013.04.035

Diercks, C S, & Yaghi, O M (2017). The atom, the molecule, and the covalent organic framework. Science, 355(6328), eaal1585. DOI: 10.1126/science.aal1585

Ding, C., Chen, Z., Cao, C., Liu, Y., & Gao, Y. (2023). Advances in Mn-based electrode materials for aqueous sodium-ion batteries. *Nano-Micro Letters*, *15*(1), 192. Advance online publication. DOI: 10.1007/s40820-023-01162-x PMID: 37555908

Djomgoue, P., & Njopwouo, D. (2013). FT-IR spectroscopy applied for surface clays. *Journal of Surface Engineered Materials and Advanced Technology*, *3*(4), 275–282. DOI: 10.4236/jsemat.2013.34037

Dong, C., Chen, Y., Yang, C., Li, P., Zhang, Y., Wang, P., Wang, Y., Fang, W., Lin, L., & Li, X. (2024). Pinning-effect single-atom NiCo alloy embedded graphene-aerogel in electro-fenton process for rapid degradation of emerging contaminants. *Applied Catalysis B: Environment and Energy*, *357*, 124286–124286. DOI: 10.1016/j.apcatb.2024.124286

Dong, S., Chen, J., Wang, K., Xiao, H., & Huang, T. (2025). Novel functional carbon dot based on Angelica residue and polyethyleneimine: A super-efficient corrosion and scale inhibitor. *Journal of Colloid and Interface Science*, *679*, 64–74. DOI: 10.1016/j.jcis.2024.09.222 PMID: 39357227

Dorozhkin, S. V. (2010). Calcium orthophosphates as bioceramics: State of the art. *Journal of Functional Biomaterials*, *1*(1), 22–107. DOI: 10.3390/jfb1010022 PMID: 24955932

Dragomirova, R., & Wohlrab, S. (2015). Zeolite membranes in catalysis: From separate units to particle coatings. *Catalysts*, *5*(4), 2161–2222. DOI: 10.3390/catal5042161

Drautz, R. (2019). Atomic cluster expansion for accurate and transferable interatomic potentials. *Physical Review. B*, *99*(1), 014104. DOI: 10.1103/PhysRevB.99.014104

Draxl, C., & Scheffler, M. (2018). NOMAD: The FAIR concept for big data-driven materials science. *MRS Bulletin*, *43*(9), 676–682. DOI: 10.1557/mrs.2018.208

Drost, J., & Clevers, H. (2018). Organoids in cancer research. *Nature Reviews. Cancer*, *18*(7), 407–418. DOI: 10.1038/s41568-018-0007-6 PMID: 29692415

Du, H., Liu, K., Xu, T., Xu, C., Lin, M., Fang, Z., Kim, S.-W., Seo, J.-Y., Chen, J., Ma, H., Hsiao, B. S., DeVetter, L. W., Piao, Z., Si, C., Chen, C., Yang, Q., Lee, S.-Y., Yao, Y., & Pan, X. (2025). Lignocellulosic Films: Preparation, Properties, and Applications. *Chemical Reviews*, *125*(24), 11666–11814. DOI: 10.1021/acs.chemrev.5c00267

Duignan, T. T. (2024). The Potential of Neural Network Potentials. *ACS Physical Chemistry Au*, *4*(3), 232–241. DOI: 10.1021/acsphyschemau.4c00004 PMID: 38800721

Dutta, A., Sahir, A. H., Tan, E., Humbird, D., Snowden-Swan, L. J., Meyer, P. A., . . . Lukas, J. (2015). *Process design and economics for the conversion of lignocellulosic biomass to hydrocarbon fuels: Thermochemical research pathways with in situ and ex situ upgrading of fast pyrolysis vapors* (No. PNNL-23823). Pacific Northwest National Laboratory.

Dutta, S., Fajal, S., & Ghosh, S. K. (2024). Heavy Metal-Based Toxic Oxo-Pollutants Sequestration by Advanced Functional Porous Materials for Safe Drinking Water. *Accounts of Chemical Research, 57*(17), 2546–2560. DOI: 10.1021/acs.accounts.4c00348 PMID: 39163541

Dutta, V., Sharma, S., Raizada, P., Thakur, V., Khan, A., Saini, V., Asiri, A., & Singh, P. (2021). An overview on WO_3 based photocatalyst for environmental remediation. *Journal of Environmental Chemical Engineering, 9*(1), 105018–105018. DOI: 10.1016/j.jece.2020.105018

E, S., He, C., Wang, J., Mao, Q., & Chen, X. (2021). Tunable organelle imaging by rational design of carbon dots and utilization of uptake pathways. *ACS Nano, 15*(9), 14465-14474. doi:https://doi.org/DOI: 10.1021/acsnano.1c04001

Edison, T. N. J. I., Atchudan, R., Sethuraman, M. G., Shim, J.-J., & Lee, Y. R. (2016). Microwave assisted green synthesis of fluorescent N-doped carbon dots: Cytotoxicity and bio-imaging applications. *Journal of Photochemistry and Photobiology. B, Biology, 161*, 154–161. DOI: 10.1016/j.jphotobiol.2016.05.017 PMID: 27236237

El Gaidoumi, A., Benabdallah, A. C., & El Bali, B. (2017). Synthesis and characterization of zeolite HS using natural pyrophyllite as new clay source. *Arabian Journal for Science and Engineering*. Advance online publication. DOI: 10.1007/s13369-017-2768-8

Elimelech, M., & Phillip, W. A. (2011). The future of seawater desalination: Energy, technology, and the environment. *Science, 333*(6043), 712–717. DOI: 10.1126/science.1200488 PMID: 21817042

Emanuele, G., Giovanni, B., Alessandro, B., & Emanuela, C. (2023). Diamond-Like Carbon (DLC) and AlCrN films onto Ti-6Al-4V substrates by Laser-Powder Bed Fusion (L-PBF): Effect of substrate heat treatment and surface finish. *Surface and Coatings Technology, 475*, 130128. DOI: 10.1016/j.surfcoat.2023.130128

Engstrand, T., Kihlström, L., Lundgren, K., Trobos, M., Engqvist, H., & Thomsen, P. (2015). Bioceramic implant induces bone healing of cranial defects. *Plastic and Reconstructive Surgery. Global Open, 3*(8), e491. DOI: 10.1097/GOX.0000000000000467 PMID: 26495204

Enkavi, G., Javanainen, M., Kulig, W., Róg, T., & Vattulainen, I. (2019). Multiscale Simulations of Biological Membranes: The Challenge To Understand Biological Phenomena in a Living Substance. *Chemical Reviews, 119*(9), 5607–5774. DOI: 10.1021/acs.chemrev.8b00538

Ernst, M. S., Tangstad, M., & Du Preez, S. P. (2024). Pre-reduction of Nchwaning manganese ore in CO/CO_2, H_2/H_2O, and H_2 atmospheres. *Minerals Engineering, 216*, 108854. Advance online publication. DOI: 10.1016/j.mineng.2024.108854

Eroglu, N., & Athanassiou, C. G. (2017). Applications of natural zeolites on agriculture and food production. *Journal of the Science of Food and Agriculture, 97*(11), 3487–3499. Advance online publication. DOI: 10.1002/jsfa.8312

Essaidi, N., Samet, B., Baklouti, S., & Rossignol, S. (2014). Science of producing geopolymers from two different Tunisian clays before and after calcination at various temperatures. *Applied Clay Science, 88–89*, 221–227. DOI: 10.1016/j.clay.2013.12.006

Faheem, A., Qin, Y., Nan, W., & Hu, Y. (2021). Advances in the Immunoassays for Detection of Bacillus thuringiensis Crystalline Toxins. *Journal of Agricultural and Food Chemistry, 69*(36), 10407–10418. DOI: 10.1021/acs.jafc.1c02195

Fajrina, N., & Tahir, M. (2018). A critical review in strategies to improve photocatalytic water splitting towards hydrogen production. *International Journal of Hydrogen Energy, 44*(2), 540–577. DOI: 10.1016/j.ijhydene.2018.10.200

Fallahzadeh, R., & Farhadian, N. (2015). Molecular dynamics simulation of lithium ion diffusion in LiCoO2 cathode material. *Solid State Ionics, 280*, 10–17. DOI: 10.1016/j.ssi.2015.07.001

Fang, S. F., Wang, M. P., Qi, W. H., & Zheng, F. (2008). Hybrid genetic algorithms and support vector regression in forecasting atmospheric corrosion of metallic materials. *Computational Materials Science, 44*(2), 647–655. DOI: 10.1016/j.commatsci.2008.05.010

Fan, Q., Lin, K., Yang, S., Guan, S., Chen, J., Feng, S., Liu, J., Liu, L., Li, J., & Shi, Z. (2020). Constructing effective TiO2 nano-coating for high-voltage Ni-rich cathode materials for lithium ion batteries by precise kinetic control. *Journal of Power Sources, 477*, 228745. DOI: 10.1016/j.jpowsour.2020.228745

Fan, X., Hou, J., Sun, D. G., Xi, S. Y., Liu, Z. H., Du, J., Luo, J. L., & Tao, C. Y. (2013). Mn-oxides catalyzed periodic current oscillation on the anode. *Electrochimica Acta, 102*, 466–471. DOI: 10.1016/j.electacta.2013.03.175

Fan, X., Yang, D. P., Ding, L. F., Du, J., & Tao, C. Y. (2015). Periodic current oscillation catalyzed by δ-MnO_2 nanosheets. *Chemphyschem : A European Journal of Chemical Physics and Physical Chemistry*, *16*(1), 176–180. DOI: 10.1002/cphc.201402623 PMID: 25382485

Farjana, S. H., Huda, N., Mahmud, M. A. P., & Lang, C. (2019). A global life cycle assessment of manganese mining processes based on EcoInvent database. *The Science of the Total Environment*, *688*, 1102–1111. DOI: 10.1016/j.scitotenv.2019.06.184 PMID: 31726541

Farrant, E. (2020). Automation of synthesis in medicinal chemistry: Progress and challenges. *ACS Medicinal Chemistry Letters*, *11*(8), 1506–1513. DOI: 10.1021/acsmedchemlett.0c00292 PMID: 32832016

Fathi, E. M., Salem, M. E., Sroor, F. M., Mohamed, F. G., Elwahy, A. H. M., Abdel-Megid, M., & Emara, M. (2025). Synthesis and evaluation of novel bis-chalcone derivatives containing a thiophene moiety as potential anticancer agents: In vitro, in silico, and mechanistic studies. *ACS Omega*, *10*(24), 25921–25937. DOI: 10.1021/acsomega.5c02394 PMID: 40584358

Feijen, E. J. P., Martens, J. A., & Jacobs, P. A. (1994). Zeolites and their mechanism of synthesis. *Studies in Surface Science and Catalysis*, *84*, 3–21. DOI: 10.1016/S0167-2991(08)64074-4

Feng, X., Yuan, X., Zhou, J., An, K., Zhu, F., Wei, X., Huang, Y., Zhang, J., Chen, L., Liu, J., Li, C., & Wei, J. (2025). A review: CNT/diamond composites prepared via CVD and its potential applications. *Materials Science in Semiconductor Processing*, *186*, 109008. DOI: 10.1016/j.mssp.2024.109008

Fernandes Machado, N. R. C., & Miotto, D. M. M. (2005). Synthesis of Na-A and -X zeolites from oil shale ash. *Fuel*, *84*(18), 2289–2294. DOI: 10.1016/j.fuel.2005.05.003

Fernando, R.-R., Victor, Z., & Fernando, G. (2023). Solid-state spark plasma sintering of super wear resistant B4C–SiC–TiB2 triplex-particulate composites. *Ceramics International*, *49*(3), 5532–5537. DOI: 10.1016/j.ceramint.2022.11.181

Field, F. R.III, Wallington, T. J., Everson, M., & Kirchain, R. E. (2017). Strategic Materials in the Automobile: A Comprehensive Assessment of Strategic and Minor Metals Use in Passenger Cars and Light Trucks. *Environmental Science & Technology*, *51*(24), 14436–14444. DOI: 10.1021/acs.est.6b06063

Finegan, D. P., Squires, I., Dahari, A., Kench, S., Jungjohann, K. L., & Cooper, S. J. (2022). Machine-learning-driven advanced characterization of battery electrodes. *ACS Energy Letters*, *7*(12), 4368–4378. DOI: 10.1021/acsenergylett.2c01996

Fourmentin, S., & Benzina, M. (2015). Synthesis of faujasite type zeolite from low grade Tunisian clay for the removal of heavy metals from aqueous waste by batch process: Kinetic and equilibrium study. *Comptes Rendus. Chimie, 18*(10), 1123–1133. DOI: 10.1016/j.crci.2015.03.013

Franco, A. A., Chotard, J. N., Loup-Escande, E., Yin, Y., Zhao, R., Rucci, A., & Ciger, J. (2020). Entering the augmented era: Immersive and interactive virtual reality for battery education and research. *Batteries & Supercaps, 3*(11), 1147–1164. DOI: 10.1002/batt.202000120

Fu, J., Yang, G., Lu, M., Fang, K., Lv, P., & Xiong, Z. (2025). Chemical simulation of high-performance CaO/La2O3 catalysts on its anti-leaching property during biodiesel production. *New Journal of Chemistry = Nouveau Journal de Chimie, 49*(26), 11180–11190. DOI: 10.1039/D5NJ01085C

Fujishima, A., & Honda, K. (1972). Electrochemical photolysis of water at a semiconductor electrode. *Nature, 238*(5358), 37–38. DOI: 10.1038/238037a0 PMID: 12635268

Furukawa, H., Cordova, K. E., O'Keeffe, M., & Yaghi, O. M. (2013). The chemistry and applications of metal-organic frameworks. *Science, 341*(6149), 1230444. DOI: 10.1126/science.1230444 PMID: 23990564

Fu, T., Monaco, F., Li, J., Zhang, K., Yuan, Q., Cloetens, P., Pianetta, P., & Liu, Y. (2022). Deep-Learning-Enabled Crack Detection and Analysis in Commercial Lithium-Ion Battery Cathodes. *Advanced Functional Materials, 32*(39), 2203070. DOI: 10.1002/adfm.202203070

Gabisa, E. W., & Aman, A. (2016). Characterization and experimental investigation of NaNO3: KNO3 as solar thermal energy storage for potential cooking application. *Journal of Solar Energy, 2016*(1), 2405094. DOI: 10.1155/2016/2405094

Galvez-Aranda, D. E., Saravanan, S., & Franco, A. A. (2025). Human-Centered Holographic Assistant using Computer Vision to Interact with Battery Manufacturing Machinery in a Predictive Manner. *Batteries & Supercaps, 8*(12), 2500292. DOI: 10.1002/batt.202500292

Galvez-Aranda, D. E., & Seminario, J. M. (2018). Simulations of a LiF solid electrolyte interphase cracking on silicon anodes using molecular dynamics. *Journal of the Electrochemical Society, 165*(3), A717–A730. DOI: 10.1149/2.0991803jes

Galvez-Aranda, D. E., & Seminario, J. M. (2022). Li-metal anode in a conventional Li-ion battery electrolyte: Solid electrolyte interphase formation using ab initio molecular dynamics. *Journal of the Electrochemical Society*, *169*(3), 030502. DOI: 10.1149/1945-7111/ac55c8

Gao, J., Hou, B., Zhu, Q., Yang, L., Jiang, X., Zou, Z., Li, X., Xu, T., Zheng, M., Chen, Y.-H., Xu, Z., Xu, H., & Yu, H. (2022). Engineered bioorthogonal POLY-PROTAC nanoparticles for tumour-specific protein degradation and precise cancer therapy. *Nature Communications*, *13*(1), 4318. Advance online publication. DOI: 10.1038/s41467-022-32050-4

Gao, W., Zhou, W., Lu, X., Liu, X., Su, H., Li, C., & Wang, H. (2023). Comprehensive utilization of steel slag: A review. *Powder Technology*, *422*, 118449. DOI: 10.1016/j.powtec.2023.118449

Gao, X., & Xu, J. (2020). Multiscale Modeling of Electro-Chemo-Mechanical Degradation in Si/C Core–Shell Anode for the Lithium-Ion Battery of High Energy Density. *Journal of Electrochemical Energy Conversion and Storage*, *18*(2), 020903. Advance online publication. DOI: 10.1115/1.4048704

Gao, Y., Yan, L., Zhao, C., Chen, M., Yang, S., Shao, G., & Mao, J. J. (2024). Accurate calculation of solid-state Li+ diffusion coefficient and kinetic activation energies for an artificial graphite anode. *Journal of the Electrochemical Society*, *171*(2), 020558. DOI: 10.1149/1945-7111/ad27b4

Garcia, G., Cabrera, S., Hedlund, J., & Mouzon, J. (2018). Selective synthesis of FAU-type zeolites. *Journal of Crystal Growth*, *489*, 36–41. DOI: 10.1016/j.jcrysgro.2018.02.022

Garcia-Segura, S., & Brillas, E. (2017). Applied photoelectrocatalysis on the degradation of organic pollutants in wastewaters. *Journal of Photochemistry and Photobiology C, Photochemistry Reviews*, *31*, 1–35. DOI: 10.1016/j.jphotochemrev.2017.01.005

Garg, H., Nagar, N., Ellamparuthy, G., Angadi, S. I., & Gahan, C. S. (2019). Bench scale microbial catalysed leaching of mobile phone PCBs with an increasing pulp density. *Heliyon*, *5*(12), e02883. DOI: 10.1016/j.heliyon.2019.e02883 PMID: 31872109

Gavilán-Arriazu, E. M., Mercer, M., Barraco, D., Hoster, H. E., & Leiva, E. P. M. (2021). Kinetic Monte Carlo simulations applied to Li-ion and post Li-ion batteries: A key link in the multi-scale chain. *Progress in Energy*, *3*(4), 042001. DOI: 10.1088/2516-1083/ac1a65

Gebara, P., Gebara, M., & Owczarek, A. (2020). Fractal structures in electrodeposition process. *Acta Physica Polonica A, 138*(2), 287–290. DOI: 10.12693/APhysPolA.138.287

Gerasimov, M., Soto, F. A., Wagner, J., Baakes, F., Guo, N., Ospina-Acevedo, F., Röder, F., Balbuena, P. B., & Krewer, U. (2023). Species Distribution During Solid Electrolyte Interphase Formation on Lithium Using MD/DFT-Parameterized Kinetic Monte Carlo Simulations. *The Journal of Physical Chemistry. C, Nanomaterials and Interfaces*, *127*(10), 4872–4886. DOI: 10.1021/acs.jpcc.2c05898

Ghasemi, S. M., & Azizi, A. (2018). Alkaline leaching of lead and zinc by sodium hydroxide: Kinetics modeling. *Journal of Materials Research and Technology*, *7*(2), 118–125. DOI: 10.1016/j.jmrt.2017.03.005

Ghosh, J., Repon, M. R., Rupanty, N. S., Asif, T. R., Tamjid, M. I., & Reukov, V. (2025). Chemical Valorization of Textile Waste: Advancing Sustainable Recycling for a Circular Economy. *ACS Omega, 10*(12), 11697–11722. DOI: 10.1021/acsomega.4c10616

Ghulam Ali, S., Khan, R., Rehman, F., Bilal, M., & Ullah, S. (2022). Thermocatalytic pyrolysis of lignin using hydrogen-loaded zeolite Y as catalyst for fuel range hydrocarbons production. *Energies*, *15*(18), 6672. DOI: 10.3390/en15186672

Giovanis, E., Akdede, S. H., & Ozdamar, O. (2021). Impact of the EU Blue Card programme on cultural participation and subjective well-being of migrants in Germany. *PLoS One*, *16*(7), e0253952. DOI: 10.1371/journal.pone.0253952

Godfrey, A. G., Masquelin, T., & Hemmerle, H. (2013). A remote-controlled adaptive medchem lab: An innovative approach to enable drug discovery in the 21st century. *Drug Discovery Today*, *18*(17-18), 795–802. DOI: 10.1016/j.drudis.2013.03.001 PMID: 23523957

Green, A. P., & Turner, N. J. (2016). Biocatalytic retrosynthesis: Redesigning synthetic routes to high-value chemicals. *Perspectives in Science*, *9*, 42–48. DOI: 10.1016/j.pisc.2016.04.106

Grey, C. P., & Hall, D. S. (2020). Prospects for lithium-ion batteries and beyond—A 2030 vision. *Nature Communications*, *11*(1), 6279. DOI: 10.1038/s41467-020-19991-4 PMID: 33293543

Grzybowski, A., Pawlikowska–Łagód, K., & Lambert, W. C. (2024). A history of artificial intelligence. *Clinics in Dermatology*, *42*(3), 221–229. DOI: 10.1016/j.clindermatol.2023.12.016 PMID: 38185196

Gualtieri, A. F., & Modena, Á. (2001). Synthesis of sodium zeolites from a natural halloysite. *Physics and Chemistry of Minerals*, *28*(2), 719–728. DOI: 10.1007/s002690100197

Guessous, G., Blake, L., Bui, A., Woo, Y., & Manzanarez, G. (2024). Disentangling the Web: An Interdisciplinary Review on the Potential and Feasibility of Spider Silk Bioproduction. *ACS Biomaterials Science & Engineering*, *10*(9), 5412–5438. DOI: 10.1021/acsbiomaterials.4c00145

Guillon, O. (2021). Ceramic materials for energy conversion and storage: A perspective. *International Journal of Ceramic Engineering & Science*, *3*(3), 100–104. DOI: 10.1002/ces2.10086

Guisnet, M., & Ribeiro, F. R. (2006). *Les zéolithes, un nanomonde au service de la catalyse*. EDP Sciences.

Gunasekara, C., Li, J., Tahmasebi, M., Law, D., Brkljac, R., & Patnaikuni, I. (2020). Geopolymer synthesis using low-grade clays. *Construction & Building Materials*, *121066*. Advance online publication. DOI: 10.1016/j.conbuildmat.2020.121066

Guo, J., Chen, X., Wu, M., Wang, D., Zhao, Y., Li, Q., &Li, M. (2024). Traditional Chinese medicine FYTF-919 (Zhongfeng Xingnao oral prescription) for the treatment of acute intracerebral haemorrhage: a multicentre, randomised, placebo-controlled, double-blind, clinical trial, 404. DOI: 10.1016/S0140-6736(24)02261-X

Guth, J., & Kessler, H. (1999). Synthesis of aluminosilicate zeolites and related silica-based materials. In *Catalysis and zeolites* (pp. 1–52). DOI: 10.1007/978-3-662-03764-5_1

Hadke, S., Kang, M.-A., Sangwan, V. K., & Hersam, M. C. (2025). Two-Dimensional Materials for Brain-Inspired Computing Hardware. *Chemical Reviews*, *125*(2), 835–932. DOI: 10.1021/acs.chemrev.4c00631

Han, F., & Wu, L. (2019). Resource utilization of electrolytic manganese residues. In Industrial solid waste recycling in Western China (pp. 127–164). DOI: 10.1007/978-981-13-8086-0_3

Han, X. W., Zhang, W. B., Li, J. J., Liu, X. Y., Yuan, X. Y., Ma, X. J., & Long, J. P. (2024). Salinity gradient energy harvesting via Na-Mn(PO_4)-based electrochemical system. *ACS Sustainable Chemistry & Engineering*, *12*(21), 8010–8021. DOI: 10.1021/acssuschemeng.3c08508

Han, Y., Gao, R., Bajpai, I., Kim, B., Yoshida, H., Nieto, A., Son, H.-W., Yun, J., Jang, B.-K., Jhung, S., Jingming, Z., Hwang, K.-H., Chen, F., Shackelford, J. F., & Kim, S. (2020). Spark plasma sintered bioceramics-from transparent hydroxyapatite to graphene nanocomposites: A review. *Advances in Applied Ceramics*, *119*(2), 57–74. DOI: 10.1080/17436753.2019.1691871

Han, Y., Tang, B., Wang, L., Bao, H., Lu, Y., Guan, C., Zhang, L., Le, M., Liu, Z., & Wu, M. (2020). Machine-learning-driven synthesis of carbon dots with enhanced quantum yields. *ACS Nano*, *14*(11), 14761–14768. DOI: 10.1021/acsnano.0c01899 PMID: 32960048

Hao, L., Zhang, J., Liu, J., Min, Y., & Chen, C. (2023). Applications of Carbon-Based Materials in Activated Peroxymonosulfate for the Degradation of Organic Pollutants: A Review. *Chemical Record : An Official Publication of the Chemical Society of Japan)*, *23*(12), e202300203. Advance online publication. DOI: 10.1002/tcr.202300203 PMID: 37639150

Haq, H., Välisuo, P., & Niemi, S. (2021). Modelling Sustainable Industrial Symbiosis. *Energies*, *14*(4), 1172. DOI: 10.3390/en14041172

Harris, A. R., & Davidson, C. I. (2005). The Role of Resuspended Soil in Lead Flows in the California South Coast Air Basin. *Environmental Science & Technology*, *39*(19), 7410–7415. DOI: 10.1021/es050642s

Hartley, S., Taitingfong, R., & Fidelman, P. (2022). The principles driving gene drives for conservation. *Environmental Science & Policy*, *135*, 36–45. DOI: 10.1016/j.envsci.2022.04.021

Hassan, M. I., Anjum, D., Mohammad, T., Alam, M., Khan, M. S., Shahwan, M., Shamsi, A., & Yadav, D. K. (2022). Integrated virtual screening and MD simulation study to discover potential inhibitors of Lyn-kinase: Targeting cancer therapy. *Journal of Biomolecular Structure & Dynamics*, *41*(20), 10558–10568. DOI: 10.1080/07391102.2022.2154849 PMID: 36495308

Hayler, J. D., Leahy, D. K., & Simmons, E. M. (2019). A Pharmaceutical Industry Perspective on Sustainable Metal Catalysis. *Organometallics*, *38*(1), 36–46. DOI: 10.1021/acs.organomet.8b00566

He, C., Xu, P., Zhang, X., & Long, W. (2022). The synthetic strategies, photoluminescence mechanisms and promising applications of carbon dots: Current state and future perspective. *Carbon*, *186*, 91–127. DOI: 10.1016/j.carbon.2021.10.002

He, D., Shu, J., Wang, R., Chen, M., Wang, R., Gao, Y., Liu, R., Liu, Z., Xu, Z., Tan, D., Gu, H., & Wang, N. (2021). A critical review on approaches for electrolytic manganese residue treatment and disposal technology: Reduction, pretreatment, and reuse. *Journal of Hazardous Materials*, *418*, 126235. Advance online publication. DOI: 10.1016/j.jhazmat.2021.126235 PMID: 34126381

He, L., Shi, L., Huang, Q., Hayat, W., Shang, Z., Ma, T., Wang, M., Yao, W., Huang, H., & Chen, R. (2021). Extraction of alumina from aluminum dross by a non-hazardous alkaline sintering process: Dissolution kinetics of alumina and silica from calcined materials. *The Science of the Total Environment*, *777*, 146123. DOI: 10.1016/j.scitotenv.2021.146123 PMID: 33684763

Hench, L. (1991). Bioceramics: From concept to clinic. *Journal of the American Ceramic Society*, *74*(7), 1487–1510. DOI: 10.1111/j.1151-2916.1991.tb07132.x

He, Q., Yu, B., Li, Z., & Zhao, Y. (2019). Density functional theory for battery materials. *Energy & Environmental Materials*, *2*(4), 264–279. DOI: 10.1002/eem2.12056

He, S., Jiang, D., Hong, M., & Liu, Z. (2021). Hazard-free treatment and resource utilisation of electrolytic manganese residue: A review. *Journal of Cleaner Production*, *306*, 127224. Advance online publication. DOI: 10.1016/j.jclepro.2021.127224

He, S., Wang, T., Liu, Z., Jiang, T., & Liu, Z. (2025). Mechanochemistry enhanced low-temperature molten salt synergistic treatment of low-grade pyrolusite and waste salts from the electrolytic manganese metal industry. *Chemical Engineering Journal*, *506*, 159338. Advance online publication. DOI: 10.1016/j.cej.2025.159338

He, X., Zhang, Y., Li, H., Zhou, M., Wang, W., Wang, R., Jiang, K., & Wang, K. (2025). Improving the reliability of classical molecular dynamics simulations in battery electrolyte design. *Journal of Energy Chemistry*, *101*, 34–41. DOI: 10.1016/j.jechem.2024.09.038

Hidalgo, M., Apachitei, G., Dogaru, D., Faraji-Niri, M., Lain, M., Copley, M., & Marco, J. (2023). Design of experiments for optimizing the calendering process in Li-ion battery manufacturing. *Journal of Power Sources*, *573*, 233091. DOI: 10.1016/j.jpowsour.2023.233091

Hoffmann, R. W. (2012). Natural product synthesis: Changes over time. *Angewandte Chemie International Edition*, *52*(1), 123–130. DOI: 10.1002/anie.201203319 PMID: 23203402

Hofmann, T., Hamar, J., Rogge, M., Zoerr, C., Erhard, S., & Schmidt, J. P. (2023). Physics-informed neural networks for state of health estimation in lithium-ion batteries. *Journal of the Electrochemical Society*, *170*(9), 090524. DOI: 10.1149/1945-7111/acf0ef

Hong, J., Kobayashi, S., Kuwabara, A., Ikuhara, Y. H., Fujiwara, Y., & Ikuhara, Y. J. M. (2021). Defect engineering and anisotropic modulation of ionic transport in perovskite solid electrolyte LixLa(1−x)/3NbO3. *ACS Nano*, *26*(12), 3559. PMID: 34200888

Hou, H. Z., Haikuo, W., Yao, T., Jiakun, W., Chao, W., Zhicai, Z., & Xiaoping, O. (2022). Research progress of spectra and properties of ultrahard carbon materials at high pressure and high temperature. *Functional Diamond*, *2*(1), 245–257. DOI: 10.1080/26941112.2022.2163857

Hou, X., Zhao, Y., Shang, Y., Shen, F., Wu, B., Chen, D., Liu, Z., Yao, M., & Liu, B. (2025). Nanostructure engineering of superhard nano-polycrystalline diamond by compressing different fullerene precursors. *Carbon*, *235*, 120078. DOI: 10.1016/j.carbon.2025.120078

Huang, D., Li, Z., Zeng, G., Zhou, C., Xue, W., Gong, X., Yan, X., Chen, S., Wang, W., & Cheng, M. (2018). Megamerger in photocatalytic field: 2D g-C_3N_4 nanosheets serve as support of 0D nanomaterials for improving photocatalytic performance. *Applied Catalysis B: Environment and Energy*, *240*, 153–173. DOI: 10.1016/j.apcatb.2018.08.071

Huang, G., Chen, L., & Fang, C. (2024). Synthesis and nitrogen content regulation of diamond in a high-pressure hydrogen-rich environment. *Chinese Physics B*, *33*(6), 068102. DOI: 10.1088/1674-1056/ad2a6d

Huang, J., Luo, Y., Weng, M., Yu, J., Sun, L., Zeng, H., Liu, Y., Zeng, W., Min, Y., & Guo, Z. (2021). Advances and applications of phase change materials (PCMs) and PCMs-based technologies. *ES Materials & Manufacturing*, *13*, 23–39. DOI: 10.30919/esmm5f458

Huang, J., Xu, Y., Wang, Q., Wang, Q., Liang, X., Wang, F., Zhang, Z., Wei, W., Zhang, B., Huang, L., Chang, J., Ma, L., Ma, T., Liang, Y., Zhang, J., Guo, J., Jiang, X., Fan, X., An, Z., & Fei, A. (2025). Foundation models and intelligent decision-making: Progress, challenges, and perspectives. *Innovation (Cambridge (Mass.))*, *6*(6), 100948. Advance online publication. DOI: 10.1016/j.xinn.2025.100948

Hua, W., Wang, S., Knapp, M., Leake, S. J., Senyshyn, A., Richter, C., Yavuz, M., Binder, J. R., Grey, C. P., Ehrenberg, H., Indris, S., & Schwarz, B. (2019). Structural insights into the formation and voltage degradation of lithium- and manganese-rich layered oxides. *Nature Communications*, *10*(1), 5365. DOI: 10.1038/s41467-019-13240-z PMID: 31772159

Huber, S. P., Zoupanos, S., Uhrin, M., Talirz, L., Kahle, L., Häuselmann, R., Gresch, D., Müller, T., Yakutovich, A. V., Andersen, C. W., Ramirez, F. F., Adorf, C. S., Gargiulo, F., Kumbhar, S., Passaro, E., Johnston, C., Merkys, A., Cepellotti, A., Mounet, N., & Pizzi, G. (2020). AiiDA 1.0, a scalable computational infrastructure for automated reproducible workflows and data provenance. *Scientific Data*, *7*(1), 300. DOI: 10.1038/s41597-020-00638-4 PMID: 32901044

Hu, D., Du, F., Cao, H., Zhou, Q., Sun, P., Xu, T., Mei, C., Hao, Q., Fan, Z., & Zheng, J. (2021). An effective strategy to control thickness of Al2O3 coating layer on nickel-rich cathode materials. *Journal of Electroanalytical Chemistry (Lausanne, Switzerland)*, *880*, 114910. DOI: 10.1016/j.jelechem.2020.114910

Hu, K., Guan, W., Bi, Y., Zhang, W., Li, L., Zhang, B., & Zhong, N. (2021). Efficacy and safety of Lianhuaqingwen capsules, a repurposed Chinese herb, in patients with coronavirus disease 2019: A multicenter, prospective, randomized controlled trial. *Phytomedicine : International Journal of Phytotherapy and Phytopharmacology*, *85*, 153242. Advance online publication. DOI: 10.1016/j.phymed.2020.153242 PMID: 33867046

Iglesias, D., Haddad, D., & Sans, V. (2022). Recent developments in process digitalisation for advanced nanomaterial syntheses. *Chemistry Methods*, *2*(12), e202200031. DOI: 10.1002/cmtd.202200031

Igwilo, C. N., Chime, T. O., Ude, C. N., Ikechukwu, O. M., Anezichukwu, A. F., Ikechukwu, N. A., Omotioma, M., Chukwudi, E. B., Chisom, M. K., Ifeanyichukwu, O. T., & Scholastica, C. O. (2025). Examining the efficiency of microbe-assisted metal extraction: A review of bio-hydrometallurgical leaching techniques. *Hybrid Advances*, *9*, 100407. DOI: 10.1016/j.hybadv.2025.100407

Ihara, L., Wu, G., Cavaleiro, A., Morina, A., & Yang, L. (2025). Environmental Impact of Physical Vapour Deposition and Plasma-Enhanced Chemical Vapour Deposition Technologies for Deposition of Diamond-like Carbon Coatings for Green Tribology. *Coatings*, *15*(2), 218. DOI: 10.3390/coatings15020218

Ipatov, A. G., & Kharanzhevskiy, E. V. (2019). The Tribological Properties of Superhard and Functional Coatings Based on Carbide and Boron Nitride. *Journal of Friction and Wear*, *40*(6), 588–592. DOI: 10.3103/S1068366619060114

Islam, I., Tandersen, D., Taimullah, A., Sata, Y. A., Hendrawan, Y., Hidayat, T., & Zulhan, Z. (2024). Hydrogen plasma smelting reduction for fast production of green ferronickel. *International Journal of Hydrogen Energy*, *94*, 300–309. DOI: 10.1016/j.ijhydene.2024.11.022

Ismail, A., Tekli, J., & Wex, B. (2025). Automating Deep Learning-Based Generation and Evaluation of De Novo Chemical Reaction with ChemRxnSAGE. *Journal of Chemical Information and Modeling*, *65*(20), 10945–10960. DOI: 10.1021/acs.jcim.5c01692

Ivanova, M., Legros, J., Poisson, T., & Jubault, P. (2021). Continuous flow synthesis of Celecoxib from 2-bromo-3,3,3-trifluoropropene. *Journal of Flow Chemistry*, *12*(2), 147–151. DOI: 10.1007/s41981-021-00205-x PMID: 34631154

Jain, A. (2024). Machine learning in materials research: Developments over the last decade and challenges for the future. *Current Opinion in Solid State and Materials Science*, *33*, 101189. DOI: 10.1016/j.cossms.2024.101189

Jain, A., Ong, S. P., Chen, W., Medasani, B., Qu, X., Kocher, M., Brafman, M., Petretto, G., Rignanese, G.-M., Hautier, G., Gunter, D., & Persson, K. A. (2015). FireWorks: A dynamic workflow system designed for high-throughput applications. *Concurrency and Computation*, *27*(17), 5037–5059. DOI: 10.1002/cpe.3505

Jain, A., Ong, S. P., Hautier, G., Chen, W., Richards, W. D., Dacek, S., Cholia, S., Gunter, D., Skinner, D., Ceder, G., & Persson, K. A. (2013). Commentary: The Materials Project: A materials genome approach to accelerating materials innovation. *APL Materials*, *1*(1), 011002. Advance online publication. DOI: 10.1063/1.4812323

Jaramillo-Fierro, X., González, S., Montesdeoca-Mendoza, F., & Medina, F. (2021). Structuring of $ZnTiO_3/TiO_2$ adsorbents for the removal of methylene blue, using zeolite precursor clays as natural additives. *Nanomaterials (Basel, Switzerland)*, *11*(4), 898. DOI: 10.3390/nano11040898

Jarvis, K. A., Deng, Z., Allard, L. F., Manthiram, A., & Ferreira, P. J. (2011). Atomic structure of a lithium-rich layered oxide material for lithium-ion batteries: Evidence of a solid solution. *Chemistry of Materials*, *23*(16), 3614–3621. DOI: 10.1021/cm200831c

Jawor, A., Jeong, B. H., & Hoek, E. M. (2009). Synthesis, characterization, and ion-exchange properties of colloidal zeolite nanocrystals. *Journal of Nanoparticle Research*, *11*(7), 1795–1803. DOI: 10.1007/s11051-009-9688-9

Jayamuthunagai, J., Flora, R. M. N., Senthilkumar, K., & Bharathiraja, B. (2025). Review on sustainable strategies for lithium recovery from spent lithium-ion batteries. *Future Batteries*, *7*, 100105. DOI: 10.1016/j.fub.2025.100105

Jayaswal, G. P., Dange, S. P., & Khalikar, A. N. (2010). Bioceramic in dental implants: A review. *Journal of Indian Prosthodontic Society*, *10*(1), 8–12. DOI: 10.1007/s13191-010-0002-4 PMID: 23204715

Jazmina, N.-C., Tomas, S.-B., Lorena, L.-C., Steven, M., Patricia, A., & Nerea, G.-P. (2022). Hot isostatic pressing of WB4 and WB4-TaB2 based ultrahard materials. *International Journal of Refractory & Hard Metals*, *109*, 105965. DOI: 10.1016/j.ijrmhm.2022.105965

Jean, D., & Jaurand, M.-C. (2018). Mesotheliomas in genetically engineered mice unravel mechanism of mesothelial carcinogenesis. *International Journal of Molecular Sciences*, *19*(8), 2191. DOI: 10.3390/ijms19082191 PMID: 30060470

Jeje, S. O., Marazani, T., Obiko, J. O., & Shongwe, M. B. (2024). Advancing the hydrogen production economy: A comprehensive review of technologies, sustainability, and future prospects. *International Journal of Hydrogen Energy*, *78*, 642–661. DOI: 10.1016/j.ijhydene.2024.06.344

Jeong, M., Lee, H., Yoon, J., & Yoon, W.-S. (2019). O3-type NaNi1/3Fe1/3Mn1/3O2 layered cathode for Na-ion batteries: Structural evolution and redox mechanism upon Na (de)intercalation. *Journal of Power Sources*, *439*, 227064. DOI: 10.1016/j.jpowsour.2019.227064

Jha, B., & Singh, D. N. (2011). A review on synthesis, characterization and industrial applications of fly ash zeolites. *The Journal of Materials Education*, *33*(1–2), 65–132.

Jiang, M., Yang, L., Zou, L., Zhang, L., Wang, S., Zhong, Z., &Li, P. (2024). A comprehensive quality evaluation for Huangqi Guizhi Wuwu decoction by integrating UPLC-DAD/MS chemical profile and pharmacodynamics combined with chemometric analysis, 319. DOI: 10.1016/j.jep.2023.117325

Jiang, J., Ye, B., Zeng, Z., Yang, X., Sun, Z., Shao, S., Feng, K., & Tan, X. (2024). Carbon Abatement and Leakage in China's Regional Carbon Emission Trading. *Environmental Science & Technology*, *58*(40), 17661–17673. DOI: 10.1021/acs.est.4c04738

Jiang, Z., Li, A., Meng, C., Chen, X., & Song, H. J. (2022). Strategies and challenges of carbon materials in the practical applications of lithium metal anode: A review. *Physical Chemistry Chemical Physics : PCCP*, *24*(43), 26356–26370. DOI: 10.1039/D2CP04032H PMID: 36286554

Jiang, Z., Qu, Z., Zhou, L., & Tao, W. (2017). A microscopic investigation of ion and electron transport in lithium-ion battery porous electrodes using the lattice Boltzmann method. *Applied Energy*, *194*, 530–539. DOI: 10.1016/j.apenergy.2016.10.125

Jiao, P. P., Xu, F. Y., Li, J. H., Duan, N., Chen, G. Y., & Jiang, L. H. (2016). The inhibition effect of SeO_2 on hydrogen evolution reaction in $MnSO_4$-$(NH_4)_2SO_4$ solution. *International Journal of Hydrogen Energy*, *41*(2), 784–791. DOI: 10.1016/j.ijhydene.2015.11.034

Jin, Z., Shi, S., Lin, J., & Jin, L. (2013). Preparation and Performance of Dynamic Friction Polishing Plate for Diamond Film. *Materials and Manufacturing Processes*, *29*(1), 20–26. DOI: 10.1080/10426914.2013.852219

Johnson, E. B. G., & Arshad, S. E. (2014). Hydrothermally synthesized zeolites based on kaolinite: A review. *Applied Clay Science*, *97–98*, 215–221. DOI: 10.1016/j.clay.2014.06.005

Jordan, M. I., & Mitchell, T. M. (2015). Machine learning: Trends, perspectives, and prospects. *Science*, *349*(6245), 255–260. DOI: 10.1126/science.aaa8415 PMID: 26185243

Joseph, I. V., Tosheva, L., Miller, G., & Doyle, A. M. (2021). FAU-type zeolite synthesis from clays and its use for the simultaneous adsorption of five divalent metals from aqueous solutions. *Materials (Basel)*, *14*(13), 3738. DOI: 10.3390/ma14133738

Juan, Y., Dai, Y., Yang, Y., & Zhang, J. (2021). Accelerating materials discovery using machine learning. *Journal of Materials Science and Technology*, *79*, 178–190. DOI: 10.1016/j.jmst.2020.12.010

Kadier, A., Ablimit, M., Tursun, H. and Ablat, N. (2025). The western transmission of traditional Chinese medicine: an investigation of the cultural elements of traditional Chinese medicine in biomedical systems of cross-Asia countries, 16. DOI: 10.3389/fphar.2025.1589275

Kaetzl, K., Riegel, M., Joseph, B., Ossenbrink, R., Gerber, H., Gwenzi, W., Morck, T., Laner, D., Heinrich, T., Kromrey, V., Friedrich, K., Wachendorf, M., & Stenchly, K. (2025). Biogenic activated carbons from conservation grassland biomass for organic micropollutants removal in municipal wastewater. *Environmental Science and Ecotechnology*, *26*, 100588. DOI: 10.1016/j.ese.2025.100588 PMID: 40599981

Kaksonen, A. H., Deng, X., Bohu, T., Zea, L., Khaleque, H. N., Gumulya, Y., Boxall, N. J., Morris, C., & Cheng, K. Y. (2020). Prospective directions for biohydrometallurgy. *Hydrometallurgy*, *195*, 105376. DOI: 10.1016/j.hydromet.2020.105376

Kalló, D. (2001). Applications of natural zeolites in water and wastewater treatment. *Reviews in Mineralogy and Geochemistry*, *45*(1), 1–67. DOI: 10.2138/rmg.2001.45.15

Kamita, M., Waweru, H., Githinji, M., Kibiro, E., & Makokha, F. (2024). Establishing a cancer registry and baseline data for Nyandarua County, Kenya: A step towards establishing a central cancer registry. *Heliyon*, *10*(20), e39056. https://doi .org/. DOI: 10.1016/j.heliyon.2024.e39056 PMID: 39492911

Kang, T., Wang, Y., Guo, F., Liu, C., Zhao, J., Yang, J., Lin, H., Qiu, Y., Shen, Y., Lu, W., & Chen, L. (2019). Self-assembled monolayer enables slurry-coating of Li anode. *ACS Central Science*, *5*(3), 468–476. DOI: 10.1021/acscentsci.8b00845 PMID: 30937374

Karaki, H., Thomitzek, M., Obermann, T., Herrmann, C., & Schröder, D. (2023). Optimizing the Microstructure and Processing Parameters for Lithium-Ion Battery Cathodes: A Use Case Scenario with a Digital Manufacturing Platform. *Energy Technology (Weinheim)*, *11*(5), 2201032. DOI: 10.1002/ente.202201032

Kato, Y., Hori, S., Saito, T., Suzuki, K., Hirayama, M., Mitsui, A., Yonemura, M., Iba, H., & Kanno, R. (2016). High-power all-solid-state batteries using sulfide superionic conductors. *Nature Energy*, *1*(4), 16030. DOI: 10.1038/nenergy.2016.30

Kaur, I., Batra, V., Bogireddy, N. K. R., Baveja, J., Kumar, Y., & Agarwal, V. (2024). Chemical- and green-precursor-derived carbon dots for photocatalytic degradation of dyes. *iScience*, *27*(2), 108920. DOI: 10.1016/j.isci.2024.108920 PMID: 38352227

Kazemian, H., Naghdali, Z., Ghaffari Kashani, T., & Farhadi, F. (2010). Conversion of high silicon fly ash to Na-P1 zeolite: Alkaline fusion followed by hydrothermal crystallization. *Advanced Powder Technology*, *21*(3), 279–283. DOI: 10.1016/j. apt.2009.12.005

Kempa, M., Bovtun, V., & Repek, D. (2024). Microwave magnetic excitations in U-type hexaferrite $Sr_4CoZnFe_{36}O_{60}$ ceramics. *Journal of Applied Physics*, *136*(7), 15. DOI: 10.1063/5.0222910

Khan, F. R., Molden, N., & Booth, A. M. (2025). As the "Wheels Come Off", Including Tire Emissions in the Plastics Treaty Would Show Drive. *Environmental Science & Technology*, *59*(50), 26914–26916. DOI: 10.1021/acs.est.5c13479

Khang, A., & Akhai, S. (2025). *E-Waste and Lithium-ion Battery Recycling Insights for Sustainable Transportation Driving Green Transportation System Through Artificial Intelligence and Automation: Approaches, Technologies and Applications*. Springer.

Kim, E. J., Kim, J. H., Kim, M.-S., Jeong, S. H., & Choi, D. H. (2021). Process Analytical Technology Tools for Monitoring Pharmaceutical Unit Operations: A Control Strategy for Continuous Process Verification. *Pharmaceutics*, *13*(6), 919. DOI: 10.3390/pharmaceutics13060919

Kim, H. H., Lee, Y. J., Park, C., Yu, S., Won, S. O., Seo, W.-S., Park, C., & Choi, W. K. (2018). Bottom-Up Synthesis of Carbon Quantum Dots With High Performance Photo- and Electroluminescence. *Particle & Particle Systems Characterization*, *35*(7), 1800080. DOI: 10.1002/ppsc.201800080

Kim, I. D. (2014). Editorial: Advances in functional ceramic materials. *Journal of Electroceramics*, *33*(1), 11–15. DOI: 10.1007/s10832-014-9970-9

Kim, I. T., Ahn, K. H., Jung, J., Jeong, Y., Shin, D. C., & Lee, Y. E. (2021). Removal of tar contents derived from lignocellulosic biomass gasification facilities using MgAl-LDH@clinoptilolite. *Catalysts*, *11*(9), 1111. DOI: 10.3390/catal11091111

Kim, K., Shim, K., Wang, Y. W., & Yang, D. (2025). Synthetic Biology Strategies for the Production of Natural Colorants and Their Non-Natural Derivatives. *ACS Synthetic Biology*, *14*(3), 662–676. DOI: 10.1021/acssynbio.4c00799

Kim, N.-Y., Kim, I., Bornamehr, B., Presser, V., Ueda, H., Lee, H.-J., Cheong, J. Y., & Jung, J.-W. (2024). Recent advances in nanoengineering of electrode-electrolyte interfaces to realize high-performance Li-ion batteries. *Energy & Environmental Materials*, *7*(3), e12622. DOI: 10.1002/eem2.12622

Kim, S.-H., Choi, K.-H., Cho, S.-J., Kil, E.-H., & Lee, S.-Y. (2013). Mechanically compliant and lithium dendrite growth-suppressing composite polymer electrolytes for flexible lithium-ion batteries. *Journal of Materials Chemistry. A, Materials for Energy and Sustainability*, *1*(16), 4949–4955. DOI: 10.1039/c3ta10612h

Kim, T., Song, W., Son, D.-Y., Ono, L. K., & Qi, Y. (2019). Lithium-ion batteries: Outlook on present, future, and hybridized technologies. *Journal of Materials Chemistry. A, Materials for Energy and Sustainability*, *7*(7), 2942–2964. DOI: 10.1039/C8TA10513H

Klawohn, S., Darby, J. P., Kermode, J. R., Csányi, G., Caro, M. A., & Bartók, A. P. (2023). Gaussian approximation potentials: Theory, software implementation and application examples. *The Journal of Chemical Physics*, *159*(17), 174108. Advance online publication. DOI: 10.1063/5.0160898 PMID: 37929869

Kochanek, S. J., Close, D. A., & Johnston, P. A. (2019). High content screening characterization of head and neck squamous cell carcinoma multicellular tumor spheroid cultures generated in 384-well ultra-low attachment plates to screen for better cancer drug leads. *Assay and Drug Development Technologies*, *17*(1), 17–36. DOI: 10.1089/adt.2018.896 PMID: 30592624

Koroviaka, Y. A., Ihnatov, A. O., Pavlychenko, A. V., Valouch, K., Rastsvietaiev, V. O., Matyash, O. V., Mekshun, M. R., & Shypunov, S. O. (2023). Studying the Performance Features of Drilling Rock Destruction and Technological Tools. *Journal of Superhard Materials*, *45*(6), 466–476. DOI: 10.3103/S1063457623060059

Ko, T. W., Finkler, J. A., Goedecker, S., & Behler, J. (2021). A fourth-generation high-dimensional neural network potential with accurate electrostatics including non-local charge transfer. *Nature Communications*, *12*(1), 398. DOI: 10.1038/s41467-020-20427-2 PMID: 33452239

Kovačević, A., Tolazzi, M., Sanadar, M., & Melchior, A. (2024). Hydrometallurgical recovery of metals from spent lithium-ion batteries with ionic liquids and deep eutectic solvents. *Journal of Environmental Chemical Engineering*, *12*(4), 113248. DOI: 10.1016/j.jece.2024.113248

Król, M. (2020). Natural vs. synthetic zeolites. *Crystals*, *10*(7), 1–8. DOI: 10.3390/cryst10070622

Krska, S. W., DiRocco, D. A., Dreher, S. D., & Shevlin, M. (2017). The evolution of chemical high-throughput experimentation to address challenging problems in pharmaceutical synthesis. *Accounts of Chemical Research*, *50*(12), 2976–2985. DOI: 10.1021/acs.accounts.7b00428 PMID: 29172435

Kuang, T., Jin, M., Lu, X., Liu, T., Vahabi, H., Gu, Z., & Gong, X. (2023). Functional carbon dots derived from biomass and plastic wastes. *Green Chemistry*, *25*(17), 6581–6602. DOI: 10.1039/D3GC01763J

Kumar, A., Indhur, R., Bux, F., & Kumari, S. (2024). Recent advances in mechanistic insights into microplastics mitigation strategies via emerging advanced oxidation processes: Legislation, challenges, and future direction. *The Science of the Total Environment*, *957*, 177150. DOI: 10.1016/j.scitotenv.2024.177150 PMID: 39486547

Kumar, A., & Lingfa, P. (2019). Sodium bentonite and kaolin clays: Comparative study on their FT-IR. *Materials Today: Proceedings*. Advance online publication. DOI: 10.1016/j.matpr.2019.10.037

Kunnumakkara, A. B., Bordoloi, D., Sailo, B. L., Roy, N. K., Thakur, K. K., Banik, K., Shakibaei, M., Gupta, S. C., & Aggarwal, B. B. (2019). Cancer drug development: The missing links. *Experimental Biology and Medicine (Maywood, N.J.), 244*(8), 663–689. DOI: 10.1177/1535370219839163 PMID: 30961357

Kunze, S., Nam, C., Kim, H., Chung, J., Hong, E., Song, J., Choi, H., & Lim, J. (2025). X-ray imaging methods for multiscale characterization of batteries. *Bulletin of the Korean Chemical Society*, *46*(4), 360–380. DOI: 10.1002/bkcs.70009

Lam, H. Y. I., Guan, J. S., Ong, X. E., Pincket, R., & Mu, Y. G. (2024). Protein language models are performant in structure-free virtual screening. *Briefings in Bioinformatics*, *25*(6), bbae480. DOI: 10.1093/bib/bbae480 PMID: 39327890

Lange, J.-P. (2021). Managing Plastic Waste—Sorting, Recycling, Disposal, and Product Redesign. *ACS Sustainable Chemistry & Engineering*, *9*(47), 15722–15738. DOI: 10.1021/acssuschemeng.1c05013

Lan, Z., Zhang, G., & Wang, X. (2016). A facile synthesis of Br-modified g-C_3N_4 semiconductors for photoredox water splitting. *Applied Catalysis B: Environment and Energy*, *192*, 116–125. DOI: 10.1016/j.apcatb.2016.03.062

LaPotin, A., Zhong, Y., Zhang, L., Zhao, L., Leroy, A., Kim, H., Rao, S. R., & Wang, E. N. (2021). Dual-stage atmospheric water harvesting device for scalable solar-driven water production. *Joule*, *5*(1), 166–182. DOI: 10.1016/j.joule.2020.09.008

Last, M. G., Voortman, L. M., & Sharp, T. H. (2023). Measuring cryo-TEM sample thickness using reflected light microscopy and machine learning. *Journal of Structural Biology*, *215*(2), 107965. DOI: 10.1016/j.jsb.2023.107965 PMID: 37100102

Latz, A., & Zausch, J. (2015). Multiscale modeling of lithium ion batteries: Thermal aspects. *Beilstein Journal of Nanotechnology*, *6*, 987–1007. DOI: 10.3762/bjnano.6.102 PMID: 25977870

Lavrinenko, V. I. (2025). Recent Developments in Diamond Processing and Surface Modification Techniques for Tailoring Properties of Diamond Grains: A Review. *Journal of Superhard Materials*, *46*(6), 462–481. DOI: 10.3103/S1063457624060054

Lee, B., Kwon, C.-Y., Suh, H.-W., Kim, Y. J., Kim, K.-I., Lee, B.-J. and Lee, J.-H. (2023). Herbal medicine for the treatment of chronic cough: a systematic review and meta-analysis, 14. DOI: 10.3389/fphar.2023.1230604

Lee, B., Yoo, J., & Kang, K. (2020). Predicting the chemical reactivity of organic materials using a machine-learning approach††Electronic supplementary information (ESI) available. See DOI: . *Chemical Science, 11*(30), 7813-7822. doi:https://doi.org/DOI: 10.1039/D0SC01328E

Lee, J. C., & Pandey, B. D. (2012). Bio-processing of solid wastes and secondary resources for metal extraction - A review. *Waste Management*, *32*(1), 3–18. DOI: 10.1016/j.wasman.2011.08.010 PMID: 21925857

Lee, S. G., Seo, J. Y., Lee, J. W., Park, W. B., Sohn, K. S., & Pyo, M. (2021). Composition-tuned lithium aluminosilicate as a new humidity-sensing ceramic material with high sensitivity. *Sensors and Actuators. B, Chemical*, *339*, 129928. DOI: 10.1016/j.snb.2021.129928

Leighton, M., & Akasapu, U. (2024). An Application of Uncertainty Quantification to Efficiency Measurements and Validating Requirements through Correlating Simulation and Physical Testing Results. *Sensors (Basel)*, *24*(15), 4867. DOI: 10.3390/s24154867 PMID: 39123913

Lei, Z., Ang, H. T., & Wu, J. (2023). Advanced in-line purification technologies in multistep continuous flow pharmaceutical synthesis. *Organic Process Research & Development*, *28*(5), 1355–1368. DOI: 10.1021/acs.oprd.2c00374

Leung, K. (2013). Electronic structure modeling of electrochemical reactions at electrode/electrolyte interfaces in lithium ion batteries. *The Journal of Physical Chemistry. C, Nanomaterials and Interfaces*, *117*(4), 1539–1547. DOI: 10.1021/jp308929a

Leung, K. (2020). DFT modelling of explicit solid–solid interfaces in batteries: Methods and challenges. *Physical Chemistry Chemical Physics : PCCP*, *22*(19), 10412–10425. DOI: 10.1039/C9CP06485K PMID: 32073055

Leyfman, Y., Pawar, S., Joshi, M., Kieft, A. V., Menon, G. P., Wilkerson, W. B., Balasubramanian, M., Jackewicz, S., Azeez, A., Chorya, H., Pai, P., Coloma, H. S., Wilson, S., & Park, C. H. (2023). Impact of a digital education platform to combat global inequity in oncology education amongst healthcare professionals in lower socioeconomic areas. *JCO Global Oncology*, *9*(Supplement_1), 140–140. DOI: 10.1200/GO.2023.9.Supplement_1.140

Li, C., Liu, H., Xu, Y., Li, W., Long, J., Meng, L., Huang Q. (2025) Curent sitation and development trend of solid waste geeration, ulization and disposal in China. *Environmental Sanitation Engineering*, 33(6): 28-36, 50.

Li, J., Hu, X., & Li, T. (2024). Simulation of Solid Electrolyte Interphase Growth for Lithium Batteries Based on Kinetic Monte Carlo. *Energy Material Advances*, *5*, 0137.

Li, J., Yu, X.-q., Xie, Y., Yang, S.-g., Zhao, L., Zhou, M. and Meng, Y. (2022). Efficacy and safety of traditional Chinese medicine treatment for idiopathic pulmonary fibrosis: An exploratory, randomized, double-blinded and placebo controlled trial, 13. DOI: 10.3389/fphar.2022.1053356

Li, J., Zhang, X., Hou, L., Liu, B.-Y., Fan, Y.-M., Zhang, Y., & Yin, X. (2025). Proteomic analysis reveals QiShen YiQi Pills ameliorates ischemia-induced heart failure through inhibition of mitochondrial fission, 138. DOI: 10.1016/j.phymed.2025.156435

Li, A. G., West, A. C., & Preindl, M. (2022). Towards unified machine learning characterization of lithium-ion battery degradation across multiple levels: A critical review. *Applied Energy*, *316*, 119030. DOI: 10.1016/j.apenergy.2022.119030

Liao, Y., Zhang, H., Peng, Y., Hu, Y., Liang, J., Gong, Z., Wei, Y., & Yang, Y. (2024). Electrolyte degradation during aging process of lithium-ion batteries: Mechanisms, characterization, and quantitative analysis. *Advanced Energy Materials*, *14*(18), 2304295. DOI: 10.1002/aenm.202304295

Li, C., Liu, W., Jiao, F., Dong, L., Liu, S., & Qin, W. (2024). Intensifying recovery of metallic aluminum in secondary aluminum dross by alkali roasting-water leaching. *Transactions of Nonferrous Metals Society of China*, *34*(12), 4063–4074. DOI: 10.1016/S1003-6326(24)66658-6

Li, D.-C., Muta, T., Zhang, L.-Q., Yoshio, M., & Noguchi, H. (2004). Effect of synthesis method on the electrochemical performance of LiNi1/3Mn1/3Co1/3O2. *Journal of Power Sources*, *132*(1), 150–155. DOI: 10.1016/j.jpowsour.2004.01.016

Li, D., Shen, C., Zheng, Y., & Xu, J. (2025). Electrochemo-Mechanical Degradation and Failure of Active Particles in High Energy Density Batteries: A Review. *Small*, *21*(8), 2407740. DOI: 10.1002/smll.202407740 PMID: 39776184

Li, D., Zhi, X., Zhang, L., Xie, F., & Wu, M. (2025). Direct regeneration of spent lithium-ion batteries: Advancing from powder to cell. *Journal of Energy Chemistry*, *111*, 767–789. DOI: 10.1016/j.jechem.2025.08.013

Li, H., Guo, S., & Zhou, H. (2021). In-situ/operando characterization techniques in lithium-ion batteries and beyond. *Journal of Energy Chemistry*, *59*, 191–211. DOI: 10.1016/j.jechem.2020.11.020

Li, H., Lin, Q., Wang, J., Hu, L., Chen, F., Zhang, Z., & Ma, C. (2024). A cost-effective sulfide solid electrolyte Li7P3S7.5O3.5 with low density and excellent anode compatibility. *Angewandte Chemie International Edition*, *63*(37), e202407892. DOI: 10.1002/anie.202407892 PMID: 38945831

Li, J., Nazari, S., Ma, X., Wei, N., & He, Y. (2025). A critical review on the advances in thermal reduction technology for recycling spent lithium-ion batteries. *Next Materials*, *8*, 100641. DOI: 10.1016/j.nxmate.2025.100641

Li, J., Niu, G., Mu, P., Yan, B., Liu, F., Zhao, S., Fang, L., & Gou, H. (2024). Amorphous diamond embedded in dense boron nitride with excellent mechanical properties. *Microstructures*, *4*(1), 2024010. DOI: 10.20517/microstructures.2023.54

Li, L., Chen, Z., Zhang, X., & Jia, Y. (2018). Divergent strategy in natural product total synthesis. *Chemical Reviews*, *118*(7), 3752–3832. DOI: 10.1021/acs.chemrev.7b00653 PMID: 29516724

Li, L., Li, Q., Li, Y., Gong, D., & Zhao, B. (2025). Research progress and challenges of molecular recognition techniques in the screening of active ingredients in traditional. *Journal of Pharmaceutical Analysis*, *15*(9), 101243. Advance online publication. DOI: 10.1016/j.jpha.2025.101243 PMID: 41069865

Li, M., Wu, D., Zhu, Y., Guan, L., Huang, Y., & Song, X. (2023). Low-temperature preparation of $La_{0.9}Ca_{0.1}CrO_3$ semiconductor ceramics by flash sintering. *Ceramics International*, *49*(17), 29364–29369. DOI: 10.1016/j.ceramint.2023.06.203

Lim, J., Jang, Y., Lee, J., Lee, C., Jbari, O., Kwon, K., & Chung, E. (2025). Hydrometallurgical process of spent lithium-ion battery recycling Part. 2 Recovery of valuable metals from the cathode active material leachates: Review and cost analysis. *Hydrometallurgy*, *236*, 106516. DOI: 10.1016/j.hydromet.2025.106516

Lin, B., Wu, X., He, X., Ye, J., Song, Y., Jiang, P., Bai, L., Liao, X., Li, Y., Huo, Y., Zheng, Q., Zhang, H., & Lin, D. (2025). High-entropy doping in NASICON cathodes: Activating the V4+/V5+ redox couple and inducing a reversible single solid-solution phase reaction for advanced sodium ion batteries. *Journal of Colloid and Interface Science*, *690*, 137299. DOI: 10.1016/j.jcis.2025.137299 PMID: 40088810

Lin, S. H., Zhou, C. X., Chen, H. L., Zhou, X. L., Yang, H. J., Sun, L. T., Zhang, L. Y., & Zhang, Y. X. (2025). Advances in hereditary colorectal cancer: How precision medicine is changing the game. *Cancers (Basel)*, *17*(21), 3461. DOI: 10.3390/cancers17213461 PMID: 41228253

Lin, X., Khosravinia, K., Hu, X., Li, J., & Lu, W. (2021). Lithium plating mechanism, detection, and mitigation in lithium-ion batteries. *Progress in Energy and Combustion Science*, *87*, 100953. DOI: 10.1016/j.pecs.2021.100953

Lin, Y., & Cai, H. (2023). Biological functions and therapeutic potential of SHCBP1 in human cancer. *Biomedicine & Pharmacotherapy = Biomédecine & Pharmacothérapie*, *160*, 114362. DOI: 10.1016/j.biopha.2023.114362 PMID: 36739763

Lin, Y., Wang, T., Gao, C., Zhang, X., Yu, W., Wang, M., Yang, C., & Zhang, J. (2025). Direct and low-temperature regeneration of degraded $LiFePO_4$ cathodes at ambient conditions using green and sustainable deep eutectic solvent. *Advanced Science (Weinheim, Baden-Württemberg, Germany)*, *12*(28), 2504683. DOI: 10.1002/advs.202504683 PMID: 40391653

Li, P., Zhao, Z., Fei, Y., Zhang, H., & Li, G. (2025). LiNO3-based electrolyte with fast kinetics for lithium metal batteries under practical conditions. *Advanced Energy Materials*, *15*(30), 2500882. DOI: 10.1002/aenm.202500882

Li, Q., Chen, Y., Yu, C., Young, L., Spector, J., & Harris, V. G. (2022). Emerging magnetodielectric materials for 5G communications: 18H hexaferrites. *Acta Materialia*, *231*, 117854. DOI: 10.1016/j.actamat.2022.117854

Li, S., Li, L., Zhao, Y., Yang, H., Tong, H., Fan, S., Wang, Z., & Xu, W. (2024). Modulating the d-band center of single-atom catalysts for efficient Li2S2-Li2S conversion in durable lithium-sulfur batteries. *Energy Storage Materials*, *70*, 103477. DOI: 10.1016/j.ensm.2024.103477

Li, T., Yang, Y., Yang, F., Li, Y., Qiao, H., Yan, Y., & Ding, X. (2025). Organoids and Organoids-on-Chip in Traditional Chinese Medicine Research: Applications, Advantages, and Future Prospects. *Cell Biology International*, *49*(10), 1233–1244. Advance online publication. DOI: 10.1002/cbin.70067 PMID: 40772545

Liu, C., Zhou, Y., Wu, G., Gao, K., Li, L., Tu, H., & Chen, Z. (2022). Sandwich-likely structured, magnetically-driven recovery, biomimetic composite penicillin G acylase-based biocatalyst with excellent operation stability. *Colloids and Surfaces. A, Physicochemical and Engineering Aspects*, *639*, 128245. DOI: 10.1016/j.colsurfa.2021.128245

Liu, J., Baeyens, J., Deng, Y., Wang, X., & Zhang, H. (2020). High temperature Mn2O3/Mn3O4 and Co3O4/CoO systems for thermo-chemical energy storage. *Journal of Environmental Management*, *267*, 110582. DOI: 10.1016/j.jenvman.2020.110582 PMID: 32364130

Liu, J., Li, R., & Yang, B. (2020). Carbon dots: A new type of carbon-based nanomaterial with wide applications. *ACS Central Science*, *6*(12), 2179–2195. DOI: 10.1021/acscentsci.0c01306 PMID: 33376780

Liu, J., Zhong, S., Zhang, F., Cao, Y., Luo, D., Liao, X., Hu, D., Wang, Y., Lu, Z., & Cheng, H. J. (2025). Stabilizing 4.6 V LiCoO2 by lithium compensative additives. *ACS Applied Materials & Interfaces*, *17*(9), 13872–13880. DOI: 10.1021/acsami.4c20095 PMID: 39970084

Liu, M., Saman, W., & Bruno, F. (2012). Review on storage materials and thermal performance enhancement techniques for high temperature phase change thermal storage systems. *Renewable & Sustainable Energy Reviews*, *16*(4), 2118–2132. DOI: 10.1016/j.rser.2012.01.020

Liu, N., Yu, J., Li, Y., Liu, Y., Jabeen, S., Cheng, Y., Zhou, L., Danilov, D. L., Notten, P. H. L., & Li, H. (2025). Multifunctional carbon dots for electrochemical capacitors reviewed: Their opportunities and challenges. *Advanced Science (Weinheim, Baden-Württemberg, Germany)*, *12*(41), e08000. DOI: 10.1002/advs.202508000 PMID: 41001802

Liu, Q., Du, C., Shen, B., Zuo, P., Cheng, X., Ma, Y., Yin, G., & Gao, Y. (2016). Understanding undesirable anode lithium plating issues in lithium-ion batteries. *RSC Advances*, *6*(91), 88683–88700. DOI: 10.1039/C6RA19482F

Liu, X., Gu, S., Zhao, Y., Zhou, G., & Li, W. (2020). BiVO4, Bi_2WO_6 and Bi_2MoO_6 photocatalysis: A brief review. *Journal of Materials Science and Technology*, *56*, 45–68. DOI: 10.1016/j.jmst.2020.04.023

Liu, X., Park, H., Ackermann, Y. S., Averous, L., Ballerstedt, H., Besenmatter, W., Blazquez, B., Bornscheuer, U. T., Branson, Y., Casey, W., de Lorenzo, V., Dong, W., Floehr, T., Godoy, M. S., Ji, Y., Jupke, A., Klankermayer, J., San Leon, D., Liu, L., & Chen, G.-Q. (2025). Exploring biotechnology for plastic recycling, degradation and upcycling for a sustainable future. *Biotechnology Advances*, *81*, 108544. Advance online publication. DOI: 10.1016/j.biotechadv.2025.108544

Liu, X., Xu, P., Zhao, J., Lu, W., Li, M., & Wang, G. (2022). Material machine learning for alloys: Applications, challenges and perspectives. *Journal of Alloys and Compounds*, *921*, 165984. DOI: 10.1016/j.jallcom.2022.165984

Liu, Y., Guo, B., Zou, X., Li, Y., & Shi, S. (2020). Machine learning assisted materials design and discovery for rechargeable batteries. *Energy Storage Materials*, *31*, 434–450. DOI: 10.1016/j.ensm.2020.06.033

Liu, Y., Ju, Z., Zhang, B., Wang, Y., Nai, J., Liu, T., & Tao, X. (2021). Visualizing the sensitive lithium with atomic precision: Cryogenic electron microscopy for batteries. *Accounts of Chemical Research*, *54*(9), 2088–2099. DOI: 10.1021/acs.accounts.1c00120 PMID: 33856759

Liu, Y., Zhao, T., Ju, W., & Shi, S. (2017). Materials discovery and design using machine learning. *Journal of Materiomics*, *3*(3), 159–177. DOI: 10.1016/j.jmat.2017.08.002

Li, W., Cui, L., Mai, J., Shi, T.-Q., Lin, L., Zhang, Z.-G., Ledesma-Amaro, R., Dong, W., & Ji, X.-J. (2022). Advances in Metabolic Engineering Paving the Way for the Efficient Biosynthesis of Terpenes in Yeasts. *Journal of Agricultural and Food Chemistry*, *70*(30), 9246–9261. DOI: 10.1021/acs.jafc.2c03917

Li, W., Rentemeister, M., Badeda, J., Jöst, D., Schulte, D., & Sauer, D. U. (2020). Digital twin for battery systems: Cloud battery management system with online state-of-charge and state-of-health estimation. *Journal of Energy Storage*, *30*, 101557. DOI: 10.1016/j.est.2020.101557

Li, W., Sun, X., & Yu, Y. (2017). Si-, Ge-, Sn-based anode materials for lithium-ion batteries: From structure design to electrochemical performance. *Small Methods*, *1*(3), 1600037. DOI: 10.1002/smtd.201600037

Li, X. T., Li, X. Y., Yang, J., Cao, Z. F., Li, C. X., Xue, J. R., Ma, X., & Wang, S. (2024). Highly efficient Mn^{2+}, Mg^{2+}, and NH_4^+ recovery from electrolytic manganese residue via leaching, solvent extraction, coprecipitation, and atmospheric oxidation. *Journal of Hazardous Materials*, *472*, 134430. Advance online publication. DOI: 10.1016/j.jhazmat.2024.134430 PMID: 38718502

Li, X., Jiang, Y., Liu, X., Shi, L., Zhang, D., & Sun, L. (2017). Direct synthesis of zeolites from a natural clay, attapulgite. *ACS Sustainable Chemistry & Engineering*, *5*(7), 6124–6130. Advance online publication. DOI: 10.1021/acssuschemeng.7b01001

Li, X., Mathur, A., Liu, A., & Liu, Y. (2023). Electrifying Carbon Capture by Developing Nanomaterials at the Interface of Molecular and Process Engineering. *Accounts of Chemical Research*, *56*(20), 2763–2775. DOI: 10.1021/acs.accounts.3c00321

Li, X., Wang, B., Cao, Y., Zhao, S., Wang, H., Feng, X., Zhou, J., & Ma, X. (2019). Water Contaminant Elimination Based on Metal–Organic Frameworks and Perspective on Their Industrial Applications. *ACS Sustainable Chemistry & Engineering*, *7*(5), 4548–4563. DOI: 10.1021/acssuschemeng.8b05751 PMID: 33659101

Li, Y. Y., Jia, Z. G., Xu, L. X., Sheng, G. H., & Wan, C. (2025). Synergistic removal of tetracycline hydrochloride by nano-MnOx/NCF composite fiber catalysts via heterogeneous peroxymonosulfate activation. *Inorganic Chemistry Communications*, *172*, 113682. Advance online publication. DOI: 10.1016/j.inoche.2024.113682

Loginov, P. A., Zhassay, U. A., Bychkova, M. Y., Petrzhik, M. I., Mukanov, S. K., Sidorenko, D. A., Orekhov, A. S., Rupasov, S. I., & Levashov, E. A. (2020). Chromium-doped Fe-Co-Ni binders for diamond cutting tools: The features of the structure, mechanical properties, and adhesion to diamond. *International Journal of Refractory & Hard Metals*, *92*, 105289. DOI: 10.1016/j.ijrmhm.2020.105289

Lombardo, T., Duquesnoy, M., El-Bouysidy, H., Årén, F., Gallo-Bueno, A., Jørgensen, P. B., Bhowmik, A., Demortière, A., Ayerbe, E., Alcaide, F., Reynaud, M., Carrasco, J., Grimaud, A., Zhang, C., Vegge, T., Johansson, P., & Franco, A. A. (2022). Artificial Intelligence Applied to Battery Research: Hype or Reality? *Chemical Reviews*, *122*(12), 10899–10969. DOI: 10.1021/acs.chemrev.1c00108

Long, X. R., Sun, K., Lai, S., Liu, Y. C., Sun, J., Chen, W. Q., Liu, R. H., He, X. Y., Zhao, S., & Huang, K. (2025). Artificial intelligence and anti-cancer drug response. *Acta Pharmaceutica Sinica. B*, *15*(7), 3355–3371. DOI: 10.1016/j.apsb.2025.05.009 PMID: 40698144

Lu, A. (2025). Elevating traditional Chinese medicine in global health research: The case for Chinese herbal medicine formulas in mainstream therapeutics, 6. DOI: 10.1016/j.xinn.2025.100811

Lu, P., Li, L., Liu, B., Cao, Z., Geng, Q., Ji, X., &Lu, C. (2024). Efficacy and safety of integrated traditional Chinese and Western medicine for rheumatoid arthritis-interstitial lung disease: A systematic review and meta-analysis, 10. DOI: 10.1016/j.heliyon.2024.e38771

Lu, C., & Weusthuis, R. A. (2025). Microbial production of medium-chain-length diols: Current stage and perspectives. *Bioresource Technology*, *435*, 132932. Advance online publication. DOI: 10.1016/j.biortech.2025.132932

Lu, H., Wang, L., Ma, X., Cheng, J., & Zhou, M. (2025). A survey of graph neural networks and their industrial applications. *Neurocomputing*, *614*, 128761. DOI: 10.1016/j.neucom.2024.128761

Lu, J. M., Dreisinger, D., & Glück, T. (2014). Manganese electrodeposition: A literature review. *Hydrometallurgy*, *141*, 105–116. DOI: 10.1016/j.hydromet.2013.11.002

Luo, W. J., Li, B., Yang, G., Xu, M. X., Pang, C. H., Kow, K. W., & Wu, T. (2024). Utilisation of electrolytic manganese residue as a sulphate activator in producing concrete blocks with high-volume fly ash. *Journal of Cleaner Production*, *434*, 139813. Advance online publication. DOI: 10.1016/j.jclepro.2023.139813

Lysogorskiy, Y., Bochkarev, A., Mrovec, M., & Drautz, R. (2023). Active learning strategies for atomic cluster expansion models. *Physical Review Materials*, *7*(4), 043801. DOI: 10.1103/PhysRevMaterials.7.043801

Lyu, K., Zhang, W. J., Chen, Y. Q., & Wang, C. Y. (2024). Energy-efficient and advanced electrowinning of metallic manganese within a novel H^+ exchange membrane cell using a Ti/IrO_2-RuO_2-SiO_2 anode. *Separation and Purification Technology*, *347*, 127489. Advance online publication. DOI: 10.1016/j.seppur.2024.127489

M., A. A. (2024). Heat Resistance of Tool Materials. *Russian Engineering Research, 44*(1), 61-69. DOI: 10.3103/S1068798X24010040

Magoda, K., Nomngongo, P.N., & Mekuto, L. (2024) Two-step bio-dissolution of metals from printed circuit boards using acidophilic iron - and sulfur-oxidizing mesophiles. *Recycling*, 9(1).

Mahmoud, A., Cézac, P., Hoadley, A. F. A., Contamine, F., & D'Hugues, P. (2017). A review of sulfide minerals microbially assisted leaching in stirred tank reactors. *International Biodeterioration & Biodegradation, 119*, 118–146. DOI: 10.1016/j.ibiod.2016.09.015

Majidi, L., Hemmat, Z., Warburton, R. E., Kumar, K., Ahmadiparidari, A., Hong, L., Guo, J., Zapol, P., Klie, R. F., Cabana, J., Greeley, J., Curtiss, L. A., & Salehi-Khojin, A. (2020). Highly Active Rhenium-, Ruthenium-, and Iridium-Based Dichalcogenide Electrocatalysts for Oxygen Reduction and Oxygen Evolution Reactions in Aprotic Media. *Chemistry of Materials, 32*(7), 2764–2773. DOI: 10.1021/acs.chemmater.9b04117

Ma, K., Su, Z. J., Zhang, Y. B., & Jiang, T. (2024). Effect mechanism of manganese on hydrogen-based reduction behavior of magnetite pellet. *Journal of Materials Research and Technology, 30*, 8304–8315. DOI: 10.1016/j.jmrt.2024.05.205

Malakpour-Permlid, A., Rodriguez, M. M., Untracht, G. R., Andersen, P. E., Oredsson, S., Boisen, A., & Zor, K. (2025). High-throughput non-homogenous 3D polycaprolactone scaffold for cancer cell and cancer-associated fibroblast mini-tumors to evaluate drug treatment response. *Toxicology Reports, 14*, 101863. DOI: 10.1016/j.toxrep.2024.101863 PMID: 39758801

Maleki, H., Aiyelabegan, H. T., Javadi, P., Abdi, F., Mirzavi, F., Behjani, Z. Z., Rizvanov, A. A., Takallu, S., Kumar, R., Barhaghtalab, R. H., Hosseinpouri, A., & Vakili-Ghartavol, R. (2025). Nanotechnology-mediated precision drug delivery strategies for breast cancer treatment. *Biomedicine & Pharmacotherapy = Biomédecine & Pharmacothérapie, 188*, 118224. DOI: 10.1016/j.biopha.2025.118224 PMID: 40456194

Mama, M., Solai, E., Capurso, T., Danlos, A., & Khelladi, S. (2025). Comprehensive review of multi-scale Lithium-ion batteries modeling: From electro-chemical dynamics up to heat transfer in battery thermal management system. *Energy Conversion and Management, 325*, 119223. DOI: 10.1016/j.enconman.2024.119223

Manea, F., Troquet, M., & Avram, L. (2006). Analyse de certains minéraux argileux. *Scientific Study & Research, 7*(3), 689–696.

Manikandan, S., Inbakandan, D., Valli Nachiyar, C., & Namasivayam, S. K. R. (2023). Towards sustainable metal recovery from e-waste: A mini review. *Sustainable Chemistry for the Environment, 2*, 100001. DOI: 10.1016/j.scenv.2023.100001

Manthiram, A., & Choi, J. (2006). Chemical and structural instabilities of lithium ion battery cathodes. *Journal of Power Sources, 159*(1), 249–253. DOI: 10.1016/j.jpowsour.2006.04.028

Manuscript, A. (2020). Framework flexibility-driven CO_2 adsorption on a zeolite. *Materials Horizons, 7*(6), 1528–1532. DOI: 10.1039/D0MH00307G

Mao, J., Miao, J., Lu, Y., & Tong, Z. (2021). Machine learning of materials design and state prediction for lithium ion batteries. *Chinese Journal of Chemical Engineering, 37*, 1–11. DOI: 10.1016/j.cjche.2021.04.009

Ma, R., Li, J., Zeng, P., Duan, L., Dong, J., Ma, Y., & Yang, L. (2024). The application of membrane separation technology in the pharmaceutical industry. *Membranes (Basel), 14*(1), 24. DOI: 10.3390/membranes14010024 PMID: 38248714

Marcilly, C. (2001). Evolution of refining and petrochemicals: What is the place of zeolites. In *Studies in Surface Science and Catalysis* (Vol. 135, pp. 37–60). Elsevier., DOI: 10.1016/S0167-2991(01)81185-X

Martinengo, B., Diamanti, E., Uliassi, E., & Bolognesi, M. L. (2024). Harnessing the 12 Green Chemistry Principles for Sustainable Antiparasitic Drugs: Toward the One Health Approach. *ACS Infectious Diseases, 10*(6), 1856–1870. DOI: 10.1021/acsinfecdis.4c00172

Marzoughi, O., & Tafaghodi, L. (2025). Hydrogen as a sustainable reducing agent for recycling electric arc furnace dust: A kinetic review. *Journal of Sustainable Metallurgy, 11*(1), 232–247. DOI: 10.1007/s40831-024-00975-3

Mathebula, N. P., Sheldon, R. A., & Bode, M. L. (2022). Lipase-catalysed enzymatic kinetic resolution of aromatic Morita-Baylis-Hillman derivatives by hydrolysis and transesterification. *Chembiochem : A European Journal of Chemical Biology, 23*(21), e202200435. DOI: 10.1002/cbic.202200435 PMID: 36049111

Matos-Pérez, C. R., White, J. D., & Wilker, J. J. (2012). Polymer Composition and Substrate Influences on the Adhesive Bonding of a Biomimetic, Cross-Linking Polymer. *Journal of the American Chemical Society, 134*(22), 9498–9505. DOI: 10.1021/ja303369p

Maximov, M., Nazarov, D., Rumyantsev, A., Koshtyal, Y., Ezhov, I., Mitrofanov, I., Kim, A., Medvedev, O., & Popovich, A. (2020). Atomic layer deposition of lithium–nickel–silicon oxide cathode material for thin-film lithium-ion batteries. *Energies*, *13*(9), 2345. DOI: 10.3390/en13092345

Ma, Y., Tayefi, S. H., Mogharabi-Manzari, M., & Luo, X. (2025). Advances in immobilized enzyme systems for enhanced microplastic biodegradation: A review. *International Journal of Biological Macromolecules*, *328*, 147656. DOI: 10.1016/j.ijbiomac.2025.147656

McCulloch, M. T., Winter, A., Sherman, C. E., & Trotter, J. A. (2024). 300 years of sclerosponge thermometry shows global warming has exceeded 1.5 °C. *Nature Climate Change*, *14*(2), 171–177. DOI: 10.1038/s41558-023-01919-7

Mei, J., Ju, C., Wang, B., Gao, R., Zhang, Y., Zhou, S., &Xu, F. (2024). The efficacy and safety of Bazi Bushen Capsule in treating premature aging: A randomized, double blind, multicenter, placebo-controlled clinical trial, 130. DOI: 10.1016/j.phymed.2024.155742

Meng, K., Bai, K., & Sun, S. (2024). Artificial intelligence driven design of cathode materials for sodium-ion batteries using graph deep learning method. *Journal of Energy Storage*, *101*, 113809. DOI: 10.1016/j.est.2024.113809

Meng, X., & Zhang, Z. (2016). Bismuth-based photocatalytic semiconductors: Introduction, challenges and possible approaches. *Journal of Molecular Catalysis A Chemical*, *423*, 533–549. DOI: 10.1016/j.molcata.2016.07.030

Mezni, M., Hamzaoui, A., Hamdi, N., & Srasra, E. (2011). Synthesis of zeolites from the low-grade Tunisian natural illite by two different methods. *Applied Clay Science*, *52*(3), 209–218. DOI: 10.1016/j.clay.2011.02.017

Michalkiewicz, B., & Koren, Z. C. (2015). Zeolite membranes for hydrogen production from natural gas: State of the art. *Journal of Porous Materials*, *22*(3), 635–646. DOI: 10.1007/s10934-015-9936-6

Miehe, C., Dal, H., Schänzel, L.-M., & Raina, A. (2016). A phase-field model for chemo-mechanical induced fracture in lithium-ion battery electrode particles. *International Journal for Numerical Methods in Engineering*, *106*(9), 683–711. DOI: 10.1002/nme.5133

Ming, W., Hu, S., Zhang, X., Ma, J., He, W., Xu, Y., Wang, H., Xu, Y., Xing, W., Chen, Z., & Li, X. (2025). Review: Biorecovery of precious metals from waste printed circuit boards. *Minerals Engineering*, *228*, 109306. DOI: 10.1016/j.mineng.2025.109306

Misaelides, P. (2011). Application of natural zeolites in environmental remediation: A short review. *Microporous and Mesoporous Materials*, *144*(1–3), 15–18. DOI: 10.1016/j.micromeso.2011.03.024

Moazzam, P., Boroumand, Y., Rabiei, P., Baghbaderani, S. S., Mokarian, P., Mohagheghian, F., Mohammed, L. J., & Razmjou, A. (2021). Lithium bioleaching: An emerging approach for the recovery of Li from spent lithium ion batteries. *Chemosphere*, *277*, 130196. DOI: 10.1016/j.chemosphere.2021.130196 PMID: 33784558

Mobarak, M. H., Mimona, M. A., Islam, M. A., Hossain, N., Zohura, F. T., Imtiaz, I., & Rimon, M. I. H. (2023). Scope of machine learning in materials research—A review. *Applied Surface Science Advances*, *18*, 100523. DOI: 10.1016/j.apsadv.2023.100523

Mohammadi, M., & Alfantazi, A. (2015). The performance of Pb-MnO_2 and Pb-Ag anodes in Mn(II)-containing sulphuric acid electrolyte solutions. *Hydrometallurgy*, *153*, 134–144. DOI: 10.1016/j.hydromet.2015.02.009

Mohammadi, M., & Alfantazi, A. (2016). Evaluation of manganese dioxide deposition on lead-based electrowinning anodes. *Hydrometallurgy*, *159*, 28–39. DOI: 10.1016/j.hydromet.2015.10.023

Mohammed, A., & Abdullah, A. (2018). *Scanning electron microscopy (SEM): A review.* Paper presented at the Proceedings of the 2018 international conference on hydraulics and pneumatics—HERVEX, Băile Govora, Romania.

Mohanty, S., Ghosh, S., Bal, B., & Das, A. P. (2018). A review of biotechnology processes applied for manganese recovery from wastes. *Reviews in Environmental Science and Biotechnology*, *17*(4), 791–811. DOI: 10.1007/s11157-018-9482-1

Mohd, R., Desforges, A., Rahman, A., & Vigolo, B. (2021). Progress in adsorption capacity of nanomaterials for carbon dioxide capture: A comparative study. *Journal of Cleaner Production*, *328*, 129553. DOI: 10.1016/j.jclepro.2021.129553

Moneim, M. A., & Ahmed, E. A. (2015). Synthesis of faujasite from Egyptian clays: Characterizations and removal of heavy metals. *Geomaterials*, *5*(2), 68–76. DOI: 10.4236/gm.2015.52007

Moore, A. R., Rosenberg, S. C., & Malek, F. M. S. (2020). RAS-targeted therapies: Is the undruggable drugged? *Nature Reviews. Drug Discovery*, *19*(8), 533–552. DOI: 10.1038/s41573-020-0068-6 PMID: 32528145

Morgan, D., & Jacobs, R. (2020). Opportunities and Challenges for Machine Learning in Materials Science. *Annual Review of Materials Research*, *50*(1), 71–103. DOI: 10.1146/annurev-matsci-070218-010015

Moro, K., Haubrich, F., Martin, M., Grimmer, M., Hoth, N., & Schippers, A. (2023). Reductive leaching behaviour of manganese and cobalt phases in laterite and manganese ores. *Hydrometallurgy*, *220*, 106101. Advance online publication. DOI: 10.1016/j.hydromet.2023.106101

Moslim, N., Beh, C., Kasim, S., & Ramakrishnan, S. (2017), Effect of composition and temperature to the HA/beta-TCP composite, in: *Regional Conference on Materials and Asean Microscopy Conference*. 1082, 12023-12027. https://doi.org/DOI: 10.1088/1742-6596/1082/1/012023

Motohashi, S., Katsuta, E., & Ban, D. (2025). Advances and challenges in drug screening for cancer therapy: A comprehensive review. *Bioengineering (Basel, Switzerland)*, *12*(12), 1315. DOI: 10.3390/bioengineering12121315 PMID: 41463612

Mozgawa, W., & Król, M. (2011). FT-IR studies of zeolites from different structural groups. *Chemik*, *65*(7), 671–674.

Mravec, D., Hudec, J., & Janotka, I. (2005). Some possibilities of catalytic and noncatalytic utilization of zeolites. *Chemical Papers*, *59*(1), 62–66.

Mukhopadhyay, A., & Sheldon, B. W. (2014). Deformation and stress in electrode materials for Li-ion batteries. *Progress in Materials Science*, *63*, 58–116. DOI: 10.1016/j.pmatsci.2014.02.001

Munro, D. H. (2019). Estimating clinical trial success rates from small molecule drug discovery and development. *Nature Reviews. Drug Discovery*, *18*, 495–496.

Muntean, E., Leba, M., & Ionica, A. C. (2025). AI-Driven Arm Movement Estimation for Sustainable Wearable Systems in Industry 4.0. *Sustainability (Basel)*, *17*(14), 6372. DOI: 10.3390/su17146372

Muravyov, M., Panyushkina, A., & Fomchenko, N. (2025). Two-step ferric and biological leaching as an environmentally friendly method for improving nickel sulfide concentrate processing. *Process Safety and Environmental Protection*, *195*, 105850. DOI: 10.1016/j.psep.2025.106850

Mustafa, B. M. (2026). Crystal symmetry effects on oxygen doping distribution, SC, and phase diagram in $YBa_2Cu_3O_7$. Journal of Superconductivity and Novel Magnetism. *Journal of Superconductivity and Novel Magnetism*, *39*(55), 152. DOI: 10.1007/s10948-026-07136-5

Musyoka, N. M., Missengue, R., Kusisakana, M., & Petrik, L. F. (2014). Conversion of South African clays into high quality zeolites. *Applied Clay Science*, *97–98*, 182–186. DOI: 10.1016/j.clay.2014.05.026

Musyoka, N. M., Petrik, L. F., Hums, E., Kuhnt, A., & Schwieger, W. (2015). Thermal stability studies of zeolites A and X synthesized from South African coal fly ash. *Research on Chemical Intermediates, 41*(2), 575–582. DOI: 10.1007/s11164-013-1211-3

Musyoka, N. M., Rambau, K. M., Manyala, N., Ren, J., Langmi, H. W., & Mathe, M. K. (2018). Utilization of waste tyres pyrolysis oil vapour in the synthesis of zeolite templated carbons (ZTCs) for hydrogen storage application. *Journal of Environmental Science and Health. Part A, Toxic/Hazardous Substances & Environmental Engineering, 53*(11), 1022–1028. DOI: 10.1080/10934529.2018.1471099

Mu, Y. G. (2010). Applications of bioceramics and its future. *Yaowu Fenxi Zazhi, 30*(7), 1343–1347. DOI: 10.16155/j.0254-1793.2010.07.058

Myrthong, B., Ansari, S., & Choudhary, R. B. (2025). Bimetallic oxide-conducting polymer composites for robust supercapacitor performance and applications: A review. *Journal of Energy Storage, 117*, 116182. DOI: 10.1016/j.est.2025.116182

Naber, J. E., & De Jong, K. P. (1994). Industrial applications of zeolite catalysis. *Studies in Surface Science and Catalysis, 84*, 219–294.

Nahum, A. B. R., Pereira, W. V. D., Martins, G. C., Dias, Y. N., Ribeiro, P. G., Salomão, G. N., Gastauer, M., Caldeira, C. F., Fernandes, A. R., & de Souza, E. S.. (2024). Properties and environmental quality of the overburden and tailings of manganese mining in the eastern Amazon. *Environmental Research, 262*, 119965. Advance online publication. DOI: 10.1016/j.envres.2024.119965 PMID: 39265760

Nandi, S., & Pumera, M. (2024). Transition metal dichalcogenide-based materials for rechargeable aluminum-ion batteries: A mini-review. *ChemSusChem, 17*(9), e202301434. DOI: 10.1002/cssc.202301434 PMID: 38212248

Nguyen, V. P., Kim, J.-H., & Lee, S.-M. (2023). Strategies of boosting synergistic adsorption-catalysis effect for high-performance lithium–sulfur batteries. *Materials Today. Energy, 38*, 101451. DOI: 10.1016/j.mtener.2023.101451

Nguyen, V.-T., Hung, C.-M., Nguyen, T.-B., Chang, J.-H., Wang, T.-H., Wu, C.-H., Lin, Y.-L., Chen, C.-W., & Dong, C.-D. (2019). Efficient Heterogeneous Activation of Persulfate by Iron-Modified Biochar for Removal of Antibiotic from Aqueous Solution: A Case Study of Tetracycline Removal. *Catalysts, 9*(1), 49. DOI: 10.3390/catal9010049

Nicholas, R. J., McGuire, M. A., Hyun, S.-H., Cullison, M. N., & Thompson, D. H. (2022). Development of an efficient, high purity continuous flow synthesis of Diazepam. *Frontiers in Chemical Engineering*, *4*, 877498. DOI: 10.3389/fceng.2022.877498

Nielsen, A., Dyer, N. L., Lechuga, C., McKee, M. D. and Dusek, J. A. (2024). Fidelity to the acupuncture intervention protocol in the ACUpuncture In The EmergencY department for pain management (ACUITY) trial: Expanding the gold standard of STRICTA and CONSORT guidelines, 13. DOI: 10.1016/j.imr.2024.101048

Nie, M., & Lucht, B. L. (2014). Role of lithium salt on solid electrolyte interface (SEI) formation and structure in lithium ion batteries. *Journal of the Electrochemical Society*, *161*(6), A1001–A1006. DOI: 10.1149/2.054406jes

Niska, J., & Loberg, B. (1994). Hot isostatic pressing of high temperature superconducting ceramics. *Materials Technology*, *9*(9-10), 211–215. DOI: 10.1080/10667857.1994.11785072

Niu, J., Wang, M., Cao, T., Cheng, X., Wu, R., Liu, H., Zhang, Y., & Liu, X. (2021). Li metal coated with Li3PO4 film via atomic layer deposition as battery anode. *Ionics*, *27*(6), 2445–2454. DOI: 10.1007/s11581-021-04030-z

Novembre, D., Gimeno, D., & Del Vecchio, A. (2021). Synthesis and characterization of Na-P1 (GIS) zeolite using a kaolinitic rock. *Scientific Reports*, *1*(1), 1–11. DOI: 10.1038/s41598-021-84383-7

Novembre, D., Sabatino, B. D. I., Gimeno, D., & Annunzio, S. G. D. (2011). Synthesis and characterization of Na-X, Na-A and Na-P zeolites and hydroxysodalite from metakaolinite. *Clay Minerals*, *46*(3), 339–354. DOI: 10.1180/claymin.2011.046.3.339

Novikov, I. S., Gubaev, K., Podryabinkin, E. V., & Shapeev, A. V. (2021). The MLIP package: Moment tensor potentials with MPI and active learning. *Machine Learning: Science and Technology*, *2*(2), 025002. DOI: 10.1088/2632-2153/abc9fe

Ogura, H., Yamamoto, T., & Kage, H. (2003). Efficiencies of CaO/H2O/Ca(OH)2 chemical heat pump for heat storing and heating/cooling. *Energy*, *28*(14), 1479–1493. DOI: 10.1016/S0360-5442(03)00119-1

Olii, M. R., Hambali, R., Djafar, L., Ichsan, I., & Saliko, M. (2024). Innovative porous mortar filters: Wastewater purification for clean water. *Water Science and Technology*, *89*(6), 1526–1538. DOI: 10.2166/wst.2024.073 PMID: 38557716

Olson, D. H.. (1970). A reinvestigation of the crystal structure of the zeolite hydrated NaX. *Journal of Physical Chemistry*, *74*(14), 2758–2764. DOI: 10.1021/j100708a006

Oluwole, O. A., Campos, E. V. R., de Oliveira, J. L., & Fraceto, L. F. (2025). Integrating the Nano–Phyto–Micro Triad for Climate-Resilient and Sustainable Agriculture. *ACS Omega*, *10*(45), 53702–53721. DOI: 10.1021/acsomega.5c07151

Omidian, H., Akhzarmehr, A., & Chowdhury, S. D. (2024). Advancements in Cellulose-Based Superabsorbent Hydrogels: Sustainable Solutions across Industries. *Gels (Basel, Switzerland)*, *10*(3), 174. Advance online publication. DOI: 10.3390/gels10030174 PMID: 38534592

Onyeachu, I. B., Chauhan, D. S., Quraishi, M. A., & Obot, I. B. (2021). Influence of hydrodynamic condition on 1,3,5-tris(4-methoxyphenyl)-1,3,5-triazinane as a novel corrosion inhibitor formulation for oil and gas industry. *Corrosion Engineering, Science and Technology*, *56*(2), 154–161. DOI: 10.1080/1478422X.2020.1827348

Ophus, C., Ercius, P., Sarahan, M., Czarnik, C., & Ciston, J. (2014). Recording and using 4D-STEM datasets in materials science. *Microscopy and Microanalysis*, *20*(S3), 62–63. DOI: 10.1017/S1431927614002037

Örkényi, R., Éles, J., Faigl, F., Vincze, P., Prechl, A., Szakács, Z., Kóti, J., & Greiner, I. (2017). Continuous synthesis and purification by coupling a multistep flow reaction with centrifugal partition chromatography. *Angewandte Chemie International Edition*, *56*(30), 8742–8745. DOI: 10.1002/anie.201703852 PMID: 28548374

Otto, D. (2019). Adoption and Diffusion of Open Educational Resources (OER) in Education: A Meta-Analysis of 25 OER-Projects. *International Review of Research in Open and Distance Learning*, *20*(5), 122–140. DOI: 10.19173/irrodl.v20i5.4472

Ouyang, J., Zhang, Z., Li, J., & Wu, C. (2024). Integrating Enzymes with Supramolecular Polymers for Recyclable Photobiocatalytic Catalysis. *Angewandte Chemie International Edition*, *63*(16), e202400105. DOI: 10.1002/anie.202400105

Owusu, P. A., & Asumadu-Sarkodie, S. (2016). A review of renewable energy sources, sustainability issues and climate change mitigation. *Cogent Engineering*, *3*(1), 1167990. DOI: 10.1080/23311916.2016.1167990

Ozdemir, O. D., & Pişkin, S. (2013). Zeolite X synthesis with different sources. *International Journal of Chemical* [IJCEBS]. *Environmental & Biological Sciences*, *1*(2), 229–232.

Paithankar, V., Devnania, D., & Nimburkar, T. A. (2025). AI-guided discovery of novel anti-inflammatory agents for cancer therapy: A new era in drug development. *Intelligent Hospitals*, *1*(1), 100007. DOI: 10.1016/j.inhs.2025.100007

Palaniappan, M. (2025). Time-dependent properties of temperature aged vinyl–based biocomposites developed using waste reinforcements. *Biomass Conversion and Biorefinery*, *15*(4), 5339–5349. DOI: 10.1007/s13399-024-05489-2

Pani, S., Dash, N., Mohapatra, B. K., & Singh, S. K. (2019). Siliceous manganese ore from eastern India: A potential resource for ferrosilicon-manganese production. *High-Temperature Materials and Processes*, *38*(2019), 425–435. DOI: 10.1515/htmp-2018-0081

Pani, S., Singh, S. K., & Mohapatra, B. K. (2016). Bog manganese ore: A resource for high manganese steel making. *Journal of the Minerals Metals & Materials Society*, *68*(6), 1525–1534. DOI: 10.1007/s11837-016-1935-9

Pathak, M., Roy, J., Patel, S., Some, S., Vyas, P., Das, N., & Shukla, P. (2021). Communicating climate change findings from IPCC reports: Insights from outreach events in India. *Climatic Change*, *168*(3), 23. DOI: 10.1007/s10584-021-03224-8

Patra, A. K. N., Jose, J. R., Sahoo, S., Chakraborty, B., & Rout, C. S. (2021). Understanding the charge storage mechanism of supercapacitors: In situ/operando spectroscopic approaches and theoretical investigations. *Journal of Materials Chemistry. A, Materials for Energy and Sustainability*, *9*(46), 25852–25891. DOI: 10.1039/D1TA07401F

Pattipaka, S., Lim, Y., Son, Y. H., Bae, Y. M., Peddigari, M., & Hwang, G. T. (2024). Ceramic-based dielectric materials for energy storage capacitor applications. *Materials (Basel)*, *17*(10), 2277. DOI: 10.3390/ma17102277 PMID: 38793340

Paudel, H. P., Syamlal, M., Crawford, S. E., Lee, Y.-L., Shugayev, R. A., Lu, P., Ohodnicki, P. R., Mollot, D., & Duan, Y. (2022). Quantum Computing and Simulations for Energy Applications: Review and Perspective. *ACS Engineering Au*, *2*(3), 151–196. DOI: 10.1021/acsengineeringau.1c00033

Pehnt, M. (2006). Dynamic life cycle assessment (LCA) of renewable energy technologies. *Renewable Energy*, *31*(1), 55–71. DOI: 10.1016/j.renene.2005.03.002

Peng, D., Wang, Q., Zang, S. H., & Mo, L. Y. (2024). Catalytic complete oxidation of ethyl acetate on MnO/MgAlO catalysts. *Kinetics and Catalysis*, *65*(4), 386–395. DOI: 10.1134/S0023158423601134

Peng, L., Guo, Y., Liu, S., He, G., & Mauzerall, D. L. (2024). Subsidizing Grid-Based Electrolytic Hydrogen Will Increase Greenhouse Gas Emissions in Coal Dominated Power Systems. *Environmental Science & Technology*, *58*(12), 5187–5195. DOI: 10.1021/acs.est.3c03045

Percival, H. J., Duncan, J. F., & Foster, P. K.H. J. P. (1974). Interpretation of the kaolinite-mullite reaction sequence. *Journal of the American Ceramic Society, 57–61*(2), 57–61. Advance online publication. DOI: 10.1111/j.1151-2916.1974.tb10813.x

Pereira, P. M., Ferreira, B., Oliveira, N., Nassar, E., Ciuffi, K., Vicente, M., Trujillano, R., Rives, V., Gil, A., Korili, S., & De Faria, E. (2018). Synthesis of zeolite A from metakaolin and its application in the adsorption of cationic dyes. *Applied Sciences (Basel, Switzerland), 8*(4), 608. DOI: 10.3390/app8040608

Perevislov, N. S. (2022). Superhard boron suboxide (B6O): Crystal structure, synthesis, properties, applications, and materials based thereon. *Mendeleev Communications, 32*(2), 152–163. DOI: 10.1016/j.mencom.2022.03.002

Pérez-Botella, E., Valencia, S., & Rey, F. (2022). Zeolites in adsorption processes: State of the art and future prospects. *Chemical Reviews, 122*(24), 17647–17695. DOI: 10.1021/acs.chemrev.2c00140

Petrich, L., Westhoff, D., Feinauer, J., Finegan, D. P., Daemi, S. R., Shearing, P. R., & Schmidt, V. (2017). Crack detection in lithium-ion cells using machine learning. *Computational Materials Science, 136*, 297–305. DOI: 10.1016/j.commatsci.2017.05.012

Petrov, T. M. I. (2012). *Synthesis of zeolite A: A review*. https://conf.uni-ruse.bg/bg/docs/cp12/9.1/9.1-5.pdf

Phelps, D. W., Parkinson, L. V., Boucher, J. M., Muncke, J., & Geueke, B. (2024). Per- and Polyfluoroalkyl Substances in Food Packaging: Migration, Toxicity, and Management Strategies. *Environmental Science & Technology, 58*(13), 5670–5684. DOI: 10.1021/acs.est.3c03702

Pichon, V., Brothier, F., & Combès, A. (2014). Aptamer-based-sorbents for sample treatment—A review. *Analytical and Bioanalytical Chemistry, 407*(3), 681–698. DOI: 10.1007/s00216-014-8129-5 PMID: 25286873

Pilania, G. (2021). Machine learning in materials science: From explainable predictions to autonomous design. *Computational Materials Science, 193*, 110360. DOI: 10.1016/j.commatsci.2021.110360

Pizzi, G., Cepellotti, A., Sabatini, R., Marzari, N., & Kozinsky, B. (2016). AiiDA: Automated interactive infrastructure and database for computational science. *Computational Materials Science, 111*, 218–230. DOI: 10.1016/j.commatsci.2015.09.013

Pley, C., van der Heijden, M. R., & Tulinius, C. (2020). A health professional's guide to the intersection of public health with intellectual property rights in trade and investment: The case of tobacco plain packaging. *Journal of Public Health Policy*, *41*(1), 52–62. DOI: 10.1057/s41271-019-00195-0

Pradhan, A., Pattnaik, G., Das, S., Acharya, B., & Patra, C. N. (2025). Advancements in lung cancer: Molecular insights, innovative therapies, and future prospects. *Medical Oncology*, *42*(9), 383. DOI: 10.1007/s12032-025-02725-1 PMID: 40719980

Prakash, M., Sakhavand, N., & Shahsavari, R. (2013). H_2, N_2, and CH_4 Gas Adsorption in Zeolitic Imidazolate Framework-95 and -100: Ab Initio Based Grand Canonical Monte Carlo Simulations. *The Journal of Physical Chemistry. C, Nanomaterials and Interfaces*, *117*(46), 24407–24416. DOI: 10.1021/jp408075y

Psioda, M. A., Xu, J. W., Jiang, Q., Ke, C. L., Yang, Z., & Ibrahim, J. G. (2021). Bayesian adaptive basket trial design using model averaging. *Biostatistics (Oxford, England)*, *22*(1), 19–34. DOI: 10.1093/biostatistics/kxz014 PMID: 31107534

Punj, S., Singh, J., & Singh, K. (2021). Ceramic biomaterials: Properties, state of the art and future prospectives. *Ceramics International*, *47*(20), 28059–28074. DOI: 10.1016/j.ceramint.2021.06.238

Putman, J. I., & Armstrong, D. W. (2022). Recent advances in the field of chiral crystallization. *Chirality*, *34*(10), 1338–1354. DOI: 10.1002/chir.23492 PMID: 35904758

Qian, J., McMurray, C. E., Mukhopadhyay, D. K., Wiggins, J. K., Vail, M. A., & Bertagnolli, K. E. (2012). Polycrystalline diamond cutters sintered with magnesium carbonate in cubic anvil press. *International Journal of Refractory & Hard Metals*, *31*, 71–75. DOI: 10.1016/j.ijrmhm.2011.09.008

Qi, K., Imparato, C., Almjasheva, O., Khataee, A., & Zheng, W. (2024). TiO2-based photocatalysts from type-II to S-scheme heterojunction and their applications. *Journal of Colloid and Interface Science*, *675*, 150–191. DOI: 10.1016/j.jcis.2024.06.204 PMID: 38968635

Rabiee, N., & Rabiee, M. (2025). Engineered Metal–Organic Frameworks for Targeted CRISPR/Cas9 Gene Editing. *ACS Pharmacology & Translational Science*, *8*(4), 1028–1049. DOI: 10.1021/acsptsci.5c00047

Raclariu-Manolică, A. C. and de Boer, H. J. (2022). DNA barcoding and metabarcoding for quality control of botanicals and derived herbal products, DOI: 10.1016/b978-0-323-85542-6.00004-410.1016/b978-0-323-85542-6.00004-4

Ragone, M., Shahabazian-Yassar, R., Mashayek, F., & Yurkiv, V. (2023). Deep learning modeling in microscopy imaging: A review of materials science applications. *Progress in Materials Science*, *138*, 101165. DOI: 10.1016/j.pmatsci.2023.101165

Randhawa, K. S. (2024). Advanced ceramics in energy storage applications: Batteries to hydrogen energy. *Journal of Energy Storage*, *98*, 113122. DOI: 10.1016/j.est.2024.113122

Ravindiran, G., Rajamanickam, S., Janardhan, G., Hayder, G., Alagumalai, A., Mahian, O., Lam, S. S., & Sonne, C. (2024). Production and modifications of biochar to engineered materials and its application for environmental sustainability: A review. *Biochar*, *6*(1), 62. DOI: 10.1007/s42773-024-00350-1

Reclaru, L., Ardelean, L., Miu, C., & Grecu, A. (2020). Are zirconia bioceramics and ceramics intended to come in contact with skin inert. *Materials (Basel)*, *13*(7), 1697–1672. DOI: 10.3390/ma13071697 PMID: 32260495

Reddy, K. R., Reddy, C. V., Nadagouda, M. N., Shetti, N. P., Jaesool, S., & Aminabhavi, T. M. (2019). Polymeric graphitic carbon nitride (g-C_3N_4)-based semiconducting nanostructured materials: Synthesis methods, properties and photocatalytic applications. *Journal of Environmental Management*, *238*, 25–40. DOI: 10.1016/j.jenvman.2019.02.075 PMID: 30844543

Reyes-Morales, Q. L., Reyes-Cruz, V. E., Trujillo-Estrada, A., Cobos-Murcia, J. A., Urbano-Reyes, G., & Pérez-Labra, M. (2021). Selenium and sulphur reactions involved in manganese reduction from sulphate solutions. *RSC Advances*, *11*(41), 25542–25550. DOI: 10.1039/D1RA03679C PMID: 35478895

Rinkel, B. L., Hall, D. S., Temprano, I., & Grey, C. P. (2020). Electrolyte oxidation pathways in lithium-ion batteries. *Journal of the American Chemical Society*, *142*(35), 15058–15074. DOI: 10.1021/jacs.0c06363 PMID: 32697590

Ríos, C., Williams, C., & Maple, M. (2007). Synthesis of zeolites and zeotypyes by hydrothermal transformation of kaolinite and metakaolinite. *Bistua*, *5*(1), 15–26.

Rodrigues, M. L. M., Leão, V. A., Gomes, O., Lambert, F., Bastin, D., & Gaydardzhiev, S. (2015). Copper extraction from coarsely ground printed circuit boards using moderate thermophilic bacteria in a rotating-drum reactor. *Waste Management*, *41*, 148–158. DOI: 10.1016/j.wasman.2015.04.001 PMID: 25899037

Rohskopf, A., Sievers, C., Lubbers, N., Cusentino, M., Goff, J., Janssen, J., & Sargsyan, K. (2023). FitSNAP: Atomistic machine learning with LAMMPS. *Journal of Open Source Software*, *8*(84), 5118. DOI: 10.21105/joss.05118

Ronchi, L., Ryzhikov, A., Nouali, H., Daou, T. J., Albrecht, S., & Patarin, J. (2018). Extra large pore opening CFI and DON-type zeosils for mechanical energy storage. *Microporous and Mesoporous Materials*, *255*, 211–219. DOI: 10.1016/j.micromeso.2017.07.039

Ruini, A., Bundgaard, A. M., Mosgaard, M. A., Løkke, S., & Pizzol, M. (2025). Normative and empirical solutions to the upcycling dilemma. *Sustainable Production and Consumption*, *54*, 466–475. DOI: 10.1016/j.spc.2025.01.022

Russell, M. G., & Jamison, T. F. (2019). Seven-step continuous flow synthesis of Linezolid without intermediate purification. *Angewandte Chemie International Edition*, *58*(23), 7678–7681. DOI: 10.1002/anie.201901814 PMID: 30964955

Salahudeen, N. (2022). *A review on zeolite: Application, synthesis and effect of synthesis parameters on product properties*. Chemistry Africa., DOI: 10.1007/s42250-022-00471-9

Saleem, S. S., Ravichandran, S., Ravichandran, S., Chinnadurai, K. R., Saravanan, A., & Vickram, S. (2025). A Comprehensive Review Molecular Origami Classical DNA Folding Principle Empowering Modern Nanobiotechnology. *ACS Applied Bio Materials*, *8*(10), 8467–8512. DOI: 10.1021/acsabm.5c01094

Salimi, E. (2020). Functionally graded calcium phosphate bioceramics: An overview of preparation and properties. *Ceramics International*, *46*(12), 19664–19668. DOI: 10.1016/j.ceramint.2020.05.077

Sánchez-López, P., Kotolevich, Y., Yocupicio-Gaxiola, R. I., Antúnez-García, J., Chowdari, R. K., Petranovskii, V., & Fuentes-Moyado, S. (2021). Recent advances in catalysis based on transition metals supported on zeolites. *Frontiers in Chemistry*, *9*, 716745. DOI: 10.3389/fchem.2021.716745

Sanchez-Rexach, E., Johnston, T. G., Jehanno, C., Sardon, H., & Nelson, A. (2020). Sustainable Materials and Chemical Processes for Additive Manufacturing. *Chemistry of Materials*, *32*(17), 7105–7119. DOI: 10.1021/acs.chemmater.0c02008

Sánchez-Trasviña, C., Flores-Gatica, M., Enriquez-Ochoa, D., Rito-Palomares, M., & Mayolo-Deloisa, K. (2021). Purification of modified therapeutic proteins available on the market: An analysis of chromatography-based strategies. *Frontiers in Bioengineering and Biotechnology*, *9*, 717326. DOI: 10.3389/fbioe.2021.717326 PMID: 34490225

Sanjabi, S., & Salehirad, A. (2025). $MgAl_2O_4$/$CoFe_2O_4$/Zeolite a magnetic ceramic nanocomposite for efficacious elimination of Methylene blue from aqueous solutions: Composition influence. *Research on Chemical Intermediates*, *51*(9), 4907–4935. DOI: 10.1007/s11164-025-05679-1

Sari, M. E. F., Suprapto, S., & Prasetyoko, D. (2018). Direct synthesis of sodalite from kaolin: The influence of alkalinity. *Indonesian Journal of Chemistry*, *18*(4), 607–613. DOI: 10.22146/ijc.25191

Sarker, I. H. (2021). Deep learning: A comprehensive overview on techniques, taxonomy, applications and research directions. *SN Computer Science*, *2*(6), 420. DOI: 10.1007/s42979-021-00815-1 PMID: 34426802

Sasaki, N., & Clevers, H. (2018). Studying cellular heterogeneity and drug sensitivity in colorectal cancer using organoid technology. *Current Opinion in Genetics & Development*, *52*, 117–122. DOI: 10.1016/j.gde.2018.09.001 PMID: 30261425

Sbaiti, M., Abbara, A., Al-Haffar, M., Budhathoki, S., Wihba, K., Cardoso Pinto, A., & Streule, M. (2024). Curricular innovations to address inequalities in global health education. *European Journal of Public Health*, *34*(Supplement_3), ckae144.1231. Advance online publication. DOI: 10.1093/eurpub/ckae144.1231

Scala, P., Serio, B., & Giudice, V. (2026). 3D In Vitro Models of the Bone Marrow Niche. *ACS Biomaterials Science & Engineering*, *12*(1), 110–127. DOI: 10.1021/acsbiomaterials.5c01421

Schmidtsdorff, S., & Schmidt, A. H. (2022). Nitrosamine impurities in pharmaceuticals: A review on the current status of the regulatory environment and analytical methods. *Journal of Pharmaceutical and Biomedical Analysis*, *209*, 114526. DOI: 10.1016/j.jpba.2021.114526

Schneider, G. (2017). Automating drug discovery. *Nature Reviews. Drug Discovery*, *17*(2), 97–113. DOI: 10.1038/nrd.2017.232 PMID: 29242609

Seavey, K. C., Khare, N. P., Liu, Y. A., Williams, T. N., & Chen, C.-C. (2003). A New Phase-Equilibrium Model for Simulating Industrial Nylon-6 Production Trains. *Industrial & Engineering Chemistry Research*, *42*(17), 3900–3913. DOI: 10.1021/ie030112+

Selicani, G. V., Mobin, M., Esposito, V., Insinga, A. R., & Haugen, A. B. (2025). Shape engineering and manufacturing of piezoceramics for energy conversion-a review. *Journal of Physics: Energy*, *7*(2), 022004. DOI: 10.1088/2515-7655/adb0ee

Selis, L. A., & Seminario, J. M. (2018). Dendrite formation in silicon anodes of lithium-ion batteries. *RSC Advances*, *8*(10), 5255–5267. DOI: 10.1039/C7RA12690E PMID: 35542415

Selis, L. A., & Seminario, J. M. (2019). Dendrite formation in Li-metal anodes: An atomistic molecular dynamics study. *RSC Advances*, *9*(48), 27835–27848. DOI: 10.1039/C9RA05067A PMID: 35530483

Selvin, T., Berglund, M., Lenhammar, L., Jarvius, M., Fryknäs, M., Larsson, R., & Andersson, C. R. (2023). Phenotypic screening platform identifies statins as enhancers of immune cell-induced cancer cell death. *BMC Cancer*, *23*(1), 164. DOI: 10.1186/s12885-023-10645-4 PMID: 36803614

Sethurajan, M., & Gaydardzhiev, S. (2021). Bioprocessing of spent lithium ion batteries for critical metals recovery - A review, Resources. *Resources, Conservation and Recycling*, *165*, 105225. DOI: 10.1016/j.resconrec.2020.105225

Sethuraman, V. A., Hardwick, L. J., Srinivasan, V., & Kostecki, R. (2010). Surface structural disordering in graphite upon lithium intercalation/deintercalation. *Journal of Power Sources*, *195*(11), 3655–3660. DOI: 10.1016/j.jpowsour.2009.12.034

Shah, A. M., Kamal, R., Wang, Q., Zhang, J., Lu, H., & Zhao, Z. K. (2025). Calcium carbonate-mediated detoxification of corn stover hydrolysate to facilitate the fermentation by Rhodosporidium toruloides. *Biomass and Bioenergy*, *200*, 108036. DOI: 10.1016/j.biombioe.2025.108036

Shahi, S., Roghani-Mamaqani, H., Hoogenboom, R., Talebi, S., & Mardani, H. (2022). Stimuli-Responsive Covalent Adaptable Hydrogels Based on Homolytic Bond Dissociation and Chain Transfer Reactions. *Chemistry of Materials*, *34*(2), 468–498. DOI: 10.1021/acs.chemmater.1c03678

Shannon, M. A., Bohn, P. W., Elimelech, M., Georgiadis, J. G., Marinas, B. J., & Mayes, A. M. (2008). Science and technology for water purification in the coming decades. *Nature*, *452*(7185), 301–310. DOI: 10.1038/nature06599 PMID: 18354474

Sharma, P., Song, J. S., Han, M. H., & Cho, C. H. (2016). GIS-NaP1 zeolite microspheres as potential water adsorption material: Influence of initial silica concentration on adsorptive and physical/topological properties. *Scientific Reports*, *6*(1), 1–26. DOI: 10.1038/srep22734

Sheldon, R. A. (2018). Metrics of Green Chemistry and Sustainability: Past, Present, and Future. *ACS Sustainable Chemistry & Engineering*, *6*(1), 32–48. DOI: 10.1021/acssuschemeng.7b03505

Sheldon, R. A., & Woodley, J. M. (2018). Role of biocatalysis in sustainable chemistry. *Chemical Reviews*, *118*(2), 801–838. DOI: 10.1021/acs.chemrev.7b00203 PMID: 28876904

Shen, K.-H., Fan, M., & Hall, L. M. (2021). Molecular dynamics simulations of ion-containing polymers using generic coarse-grained models. *Macromolecules*, *54*(5), 2031–2052. DOI: 10.1021/acs.macromol.0c02557

Shen, W., He, W., Zhong, M., Zhang, J., & Guo, S. (2025). Heat generation power of graphite anode during electrochemical lithiation. *Journal of Power Sources*, *639*, 236702. DOI: 10.1016/j.jpowsour.2025.236702

Shen, Y., & Zhang, S. (2025). Recent advances in sustainable reductive roasting for spent lithium-ion batteries recycling. *Energy Storage Materials*, *79*, 104358. DOI: 10.1016/j.ensm.2025.104358

Shi, H., Luo, Y., Deng, Y., Dai, J., Zhang, J., & Ou, L. (2025). Recovery of valuable metals from spent lithium-ion batteries based on a green and efficient leaching system of dimethyl-β-propionic acid thiophene (DMPT). *Hydrometallurgy*, *237*, 106551. DOI: 10.1016/j.hydromet.2025.106551

Shirk, B. D., Heichel, D. L., Eccles, L. E., Rodgers, L. I., Lateef, A. H., Burke, K. A., & Stoppel, W. L. (2024). Modifying Naturally Occurring, Nonmammalian-Sourced Biopolymers for Biomedical Applications. *ACS Biomaterials Science & Engineering*, *10*(10), 5915–5938. DOI: 10.1021/acsbiomaterials.4c00689

Shi, Y., Zhen, X., Zhang, Y., Li, Y., Koo, S., Saiding, Q., Kong, N., Liu, G., Chen, W., & Tao, W. (2024). Chemically Modified Platforms for Better RNA Therapeutics. *Chemical Reviews*, *124*(3), 929–1033. DOI: 10.1021/acs.chemrev.3c00611

Shodiev, A., Primo, E., Arcelus, O., Chouchane, M., Osenberg, M., Hilger, A., Manke, I., Li, J., & Franco, A. A. (2021). Insight on electrolyte infiltration of lithium ion battery electrodes by means of a new three-dimensional-resolved lattice Boltzmann model. *Energy Storage Materials*, *38*, 80–92. DOI: 10.1016/j.ensm.2021.02.029

Shokri, Z., Mousavi, S. M., & Omidkhah, M. R. (2025). Advancing sustainable lithium recovery from spent laptop lithium-ion battery cathodes through high pulp density bioleaching followed by Li_2CO_3 precipitation: Process intensification and environmental insights. *Journal of Environmental Chemical Engineering*, *13*(3), 117014. DOI: 10.1016/j.jece.2025.117014

Short, P. L. (2007). New Regulatory Era in Europe Begins: European Chemicals Agency opens does as companies prepare for new regulations. *Chemical and Engineering News*, *85*(33), 33–35. DOI: 10.1021/cen-v085n033.p033

Shu, J. C., Liu, R. L., Liu, Z. H., Qiu, J., Chen, H. L., & Tao, C. Y. (2017). Simultaneous removal of ammonia nitrogen and manganese from wastewater using nitrite by electrochemical method. *Environmental Technology, 38*(3), 370–376. DOI: 10.1080/09593330.2016.1194482 PMID: 27249226

Shu, Y., Yang, Z. Y., Pan, W. K., Xiang, D. X., Wu, J. G., Gao, L. H., Jiao, X., He, H., Zhang, Z. X., & Chen, G. (2025). Achieving carbon neutrality through conversion of CO_2 to cyclic carbonates using bifunctional benzimidazolium-based pyridine hypercrosslinked polymers. *Applied Surface Science, 686*, 162164. Advance online publication. DOI: 10.1016/j.apsusc.2024.162164

Siegel, H. G., Nason, S. L., Warren, J. L., Prunas, O., Deziel, N. C., & Saiers, J. E. (2023). Investigation of Sources of Fluorinated Compounds in Private Water Supplies in an Oil and Gas-Producing Region of Northern West Virginia. *Environmental Science & Technology, 57*(45), 17452–17464. DOI: 10.1021/acs.est.3c05192

Simić, S., Zukić, E., Schmermund, L., Faber, K., Winkler, C. K., & Kroutil, W. (2022). Shortening Synthetic Routes to Small Molecule Active Pharmaceutical Ingredients Employing Biocatalytic Methods. *Chemical Reviews, 122*(1), 1052–1126. DOI: 10.1021/acs.chemrev.1c00574

Singh, K., Gupta, J. K., Jain, D., Kumar, S., Singh, T. and Saha, S. (2024). Exploring the ancient wisdom and modern relevance of Chinese medicine: A comprehensive review, 11. DOI: 10.1016/j.prmcm.2024.100448

Singh, P., Shandilya, P., Raizada, P., Sudhaik, A., Rahmani-Sani, A., & Hosseini–Bandegharaei, A. (2018). Review on various strategies for enhancing photocatalytic activity of graphene based nanocomposites for water purification. *Arabian Journal of Chemistry, 13*(1), 3498–3520. DOI: 10.1016/j.arabjc.2018.12.001

Singh, S., Santhiya, D., & Sharma, J. G. (2025). An overview of bioprocess based recovery and recycling techniques for battery waste management: Transitioning from environmental hazard to sustainability. *Minerals Engineering, 233*, 109598. DOI: 10.1016/j.mineng.2025.109598

Siu, P. M., Yu, D. J., Yu, A. P., Recchia, F., Li, S. X., Chan, R. N. Y., &Irwin, M. R. (2025). Tai chi or cognitive behavioural therapy for treating insomnia in middle aged and older adults: randomised non-inferiority trial, 391. DOI: 10.1136/bmj-2025-084320

Soboyejo, W., Obayemi, J., Annan, E., Ampaw, E. K., Daniels, L., & Rahbar, N. (2015). Review of high temperature ceramics for aerospace applications. *Advanced Materials Research, 1132*, 385–407.. DOI: 10.4028/www.scientific.net/AMR.1132 .385

Solomon, M. B., Rabha, S. S., Fimbres-Weihs, G., Goyal, H., Taghikhah, F. R., Varghese, J. J., Wenger, S. R., Liang, W., Kearns, E. R., Huang, J., Kaisare, N. S., & D'Alessandro, D. M. (2024). Decarbonization in Australia and India: Bilateral Opportunities and Challenges for the Net Zero Transformation. *ACS Engineering Au*, *4*(3), 295–311. DOI: 10.1021/acsengineeringau.3c00060

Song, G., Sun, C., Madadi, M., Dou, S., Yan, J., Huan, H., Aghbashlo, M., Tabatabaei, M., Sun, F., & Ashori, A. (2024). Dual assistance of surfactants in glycerol organosolv pretreatment and enzymatic hydrolysis of lignocellulosic biomass for bioethanol production. *Bioresource Technology*, *395*, 130358. DOI: 10.1016/j.biortech.2024.130358 PMID: 38253243

Song, X. S., Shen, Q. C., Liu, B., Li, J., & Zhu, D. W. (2024). Research on the use of low-selenium additives in the electrodeposition of manganese metal and co-production of electrolytic manganese. *Electrochemistry (Tokyo, Japan)*, *92*(7), 077004. Advance online publication. DOI: 10.5796/electrochemistry.24-00045

Sorokina, S. A., Vorob'ev, R. A., Gorshunov, M. G., & Chernigin, M. A. (2023). Use of chromium-manganese corrosion-resistant steel in district heating network applications. *Metal Science and Heat Treatment*, *65*(3–4), 159–166. DOI: 10.1007/s11041-023-00908-z

Srilai, S., Tanwongwan, W., Onpecth, K., Wongkitikun, T., & Panpiemrasda, K. (2020). Synthesis of zeolite A from bentonite via hydrothermal method: The case of different base solution. In *The Second Materials Research Society of Thailand International Conference* (Vol. 060006). https://doi.org/DOI: 10.1063/5.0025043

Stefanache, A., Marcinschi, A., Marin, G.-A., Mitran, A.-M., Lungu, I. I., Miftode, A. M., & Hancianu, M. (2025). Green chemistry approaches in pharmaceutical synthesis: Sustainable methods for drug development. *AppliedChem*, *5*(2), 22. DOI: 10.3390/appliedchem5020013

Subbiah, G., Jena, A., Kalidhas, A. M., Kele, V., Aadiwal, V., Awasthi, K., & K, K. P. (2025). Sustainable Food Production and Valorization of Food and Agricultural Waste: Recent Advances and Future Prospects. *ACS Food Science & Technology*, *5*(7), 2549–2562. DOI: 10.1021/acsfoodscitech.5c00403

Su, G., Ong, H. C., Mohd Zulkifli, N. W., Ibrahim, S., Chen, W. H., Chong, C. T., & Ok, Y. S. (2022). Valorization of animal manure via pyrolysis for bioenergy: A review. *Journal of Cleaner Production*, *343*, 130965. DOI: 10.1016/j.jclepro.2022.130965

Sulley, G. A., Raush, J., Montemore, M. M., & Hamm, J. (2024). Accelerating high-entropy alloy discovery: Efficient exploration via active learning. *Scripta Materialia*, *249*, 116180. DOI: 10.1016/j.scriptamat.2024.116180

Sun, H., Zhang, H., Ang, E. L., & Zhao, H. (2018). Biocatalysis for the synthesis of pharmaceuticals and pharmaceutical intermediates. *Bioorganic & Medicinal Chemistry, 26*(7), 1275–1284. DOI: 10.1016/j.bmc.2017.06.043 PMID: 28709846

Sun, J., Hou, S., Guo, Y., Zhou, J., Cui, Y., & Bao, J. (2024). Preparation of steel slag-based cementitious materials compounded with GBFS and phosphogypsum: Mix design, hydration, and microstructure characteristic. *Journal of Thermal Analysis and Calorimetry, 149*(23), 13785–13798.

Sun, Q., Wang, N., Bai, R., Hui, Y., Zhang, T., Do, D. A., Zhang, P., Song, L., Miao, S., & Yu, J. (2019). Synergetic effect of ultrasmall metal clusters and zeolites promoting hydrogen generation. *Advanced Science (Weinheim, Baden-Württemberg, Germany), 6*(10), 1802350. DOI: 10.1002/advs.201802350

Sun, Q., Wang, N., & Yu, J. (2021). Advances in catalytic applications of zeolite-supported metal catalysts. *Advanced Materials, 33*(51), 2104442. DOI: 10.1002/adma.202104442

Sun, Y. Q., & Ma, H. Y. (2023). Application of three-dimensional cell culture technology in screening anticancer drugs. *Biotechnology Letters, 45*(9), 1073–1092. DOI: 10.1007/s10529-023-03410-x PMID: 37421554

Sun, Y., Yang, T., Ji, H., Zhou, J., Wang, Z., Qian, T., & Yan, C. (2020). Boosting the optimization of lithium metal batteries by molecular dynamics simulations: A perspective. *Advanced Energy Materials, 10*(41), 2002373. DOI: 10.1002/aenm.202002373

Sun, Z., Sun, G. Y., Salazar-Tio, R., Fager, A., & Crouse, B. (2025). A comprehensive review of CO_2 geological storage projects with regulatory requirements of computational modeling. *Fuel, 386*, 134209. Advance online publication. DOI: 10.1016/j.fuel.2024.134209

Surana, M., Pattanayak, D. S., Yadav, V., Singh, V. K., & Pal, D. (2024). An insight decipher on photocatalytic degradation of microplastics: Mechanism, limitations, and future outlook. *Environmental Research, 247*, 118268. DOI: 10.1016/j.envres.2024.118268 PMID: 38244970

Surana, P., Singh Dhull, K., Arya, A., Samreen, S., Rajan, M., & Parihar, A. S. (2024). Bio-ceramics application in dentistry. *Bioinformation, 20*(2), 136–139. DOI: 10.6026/973206300200136 PMID: 38497078

Surin, A. M., Sharipov, R. R., Krasil'nikova, I. A., Boyarkin, D. P., Lisina, O. Y., Gorbacheva, L. R., Avetisyan, A. V., & Pinelis, V. G. (2017). Disruption of functional activity of mitochondria during MTT assay of viability of cultured neurons. *Biochemistry. Biokhimiia*, *82*(6), 737–749. DOI: 10.1134/S0006297917060104 PMID: 28601083

Susarla, N., & Ahmed, S. (2019). Estimating cost and energy demand in producing lithium hexafluorophosphate for Li-ion battery electrolyte. *Industrial & Engineering Chemistry Research*, *58*(9), 3754–3766. DOI: 10.1021/acs.iecr.8b03752

Su, Z. H., Guo, Z. H., Xie, H. Y., Qu, M. Z., Wang, H., & Peng, G. C. (2024). Multifunctional surface modification to enhance the electrochemical performance of lithium-rich manganese-based cathode materials. *Electrochimica Acta*, *500*, 144727. Advance online publication. DOI: 10.1016/j.electacta.2024.144727

Szwabowski, G. L., Cole, J. A., Baker, D. L., & Parrill, A. L. (2023). Structure-based pharmacophore modeling 1: Automated random pharmacophore model generation. *Journal of Molecular Graphics & Modelling*, *121*, 108429. DOI: 10.1016/j.jmgm.2023.108429 PMID: 36804368

Tabone, M. D., Cregg, J. J., Beckman, E. J., & Landis, A. E. (2010). Sustainability Metrics: Life Cycle Assessment and Green Design in Polymers. *Environmental Science & Technology*, *44*(21), 8264–8269. DOI: 10.1021/es101640n

Tahmasebi, Y. M. (2021). Alkali activation of mechanically activated low-grade clay. *Journal of Sustainable Cement-Based Materials*, *5*(10), 272–288. DOI: 10.1080/21650373.2020.1838355

Tang, B., Yang, F., Chen, C., Shi, C., Deng, W., Li, J., & Jiang, L. (2023). A novel lead-based pseudo-dimensional stable anode toward efficient and clean extraction of metallic manganese. *Journal of Cleaner Production*, *386*, 135806. Advance online publication. DOI: 10.1016/j.jclepro.2022.135806

Tang, H., Wu, Y., Cai, Y., Wang, F., Lin, Z., & Pei, Y. (2022). Design of power lithium battery management system based on digital twin. *Journal of Energy Storage*, *47*, 103679. DOI: 10.1016/j.est.2021.103679

Tang, M., He, D., Wang, W., Wang, H., Xu, C., Li, F., & Guan, J. (2012). Superhard solid solutions of diamond and cubic boron nitride. *Scripta Materialia*, *66*(10), 781–784. DOI: 10.1016/j.scriptamat.2012.02.006

Tang, Y., Zhang, Z., Jin, Y., Gao, S., Huang, D., Xu, S., & Zou, R. J. (2022). Kinetic and thermodynamic insights into advanced energy storage mechanisms of battery-type bimetallic metal–organic frameworks. *Chemistry of Materials, 34*(23), 10338–10346. DOI: 10.1021/acs.chemmater.2c01798

Tao, M., Qiu, X. P., & Lu, D. M. (2024). Life cycle assessment of electrolytic manganese metal production. *The Science of the Total Environment, 949*, 174862. Advance online publication. DOI: 10.1016/j.scitotenv.2024.174862 PMID: 39038680

Tavoosi, M. (2023). The precipitation of $PbFe_{12}O_{19}$ hard magnetic phase in PbO-SiO_2-B_2O_3-Fe_2O_3 glass-ceramics. Journal of Non-Crystalline Solids: A Journal Devoted to Oxide, Halide, Chalcogenide and Metallic Glasses, Amorphous Semiconductors, Non-Crystalline Films. *Journal of Non-Crystalline Solids, 615*, 122402. DOI: 10.1016/j.jnoncrysol.2023.122402

Teel, T., Zoelle, A., Williams, J. M., Gavvalapalli, N., Woods, M., Hancu, D., Fout, T., & Homsy, S. (2026). Insights from Initial Engineering Designs of Point Source Capture at Industrial Facilities. *Industrial & Engineering Chemistry Research, 65*(3), 1751–1770. DOI: 10.1021/acs.iecr.5c03762

Teo, K. S. K., Kondo, K., Khattab, S. M. R., Watanabe, T., Nagata, T., & Katahira, M. (2024). Enhancing Bioethanol Production from Rice Straw through Environmentally Friendly Delignification Using Versatile Peroxidase. *Journal of Agricultural and Food Chemistry, 72*(5), 2657–2666. DOI: 10.1021/acs.jafc.3c07998 PMID: 38288662

Theerthagiri, J., Lee, S. J., Karuppasamy, K., Arulmani, S., Veeralakshmi, S., Ashokkumar, M., & Choi, M. Y. (2021). Application of advanced materials in sono-photocatalytic processes for the remediation of environmental pollutants. *Journal of Hazardous Materials, 412*, 125245. DOI: 10.1016/j.jhazmat.2021.125245 PMID: 33545645

Tokranov, A., Sheldon, B. W., Li, C., Minne, S., & Xiao, X. (2014). In situ atomic force microscopy study of initial solid electrolyte interphase formation on silicon electrodes for Li-ion batteries. *ACS Applied Materials & Interfaces, 6*(9), 6672–6686. DOI: 10.1021/am500363t PMID: 24670933

Tom, G., Schmid, S. P., Baird, S. G., Cao, Y., Darvish, K., Hao, H., Lo, S., Pablo-García, S., Rajaonson, E. M., Skreta, M., Yoshikawa, N., Corapi, S., Akkoc, G. D., Strieth-Kalthoff, F., Seifrid, M., & Aspuru-Guzik, A. (2024). Self-Driving Laboratories for Chemistry and Materials Science. *Chemical Reviews, 124*(16), 9633–9732. DOI: 10.1021/acs.chemrev.4c00055

Tompsett, G. A., Conner, C., & Yngvesson, K. S. (2006). Microwave synthesis of nanoporous materials. *Chemphyschem : A European Journal of Chemical Physics and Physical Chemistry*, *7*(2), 296–319. DOI: 10.1002/cphc.200500449

Tong, Q., Gao, P., Liu, H., Xie, Y., Lv, J., Wang, Y., & Zhao, J. (2020). Combining Machine Learning Potential and Structure Prediction for Accelerated Materials Design and Discovery. *The Journal of Physical Chemistry Letters*, *11*(20), 8710–8720. DOI: 10.1021/acs.jpclett.0c02357 PMID: 32955889

Tran, Y. H. T., An, K., Vu, D. T. T., & Song, S.-W. (2025). High-voltage electrolyte and interface design for mid-nickel high-energy Li-ion batteries. *ACS Energy Letters*, *10*(1), 356–370. DOI: 10.1021/acsenergylett.4c02860

Travlou, N. A., Giannakoudakis, D. A., Algarra, M., Labella, A. M., Rodríguez-Castellón, E., & Bandosz, T. J. (2018). S- and N-doped carbon quantum dots: Surface chemistry dependent antibacterial activity. *Carbon*, *135*, 104–111. DOI: 10.1016/j.carbon.2018.04.018

Troncoso, J. F., Zanotto, F. M., Galvez-Aranda, D. E., Zapata Dominguez, D., Denisart, L., & Franco, A. A. (2025). The ARTISTIC battery manufacturing digitalization initiative: From fundamental research to industrialization. *Batteries & Supercaps*, *8*(1), e202400385. DOI: 10.1002/batt.202400385

Under pressure, Japan's chemical giants pivot. (2025). *C&EN Global Enterprise*, *103*(10), 29-31. DOI: 10.1021/cen-10310-feature3

Vaiani, L., Boccaccio, A., Uva, A. E., Palumbo, G., Piccininni, A., Guglielmi, P., & Ballini, A. (2023). Ceramic materials for biomedical applications: An overview on properties and fabrication processes. *Journal of Functional Biomaterials*, *14*(3), 146–152. DOI: 10.3390/jfb14030146 PMID: 36976070

Vandermause, J., Torrisi, S. B., Batzner, S., Xie, Y., Sun, L., Kolpak, A. M., & Kozinsky, B. (2020). On-the-fly active learning of interpretable Bayesian force fields for atomistic rare events. *npj Computational Materials*, *6*(1), 20. DOI: 10.1038/s41524-020-0283-z

Vandghanooni, S., Eskandani, M., Barar, J., & Omidi, Y. (2018). Bispecific therapeutic aptamers for targeted therapy of cancer: A review on cellular perspective. *Journal of Molecular Medicine : Official Organ of the "Gesellschaft Deutscher Naturforscher und Ärzte"*, *96*(9), 885–902. DOI: 10.1007/s00109-018-1669-y PMID: 30056527

Vassilios, Z. (2004). Influence of compaction and granulation processes on the properties of soft magnetic ceramics. *Journal of the American Ceramic Society*, *87*(5), 824–829. DOI: 10.1111/j.1551-2916.2004.00824.x

Viano, R., Milani, L., Orkenyi, R., & Adamo, A. (2023). Continuous separation and purification of chemical species by continuous liquid–liquid partition chromatography implemented with membrane-based phase separators. *Organic Process Research & Development, 27*(12), 2296–2308. DOI: 10.1021/acs.oprd.3c00239

Vigil, R., Villa, D., Goiti, E., Ocejo, M., & Gim, R. G. (2019). Synthesis of zeolite type analcime from industrial wastes. *Microporous and Mesoporous Materials, 293*, 109817. DOI: 10.1016/j.micromeso.2019.109817

Waikhom, L., & Patgiri, R. (2023). A survey of graph neural networks in various learning paradigms: Methods, applications, and challenges. *Artificial Intelligence Review, 56*(7), 6295–6364. DOI: 10.1007/s10462-022-10321-2

Wan, X, Xiao, Z, Tian, Y, Chen, M, Liu, F, Wang, D, Liu, Y, Bartolo, PJDS, Yan, C, Shi, Y, Zhao, RR, Qi, HJ, & Zhou, K (2024). Recent Advances in 4D Printing of Advanced Materials and Structures for Functional Applications. Advanced materials (Deerfield Beach, Fla.), 36(34), e2312263. DOI: 10.1002/adma.202312263

Wan, B., Insinga, A. R., Dash, A., & Grivel, J. C. (2024). Application of spark plasma sintering to the manufacture of $La_{2-x}Sr_xCuO_4$ superconducting ceramics. *Ceramics International, 50*(23), 51421–51428. DOI: 10.1016/j.ceramint.2024.10.058

Wan, C. P., Hu, X. M., Yang, P. X., Li, M. X., Nisar, M. F., Zhou, Z. H., Gan, Y. D., Cai, Y., & Liu, Z. H. (2025). Screening of theaflavins that inhibit proliferation of nasopharyngeal carcinoma cells through metabolomics approaches. *Food Science & Nutrition, 13*(8), e70642. DOI: 10.1002/fsn3.70642 PMID: 40755495

Wang, L, & Zhu, W (2024). Organic Donor-Acceptor Systems for Photocatalysis. Advanced science (Weinheim, Baden-Wurttemberg, Germany), 11(10), e2307227. DOI: 10.1002/advs.202307227

Wang, M., Yang, Q., Li, Y., Zhao, Y., Zou, J., Luan, F., &Liu, F. (2025). Therapeutic potential of traditional Chinese medicine and mechanisms for the treatment of type 2 diabetes mellitus, 20. DOI: 10.1186/s13020-025-01222-x

Wang, M., Yin, F., Kong, L., Yang, L., Sun, H., Sun, Y., &Wang, X. (2024). Chinmedomics: a potent tool for the evaluation of traditional Chinese medicine efficacy and identification of its active components, 19. DOI: 10.1186/s13020-024-00917-x

Wang, Y., Jiang, Y., Li, M., Xiao, Y., Zhao, Q., Zeng, J., &Wu, X. (2025). Rosavin derived from Rhodiola alleviates colitis in mice through modulation of Th17 differentiation, 136. DOI: 10.1016/j.phymed.2024.156318

Wang, A., O'Kane, S., Planella, F. B., Le Houx, J., O'Regan, K., Zyskin, M., & Howey, D. A. (2022). Review of parameterisation and a novel database (LiionDB) for continuum Li-ion battery models. *Progress in Energy*, *4*(3), 032004. DOI: 10.1088/2516-1083/ac692c

Wang, B., Guo, W., & Fu, Y. (2022). High-performance garnet solid-state battery enabled by improved interfaces. *Journal of Power Sources*, *542*, 231798. DOI: 10.1016/j.jpowsour.2022.231798

Wang, F., Long, G. C., & Zhou, J. L. (2024). Enhanced green remediation and refinement disposal of electrolytic manganese residue using air-jet milling and horizontal-shaking leaching. *Journal of Hazardous Materials*, *465*, 133419. Advance online publication. DOI: 10.1016/j.jhazmat.2023.133419 PMID: 38183942

Wang, F., Zhai, Z., Zhao, Z., Di, Y., & Chen, X. (2024). Physics-informed neural network for lithium-ion battery degradation stable modeling and prognosis. *Nature Communications*, *15*(1), 4332. DOI: 10.1038/s41467-024-48779-z PMID: 38773131

Wang, G., Wang, C., Zhang, X., Li, Z., Zhou, J., & Sun, Z. (2024). Machine learning interatomic potential: Bridge the gap between small-scale models and realistic device-scale simulations. *iScience*, *27*(5), 109673. DOI: 10.1016/j.isci.2024.109673 PMID: 38646181

Wang, H. (2023). Quantitative drug safety and benefit-risk evaluation: Practical and cross-disciplinary approaches. *The American Statistician*, *77*(1), 111–112. DOI: 10.1080/00031305.2022.2160592

Wang, H., Ben, J., Zhang, S., Wang, K., & Xie, P. (2025). Green Methanol from CO_2 Hydrogenation at Industrial Scale: Progress, Challenges, and Perspectives. *Chem & Bio Engineering*. Advance online publication. DOI: 10.1021/cbe.5c00113

Wang, H., Mao, Y., Xu, P., Ding, Y., Yang, H., Li, J.-F., Gu, Y., Han, J., Zhang, L., & Mao, B.-W. (2025). Scalable copper current collectors with precisely engineered lithiophilic alloy "skins" for durable lithium-metal batteries. *Energy & Environmental Science*, *18*(5), 2622–2633. DOI: 10.1039/D4EE05862C

Wang, J., & Qi, F. (2020). Traditional Chinese medicine to treat COVID-19: The importance of evidence-based research. *Drug Discoveries & Therapeutics*, *14*(3), 149–150. DOI: 10.5582/ddt.2020.03054 PMID: 32669523

Wang, J., & Sun, X. (2015). Olivine LiFePO4: The remaining challenges for future energy storage. *Energy & Environmental Science*, *8*(4), 1110–1138. DOI: 10.1039/C4EE04016C

Wang, K. T., Yi, M., Sun, Y. X., Yang, Y., Lu, C. J., Fujita, T., & Cui, X. M. (2025). Fabrication of electrolytic manganese residue/granulated blast furnace slag-based geopolymer microspheres and enhanced synergic adsorption of Cr(III)-containing wastewater. *Separation and Purification Technology*, *355*, 129555. Advance online publication. DOI: 10.1016/j.seppur.2024.129555

Wang, L., Zhou, W. W., Lin, X. F., & Xie, Y. L. (2024). TiO_2-doping boosting resource utilization of high-manganese electrolytic manganese residue for superior performance glass-ceramics. *Ceramics International*, *50*(23), 49889–49905. DOI: 10.1016/j.ceramint.2024.09.333

Wang, M., Meng, Y., Xu, Y., Chen, N., Chuai, M., Yuan, Y., Sun, J., Liu, Z., Zheng, X., Zhang, Z., Li, D., & Chen, W. (2023). Aqueous all-manganese batteries. *Energy & Environmental Science*, *16*(11), 5284–5293. DOI: 10.1039/D3EE01679J

Wang, M., Xu, J., Li, T., Liu, H., & Qu, L. (2025). Comprehensive Performance Evaluation of Steel Slag-Slag-Desulfurization Gypsum Ternary Solid Waste Cementitious Material Based on Principal Component Analysis. *Buildings (Basel, Switzerland)*, *15*(4), 645. DOI: 10.3390/buildings15040645

Wang, R. C., Zhang, S. L., Guo, Y. J., Liu, J. Y., Zhang, F. L., & Dang, C. E. (2024). Techno-economic-environmental analysis based on life cycle carbon accounting and pollutants audit for cleaner production of electrolytic manganese. *Process Safety and Environmental Protection*, *191*, 1010–1023. DOI: 10.1016/j.psep.2024.09.035

Wang, R. F., Yuan, S., Li, Y. J., Gao, P., & Li, R. (2024). Hydrogen-based mineral phase transformation mechanism investigation of pyrolusite ore. *International Journal of Minerals Metallurgy and Materials*, *31*(11), 2445–2457. DOI: 10.1007/s12613-023-2819-y

Wang, R., Mitchell, J. B., Gao, Q., Tsai, W.-Y., Boyd, S., Pharr, M., Balke, N., & Augustyn, V. (2018). Operando atomic force microscopy reveals mechanics of structural water driven battery-to-pseudocapacitor transition. *ACS Nano*, *12*(6), 6032–6039. DOI: 10.1021/acsnano.8b02273 PMID: 29767999

Wang, S., Chen, F., Zhuang, G., Wei, K., Chen, T., Zhang, X., Chen, C., & Du, P. (2023). Synthesis of an all-carbon conjugated polymeric segment of carbon nanotubes and its application for lithium-ion batteries. *Nano Research*, *16*(7), 10342–10347. DOI: 10.1007/s12274-023-5530-4

Wang, S., Shen, T., Yang, C., Luo, G., & Wang, D. (2023). Engineering Iridium-Based Oxygen Evolution Reaction Electrocatalysts for Proton Exchange Membrane Water Electrolyzers. *ACS Catalysis*, *13*(13), 8670–8691. DOI: 10.1021/acscatal.3c01511

Wang, X., Faungnawakij, K., & Chareonpanich, M. (2019). Editorial: Photocatalysis – From Solar Power to Sustainable Chemical Production. *ChemCatChem, 11*(24), 5838–5841. DOI: 10.1002/cctc.201902248

Wang, X., Li, X., Fan, H., & Ma, L. (2022). Solid electrolyte interface in Zn-based battery systems. *Nano-Micro Letters, 14*(1), 205. DOI: 10.1007/s40820-022-00939-w PMID: 36261666

Wang, X., Wu, X., Lu, K., & Ye, J. (2025). Influence of cemented carbide substrate microstructure on CVD diamond coating characteristics. *Diamond and Related Materials, 152*, 111886. DOI: 10.1016/j.diamond.2024.111886

Wang, Z., Tang, Y., Zhang, L., Li, M., Shan, Z., & Huang, J. (2020). In situ TEM observations of discharging/charging of solid-state lithium-sulfur batteries at high temperatures. *Small, 16*(28), 2001899. DOI: 10.1002/smll.202001899 PMID: 32519445

Wang, Z., Wang, Z., Liu, Y., Li, B., Zhao, H., Guo, Q., Ma, H., & Jia, X. (2024). Properties of boron-doped HPHT diamond single crystals grown in a Fe-Ti-B-C system. *Diamond and Related Materials, 145*, 111073. DOI: 10.1016/j.diamond.2024.111073

Warner, J. T. (2024). *The handbook of lithium-ion battery pack design: Chemistry, components, types, and terminology*. Elsevier.

Warzybok, M. (2018). Synthesis of kaolin-based zeolite Y and its application for adsorption of two carbonyl compound gases. *Czasopismo Inżynierii Lądowej, Środowiska i Architektury, 65*(3), 13–26. DOI: 10.7862/rb.2018.2

Weeranoppanant, N. (2019). Enabling tools for continuous-flow biphasic liquid–liquid reaction. *Reaction Chemistry & Engineering, 4*(2), 235–243. DOI: 10.1039/C8RE00230D

Weeranoppanant, N., & Adamo, A. (2019). In-line purification: A key component to facilitate drug synthesis and process development in medicinal chemistry. *ACS Medicinal Chemistry Letters, 11*(1), 9–15. DOI: 10.1021/acsmedchemlett.9b00491 PMID: 31938456

Wei Seh, Z., Li, W., Cha, J. J., Zheng, G., Yang, Y., McDowell, M. T., Hsu, P.-C., & Cui, Y. (2013). Sulphur–TiO2 yolk–shell nanoarchitecture with internal void space for long-cycle lithium–sulphur batteries. *Nature Communications, 4*(1), 1331. DOI: 10.1038/ncomms2327 PMID: 23299881

Wei, C., Tan, L., Zhang, Y., Wang, Z., Feng, J., & Qian, Y. (2022). Towards better Mg metal anodes in rechargeable Mg batteries: Challenges, strategies, and perspectives. *Energy Storage Materials, 52*, 299–319. DOI: 10.1016/j.ensm.2022.08.014

Wei, D., Wang, Z., Lin, H., Yin, X. P., & Wang, Y. (2026). Research Progress in Artificial Intelligence-Assisted Preparation of High-Quality Biomaterials. *ACS Omega*, acsomega.5c09392. Advance online publication. DOI: 10.1021/acsomega.5c09392

Weng, W., Lan, Y., Zhang, R., Chi, X., Tan, W., Tang, D., Chen, H., & Zhong, S. (2024). Gluconic acid as the supplementary additive realizing high-efficiency production of metal manganese with decreased dosage of toxic selenium dioxide. *Electrochimica Acta*, *499*, 144708. Advance online publication. DOI: 10.1016/j.electacta.2024.144708

Wen, T., Bu, Y., & Zhang, H. (2025). Research advances and futureprospects of artificial intelligence technology in the integratedmanagement of solid waste. *Journal of Ecology and Rural Environment*, *41*(9), 1122–1133.

Wen, Y., Zhang, H., Li, K., Wang, H., & Li, Z. (2024). Bi-modified samarium cuprate ($Bi_xSm_{2-x}CuO_4$ ($0 \leq x \leq 0.1$)) thermosensitive ceramics with low sintering temperature. *Journal of Materials Science Materials in Electronics*, *35*(4), 266–272. DOI: 10.1007/s10854-024-12019-3

Wiesinger, H., Shalin, A., Huang, X., Siegrist, A., Plinke, N., Hellweg, S., & Wang, Z. (2024). LitChemPlast: An Open Database of Chemicals Measured in Plastics. *Environmental Science & Technology Letters*, *11*(11), 1147–1160. DOI: 10.1021/acs.estlett.4c00355

Williams, D. B., & Carter, C. B. (1996). *The transmission electron microscope Transmission electron microscopy: a textbook for materials science*. Springer. DOI: 10.1007/978-1-4757-2519-3

Wissinger, J. E., Visa, A., Saha, B. B., Matlin, S. A., Mahaffy, P. G., Kummerer, K., & Cornell, S. (2021). Integrating Sustainability into Learning in Chemistry. *Journal of Chemical Education*, *98*(4), 1061–1063. DOI: 10.1021/acs.jchemed.1c00284

Woeldgen, E., Teoh, R., Stettler, M. E. J., & Malina, R. (2025). Sustainable Aviation Fuel Deployment Strategies in Europe: Supply Chain Implications and Climate Benefits. *Environmental Science & Technology*, *59*(25), 12447–12457. DOI: 10.1021/acs.est.5c02364

Wu, H., Tian, J., Dai, D., Liao, J., Wang, X., Wei, X., &Tong, X. (2020). Efficacy and safety assessment of traditional Chinese medicine for metabolic syndrome, 8. DOI: 10.1136/bmjdrc-2020-001181

Wu, L., Wu, Q., Du, X., Ling, M. and Tong, H. (2025). Revolutionizing pharmacological research of traditional Chinese medicine with single-cell omics technologies, 186. DOI: 10.1016/j.fitote.2025.106846

Wu, Y.-Z., Zhang, Q., Wei, X.-H., Jiang, C.-X., Li, X.-K., Shang, H.-C. and Lin, S. (2022). Multiple anti-inflammatory mechanisms of Zedoary Turmeric Oil Injection against lipopolysaccharides-induced acute lung injury in rats elucidated by network pharmacology combined with transcriptomics, 106. DOI: 10.1016/j.phymed.2022.154418

Wu, B. X., Wu, Z., Hou, Y.-Y., Fang, Z.-X., Deng, Y., Wu, H.-T., & Liu, J. (2023). Application of three-dimensional (3D) bioprinting in anti-cancer therapy. *Heliyon*, *9*(10), e20475. DOI: 10.1016/j.heliyon.2023.e20475 PMID: 37800075

Wu, B., Lochala, J., Taverne, T., & Xiao, J. (2017). The interplay between solid electrolyte interface (SEI) and dendritic lithium growth. *Nano Energy*, *40*, 34–41. DOI: 10.1016/j.nanoen.2017.08.005

Wu, C., Yang, Y., Zhang, Y., Xu, H., He, X., Wu, X., & Chou, S. (2024). Hard carbon for sodium-ion batteries: Progress, strategies and future perspective. *Chemical Science (Cambridge)*, *15*(17), 6244–6268. DOI: 10.1039/D4SC00734D PMID: 38699270

Wu, D. Y., Zhang, Z. H., & Wang, Z. H. (2024). Impact of manganese mining on potentially toxic elements pollution and bioaccumulation in the Xiaojiang River. *Environmental Geochemistry and Health*, *46*(10). Advance online publication. DOI: 10.1007/s10653-024-02171-4 PMID: 39167155

Wu, D., Housel, L. M., King, S. T., Mansley, Z. R., Sadique, N., Zhu, Y., Ma, L., Ehrlich, S. N., Zhong, H., Takeuchi, E. S., Marschilok, A. C., Bock, D. C., Wang, L., & Takeuchi, K. J. (2022). Simultaneous elucidation of solid and solution manganese environments via multiphase operando extended X-ray absorption fine structure spectroscopy in aqueous Zn/MnO2 batteries. *Journal of the American Chemical Society*, *144*(51), 23405–23420. DOI: 10.1021/jacs.2c09477 PMID: 36513373

Wu, J.-G., Yan, Y., Zhang, D.-X., Liu, B.-W., Zheng, Q.-B., Xie, X.-L., Liu, S.-Q., Ge, S.-X., Hou, Z.-G., & Xia, N.-S. (2021). Machine learning for structure determination in single-particle cryo-electron microscopy: A systematic review. *IEEE Transactions on Neural Networks and Learning Systems*, *33*(2), 452–472. DOI: 10.1109/TNNLS.2021.3131325 PMID: 34932487

Wu, J., Guo, S., Li, J., & Zeng, D. (2016). Big data meet green challenges: Big data toward green applications. *IEEE Systems Journal*, *10*(3), 888–900. DOI: 10.1109/JSYST.2016.2550530

Wu, J., Li, J., & Yao, X. (2025). Exploring the potential of halide electrolytes for next-generation all-solid-state lithium batteries. *Advanced Functional Materials*, *35*(10), 2416671. DOI: 10.1002/adfm.202416671

Wu, T., Ma, B., An, Y., Chen, Y., & Wang, C. (2024). Improvement of manganese electrolytic process and secondary resources recovery of manganese: A review. *Process Safety and Environmental Protection, 186*, 895–909. DOI: 10.1016/j.psep.2024.03.097

Wu, W., Zhao, B., Liang, Q., Lei, M., Zhang, B., & Zhang, F. (2025). Effect of direct alloying of manganese ore-cored wires on the composition and quality of molten steel. *Ironmaking & Steelmaking, 52*(7), 739–751. DOI: 10.1177/03019233241307487

Wu, X. W., Yamamura, T., Ohta, S., Zhang, Q. X., Lv, F. C., Liu, C. M., Shirasaki, K., Satoh, I., Shikama, T., Lu, D., & Liu, S. Q. (2011). Acceleration of the redox kinetics of VO2+/VO2+ and V3+/V2+ couples on carbon paper. *Journal of Applied Electrochemistry, 41*(10), 1183–1190. DOI: 10.1007/s10800-011-0343-7

Wu, X., Li, J., Deng, C., Yang, L., Lv, J., & Fu, L. (2022). Novel carbon dots as effective corrosion inhibitor for N80 steel in 1 M HCl and CO_2-saturated 3.5 wt% NaCl solutions. *Journal of Molecular Structure, 1250*, 131897. DOI: 10.1016/j.molstruc.2021.131897

Wu, Y., Guo, L., & She, Y. (2022). Insight on the corrosion inhibition performance of psidium guajava linn leaves extract. *Journal of Molecular Liquids, 346*, 117858. DOI: 10.1016/j.molliq.2021.117858

Wu, Y., Zeng, L., Zheng, Y., You, F., & Liu, X. (2025). Pectin-derived nitrogen-doped carbon dots as green corrosion inhibitors for N80 steel in 1 M sulfuric acid solution. *Applied Surface Science, 687*, 162245. DOI: 10.1016/j.apsusc.2024.162245

Xiang, D. X., He, A. N., Zhou, R., Wang, Y. G., Xiao, X. Y., Gong, T., Kang, W. Y., Lin, X. L., Wang, X. C., Liu, L. X., Chen, Y. G., Gao, S. R., & Liu, Y. B. (2024). Building consensus on the application of organoid-based drug sensitivity testing in cancer precision medicine and drug development. *Theranostics, 14*(8), 3300–3316. DOI: 10.7150/thno.96027 PMID: 38855182

Xiang, L., Chen, Z., Wei, S., & Zhou, H. (2022). Global Trade Pattern of Traditional Chinese Medicines and China's Trade Position. *Frontiers in Public Health, 10*, 865887. Advance online publication. DOI: 10.3389/fpubh.2022.865887 PMID: 35602159

Xiao, W., Zhang, M., Zhao, D., Meng, F., Tang, Q., Hu, L., &Chen, W. (2025). TC-MKD: From ancient wisdom to modern insights-A comprehensive platform for traditional Chinese medicine knowledge discovery, 15. DOI: 10.1016/j.jpha.2025.101297

Xiaoran, Z., Jinyu, S., Xiangzhuo, X., Chutong, Z., Yonghao, S., & Xiaobing, L. (2024). Significant enhancement of electrical conductivity in boron-doped diamond through HPHT post-annealing treatment. *International Journal of Refractory & Hard Metals, 119*, 106509. DOI: 10.1016/j.ijrmhm.2023.106509

Xiao, X., Zheng, L., Li, Q., Chen, C., He, D., Wei, X., Hu, Z., Yang, J., & Zhan, G. (2025). HPHT welding of CVD polycrystalline diamond films. *Journal of the European Ceramic Society, 45*(7), 117246. DOI: 10.1016/j.jeurceramsoc.2025.117246

Xie, Y., Ren, L., Zhu, X., Gou, X., & Chen, S. (2018). Physical and chemical treatments for removal of perchlorate from water-A review. *Process Safety and Environmental Protection, 116*, 180–198. DOI: 10.1016/j.psep.2018.02.009

Xie, Z. N., Liu, Z. H., Zhang, X. J., Yang, L. R., Chang, J., & Tao, C. Y. (2019). Electrochemical oscillation on anode regulated by sodium oleate in electrolytic metal manganese. *Journal of Electroanalytical Chemistry (Lausanne, Switzerland), 845*, 13–21. DOI: 10.1016/j.jelechem.2019.05.012

Xie, Z., Liu, Z., Tao, C., Li, C., & Chang, J. (2022). Production of electrolytic manganese metal using a new hyperchaotic circuit system. *Journal of Materials Research and Technology, 18*, 4804–4815. DOI: 10.1016/j.jmrt.2022.04.141

Xing, Y., Wang, M., Yuan, Y., Hu, J., Wang, Z., Sun, Z., &Mao, T. (2025). Gut microbiota-derived butyrate mediates the anticolitic effect of indigo supplementation through regulating CD4+ T cell differentiation, 4. DOI: 10.1002/imt2.70040

Xi, Z., Dai, R., Ze, Y., Jiang, X., Liu, M., & Xu, H. (2025). Traditional Chinese medicine in lung cancer treatment. *Molecular Cancer, 24*(1), 57. Advance online publication. DOI: 10.1186/s12943-025-02245-6 PMID: 40001110

Xu, J., Li, H., Sze, D. M.-y., Chan, V. W. S. and Yang, A. W. H. (2025). Effectiveness of qigong and Tai Chi for quality of life in patients with cancer: an umbrella review and meta-analysis, 25. DOI: 10.1186/s12906-025-04875-1

Xue, Z., He, D., & Xie, X. (2015). Poly(ethylene oxide)-based electrolytes for lithium-ion batteries. *Journal of Materials Chemistry. A, Materials for Energy and Sustainability, 3*(38), 19218–19253. DOI: 10.1039/C5TA03471J

Xu, G., Jiang, M., Li, J., Xuan, X., Li, J., Lu, T., & Pan, L. (2024). Machine learning-accelerated discovery and design of electrode materials and electrolytes for lithium ion batteries. *Energy Storage Materials, 72*, 103710. DOI: 10.1016/j.ensm.2024.103710

Xu, H., Ouyang, S., Liu, L., Reunchan, P., Umezawa, N., & Ye, J. (2014). Recent advances in TiO_2-based photocatalysis. *Journal of Materials Chemistry. A, Materials for Energy and Sustainability, 2*(32), 12642–12642. DOI: 10.1039/C4TA00941J

Xu, J., Gao, N.-y., Deng, Y., & Xia, S. (2013). Nanoscale iron hydroxide-doped granular activated carbon (Fe-GAC) as a sorbent for perchlorate in water. *Chemical Engineering Journal*, *222*, 520–526. DOI: 10.1016/j.cej.2012.07.141

Xu, J., Gao, N.-y., Zhao, D.-y., Liu, Z., Tang, M., Chen, Y., & Zhu, Y. (2016). Different iron salt impregnated granular activated carbon (Fe-GAC) for perchlorate removal: Characterization, performance and mechanism. *Colloids and Surfaces. A, Physicochemical and Engineering Aspects*, *509*, 99–108. DOI: 10.1016/j.colsurfa.2016.08.039

Xu, L., Liu, W., Hu, Y., & Luo, L. (2020). Stress-resilient electrode materials for lithium-ion batteries: Strategies and mechanisms. *Chemical Communications (Cambridge)*, *56*(87), 13301–13312. DOI: 10.1039/D0CC05359G PMID: 33034589

Xu, X. P., Liu, Y., Gong, Q., Ma, L., Wei, W., Zhao, L. Q., & Luo, Z. B. (2024). PARP1 promotes EGFR-TKI drug resistance via PI3K/AKT pathway in non-small-cell lung cancer. *Cancer Chemotherapy and Pharmacology*, *94*(2), 209–221. DOI: 10.1007/s00280-024-04668-2 PMID: 38609654

Xu, Y., Jia, H., Gao, P., Galvez-Aranda, D. E., Beltran, S. P., Cao, X., & Li, S. (2023). Direct in situ measurements of electrical properties of solid–electrolyte interphase on lithium metal anodes. *Nature Energy*, *8*(12), 1345–1354. DOI: 10.1038/s41560-023-01361-1 PMID: 38249622

Yaashikaa, P. R., Priyanka, B., Kumar, P. S., Karishma, S., Jeevanantham, S., & Indraganti, S. (2022). A review on recent advancements in recovery of valuable and toxic metals from e-waste using bioleaching approach. *Chemosphere*, *287*, 132230. DOI: 10.1016/j.chemosphere.2021.132230 PMID: 34826922

Yang, H., Liu, X., Guo, H., He, Y., Shi, J., Liu, H., Hao, J., & Liu, S. (2025). Effect of laser surface remelting on the microstructure and properties of high manganese steel coating. *Surface and Coatings Technology*, *496*, 131667. Advance online publication. DOI: 10.1016/j.surfcoat.2024.131667

Yang, J., Li, C. B., Zhang, Q., Huang, K. Y., Liu, P. Q., Liu, Z. H., Li, Z. N., Meng, T., Zhang, X., Wu, J. Y., Tao, C., & Zheng, G. (2023). A new approach of electrolytic metal manganese with lower energy consumption and fewer spherical dendrites based on a hyperchaotic circuit. *International Journal of Bifurcation and Chaos in Applied Sciences and Engineering*, *33*(14), 2350173. Advance online publication. DOI: 10.1142/S0218127423501730

Yang, J., Li, C., Zhang, Q., Wu, Z., Liu, P., Liu, Z., Tao, C., Zheng, G., Yang, Y., & Wei, H. (2025). Improving the current efficiency of electrodeposited manganese metal using variable current electrolysis and its mechanism. *Chemical Engineering Journal, 503*, 158195. Advance online publication. DOI: 10.1016/j.cej.2024.158195

Yang, J., Wu, Z., Li, C., Zhang, Q., Liu, P., Zheng, G., Liu, Z., Li, J., Tao, C., Yang, Y., & Wei, H. (2025). Enhancing manganese electrodeposition efficiency at high current densities utilizing a novel variable gradient chaotic current technique. *Journal of Cleaner Production, 503*, 145373. Advance online publication. DOI: 10.1016/j.jclepro.2025.145373

Yang, L. Z., Feng, Z., Hu, H., Wei, G. S., Xue, B. T., Guo, Y. F., & Jiang, T. (2025). A review on production and application of direct reduced iron in gas-based shaft furnace-electric arc furnace route. *Journal of Iron and Steel Research International, 32*(3), 485–518. DOI: 10.1007/s42243-024-01377-1

Yang, L., Lu, X., Wang, S., Wang, J., Guan, X., Guan, X., & Wang, G. (2020). Designed synthesis of nickel–cobalt-based electrode materials for high-performance solid-state hybrid supercapacitors. *Nanoscale, 12*(3), 1921–1938. DOI: 10.1039/C9NR08156A PMID: 31907496

Yang, S., Zhang, F., Ding, H., He, P., & Zhou, H. (2018). Lithium Metal Extraction from Seawater. *Joule, 2*(9), 1648–1651. DOI: 10.1016/j.joule.2018.07.006

Yang, W., Liu, Y., Li, R., Lv, J., Zhang, Y., Ren, Y., Yuan, Z., & Zhang, Z. (2025). Toward sustainable and scalable synthesis of ibuprofen: Integrative insights into batch and continuous flow strategies. *AIChE Journal. American Institute of Chemical Engineers*. Advance online publication. DOI: 10.1002/aic.70198

Yang, Y., Liu, G., Zhang, Y., Feng, S., Xu, D., & Chen, Y. M. (2025). Advances in Gel-Based Solar Interfacial Evaporators: Advantages, Performance and Applications. *ACS Energy Letters, 10*(12), 6123–6150. DOI: 10.1021/acsenergylett.5c02413

Yang, Z., Li, Y., & Zhang, G. (2024). Degradation of microplastic in water by advanced oxidation processes. *Chemosphere, 357*, 141939. DOI: 10.1016/j.chemosphere.2024.141939 PMID: 38621489

Yang, Z., Yan, R., Han, J., Wu, T., Wu, Q., Wei, G., Fu, Y., & Wen, M. (2025). Oxygen bridges of CoTe2/Co-O-NC enhancing adsorption-catalysis of polysulfide for stable lithium–sulfur batteries. *Advanced Functional Materials, 35*(13), 2417834. DOI: 10.1002/adfm.202417834

Yan, Y., Hai, F., Wang, B., Cao, W., Li, M., Wang, C., Li, N., & Zhao, D. (2025). Machine learning accelerates high-voltage electrolyte discovery for lithium metal batteries. *Energy Storage Materials*, *79*, 104312. DOI: 10.1016/j.ensm.2025.104312

Yao, D., Yang, H., Chen, H., & Williams, P. T. (2018). Investigation of nickel-impregnated zeolite catalysts for hydrogen/syngas production from the catalytic reforming of waste polyethylene. *Applied Catalysis B: Environment and Energy*, *227*, 477–487. DOI: 10.1016/j.apcatb.2018.01.050

Yao, D., Yang, H., Chen, H., & Williams, P. T. (2021). Hydrogen production by pyrolysis–nonthermal plasma/catalytic reforming of waste plastic over different catalyst support materials. *Energy & Fuels*, *35*(10), 8280–8293. DOI: 10.1021/acs.energyfuels.1c04186

Yao, N., Chen, X., Fu, Z.-H., & Zhang, Q. (2022). Applying classical, ab initio, and machine-learning molecular dynamics simulations to the liquid electrolyte for rechargeable batteries. *Chemical Reviews*, *122*(12), 10970–11021. DOI: 10.1021/acs.chemrev.1c00904 PMID: 35576674

Yao, X., Hafiz, H., Chen, L., Cai, W., Mao, Z., & Hu, S. (2025). Coarse-grained molecular dynamics modelling of lithium-ion battery electrode drying: A sensitivity analysis. *Journal of Manufacturing Processes*, *144*, 54–59. DOI: 10.1016/j.jmapro.2025.03.117

Yashnik, S. A., Surovtsova, T. A., & Salnikov, A. V. (2025). Low-loaded Pt–Pd diesel oxidation catalyst: MnO loading effects. *Topics in Catalysis*, *68*(3–4), 357–378. DOI: 10.1007/s11244-024-02023-6

Yerumoh, E., Cardenas-Sierra, I., Fernandez, F., & Franco, A. A. 3D Resolved Computational Modeling to Simulate the Electrolyte Wetting of a Lithium-Ion Battery Cell with 18650 Format. *Batteries & Supercaps, n/a*(n/a), 2500434. doi:DOI: 10.1002/batt.202500434

Ye, W. Q., Xu, F. Y., Jiang, L. H., Duan, N., Li, J. H., Zhang, F. L., Zhang, G., & Chen, L. J. (2021). A novel functional lead-based anode for efficient lead dissolution inhibition and slime generation reduction in zinc electrowinning. *Journal of Cleaner Production*, *284*, 124767. Advance online publication. DOI: 10.1016/j.jclepro.2020.124767

Ying, H., & Han, W.-Q. (2017). Metallic Sn-based anode materials: Application in high-performance lithium-ion and sodium-ion batteries. *Advanced Science (Weinheim, Baden-Württemberg, Germany)*, *4*(11), 1700298. DOI: 10.1002/advs.201700298 PMID: 29201624

Yin, J., Wu, Z., Fang, K., Zhu, Y., Zhang, K., Zhang, H., Chen, Y., Li, L., Fan, L., Dong, K., Zheng, L., Wang, Q., Huang, H., Zhang, J., Qiao, Y., & Sun, S.-G. (2025). Reassessing anionic redox in conventional layered oxide cathodes for Li-ion batteries: Ionic and covalent mechanisms. *Chemical Science (Cambridge)*, *16*(19), 8268–8281. DOI: 10.1039/D5SC00429B PMID: 40213374

Yin, S. Y., Yu, Y., Wu, N., Zhuo, M. L., Wang, Y. M., Niu, Y. J., Ni, Y. Q., Hu, F., Ding, C. M., Liu, H. S., Cheng, X. H., Peng, J., Li, J., He, Y., Li, J. X., Wang, J. Y., Zhang, H. S., Zhai, X. Y., Liu, B., & Xi, J. Z. J. (2024). Patient-derived tumor-like cell clusters for personalized chemo- and immunotherapies in non-small cell lung cancer. *Cell Stem Cell*, *31*(5), 717–733. DOI: 10.1016/j.stem.2024.03.008 PMID: 38593797

Yin, X.-T., You, E.-M., Zhou, R.-Y., Zhu, L.-H., Wang, W.-W., Li, K.-X., Wu, D.-Y., Gu, Y., Li, J.-F., Mao, B.-W., & Yan, J.-W. (2024). Unraveling the energy storage mechanism in graphene-based nonaqueous electrochemical capacitors by gap-enhanced Raman spectroscopy. *Nature Communications*, *15*(1), 5624. DOI: 10.1038/s41467-024-49973-9 PMID: 38965231

Yoldi, M., Korili, S. A., & Gil, A. (2019). Zeolite synthesis from industrial wastes. *Microporous and Mesoporous Materials*, *287*, 183–191. DOI: 10.1016/j.micromeso.2019.06.009

Young, J., & Smeu, M. (2024). *Applications of Ab Initio Molecular Dynamics for Modeling Batteries Computational Design of Battery Materials*. Springer.

Yuan, H., Read, J. A., & Wang, Y. (2019). Capacity loss of non-aqueous Li-Air battery due to insoluble product formation: Approximate solution and experimental validation. *Materials Today. Energy*, *14*, 100360. DOI: 10.1016/j.mtener.2019.100360

Yuan, R., Zhang, J., He, L., & Zuo, J.-M. (2021). Training artificial neural networks for precision orientation and strain mapping using 4D electron diffraction datasets. *Ultramicroscopy*, *231*, 113256. DOI: 10.1016/j.ultramic.2021.113256 PMID: 33773843

Yue, J., Hao, D., Liu, S., Yu, J., Meng, L., Lv, J., & Guo, J. (2024). Research progress of traditional Chinese medicine in the treatment of allergic rhinitis. *Heliyon*, *10*(7), e29262. Advance online publication. DOI: 10.1016/j.heliyon.2024.e29262 PMID: 38617960

Yun, D. Y., Wu, X. H., Chen, X., Yang, Y. H., Shang, Y. J., Liu, S. P., Gunasekeran, D. V., Lin, D. R., Liu, L. X., Zhao, L. Q., Chen, W. B., Chen, J. J., Jin, L., Wu, Y. X., Hong, J. M., Hu, W. L., Lin, Z. Z., Cheung, C. Y., Xie, P. C., & Lin, H. T. (2025). An artificial intelligence and blockchain technology-based data management framework for multicenter randomized controlled trials. *Science Bulletin*, *70*(6), 856–860. DOI: 10.1016/j.scib.2025.01.041 PMID: 39933985

Yu, P., Liu, Q., Zuo, L., Zhang, X., Yin, C., & Zhou, H. (2024). Water-mediated one-pot multi-step synthesis of chiral 1,3-diarylpropan-1-ols by the asymmetric hydrofunctionalisation of simple alkynes. *Green Chemistry*, *26*(7), 4074–4078. DOI: 10.1039/D3GC04945K

Yu, X., & Manthiram, A. (2018). Electrode–electrolyte interfaces in lithium-based batteries. *Energy & Environmental Science*, *11*(3), 527–543. DOI: 10.1039/C7EE02555F

Zakzeski, J., Bruijnincx, P. C. A., Jongerius, A. L., & Weckhuysen, B. M. (2010). The Catalytic Valorization of Lignin for the Production of Renewable Chemicals. *Chemical Reviews*, *110*(6), 3552–3599. DOI: 10.1021/cr900354u

Zarrabian, M., Adhikary, L., Nita, M., Sriyanka, L., & Sherif, S. M. (2025). Toxicological and Functional Assessment of Minicell-Encapsulated dsRNA on Biocontrol Agents in Agriculture. *ACS Environmental Au*, *5*(4), 427–441. DOI: 10.1021/acsenvironau.5c00067

Zeng, S., Wang, R., Zhang, Z., & Qiu, S. (2016). Solventless green synthesis of sodalite zeolite using diatomite as silica source by a microwave heating technique. *Inorganic Chemistry Communications*, *70*, 168–171. DOI: 10.1016/j.inoche.2016.06.013

Zhang, W., Wu, P., Song, Y., Liang, D., Meng, G. and Zeng, H. (2025). Immunomodulatory mechanism of Huangqi-Guizhi-Wuwu Decoction in alleviating autoimmune arthritis based on network pharmacology and experimental validation, 143. DOI: 10.1016/j.phymed.2025.156644

Zhang, F., Gallagher, K. S., Deng, M., Liu, H., Orvis, R., & Xuan, X. (2025). Assessing the Policy Gaps for Achieving China's Carbon Neutrality Target. *Environmental Science & Technology*, *59*(34), 18124–18133. DOI: 10.1021/acs.est.4c12478

Zhang, L., Yang, S., Fu, W., Cui, Y., Wang, J., Zhao, D., Yang, C., Wang, X., & Cao, B. (2022). Plasma-induced ε-MnO2 based aqueous zinc-ion batteries and their dissolution-deposition mechanism. *Journal of Materials Science and Technology*, *127*, 206–213. DOI: 10.1016/j.jmst.2022.03.028

Zhang, P. J., Wang, X. Y., Cen, X. F., Zhang, Q., Fu, Y. Y., Mei, Y. Q., Wang, X. R., Wang, R. Y., Wang, J. J., Ouyang, H. W., Liang, T. B., Xia, H. G., Han, X. P., & Guo, G. J. (2025). A deep learning framework for in silico screening of anticancer drugs at the single-cell level. *National Science Review, 12*(2), nwae451. DOI: 10.1093/nsr/nwae451 PMID: 39872221

Zhang, Q., Yang, L. T., Chen, Z., & Li, P. (2018). A survey on deep learning for big data. *Information Fusion, 42*, 146–157. DOI: 10.1016/j.inffus.2017.10.006

Zhang, R., Motes, B., Tan, S., Lu, Y., Shih, M.-C., Hao, Y., Yang, K., Srinivasan, S., Bawendi, M. G., & Bulović, V. (2025). Machine Learning Prediction of Organic–Inorganic Halide Perovskite Solar Cell Performance from Optical Properties. *ACS Energy Letters, 10*(4), 1714–1724. DOI: 10.1021/acsenergylett.4c03592

Zhang, S. Q., & Li, J. C. (2021). An introduction to traditional Chinese medicine, including acupuncture. *The Anatomical Record : Advances in Integrative Anatomy and Evolutionary Biology, 304*(11), 2359–2364. DOI: 10.1002/ar.24782 PMID: 34626156

Zhang, S., Gu, K., Lu, B., Han, J., & Zhou, J. (2024). Hydrometallurgical Processes on Recycling of Spent Lithium-lon Battery Cathode: Advances and Applications in Sustainable Technologies. *Wuli Huaxue Xuebao, 40*(10), 2309028. DOI: 10.3866/PKU.WHXB202309028

Zhang, W. J., Zhuang, A., Gu, P., Zhou, H. F., & Fan, X. Q. (2016). A review of the three-dimensional cell culture technique: Approaches, advantages and applications. *Current Stem Cell Research & Therapy, 11*(4), 370–380. DOI: 10.2174/1574888X11666160217154007 PMID: 26899396

Zhang, X., He, X., Fang, Z., Yu, S., You, L., Zhu, S., & Zhang, B. (2025). Analyzing the highlights and trends of traditional Chinese medicine academic development based on the “Top 10 Academic Advances in TCM from 2020 to 2023&rdquo. *Zhongguo Ke Xue : Sheng Ming Ke Xue = Scientia Sinica Vitae, 55*(2), 267–281. DOI: 10.1360/SSV-2024-0293

Zhang, Y., Li, J., Gao, L., & Wang, M. (2020). Nitrate based nanocomposite thermal storage materials: Understanding the enhancement of thermophysical properties in thermal energy storage. *Solar Energy Materials and Solar Cells, 216*, 110727. DOI: 10.1016/j.solmat.2020.110727

Zhang, Y., Lu, P., Qin, H., Zhang, Y., Sun, X., Song, X., Liu, Z., Peng, H., Liu, Y., Nwafor, E. O., Li, J., & Liu, Z. (2021). Traditional Chinese medicine combined with pulmonary drug delivery system and idiopathic pulmonary fibrosis: Rationale and therapeutic potential. *Biomedicine & Pharmacotherapy = Biomédecine & Pharmacothérapie*, *133*, 111072. Advance online publication. DOI: 10.1016/j.biopha.2020.111072 PMID: 33378971

Zhang, Y., Wu, Z., Hong, Z., Gu, X., & Xu, N. (2012). Hydrogen-selective zeolite membrane reactor for low temperature water gas shift reaction. *Chemical Engineering Journal*, *197*, 314–321. DOI: 10.1016/j.cej.2012.05.064

Zhang, Y., Zhao, C., & Guo, Z. (2019). Simulation of crack behavior of secondary particles in Li-ion battery electrodes during lithiation/de-lithiation cycles. *International Journal of Mechanical Sciences*, *155*, 178–186. DOI: 10.1016/j.ijmecsci.2019.02.042

Zhang, Z., Smith, K., Jervis, R., Shearing, P. R., Miller, T. S., & Brett, D. J. L. (2020). Operando electrochemical atomic force microscopy of solid–electrolyte interphase formation on graphite anodes: The evolution of SEI morphology and mechanical properties. *ACS Applied Materials & Interfaces*, *12*(31), 35132–35141. DOI: 10.1021/acsami.0c11190 PMID: 32657567

Zhao, D., Liu, X., Zhang, R., Xiao, X., & Li, J. (2021). Preparation of two types of silver-doped fluorescent carbon dots and determination of their antibacterial properties. *Journal of Inorganic Biochemistry*, *214*, 111306. DOI: 10.1016/j.jinorgbio.2020.111306 PMID: 33221600

Zhao, M., Che, Y., Gao, Y., & Zhang, X. (2024). Application of multi-omics in the study of traditional. *Frontiers in Pharmacology*, *15*, 1431862. Advance online publication. DOI: 10.3389/fphar.2024.1431862 PMID: 39309011

Zhao, Y. N., Zhu, G. C., & Cheng, Z. (2010). Thermal analysis and kinetic modeling of manganese oxide ore reduction using biomass straw as reductant. *Hydrometallurgy*, *105*(1–2), 96–102. DOI: 10.1016/j.hydromet.2010.08.004

Zhao, Y.-Z., Dai, Y.-Z., & Nie, K. (2022). Research Progress on the Antiemetic Effect of Traditional Chinese Medicine Against Chemotherapy-Induced Nausea and Vomiting: A Review. *Frontiers in Pharmacology*, *12*, 790784. Advance online publication. DOI: 10.3389/fphar.2021.790784 PMID: 35222008

Zhao, Z. Y., Pei, Z. Y., Hu, A. X., Zhang, Y. H., & Chen, J. (2023). Analysis of incentive policies and initiatives on orphan drug development in China: Challenges, reforms and implications. *Orphanet Journal of Rare Diseases*, *18*(1), 220. DOI: 10.1186/s13023-023-02684-8 PMID: 37501126

Zhao, Z., Xu, B., & Tian, Y. (2016). Recent Advances in Superhard Materials. *Annual Review of Materials Research, 46*(1), 383–406. DOI: 10.1146/annurev-matsci-070115-031649

Zhao, Z., Zuo, H., Fang, S., Duan, M., Zhao, D., & Chen, T. (2025). Hydrophobic metal–organic frameworks and membranes for enhanced carbon capture. *Chemical Engineering Journal, 526*, 170657. DOI: 10.1016/j.cej.2025.170657

Zheng, Q., Song, Y., Huang, W., Yang, J., Li, T., & Xu, Y. J. (2023). Challenges and strategies towards the interface between lithium anode and Li10GeP2S12 electrolyte in all-solid-state lithium metal batteries. *Energy Storage Materials, 63*, 103038. DOI: 10.1016/j.ensm.2023.103038

Zheng, S., Feng, L., Hu, Z., Li, J., Zhu, H., & Ma, X. (2023). Study on the corrosion inhibition of biomass carbon quantum dot self- aggregation on Q235 steel in hydrochloric acid. *Arabian Journal of Chemistry, 16*(4), 104605. DOI: 10.1016/j.arabjc.2023.104605

Zhong, H. D., Zhang, Q., Liu, Z. H., Du, J., & Tao, C. Y. (2023). Ti/TiO_2 anodes for efficient electrodeposition of manganese metal and anode slime generation reduction. *ACS Omega, 8*(41), 38469–38480. DOI: 10.1021/acsomega.3c05273 PMID: 37867691

Zhou, T., Yuk, H., Hu, F., Wu, J., Tian, F., Roh, H., Shen, Z., Gu, G., Xu, J., Lu, B., & Zhao, X. (2023). 3D printable high-performance conducting polymer hydrogel for all-hydrogel bioelectronic interfaces. *Nature Materials, 22*(7), 895-+. DOI: 10.1038/s41563-023-01569-2

Zhou, Y., Wu, Q., Ren, C., Jiang, H., Huang, X., Chen, J., &Pei, J. (2025). Construction and application of a precise evaluation method for the quality of traditional Chinese medicine based on "target-combined quality evaluation" using safflower as an example, 16. DOI: 10.3389/fphar.2025.1554895

Zhou, H., Cao, H. B., Que, Y. T., Chen, Y. F., Fu, J. J., Wang, H. Y., Zheng, Y. Y., He, X., Yang, Y. X., & Liu, J. (2024). Hierarchical Cu dendrites@CuO multifunctional nanowire mesh for solar-thermal clean water production. *Desalination, 572*, 117114. Advance online publication. DOI: 10.1016/j.desal.2023.117114

Zhou, H., Xue, J., Che, Q. L., Kong, L. F., Zhang, N. N., Tao, C. Y., & Fan, X. (2023). Rollable and ventilated net-based solar thermal water evaporator for casting on water surface. *Small Structures, 4*(7), 2300008. Advance online publication. DOI: 10.1002/sstr.202300008

Zhou, H., Zhang, N. N., Bai, H., Ming, H., Zhang, Q., Du, J., Fan, X., & Tao, C. Y. (2020). A pulse modulatable self-oscillation kinetics for water oxidation at large current on manganese catalyst. *Electrochimica Acta*, *337*, 135798. Advance online publication. DOI: 10.1016/j.electacta.2020.135798

Zhou, Y. Y., Yin, G. D., Zeng, X., Zhao, J. F., & Yao, G. D. (2022). Potential application of carbohydrate biomass in hydrometallurgy: One-pot reduction of metal oxides/salts under mild hydrothermal conditions. *RSC Advances*, *12*(32), 20747–20754. DOI: 10.1039/D2RA01493A PMID: 35919188

Zhu, H., Hu, B., & Yang, F. (2021). Removal of sulfadiazine by polyamide nanofiltration membranes: Measurement, modeling, and mechanisms. *Membranes (Basel)*, *11*(2), 143. DOI: 10.3390/membranes11020104 PMID: 33540550

Zhu, S. J., Hinman, R. S., Nelligan, R. K., Li, P., De Silva, A. P., Harrison, J., & Bennell, K. L. (2026). Online Unsupervised Tai Chi Intervention for Knee Pain and Function in People With Knee Osteoarthritis. *JAMA Internal Medicine*, *186*(1), 15. Advance online publication. DOI: 10.1001/jamainternmed.2025.5723 PMID: 41143827

Zhu, X., Dong, H., Liu, Y., Feng, Y.-H., Tang, Y., Yu, L., Xu, S.-W., Wei, G.-X., Sun, S., Liu, M., Xiao, B., Xu, R., Xiao, Y., Chou, S., & Wang, P.-F. (2024). Mitigating mechanical stress by the hierarchical crystalline domain for high-energy P2/O3 biphasic cathode materials. *ACS Nano*, *18*(46), 32003–32015. DOI: 10.1021/acsnano.4c09918 PMID: 39523605

Zhu, X., Yang, J., Yang, Y., Huang, Q., & Liu, T. (2022). Pyrometallurgical process and multipollutant co-conversion for secondary aluminum dross: A review. *Journal of Materials Research and Technology*, *21*, 1196–1211. DOI: 10.1016/j.jmrt.2022.09.089

Zmuidzinaviciene, N., Griskonis, E., & Sulcius, A. (2023). The corrosion of Mn coatings electrodeposited from a sulphate bath with Te(VI) additive and influence of phosphate post-treatment on corrosion resistance. *Coatings*, *13*(9), 1617. Advance online publication. DOI: 10.3390/coatings13091617

Zorn, M., Ionescu, C., Klohs, D., Zähl, K., Kisseler, N., Daldrup, A., & Flamme, S. (2022). An approach for automated disassembly of lithium-ion battery packs and high-quality recycling using computer vision, labeling, and material characterization. *Recycling*, *7*(4), 48. DOI: 10.3390/recycling7040048

Zulkharnay, R., & May, P. W. (2024). Applications of diamond films: A review. *Functional Diamond*, *4*(1), 2410160. DOI: 10.1080/26941112.2024.2410160

Zuo, T.-T., Liu, J., Zan, K., Liu, L.-N., Wang, Q., Wang, Z., &Ma, S.-C. (2024). Bioaccessibility and bioavailability of exogenous and endogenous toxic substances in traditional Chinese medicine and their significance in risk assessment, 208. DOI: 10.1016/j.phrs.2024.107388

About the Contributors

Lei Guo received his PhD in Chemical Engineering and Technology from Chongqing University, China, in 2015. He is currently working as a professor at the Department of Materials and Chemical Engineering, Tongren University, China. He has conducted a series of studies in various areas of the field of physical chemistry: electrochemistry, corrosion science, density functional theory (DFT), as well as molecular dynamic simulation. He has authored over 350 papers published in international journals, 9 books, and 20 patents. He is also a member of the editorial boards of 7 international journals. He recently focuses on metal-air batteries, organic corrosion inhibitors, and the first principle calculations.

Yan Tan received her PhD in Optoelectronics and Nanostructure Science in the Graduate School of Science and Technology of Shizuoka University, Japan in 2019. She is currently working as an Associate Professor at the Department of Materials and Chemical Engineering, Tongren University, China. She has conducted a series of studies in the field of materials science: vapor deposition polymerization of thin films, carbon source materials and electrospinning functional materials. She has authored more than 10 papers in international journals, and 1 book. She recently focuses on organic corrosion inhibitors and antimicrobial materials.

Amir Adam is an accomplished researcher with a diverse expertise in sustainable energy applications and advanced material design. He received his Ph.D. in Chemical Engineering and Technology from Xiamen University, China in 2020. He obtained a Postdoc fellow from Zhejiang University of Technology, where his research focused on the development of advanced electrocatalysis for energy conversion devices. He is currently a full-time researcher in the Department of Material and Chemical Engineering, Tongren University. His research interests include electro/photo catalysis, Fuel cells, energy conversion devices, and metal-air batteries. His work aims to advance sustainable energy technologies and contribute to innovative solutions for clean energy production. His contributions to the field are evident through his publications and research outcome.

***

Imen Amri focusing on improving water treatment technologies by synthesizing and characterizing functional materials, particularly zeolites and activated carbon derived from clays and waste, for the adsorption and degradation of pollutants such as pharmaceuticals, volatile organics, and dyes. Experience includes laboratory-scale treatment design, extensive use of analytical techniques (e.g., UV-Vis spectroscopy, TOC analysis, gas chromatography, XRD, FTIR, and SEM), and collaborative research under EU Horizon projects such as SOLWATER.

Du Yun received his Bachelor's degree in Materials Chemistry from Jilin Institute of Chemical Technology in 2012, and subsequently earned his Ph.D. in 2025 under the supervision of Professor Qingji Xie at Hunan Normal University. He is currently working in the Department of Physicochemical Analysis at the Changsha Center for Disease Control and Prevention, where he is dedicated to developing innovative chemical and biosensing approaches to address critical challenges in practical applications. His research primarily focuses on designing novel analytical strategies for the detection of environmental contaminants—such as heavy metals and nitrite—and the analysis of food components, with the aim of advancing public health and food safety.

Noureddine Hamdi (Eng. Dr.) Full professor at the Higher Institute of Sciences and Techniques of Waters of Gabès (ISSTEG), his main specialty is materials sciences and environment. He has more than 15 years of professional experience in the field of higher education and scientific research. From 2014 to 2020 he is the director of the Higher Institute of Water Sciences and Techniques of Gabes (Tunisia). He has participated in several international projects (RE-MED, ARIMENT2, PRIMA, VRR12/24) and has been a Fulbright Lecturer at Case Western Reserve University (USA) in 2016. His research interests cover the field of composites materials, new materials for wastewater treatment, new minerals materials (Geopolymer, Plaster and ceramic membrane). Noureddine has served on many conference committees in the field of materials engineering, environment and water, green engineering. He has written more than 90 scientific articles in international journals. He has supervised 10 Doctoral Thesis, 40 Master. H-index 24

Chen Hao received his Ph.D. in Analytical Chemistry from Nanjing University in 2008. He is currently a Professor and Master's Supervisor at the School of

Chemistry and Chemical Engineering, JiangSu University. His primary research interests include the design, development, and application of electrode materials for supercapacitors, as well as the development and application of electrochemical biosensors. In recent years, he has published more than 80 peer-reviewed SCI papers in internationally recognized journals, including Analytical Chemistry, Chemical Engineering Journal, Desalination, Inorganic Chemistry Frontiers, Journal of Colloid and Interface Science, Industrial Crops and Products, Journal of Cleaner Production, and Journal of Hazardous Materials. His work has been cited over 6,000 times, with an h-index of 42, demonstrating significant academic and industrial impact. He teaches undergraduate courses such as Materials Chemistry and Applied Electrochemistry, as well as a graduate course, Testing and Characterization of Polymer Materials.

Huang Jiming is a professor and master's supervisor at the School of Materials and Chemical Engineering of Tongren University. He holds a Ph.D. in Engineering from Hubei University (2018-2021). His primary research focuses on functional materials and water pollution control chemistry. He has published significant research in reputable journals, such as designing S-scheme heterojunctions of MIL-101(Fe)/WO3 for efficient photocatalytic degradation of antibiotics under visible light. Currently, he leads a National Natural Science Foundation of China project investigating the mechanisms of antibiotic degradation using heterogeneous catalysts coupled with peroxymonosulfate. His contact email is chyhjm@gztrc.edu.cn

Wu Liu, PhD in Chemistry from Hunan Normal University, is currently a researcher at Hunan University of Medicine. His research centers on organic electrochemical transistor (OECT) biosensors, with a core focus on the molecular design, synthetic modification and performance optimization of high-performance channel materials, as well as systematic investigations into OECT device structure design, fabrication process improvement and overall performance promotion.

Chenglong Lu Specializing in the research and development of new energy ceramic materials and devices, as well as eco-friendly ceramic materials.

Zhang-Hui Lu is currently a professor at Jiangxi Normal University and Vice Dean of the School of Chemistry and Materials Science at Jiangxi Normal University. He received his PhD degree in Chemical Science and Engineering in 2011 from Kobe University (Japan) under the supervision of Prof. Qiang Xu. He is the winner of First Prize in the Natural Science Award of Jiangxi Province in 2020 and has published more than 180 papers with citations > 11000 and a h-index of 62. His current research interest is mainly focused on synthesizing micro/nanostructured materials and exploring their applications in the area of heterogeneous catalysis and electro-catalysis for clean energy and enviornment.

Pan Xiaoguang, Male. His main research focuses on two-dimensional materials and inorganic non-metallic functional materials.

Luis A. Selis received his Bachelor of Science in Electronic Engineering from National University of Engineering (Peru) and Master of Science in Electrical Engineering from Texas A&M University (USA). He is a researcher specializing in computational chemistry and materials modeling, with a focus on lithium-ion batteries. His work combines ab initio simulations, classical molecular dynamics, and machine-learned interatomic potentials to study solid–liquid interfaces, lithium dendrite formation, and predict material properties. Currently, his research aims to integrate physics-based models with machine learning to accelerate the discovery of functional materials for energy storage applications.

Yuru Shen graduated with his first degree in chemistry from Hunan City University in 2012, and then received his Ph.D. (2023) degrees under the direction of Professor Qingji Xie from Hunan Normal University. He worked in Tongren University in China now. He is currently engaged in developing innovative chemical and biosensing approaches aimed at addressing practical challenges. His work primarily involves creating novel analytical strategies for the detection of environmental contaminants and the analysis of food components.

Sujitraj Sheth is currently a Foreign Teacher at the National Key Laboratory of Green Pesticide, Guizhou University, China. He obtained his PhD in Chemistry from University Paris Sud XI, France in 2014. He has extensive research experience as a Postdoctoral Fellow at the Dalian Institute of Chemical Physics (CAS), where he received the prestigious CAS-PIFI fellowship, and at Jiangnan University. His research interests include Synthetic Organic/Inorganic Chemistry, Photo-catalysis, Bio-sensors, and Plant Protection.

Kai Wang is an Associate Professor at the School of Materials and Chemical Engineering, Tongren University, China. He received his Ph.D. in Biochemistry and Molecular Biology from Nanjing University in 2016. His research interests include functional polymer materials, antifouling and antibacterial coatings, and corrosion inhibition materials, with a focus on controlled polymerization techniques such as RAFT and ATRP. He has participated in multiple research projects supported by the National Natural Science Foundation of China and has published over ten peer-reviewed journal articles. His current work bridges materials chemistry and pharmaceutical engineering, contributing to the development of advanced synthesis and purification strategies for pharmaceutical intermediates.

Index

A

B

C

D

E

F

G

H

I

L

M

N

P

www.ingramcontent.com/pod-product-compliance
Lightning Source LLC
LaVergne TN
LVHW061529290926
849162LV00004B/729